AF322917

PHYSICS AND CHEMISTRY OF UPPER ATMOSPHERES

ASTROPHYSICS AND SPACE SCIENCE LIBRARY

A SERIES OF BOOKS ON THE RECENT DEVELOPMENTS
OF SPACE SCIENCE AND OF GENERAL GEOPHYSICS AND ASTROPHYSICS
PUBLISHED IN CONNECTION WITH THE JOURNAL
SPACE SCIENCE REVIEWS

VOLUME 35

PHYSICS AND CHEMISTRY OF UPPER ATMOSPHERES

PROCEEDINGS OF A SYMPOSIUM ORGANIZED BY
THE SUMMER ADVANCED STUDY INSTITUTE, HELD AT
THE UNIVERSITY OF ORLÉANS, FRANCE,
JULY 31 – AUGUST 11, 1972

Edited by

B. M. McCORMAC

Lockheed Palo Alto Research Laboratory, Palo Alto, Calif., U.S.A.

D. REIDEL PUBLISHING COMPANY

DORDRECHT-HOLLAND / BOSTON-U.S.A.

Library of Congress Catalog Card Number 72–92533

ISBN 90 277 0283 7

Published by D. Reidel Publishing Company,
P.O. Box 17, Dordrecht, Holland

Sold and distributed in the U.S.A., Canada, and Mexico
by D. Reidel Publishing Company, Inc.
306 Dartmouth Street, Boston,
Mass. 02116, U.S.A.

Printed in The Netherlands by D. Reidel, Dordrecht

PREFACE

This book contains the lectures presented at the Summer Advanced Study Institute, 'Physics and Chemistry of Upper Atmospheres' which was held at the University of Orléans, Orléans, France, during the period July 31 through August 11, 1972. One-hundred thirty nine persons from 14 different countries attended the Institute.

The authors and the publisher have made a special effort for rapid publication of an up-to-date status of the particles, fields, and processes in the earth's magnetosphere, which is an ever changing area. Special thanks are due to the lecturers for their diligent preparation and excellent presentations. The individual lectures and the published papers were deliberately limited; the authors' cooperation in conforming to these specifications is greatly appreciated. The contents of the book are organized by subject area rather than in the order in which papers were presented during the Institute. Many thanks are due to Warren Berning, Donald M. Hunten, Edward Llewellyn, J. Ortner, Henry Rishbeth, Harold I. Schiff, Lance Thomas, Alister Vallance Jones, and Gilbert Weill, who served as session chairmen during the Institute and contributed greatly to its success by skillfully directing the discussion period in a stimulating manner after each lecture.

Many persons contributed to the success of the Institute. The co-chairman, Dr Gilbert Weill, was most helpful during all phases of the preparation, planning, and conduct of the Institute. Drs Rolf Boström, Alv Egeland, Donald M. Hunten, Günther Lange-Hesse, M. Nicolet, J. Ortner, Harold I. Schiff, Erwin R. Schmerling, Lance Thomas, and Alister Vallance Jones were especially helpful in preparing the technical program.

Dr J. Hieblot, Director of the Groupe de Recherches Ionospheriques, played a most important role in helping to arrange the facilities in Orleans and in providing visual aids, recording equipment, office support, and projectionists. The assistant editor, Mrs Diana R. McCormac, checked the manuscripts and proofs, and worked hard to achieve a uniform style in this book.

Direct financial support was provided the Institute by the Army Research Office, Centre National de la Recherche Scientifique, Lockheed Palo Alto Research Laboratory, Advanced Research Projects Agency, Defense Nuclear Agency, and the Office of Naval Research. The European Space Research Organization assisted a number of attendees.

BILLY M. MCCORMAC

Palo Alto,
November 1972

TABLE OF CONTENTS

PART I

STRUCTURE AND COMPOSITION OF THE ATMOSPHERE

NEUTRAL AIR DENSITY AND COMPOSITION

U. VON ZAHN

Physikalisches Institut, Universität Bonn, Bonn, Germany (F.R.G.)

1. Observational Facts

1.1. DENSITY

The thermosphere (85 to 500 km) is heated primarily by the following processes:
 (a) absorption of solar UV, EUV, and X-rays ($\lambda = 1700$ to 8 Å);
 (b) dissipation of solar wind energy deposited into the thermosphere through solar wind magnetosphere upper atmosphere interactions (including absorption of accelerated solar wind particles); and
 (c) dissipation of tidal and gravity waves entering the thermosphere from below.

Common to all three energy sources is their strong temporal and/or spatial variability of source strength. This leads to correspondingly large variations in the temperature and total density of thermospheric air.

By far the most comprehensive set of data on upper atmosphere densities and their variations has been collected by observations of satellite drag. It is largely through these efforts that the following 6 different patterns of density variations have been clearly recognized: (1) the 11 yr solar cycle variation, (2) the seasonal-latitudinal variations, (3) the semi-annual variation, (4) the 27 day density variation, (5) the diurnal density variation, and (6) variations with geomagnetic activity. In addition *in situ* measurements by means of satellite- borne density gauges and accelerometers have shown (7) rapid density fluctuations probably connected with gravity waves.

The observational evidence for these variations is very extensive and has been reviewed by various authors (Priester *et al.*, 1967; Jacchia, 1970; Roemer, 1971). Even though the scope of this paper does not allow a detailed discussion of each of these variations the following remarks on some more recent developments are thought to be appropriate.

The *seasonal-latitudinal variations* are most pronounced for He, giving rise to the so-called He bulge. They will be discussed later in connection with composition measurements.

The *semi-annual* density variation is the least understood effect among the 7 variations. Satellite drag measurements also increasingly help to investigate this variation in the lower thermosphere even though the orbital lifetime is quite short of any satellite having a perigee about or below 150 km altitude. Total densities can be deduced both from observations of the changes in orbital period (Champion *et al.*, 1972) as well as from onboard accelerometers (Champion and Marcos, 1972). One of the important results has been the confirmation that the semi-annual variation in density extends with a sizable amplitude downwards into the lowest parts of the thermosphere (King-Hele and Walker, 1969). Analysis of data obtained from rocket grenade and pitot-static tube experiments led Cook (1969) to postulate a semi-annual variation of $\pm 12\%$ in density at a height of 90 km. The amplitude was later revised downwards to $\pm 7\%$ by Jacchia (1971a). In a recent analysis Volland *et al.* (1972) argue that the semi-annual variation may be explained by semi-annually varying Joule heating at thermospheric altitudes and by dissipation of tidal and gravity wave energy within the lower thermosphere.

The response of the neutral atmosphere to *geomagnetic* activity has now been measured by satellite-borne instrumentation (Taeusch *et al.*, 1971; Anderson and Sharp, 1972) with considerably higher spatial and temporal resolution than was possible earlier by satellite drag measurements (Roemer, 1971). From these data the picture is clearly emerging that first the time lag between a given magnetic disturbance and the atmospheric response at high latitudes is very short, apparently less than 1 hr, and second, that the *major* effects of a magnetic storm on the upper neutral atmosphere are quite localized and confined to magnetic latitudes above $50°$.

The *diurnal* variation exhibits a well pronounced density maximum at about 1400 hr LT and a temperature maximum between 1600 and 1700 hr LT. The phase of the density maximum and the phase delay between temperature and density variation are of great theoretical interest and form a cornerstone for testing theoretical model atmospheres. The phase difference between density and temperature has not yet found a commonly accepted and quantitative explanation. Fore more detailed discussions of this effect the reader is referred to Rishbeth (1969), Chandra and Stubbe (1970), Blum *et al.* (1971), and Mayr and Volland (1972).

1.2. COMPOSITION

1.2.1. *Determining Processes*

The composition of the upper atmosphere is strongly influenced by
    (a) diffusive separation above the turbopause (~ 100 km),
    (b) photochemical processes (like photodissociation, ion molecule reaction), and
    (c) dynamical processes, in particular vertical transport.
Due to diffusive separation the composition becomes strongly altitude dependent above 100 km. The other two effects, if strong enough, will cause deviations from diffusive equilibrium conditions. Therefore composition studies give valuable insights into the interplay between the three processes mentioned above in the upper atmosphere.

1.2.2. *Average Values*

Even though it is well recognized that the upper atmosphere exhibits strong variations in density as well as in composition it is felt that a first step towards an improved understanding of the state of the upper atmosphere is establishing reasonable average values for the composition of neutral air in the thermosphere.

For the *altitude of 150 km* von Zahn (1970) has made an extensive comparison between individual number density measurements for N_2, O_2, O, and Ar, and with total density values ϱ determined from satellite drag observations. He showed that the sum of the number density values yields a total mass density considerably lower than the one obtained from satellite drag. In order to solve this discrepancy he recommended to discard the O measurements obtained by rocket-borne mass spectrometers. As a result von Zahn proposed the following values (see Table I) for upper air composition at 150 km altitude.

Subsequently Jacchia in a recent model atmosphere (Jacchia, 1971b; =J71) strove to approach these values, as can be seen from Table I.

TABLE I

Average composition and density at 150 km altitude

Parameter	von Zahn (1970)	Jacchia (1971b) ($T_\infty = 900$ K)
$n(N_2)$ [cm^{-3}]	26 $\times 10^9$	25.4 $\times 10^9$
$n(O)$ [cm^{-3}]	23 $\times 10^9$	23.6 $\times 10^9$
$n(O_2)$ [cm^{-3}]	2.5 $\times 10^9$	2.68 $\times 10^9$
$n(Ar)$ [cm^{-3}]	0.05 $\times 10^9$	0.04 $\times 10^9$
ϱ [g·cm^{-3}]	1.96 $\times 10^{-12}$	1.95 $\times 10^{-12}$

Since publication of the J71 model the University of Bonn group performed a number of sounding rocket experiments in which the composition of neutral air in the lower thermosphere was measured by means of mass spectrometers with a newly developed He cooled ion source. The results of these flights (Offermann and von Zahn, 1971; Grossmann and Offermann, 1972; Offermann, 1972) fully support the high O values contained in J71.

For the *altitude of 400 km* Taeusch and Carignan (1972) have studied the average composition for a 1 yr period starting July 1969 using compositional data obtained from a mass spectrometer experiment onboard OGO-6. Their average values (for the latitude belt between $\pm 50°$) are given in Table II and they also compare with the relevant values of the J71 model. Even though there is good agreement with respect to the total density ϱ there is an apparently significant difference in the N_2 number density between the OGO-6 measurements and J71.

With the assumption of diffusive equilibrium one can show that the ratio of temperatures at 150 and 400 km is uniquely determined by the composition at these altitudes

TABLE II

Average composition and density at 400 km altitude
(data between $\pm 50°$ latitude)

Parameter	Taeusch and Carignan (1972)	Jacchia (1971b)
$n(N_2)$ [cm^{-3}]	1.29 $\times 10^7$	6.49 $\times 10^6$
$n(O)$ [cm^{-3}]	1.63 $\times 10^8$	1.65 $\times 10^8$
ϱ [g·cm^{-3}]	4.93 $\times 10^{-15}$	4.72 $\times 10^{-15}$
T_∞ [K]	1166	1065

from

$$\left(\frac{T_{150}}{T_{400}}\right)^{16-28} = \frac{\left(n(O)_{150}\right)^{28}}{\left(n(N_2)_{150}\right)^{16}} \cdot \frac{\left(n(N_2)_{400}\right)^{16}}{\left(n(O)_{400}\right)^{28}}. \tag{1}$$

Introducing the measured values of Tables I and II into Equation (1) yields a temperature ratio of

$$T_{150}/T_{400} = 0.246. \tag{2}$$

Assuming a temperature $T_{150} = 600$ K, one obtains from Equation (2) an average temperature $T_{400} = 2443$ K. Because this value is unacceptably high it is concluded that the two sets of average composition in Tables I and II are not compatible indeed. In which way this discrepancy will be solved remains to be seen.

1.2.3. *Variations in Composition*

Within the scope of this brief paper it is impossible to deal with all the possible temporal and spatial variations of the various constituents. Hence only some of the most prominent variations in composition will be dealt with.

(a) Data obtained from satellite-borne mass spectrometers have clearly shown two interesting effects: First very strong changes of neutral atmospheric composition in response to geomagnetic activity have been observed by Taeusch *et al.* (1971). Variations in the O/N_2 ratio suggest dynamic processes that cause a circulation in the atmosphere that is upward at the pole with subsidence at the equator.

The second effect is a considerable control of the *geomagnetic* field over *neutral* atmospheric composition. Clear evidence for this has been given with respect to the He distribution at high latitudes by Reber *et al.* (1971) and with respect to the N_2 distribution over the geomagnetic equator by Philbrick and McIsaac (1971), and Hedin and Mayr (1971) through mass spectrometer measurements onboard the satellites OGO-6, OV3-6, and OGO-6, respectively. These observations are consistent with the idea that the global distribution of minor constituents is rather sensitive to thermospheric winds. They in turn are controlled by ion concentrations and drift velocities ('ion drag') which are geomagnetically controlled.

(b) The strong seasonal-latitudinal variations in thermospheric He density (He bulge) have attracted further attention. Helium can be used as a tracer for dynamical processes due to its chemical inertness and its atomic weight which is much smaller than the mean molecular weight of air in the thermosphere. Recent rocket measurements in the lower thermosphere (Hickman and Nier, 1972; Schneppe, 1972), satellite measurements in the upper thermosphere (Reber *et al.*, 1971), and satellite drag measurements in the exosphere (Keating *et al.*, 1972) clearly depict these compositional changes as the most prominent compositional variation in the upper atmosphere. Following a suggestion by Johnson and Gottlieb (1970) these variations are interpreted as being due to a global circulation system transporting air from the summer pole region towards the winter pole region and giving rise to vertical air flow at high latiudes.

For altitudes near 1000 km, where He is the dominant constituent, a large and remarkably systematic *asymmetry* in density between the northern and southern hemisphere has been discovered recently by Keating *et al.* (1972) by analysis of satellite drag data. The phenomenon may be indicative of a basic asymmetry in heating of the lower thermosphere causing a stronger meridional outflow from the southern hemisphere in December (local summer) than the corresponding outflow from the northern hemisphere in June (local summer). The asymmetry may also be related to systematic differences in the turbopause altitudes of the two hemispheres.

(c) Particular attention is paid to regular compositional changes at turbopause altitudes. These variations are important as input data for atmospheric models because they will, if real, propagate upwards through the whole thermosphere and exosphere.

Barlier *et al.* (1971) noted systematic differences between exospheric temperatures deduced from incoherent-scatter measurements and from an analysis of satellite drag data. They have interpreted these differences as being due to a seasonal variation by $\pm 23\%$ of the O concentration at 120 km (maximum value attained in January–February).

Variations of O_2 concentration in the lower thermosphere have been inferred from satellite occultation measurements by Roble and Norton (1972) (seasonal variations) and by May (1972) (latitudinal variations).

1.2.4. *Minor Constituents*

Trace gases like NO, O_3, H_2O, and CO_2 are of paramount importance in the lower thermosphere for ionospheric processes and also for the energy balance of the neutral atmosphere. Their measurement, however, still remains a great challenge. The optical measurements of mesospheric NO concentrations by Barth (1964) and Meira (1971) are not yet universally accepted (see for example Strobel, 1972). Offermann *et al.* (1972), on the other hand, found reasonable agreement between mass spectrometric determined NO densities above 140 km altitude and model predictions by Strobel (1971).

No reliable H_2O measurement is available above the stratopause.

Carbon dioxide was found well mixed up to the turbopause and in approximate diffusive equilibrium higher up (Offermann and von Zahn, 1971; Offermann, 1972).

First measurements of O_3 in the altitude region 80 to 100 km have been obtained by use of mass spectrometers by Philbrick and Faucher (1972), and Grossmann and Offermann (1972). Ozone densities were also determined from OAO-2 stellar occultation data by Hays and Roble (1972). In all three cases the amounts of O_3 found were higher than predicted by appropriate models. For the mass spectrometer results, however, further data analysis is required to really confirm the atmospheric origin of the measured O_3. Ozone densities at 90 km derived from $O_2\,(^1\Delta_g)$ measurements by Good (1972) are more in line with present theoretical expectations.

2. Atmospheric Modeling

2.1. Physical models

A large bulk of data on the state and the behavior of the upper atmosphere above

120 km has been accumulated in the past 15 yr. In order to allow a comparison of results obtained at different times, at different locations, and for various parameters it is necessary to develop models for the upper atmosphere which can be used to normalize or calculate temperature, density, and compositional values under various geophysical conditions. A first step in this direction is the construction of strictly empirical models derived from a fitting process to all available observational data with due regard to basic assumptions like maintenance of hydrostatic and diffusive equilibrium. The most advanced model of this type is J71. A global empirical model of thermospheric composition based on OGO-6 mass spectrometer measurements has been published recently by Hedin *et al.* (1972).

A second and much more involved step is the attempt to produce theoretical models of the upper atmosphere from basic physical assumptions only, which requires a simultaneous integration of the complete system of the three-dimensional hydrodynamic equations including all the forces acting on the atmosphere and all the sources and sinks of energy. A theoretical model which describes the observed state of the upper atmosphere on a global scale with satisfactory accuracy does not yet exist, but partial solutions may be obtained by various simplifications.

An early attempt to formulate a purely theoretical model was made by Harris and Priester (1962). An important result of their work is that it is impossible to obtain agreement between measurements of the diurnal density variation and model predictions as long as absorption of solar radiation is considered the sole energy source for heating the upper atmosphere. Because there was no other heat source quantitatively known, Harris and Priester had to revert to a semi-empirical model by introducing an *ad hoc* 'second heat source'. In this way the model of Harris and Priester became quite successful in predicting the physical parameters of the upper atmosphere in accordance with the available observational data. The CIRA 1965 model values above 120 km were based on a further refinement of this approach.

The artificial 'second heat source' of Harris and Priester, however, plagued aeronomists for the next 10 yr. Today it is widely held that the energy balance of the neutral upper atmosphere is not only determined by absorption of solar radiation, but that there are *significant* contributions to the energy balance by dissipation of tidal and gravity wave energy, dissipation of solar wind energy through magnetospheric upper atmospheric interactions and by a whole family of transport processe-acting in the thermosphere. Furthermore the time-independent boundary conditions at 120 km altitude in the Harris and Priester model have been replaced in recens models either by more realistic time-dependent boundary conditions or by timet independent conditions at lower altitudes (90 km in case of J71). Through these means the 'second heat source' has now dissolved into a number of different effects. An excellent review of the prevailing situation with respect to theoretical models of the upper atmosphere has been given by Blum *et al.* (1971).

2.2. PHOTOCHEMICAL MODELS

A somewhat different class of models has been developed in studies of the chemical composition of the upper mesosphere and lower thermosphere. These models place major emphasis on the photochemistry of an atmosphere containing O, H, and N compounds in order to predict the altitude distribution of minor neutral constituents together with their diurnal and latitudinal variations. The main interest centers in the 60 to 120 km altitude region.

Bates and Nicolet (1950) initiated intensive studies of the photochemistry of an O-H atmosphere. Hunt (1966) demonstrated that above 50 km the assumption of photochemical equilibrium becomes invalid. Colegrove *et al.* (1965) introduced vertical eddy and molecular diffusion into photochemical models but included only a rather simple set of photochemical reactions in their model. This approach was improved by Hesstvedt (1968) who included in his steady state calculations a large set of chemical reactions pertinent to an O-H atmosphere.

Shimazaki and Laird (1970, 1972) went even further by including N and its oxides into a set of 38 photochemical reactions and by solving the time-dependent equations of motion and continuity. George *et al.* (1972) followed the approach taken by Shimazaki and Laird, but included in their models a different height dependence of the eddy diffusion coefficient and some mathematical refinements.

These models have tremendously improved our understanding of the behavior of minor constituents, radicals, and excited species in the upper mesosphere and lower thermosphere. The strong influence of dynamical processes on the concentrations and altitude profiles of most neutral constituents has become quite evident and this situation provides a close parallel to the previously discussed physical models for the thermosphere and exosphere. There is, however, an important difference in the development of upper atmospheric models and photochemical calculations for the mesopause region. For the upper atmosphere there exists now a great wealth of observational data. This will in general allow immediate and quantitative testing of any newly developed theoretical model atmosphere. On the other hand measurements of minor constituents at mesopause altitudes are still so scarce that only very general features of the relevant models can be checked (like absolute number densities and altitude profiles for O). This situation can only be improved by collection of more data about the state and behavior of this highly interesting 'mixing region' between the lower and upper atmosphere.

References

Anderson, A. D. and Sharp, G. W.: 1972, *J. Geophys. Res.* **77**, 1878.
Barlier, F., Perret, D., and Jaeck, C.: 1971, *J. Geophys. Res.* **76**, 7797.
Barth, C. A.: 1964, *J. Geophys. Res.* **69**, 3301.
Bates, D. R. and Nicolet, M.: 1950, *J. Geophys. Res.* **55**, 301.
Blum, P., Harris, J., and Priester, W.: 1971, *CIRA 1970*, in press.
Champion, K. S. W. and Marcos, F. A.: 1972, Paper, XV Plenary Meeting of COSPAR.
Champion, K. S. W., Marcos, F. A., and Schweinfurth, R. A.: 1972, Paper, XV Plenary Meeting of COSPAR.
Chandra, S. and Stubbe, P.: 1970, *Planetary Space Sci.* **18**, 1021.
Colegrove, F. D., Hanson, W. B., and Johnson, F. S.: 1965, *J. Geophys. Res.* **70**, 4931.
Cook, G. E.: 1969, *Nature* **222**, 969.
George, J. D., Zimmerman, S. P., and Keneshea, T. J.: 1972, *Space Res.* **12**, 695.

Good, R. E.: 1972, *Trans Am. Geophys. Union* **53**, 463.

Grossmann, U. and Offermann, D.: 1972, Paper, XV Plenary Meeting of COSPAR.

Harris, I. and Priester, W.: 1962, *J. Atmospheric Sci.* **19**, 286.

Hays, P. B. and Roble, R. G.: 1972, *Trans. Am. Geophys. Union* **53**, 466.

Hedin, A. E. and Mayr, H. G.: 1971, *Trans. Am. Geophys. Union* **52**, 872.

Hedin, A. E., Mayr, H. G., Reber, C. A., Carignan, G. R., and Spencer, N. W.: 1972, Paper, XV Plenary Meeting of COSPAR.

Hesstvedt, E.: 1968, *Geofys. Publik.* **27**, 1.

Hickman, D. R. and Nier, A. O.: 1972, *J. Geophys. Res.* **77**, 2871.

Hunt, B. G.: 1966, *J. Geophys. Res.* **71**, 1385.

Jacchia, L. G.: 1970, *Space Res.* **10**, 367.

Jacchia, L. G.: 1971a, *J. Geophys. Res.* **76**, 4602.

Jacchia, L. G.: 1971b, Special Report 332, Smithsonian Astrophysical Observatory.

Johnson, F. S. and Gottlieb, B.: 1970, *Planetary Space Sci.* **18**, 1707.

Keating, G. M., McDougal, D. S., Prior, E. J., and Levine, J. S.: 1972, Paper, XV Plenary Meeting of COSPAR.

King-Hele, D. G. and Walker, D. M. C.: 1969, *Planetary Space Sci.* **17**, 2027.

May, B. R.: 1972, Paper, XV Plenary Meeting of COSPAR.

Mayr, H. G. and Volland, H.: 1972, *J. Geophys. Res.* **77**, 2359.

Meira, L. G., Jr.: 1971, *J. Geophys. Res.* **76**, 202.

Offermann, D.: 1972, *J. Geophys. Res.* 6284.

Offermann, D. and von Zahn, U.: 1971, *J. Geophys. Res.* **76**, 2520.

Offermann, D. and Grossmann, K. U.: 1972, *Space Res.* **12**, 665.

Offermann, D., Pelka, K., and von Zahn, U.: 1972, *Int. J. Mass Spectr. Ion Phys.* **8**, 391.

Philbrick, C. R. and McIsaac, J. P.: 1971, Paper, XIV Plenary Meeting of COSPAR.

Philbrick, C. R. and Faucher, G. A.: 1972, *Trans. Am. Geophys. Union* **53**, 463.

Priester, W., Roemer, M., and Volland, H.: 1967, *Space Sci. Rev.* **6**, 707.

Reber, C. A., Harpold, D. N., Horowitz, R., and Hedin, A. E.: 1971, *J. Geophys. Res.* **76**, 1845.

Rishbeth, H.: 1969, *Ann. Geophys.* **25**, 495.

Roble, R. G. and Norton, R. B.: 1972, *J. Geophys. Res.*, 3524.

Roemer, M.: 1971, in F. Verniani (ed.), *Physics of the Upper Atmosphere*, Editrice Compositori, Bologna, p. 229.

Schneppe, G.: 1972, Forschungsbericht 72-29 Bundesministerium Fuer Bildung und Wissenschaft, Bonn.

Shimazaki, T. and Laird, A. R.: 1970, *J. Geophys. Res.* **75**, 3221.

Shimazaki, T. and Laird, A. R.: 1972, *J. Geophys. Res.* **77**, 276.

Strobel, D. F.: 1971, *J. Geophys. Res.* **76**, 2441.

Strobel, D. F.: 1972, *J. Geophys. Res.* **77**, 1337.

Taeusch, D. R. and Carignan, G. R.: 1972, *J. Geophys. Res.*, **77**, 4870.

Taeusch, D. R., Carignan, G. R., and Reber, C. A.: 1971, *J. Geophys. Res.* **76**, 8313.

Volland, H., Wulf-Mathies, C., and Priester, W.: 1972, *J. Atmospheric Terrest. Phys.* **34**, 1053.

von Zahn, U.: 1970, *J. Geophys. Res.* **75**, 5517.

NEUTRAL WIND STRUCTURE IN THE THERMOSPHERE DURING QUIET AND DISTURBED GEOMAGNETIC PERIODS

D. REES

Dept. of Physics, University College, London

1. Introduction

The neutral wind system in the thermosphere, which for the purposes of this discussion will be taken to be the atmosphere above an altitude of about 100 km, has been the subject of experimental investigation since about 1950 by a number of ground-based rocket and satellite techniques. A very considerable amount of data concerning its structure and movements has been amassed.

Early ground-based radio techniques – radio meteor (Greenhow and Neufeld, 1961) and spaced receiver ionospheric backscatter (Briggs *et al.*, 1950) – provided data of limited time and altitude range and resolution, while rocket vapor cloud experiments (Rosenberg *et al.*, 1963) were limited to twilight or nighttime measurements up to about 140 to 150 km. Since about 1967 several incoherent scatter sounders have been commissioned (Vasseur, 1969; Evans *et al.*, 1970; Woodman, 1970) and provided measurements of ion and electron density, temperature and velocities, enabling computation of the thermospheric neutral wind.

King-Hele (1964) presented evidence, from the rate of change of satellite orbital elements (particularly inclination), that the atmosphere above 200 km was rotating appreciably faster than the earth (super-rotation). Later investigations (King-Hele, 1972a) and other satellite measurements using accelerometers (DeVries, 1972) have indicated substantial time and spatial variations of the wind. Ground-based observations of the thermal broadening and Doppler shift of the 0 1 6300Å night-glow (and auroral) emission have also been used to determine the wind (Armstrong and Bull, 1970).

Despite measurements by these and other techniques, there is no comprehensive experimental description of the global neutral wind system above 100 km. There is very little information about tidal winds, which can be driven by solar heating and lunar/solar gravitational forces, or about the vertical movements, and generation, propagation and dissipation of gravity waves, which may directly, or indirectly, by modifying the stability of the atmosphere especially near the turbopause, change the vertical transportation rates of heat and atmospheric constituents. Both gravity waves and tides may effectively transport energy upward in the atmosphere from their levels of generation to altitudes where their energy is dissipated by the action of molecular viscosity (above 120 km) or eddy viscosity or turbulence (85 to 105 km).

This paper will discuss two aspects of the thermospheric neutral wind system which have been reasonably well documented by experimental data. These are the seasonal, latitudinal, and local time variations of the wind above 200 km altitude at mid-latitude associated with global transportation of solar EUV heating, and impulsive acceleration and heating processes associated with the response of the high latitude thermosphere to geomagnetic activity. The last section will briefly discuss the application of newly developed techniques which will allow measurements of thermospheric wind to be made throughout the day and night.

2. Neutral Wind Behavior Above 200 km at Mid-Latitude

Rishbeth (1972) has compared several recently computed models of the mid-latitude thermospheric wind with meridional wind measurements deduced from incoherent scatter sounders (principally the system at St. Santin, 45° N). He has also shown that equatorward winds during the night given by the models would help to maintain nighttime F region ionization by moving ions and electrons upward to regions of

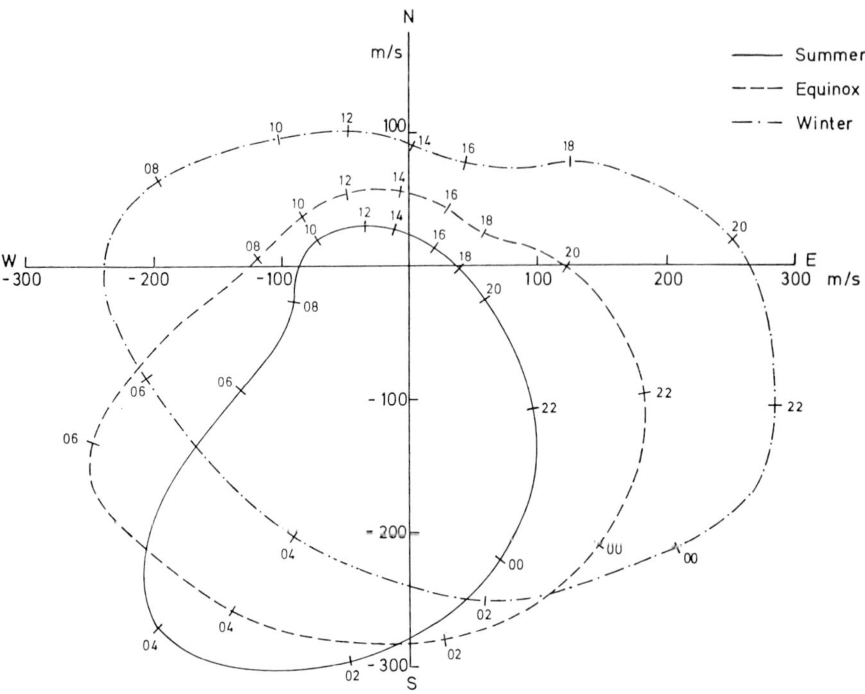

Fig. 1. Locus of wind velocity vector as a function of local time at a height of 300 km at 40°N during medium solar activity for summer, equinox and winter (Cho and Yeh, 1970a).

lower combination rates. Figure 1 shows the computed models of Cho and Yeh (1970a) of the mid-latitude thermospheric wind at a mean level of solar activity and for summer, winter, and equinox conditions, and Figure 2 compares meridional wind observations at St. Santin with model predictions (Amayenc and Vasseur, 1972).

2.1. MERIDIONAL WINDS

Both model and incoherent scatter data show that, independent of season, there is a

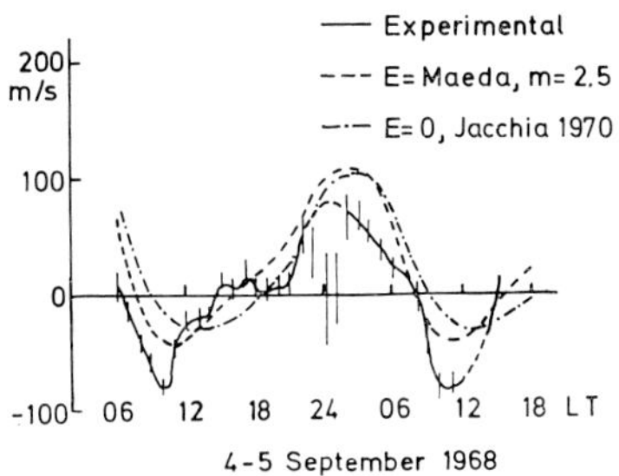

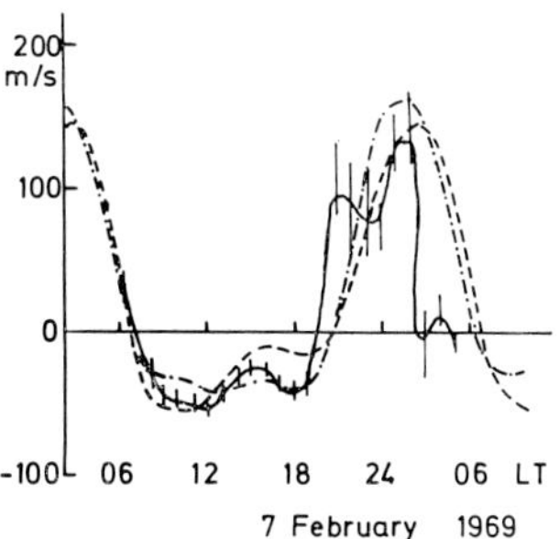

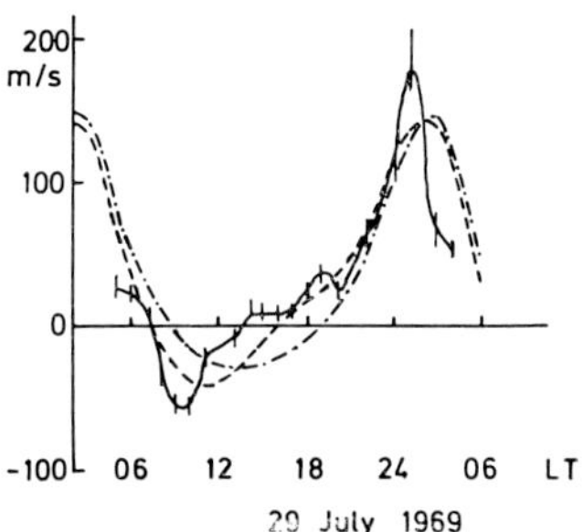

Fig. 2.   Diurnal variation of experimental and theoretical N–S winds (equatorward positive) at 250 km at St. Santin showing the effect of the neutral atmosphere density model (Jacchia, 1970) and electric fields (Maeda, 1963) for an equinox day, a winter day and a summer day (Amayenc and Vasseur, 1972).

predominant poleward wind during the day and a much stronger equatorward wind during the night. Most models have used the global pressure distribution of Jacchia (1965) as the principal driving force, although some of the latest models have used those of Jacchia (1970) or Jacchia and Slowey (1967). The principal constraint on the pressure gradient driven winds has been the ion drag on the neutral gas. Viscosity forces, Coriolis, and nonlinear terms of the equation of motion will act as relatively minor modifying effects. Other components of the winds such as atmospheric tides or gravity waves have not been included in the models, although Volland (1969) considers that they may be significant up to 300 km. The direction of the computed wind therefore follows approximately that of the Jacchia model pressure gradient, while the speed is largely controlled by the assumed electron density model. It is for the former reason that daytime winds are poleward and nighttime winds equatorward, and the latter reason that the nighttime wind speed (typically 300 m s^{-1}) corresponding to low electron density and ion drag greatly exceeds the daytime wind speed (typically 75 m s^{-1}) corresponding to high electron density and ion drag. Also for the latter reason, the Cho and Yeh (1970a) daytime wind speeds are higher in the winter (100 m s^{-1}) than in the summer (30 m s^{-1}).

A global electric field, associated with the Sq ionospheric current system, has been considered by Amayenc and Vasseur (1972) and Cho and Yeh (1970b). Both show that electric fields, by inducing a large-scale ion drift, can modify the wind system. Matsushita (1971, 1972) has concluded that present models of the mid-latitude ionospheric electric field are inadequate and the particular electric field model derived by Maeda (1955, 1963) used by the above authors is probably erroneous. At present therefore the inclusion of a global ionospheric electric field will make little improvement to the resulting wind model.

Johnson and Gottlieb (1970) have shown that wind systems derived from the Jacchia model create an energy deficit at high latitude, particularly in the winter hemisphere, due to the high equatorial wind speeds at night. D. Rees (1971a, 1972) has concluded that winter polar temperatures under very quiet geomagnetic conditions are, in fact, considerably lower than those of the Jacchia models. Figure 3 shows latitudinal temperature profiles at 170 km at local midnight for various levels of geomagnetic activity. Further evidence in favor of a cold winter pole was presented by D. Rees (1972) and Lloyd et al. (1972) from twilight measurements at mid-latitude (Figure 4). They found a mean meridional wind component of about 50 m s^{-1} blowing from the summer to the winter pole, probably sufficient to overcome the energy balance problem. There is a serious disagreement, however, between the rocket results and incoherent scatter measurements which indicate 300 m s^{-1} equatorward winds during the night in both winter and summer (Amayenc and Vasseur, 1972; Evans, 1972).

One midday neutral wind measurement near 200 km using vapor releases has recently been reported (D. Rees et al., 1972). The measurement, corresponding to a wind of 35 m s^{-1} directed about 40° west of north in the northern hemisphere, agrees well with Cho and Yeh's (1970a) summer model and is also in approximate agreement with the incoherent scatter data of Amayenc and Vasseur (1972). In the summer hemisphere near midnight, D. Rees (1972) and D. Rees et al. (1973) have reported some measurements made at 68°N. The observations showed equatorward wind components of the order of 150 m s^{-1} – much lower than the 300 m s^{-1} predicted by the various models.

2.2. ZONAL WINDS

King-Hele (1964, 1972a, 1972b) has calculated, from the rate of change of orbital inclination of several satellites, that the atmosphere between 200 and 500 km altitude is rotating faster than the earth. This 'super-rotation' may maximize in the afternoon or evening and reach a value of about 150 m s^{-1} near 300 km.

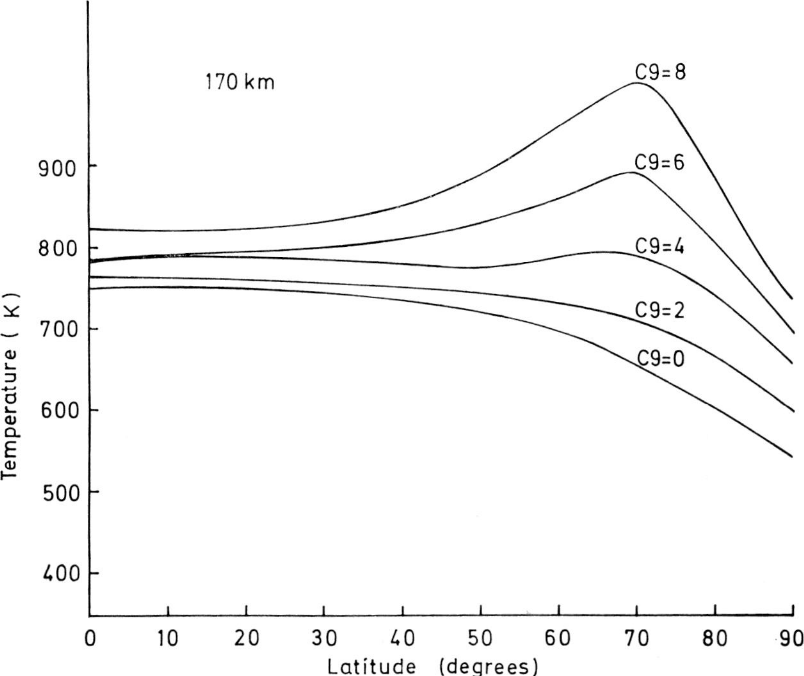

Fig. 3.   Latitudinal variation of temperature at 170 km as a function of geomagnetic activity for the midnight meridian (Lloyd *et al.*, 1972; D. Rees *et al.*, 1973).

Lloyd *et al.* (1972) found, from about forty mid-latitude twilight chemical release wind measurements (Figure 4), a mean diurnal eastward wind component for the winter hemisphere, which increased with height from 25 m s^{-1} at 200 km to 70 m s^{-1} above 220 km and was stronger in the evening than in the morning. In the summer hemisphere the mean diurnal zonal component was slightly westward (10 m s^{-1}) at 200 km and slightly eastward (10 m s^{-1}) above 220 km. However, the summer evening winds were also eastward while the morning winds were westward. The mean annual zonal component was nearly zero at 200 km and about 30 m s^{-1} eastward above 220 km, considerably less than the 70 m s^{-1} at 230 km given by King-Hele. From the difference in behavior of the wind in the winter and summer hemispheres Lloyd *et al.* (1972) concluded that the mean eastward winds of the winter hemisphere during quiet geomagnetic conditions could be due to a geostrophic wind (King-Hele, 1964) and reflect a mean polar temperature perhaps 200 to 300 K lower than the mean equatorial temperature at 150 to 200 km altitude.

3. Effects of Geomagnetic Activity on Thermospheric Winds

3.1. MORPHOLOGY OF THE AURORAL OVAL AND POLAR CAP

From the limited sample of *in situ* measurements of thermospheric wind velocity and

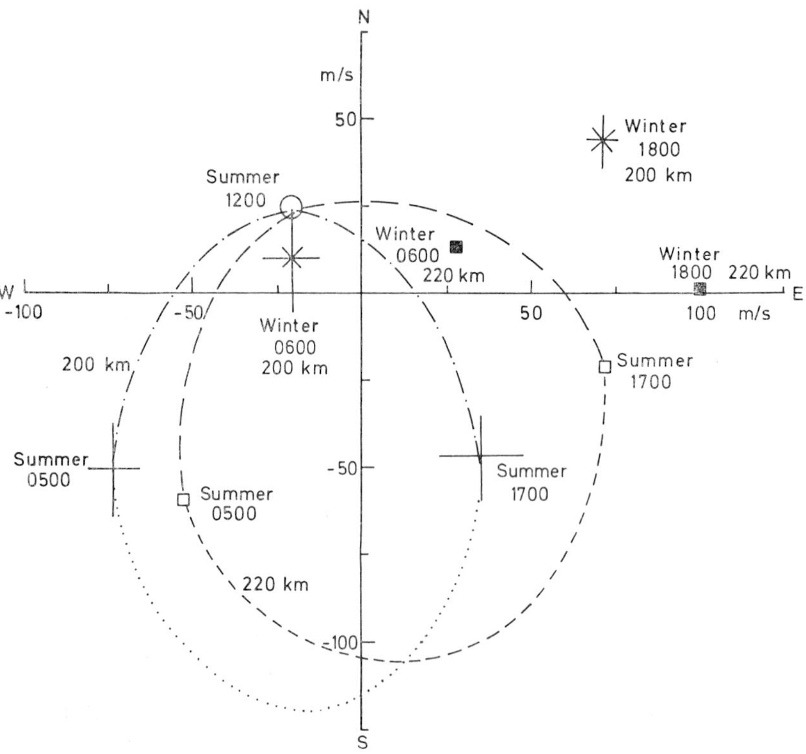

Fig. 4. Mid-latitude winds at 200 and 220 km for summer and winter morning and evening twilight (Lloyd *et al.*, 1972).

temperature at high latitudes (Haerendel, 1970; Wescott *et al.*, 1970; Meriwether *et al.*, 1971; D. Rees, 1971a, 1972; Andreeva *et al.*, 1972) there is a clear experimental picture of a high probability that elevated neutral gas temperatures and wind speeds will occur following even brief periods of moderate to severe geomagnetic activity (Jacchia and Slowey, 1964; Blamont and Luton, 1971; Jacchia, 1971; Roemer, 1971; D. Rees, 1971b, 1972). A detailed understanding of the complex interactions in the auroral and polar thermosphere is, however, dependent on the global structure of five quantities, about which there is only limited information at present: (i) electron and proton precipitation fluxes and energy spectra; (ii) auroral and polar ionospheric electric fields; (iii) vertical profiles of electron density and thus ionospheric conductivity; (iv) undisturbed polar neutral thermospheric structure; and (v) the spatial and temporal variations of the first three parameters throughout a typical auroral substorm. Of the above parameters, the best mapped at present is the auroral and polar electric field. Matsushita (1972) has reviewed existing work which is depicted schematically in Figure 5. Despite its limitations it will serve to describe the basic processes associated with the dumping of energy from the magnetosphere into the atmosphere.

Figure 5 shows the auroral oval at an average level of geomagnetic activity ($K_p \sim 3$). During very quiet periods ($K_p = 0$ or 1) the auroral oval will contract 3 to 5° polewards, while during very disturbed geomagnetic conditions ($K_p > 6$) it will expand about 5° equatorwards. The ionospheric electric fields of the oval are generally about 30 to 50 mV m^{-1}, northward before midnight and southward after midnight, while the dawn to dusk polar cap field is between 10 and 30 mV m^{-1}.

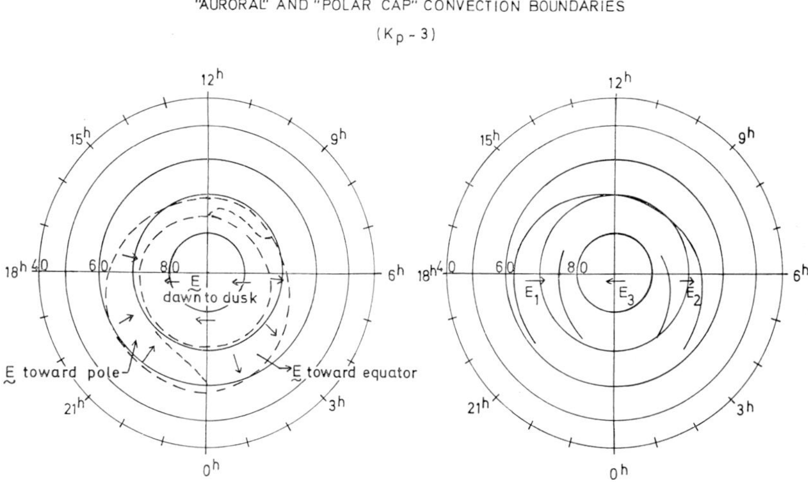

Fig. 5.   Average configuration of electric fields in the polar and auroral regions (Matsushita, 1972). *Left* – ground-based and rocket techniques; *right* – OGO 6 (after Heppner, 1969).

Satellite observations (O'Brien, 1964; Hess, 1968; Heppner, 1969) have shown that even under quiet geomagnetic conditions the auroral oval is characterized by continuous energetic particle precipitation and enhanced ionospheric electric fields and currents (the auroral electrojet). During periods of high geomagnetic activity the auroral electrojet and particle precipitation rates become greatly intensified. During magnetospheric substorm events a wide range of spectacular geophysical phenomena occur particularly in the nighttime sector. During the most intense substorms the auroral electrojet may increase from an integrated intensity during very quiet geomagnetic conditions of about 10^3 A to 10^6 A, driven by a meridional electric potential across the auroral oval of about 3×10^4 V, when the total global power dissipation of the auroral electrojet may approach 10^{11} W. Particle precipitation in the most intense aurora deposits up to 3 W m^{-2} in the atmosphere above 100 km. Worldwide, the particle precipitation may increase from $< 10^9$ W during quiet geomagnetic conditions to $> 10^{11}$ W during the most intense substorm. The regions of the thermosphere into which auroral energy is dumped react violently to the perturbations. The following section will deal with some of the main features of this reaction – the atmospheric substorm.

3.2. HEATING AND ACCELERATION MECHANISMS IN THE AURORAL AND POLAR THERMOSPHERE

Although energy from the magnetosphere is continuously being dumped into the auroral ovals, it is only during moderate to severe geomagnetic disturbances ($K_p > 3$) that the energy input will be enough to generate significantly observable atmospheric effects. These effects are:

(i) Vertical convection and horizontal advection due to temperature and pressure gradients caused by heating due to energetic particle precipitation (M. H. Rees, 1971; Kennel and Rees, 1972) and Joule losses of the auroral electrojet (Cole 1962, 1971).

(ii) Horizontal winds, initially zonal in the auroral oval, due to the momentum transfer via ion drag from the ions of the auroral electrojet to the neutral gas (Axford and Hines, 1961; D. Rees, 1971a).

(iii) Gravity wave generation (Hines, 1965).

During disturbed periods all three processes will occur simultaneously although the relative importance of each will vary at different phases of individual substorms and from one substorm to another. Additionally, it is not possible to model one of the processes adequately without at least qualitatively considering the others. The effects of the three processes can be conveniently described by dealing with two separate regions of the atmosphere: the region from 100 to 150 km – the auroral E region; and the region above 200 km.

3.3. NEUTRAL WIND RESPONSE UP TO 150 KM

D. Rees (1971c, 1972) (Figure 6) has reported a correlation for twilight wind measurements in Northern Scandinavia (68°N) between 130 and 150 km altitude given by

$$V_Y = + 2.5 (\pm 0.5) \, \Delta X,$$

where ΔX is the mean northward ground magnetic perturbation for the 2 hr prior to the experiment in gamma and V_Y is the westward neutral wind component in m s^{-1}. The correlation is attributed to drag on the neutral atmosphere by ions in the auroral oval driven by the electric fields with typical speeds of 1 km s^{-1}, westward in the evening and eastward in the morning. The correlation holds accurately for winter and equinox evening positive bay disturbances of 200 to 300γ which induce westward winds of the order of 500 m s^{-1} between 130 and 150 km. D. Rees *et al.* (1973) (Figure 6) have indicated that, for intense negative bay disturbances, the constant of proportionality is much reduced and for negative bays of $> 1000 \, \gamma$ the resulting eastward neutral wind is more adequately but approximately represented by

$$V_Y = + 0.5 \, \Delta X.$$

The change of the proportionality for severe negative bays is attributed to much of the ionospheric current flowing below 110 km, due to the higher energy electron spectrum associated with negative bay disturbances. Below 110 km the Hall drift of ions in an electric field is impeded and therefore the neutral gas is not accelerated

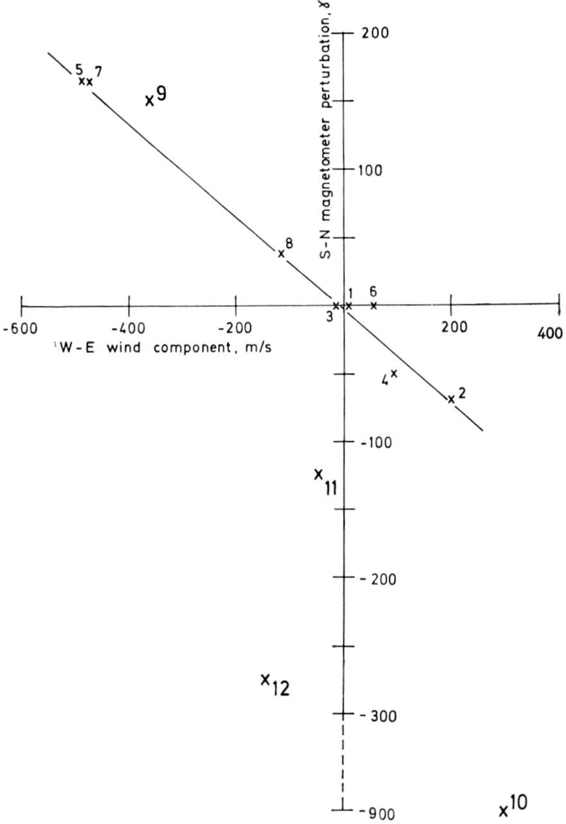

Fig. 6.   W–E neutral wind component (near 150 km generally) plotted against S–N ground magnetic perturbation (mean of 2 hr period prior to launch) for 68°N (D. Rees *et al.*, 1973).

proportionately to the strength of the ionospheric current and therefore magnetic disturbance. Due to the higher atmospheric density below 110 km local temperature changes are also relatively reduced. A mechanism which may enhance the effectiveness of westward evening wind generation during positive bays is simply the co-rotation of the wind and the substorm itself (relative to the earth's surface) which will tend to keep a particular parcel of air in a region of acceleration, i.e. the eastward electrojet of the auroral oval. On the other hand, an eastward gas motion during a morning negative bay will be counter-rotating with respect to its accelerating source.

The Heiss Island (80°N) wind data of Andreeva *et al.* (1972) show no clear correlation with magnetic perturbation, which may be partially attributed to the higher geomagnetic latitude of Heiss Island. During disturbed periods Heiss Island may be in the polar cap and subject to comparatively lower, or sporadic, auroral activity. Accompanied by lower electric fields and conductivity, wind acceleration would thus be less efficient than in the auroral oval. Also, D. Rees *et al.* (1973) have indicated that, during the summer period, defined as the time when the local ionosphere (at high latitude) is continuously sunlit, the lack of a strong differential between auroral and non-auroral E region electron density concentration may effect a coupling of the ion drag acceleration process to a larger volume of the atmosphere and thus reduce local wind speeds. This coupling may well be enhanced by the thermospheric wind circulation of the undisturbed summer pole (perhaps also the mechanism discussed by Fedder and Banks (1972) which implies equatorward winds of the order of 100 to 200 m s^{-1} in the auroral zone near midnight even in the absence of significant magnetic activity (see D. Rees *et al.*, 1973).

No strong correlation has been found between meridional winds up to 150 km altitude and geomagnetic activity comparable with that for the zonal wind. Since the temperature increases following geomagnetic heating are smaller below 150 km than above 150 km (D. Rees, 1971a; D. Rees *et al.*, 1973), the absence of high meridional winds probably reflects lower meridional temperature and pressure gradients.

3.4. Wind observations above 200 km

Several high latitude measurements have been reported by Haerendel (1970), Wescott *et al.* (1969, 1970), and Meriwether *et al.* (1971). Their results show a high probability of observing neutral wind speeds of the order of 300 to 600 m s^{-1} following ground-level disturbances of the order of a few hundred gamma. In general, the results show that ion drag is again an important process, particularly in the evening. However, many results past midnight also show high equatorward wind speeds. These winds probably reflect the intense heating in the midnight sector associated with negative bay events when the heating at lower altitude is adequate to support upward convection. Meridional winds in particular can only be maintained over a significant time and spatial extent if vertical convection due to heating can maintain the continuity requirements of the divergent equatorward flow.

If the heating of the disturbed F region of the auroral oval extends into the polar cap, as the observations of Blamont and Luton (1971) suggest, then a modest vertical convection of about 5 m s^{-1} estimated by M. H. Rees (1971) would be adequate to support an equatorward wind of several hundred meters per second across much of the midnight to dawn sector of the auroral oval. It is probable that only measurements of the divergence of the horizontal wind system will be able to confirm experimentally the presence of vertical winds of the order that M. H. Rees has computed.

3.5. Mid-latitude winds during disturbed periods

Smith (1968) and D. Rees (1972) have reported twilight mid-latitude thermospheric winds following moderate to severe geomagnetic disturbances ($K_p > 5$) in excess of 200 m s^{-1} between 150 and 200 km altitude. These measurements are consistent with the Doppler shift airglow observations of Nagy *et al.* (1971) and Armstrong and Bull (1970) of equatorward winds of about 300 m s^{-1} near midnight. The continuity of the mid-latitude winds with a high-latitude source has not been established experimentally, but it is probable that at least to 40° latitude there is a direct continuity with the winds of the disturbed auroral and polar regions.

Forbes (1972) has analyzed the latitudinal transport of energy by equatorward thermospheric winds during disturbed geomagnetic periods. He has shown that an equatorward wind of about 300 to 500 m s^{-1} can transport enough heat from the auroral and polar regions to account fully for the mid-latitude geomagnetic heating effect. Substantial enhancements of this meridional transport mechanism will occur due to mid-latitude dissipation of gravity waves generated in the auroral oval (Hines, 1965) and also the eventual dissipation via molecular viscosity of the kinetic energy of winds generated in the auroral oval.

4. Summary

There are many large areas of uncertainty in our understanding of atmospheric dynamics. Some of the more urgent problems which can be examined in the near future are

(a) Latitudinal temperature structure as a function of geomagnetic activity throughout the thermosphere, particularly during the night.

(b) Series of rocket measurements at all altitudes above 100 km throughout 24 hr. These measurements require the application of two newly developed techniques. First, daytime wind measurements using photometric ground or aircraft based observations of Li releases. Second, nighttime wind measurements above the ceiling at 150 km of the trimethyl aluminium trail technique, using a dye-tuned laser radar system. These new measurements should be coordinated where possible with the use of incoherent scatter sounders for comparative measurements, both to resolve discrepancies of the nighttime meridional wind, and also to obtain an adequate mid-latitude electric field model.

(c) At high latitude similar series of rocket measurements to those at mid-latitude are required. However, two additional problems need to be tackled simultaneously: first to obtain an adequate model of the time and spatial variation of the auroral energy sources, and second to observe the continuity of the wind systems associated with disturbed geomagnetic conditions by means of simultaneous measurements at several situations. It is likely that the latter objective can only be achieved in coordination with suitably placed ground instrumentation such as incoherent scatter sounders.

Acknowledgments

I am pleased to acknowledge the invaluable assistance provided by Miss A. Harris, Miss J. Norman and Miss B. Waters of the Department of Physics, University College London in the preparation of figures and presentation of the manuscript. I would also like to thank Dr J. Heppner and Dr J. Meriwether for discussions on their unpublished data.

References

Amayenc, P. and Vasseur, G.: 1972, *J. Atmospheric Terrest. Phys.* **34**, 351.

Andreeva, L. A., Katasyev, L. A., and Uvarov, D. B.: 1972, *Phil. Trans. Roy. Soc. London* **A271**, 559.
Armstrong, E. B. and Bull, J. A.: 1970, *Planetary Space Sci.* **18**, 784.
Axford, W. I. and Hines, C. O.: 1961, *Can. J. Phys.* **39**, 1433.
Blamont, J. E. and Luton, J. M.: 1971, Paper presented at COSPAR Meeting, Seattle.
Briggs, B. H., Phillips, G. J., and Shinn, D. H.: 1950, *Proc. Phys. Soc.* **B63**, 106.
Cho, H. R. and Yeh, K. C.: 1970a, *Radio Sci.* **5**, 881.
Cho, H. R. and Yeh, K. C.: 1970b, *Ann. Geophys.* **26**, 801.
Cole, K. D.: 1962, *Aust. J. Phys.* **15**, 223.
Cole, K. D.: 1971, *Planetary Space Sci.* **19**, 59.
DeVries, L. L.: 1972, in S. A. Bowhill, L. D. Jaff, and M. J. Rycroft (eds.), *Space Res.* **12**, Akademie-Verlag, Berlin, p. 867.
Evans, J. V.: 1972, *J. Atmospheric Terrest. Phys.* **34**, 175.
Evans, J. V., Brockelman, R. A., Julian, R. F., Reid, W. A., and Carpenter, L. A.: 1970, *Radio Sci.* **5**, 27.
Fedder, J. A. and Banks, P. M.: 1972, *J. Geophys. Res.* **77**, 2328.
Forbes, J. M.: 1972, private communication.
Greenhow, J. S. and Neufeld, E. L.: 1961, *Quart. J. Roy. Meteorol. Soc.* **87**, 472.
Haerendel, G.: 1970, in *The Upper Atmosphere*, Part IV of *Solar-Terrestrial Physics/1970* (ed. by E. R. Dyer), D. Reidel Publ. Company, Dordrecht-Holland, p. 87.
Heppner, J. P.: 1969, in B. M. McCormac and A. Omholt (eds.), *Atmospheric Emissions*, Van Nostrand Reinhold Company, New York, p. 251.
Hess, W. N.: 1968, *The Radiation Belt and Magnetosphere*, Blaisdell Publ. Co., Waltham, Mass.
Hines, C. O.: 1965, *J. Geophys. Res.* **70**, 177.
Jacchia, L. G.: 1965, *Smithsonian Contrib. Astrophys.* **8**, 215.
Jacchia, L. G.: 1970, Smithsonian Astrophys. Obs. Spec. Rep. No. 313.
Jacchia, L. G.: 1971, Smithsonian Astrophys. Obs. Spec. Rep. No. 332.
Jacchia, L. G. and Slowey, J.: 1964, *J. Geophys. Res.* **69**, 905.
Jacchia, L. G. and Slowey, J.: 1967, in R. L. Smith-Rose (ed.), *Space Res.* **10**, North-Holland Publ. Co., Amsterdam, p. 1077.
Johnson, F. S. and Gottlieb, B.: 1970, *Planetary Space Sci.* **18**, 1707.
Kennel, C. F. and Rees, M. H.: 1972, *J. Geophys. Res.* **77**, 2294.
King-Hele, D. G.: 1964, *Planetary Space Sci.* **12**, 835.
King-Hele, D. G.: 1972a, Paper presented at COSPAR Meeting, Madrid.
King-Hele, D. G.: 1972b, in S. A. Bowhill, L. D. Jaff, and M. J. Rycroft (eds.), *Space Res.* **12**, Akademie-Verlag, Berlin, p. 847.
Lloyd, K. H., Low, C. H., McAvaney, B. J., Rees, D., and Roper, R. G.: 1972, *Planetary Space Sci.* **20**, 761.
Maeda, H.: 1955, *J. Geomagn. Geoelect.* **7**, 121.
Maeda, H.: 1963, *Proc. of the Internat. Conf. on the Ionosphere*, Inst. of Physics and the Physical Society, London, p. 187.
Matsushita, S.: 1971, *Radio Sci.* **6**, 279.
Matsushita, S.: 1972, Paper presented at COSPAR Meeting, Madrid.
Meriwether Jr., J. W., Stolarik, J. D., Heppner, J. P., and Wescott, E. M.: 1971, Program 52nd Ann. Meeting AGU Washington, p. 30.
Nagy, A. F., Hayes, P. B., and McWatters, K.: 1971, in K. Ya. Kondratyev, M. J. Rycroft, and C. Sagan (eds.), *Space Res.* **11**, Akademie-Verlag, Berlin, p. 919.
O'Brien, B. J.: 1964, *J. Geophys. Res.* **69**, 1.
Rees, D.: 1971a, *J. Brit. Interplan. Soc.* **24**, 643.
Rees, D.: 1971b, *Planetary Space Sci.* **19**, 233.
Rees, D.: 1971c, *J. Brit. Interplan. Soc.* **24**, 233.
Rees, D.: 1972, *Phil. Trans. Roy. Soc. London* **A271**, 563.
Rees, D., Neal, M. P., Low, C. H., Hind, A. D., Burrows, K., and Fitchew, R. S.: 1972, *Nature* **240**, 32.
Rees, D., Aggson, T. L., Burrows, K., Haerendel, G., and Wilson, J. W. G.: 1973, in M. J. Rycroft and S. K. Runcorn (eds.), *Space Res.* **13**, in press.
Rees, M. H.: 1971, Presented at IUGG Symposium, Moscow.
Rishbeth, H.: 1972, *J. Atmospheric Terrest. Phys.* **34**, 1.

Roemer, M.: 1971, in K. Ya. Kondratyev, M. J. Rycroft, and C. Sagan (eds.), *Space Res.* **11**, Akademie-Verlag, Berlin, p. 965.
Rosenberg, N. W., Golomb, D., and Allen, E. F.: 1963, *J. Geophys. Res.* **68**, 5895.
Smith, L. B.: 1968, *J. Geophys. Res.* **73**, 4959.
Vasseur, G.: 1969, *J. Atmospheric Terrest. Phys.* **31**, 397.
Volland, H.: 1969, *Planetary Space Sci.* **17**, 1581.
Wescott, E. M., Stolarik, J. D., and Heppner, J. P.: 1969, *J. Geophys. Res.* **74**, 3469.
Wescott, E. M., Stolarik, J. D., and Heppner, J. P.: 1970, in B. M. McCormac (ed.), *Particles and Fields in the Magnetosphere*, D. Reidel Publishing Company, Dordrecht-Holland, p. 229.
Woodman, R. F.: 1970, *J. Geophys. Res.* **75**, 6249.

THE DISTRIBUTION OF MINOR CONSTITUENTS IN THE STRATOSPHERE AND LOWER MESOSPHERE

E. A. MARTELL

National Center for Atmospheric Research, Boulder, Colo. 80302, U.S.A.*

1. Introduction

Experimental determination of the distribution of minor constituents in the middle atmosphere between 20 and 80 km altitude, a region accessible only by balloons and rockets, is a neglected area of atmospheric research. The distributions of gaseous minor constituents in this region are governed by many processes: (a) the pattern and rate of troposphere-stratosphere and stratosphere-mesosphere exchange, (b) transport processes which influence the meridional and vertical redistribution of constituents at these levels, and (c) chemical reactions in the stratosphere and mesosphere. The distribution of particulate constituents in the stratosphere and mesosphere is further complicated by the influence of particle growth and sedimentation processes.

Observations of the stratospheric distribution of unique radioactive tracers and fission products from nuclear explosions in the atmosphere have provided information on residence times and transport processes in the middle atmosphere and the most pertinent literature is briefly reviewed, below. In addition, recent experimental results from balloon and rocket sampling which provide information on the altitude distribution of H_2O, CH_4, H_2, CO_2 and other minor constituents in the stratosphere and lower mesosphere are summarized and discussed. The atmospheric nomenclature used throughout this paper is based on the thermal structure of the atmosphere and was recommended by the International Union of Geodesy and Geophysics at Helsinki in 1960 (shown in Figure 1).

2. Transport Processes

The complex circulation processes which govern the redistribution of minor constituents within the stratosphere and mesosphere, and the patterns and rates of troposphere-stratosphere exchange and stratosphere-mesosphere exchange have been clarified by recent meteorological and tracer studies. Tropospheric air rises slowly across the cold ($193 \pm 5K$) tropical tropopause where aerosols and H_2O are largely removed by ice particle growth and sedimentation. Transport in this region has been discussed by Brewer (1949) and Reed and Vlcek (1969). Within the lower stratosphere atmospheric constituents are spread poleward by the competing processes of eddy mixing and slow mean motions. Rates of meridional and vertical eddy mixing in the stratosphere vary widely in time and space and are highest at the higher latitudes in winter (Newell *et al.*, 1966). The redistribution of nuclear bomb debris at ballooning

---

* The National Center for Atmospheric Research is sponsored by the National Science Foundation.

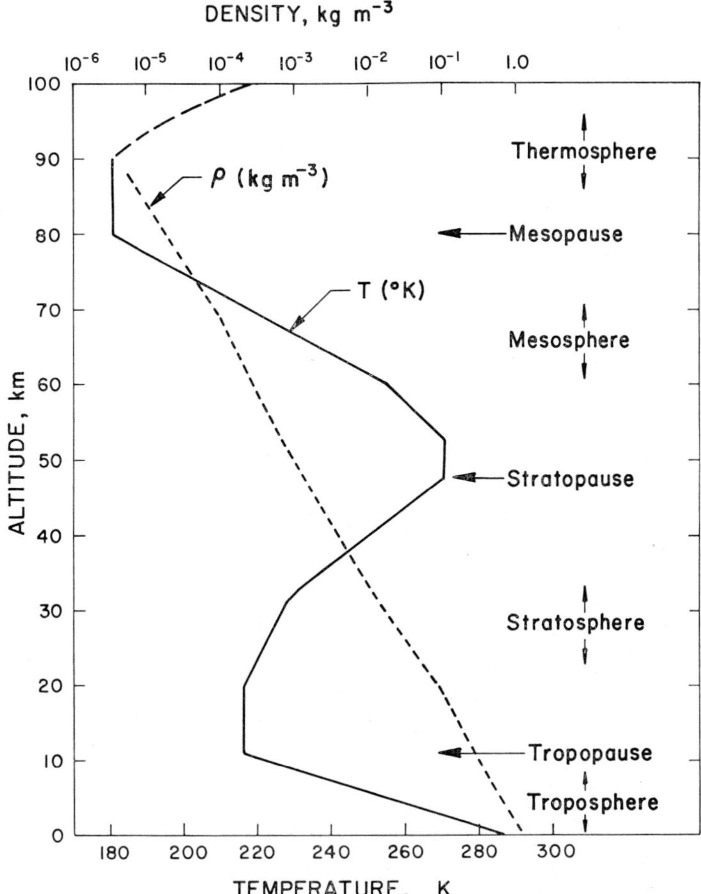

Fig. 1. Temperature and density profile according to the U.S. Standard Atmosphere (1962) and the IUGG nomenclature (1960).

altitudes indicates that poleward transport dominates the circulation between 23 and 41 km altitude (Martell, 1968, 1970a). Thus the stratospheric circulation clearly involves a well defined meridional cell within each hemisphere, with upward flow in the tropical statosphere, poleward transport in the middle stratosphere, and downward transport at high latitudes. Periodically, this cell is highly disturbed, principally by biennial wind reversals in the tropical stratosphere, by the formation and breakup of the polar vortex each year and by sudden warmings of the stratosphere in winter (Wallace and Newell 1966; Dickinson, 1972). Eddy diffusion rates in the stratosphere, mesosphere, and lower thermosphere are discussed elsewhere (Brewer, 1949; Kellogg, 1964; Colegrove *et al.*, 1965; Johnson and Wilkins, 1965; Martell, 1970a). Because of the stratospheric meridional circulation and its periodic disturbances, the stratospheric distribution of minor constituents cannot be explained by vertical eddy mixing and chemical processes alone.

The mean residence time of minor constituents at various stratospheric levels increases with appreciable altitude, with estimates of about 1 mo in the lower stratosphere between the polar tropopause and 15 km, about 1 to 2 yr near 20 km altitude, and about 4 to 20 yr in the upper stratosphere (Martell, 1970a). Stratospheric residence time estimates have been based on particulate radioactive tracers (Kalkstein, 1962; Salter, 1964; Telegadas and List, 1964; Bhandari and Lal and Rama, 1966; Leipunskii *et al.*, 1970) which are influenced by particle growth and sedimentation processes. Residence times applicable for H_2O and trace gases will be somewhat longer, particularly in the tropical stratosphere where slow mean motions upward are involved.

High altitude tracer experiments have demonstrated that constituents of the mesosphere mix downward through the stratosphere selectively at high latitudes in winter, in keeping with the meridional cell described above. There also is evidence to suggest that transport from the stratosphere into the mesosphere occurs primarily at high latitudes. Enhanced vertical eddy mixing is known to take place within the polar westerly regime (Newell *et al.*, 1966). In addition, prominent enhancements of Li were observed near the mesopause in the northern hemisphere during 1960, 1961, and 1962 (Gault and Hunten, 1963; Sullivan and Hunten, 1964). In this period Soviet thermonuclear debris in the Arctic stratosphere provided the only Li source which can explain these three enhancements. These twilight Li enhancements suggest that air from the Arctic upper stratosphere is injected into the lower mesosphere each year when the westerly vortex is formed. The time delay between onset of the polar vortex, in late September, and the observed twilight Li maxima, in mid-November, is consistent with the time required for Li transport from the lower mesosphere to the mesopause by vertical eddy mixing. On this basis, H_2O carried up from the troposphere in nuclear debris clouds from the 1961 and 1962 Soviet tests can explain the occurrence of the spectacular noctilucent cloud displays observed in the northern hemisphere during 1963, 1964, and 1965.

These transport processes significantly influence the distribution of minor constituents in the middle atmosphere. The long residence times for constituents in the stagnant atmospheric layer between 20 and 50 km altitude allow appreciable time for the partial or complete oxidation of gaseous constituents (CH_4, H_2, SO_2, NO_2, etc.) which mix up from the troposphere. The meridional circulation cell in the stratosphere in each hemisphere, and the periodic perturbations of this cell, give rise to variations in the altitude and latitude distribution of minor constituents in the stratosphere. The stratosphere-mesosphere exchange of air when the polar vortex is formed, suggested above, also may introduce significant seasonal or periodic changes in the concentration of minor constituents in the mesosphere.

3. Rocket Sampling Results

3.1. SAMPLING METHOD

A rocketborne cryogenic air sampler, developed by the author and his associates at the National Center for Atmospheric Research, successfully retrieved a large, representative air sample between 44 and 62 km altitude in NASA Aerobee 4.130 UA at White Sands, September 4, 1968. The sampling technique and some of the results from the analysis of this stratopause air sample have been reported elsewhere (Martell, 1970b; Scholz *et al.*, 1970; Bieri *et al.*, 1970; Ehhalt *et al.*, 1972) and are summarized below.

The cryogenic air sampler, illustrated schematically in Figure 2, employs a simple normal shock diffuser inlet and collects a column of air at supersonic speeds during

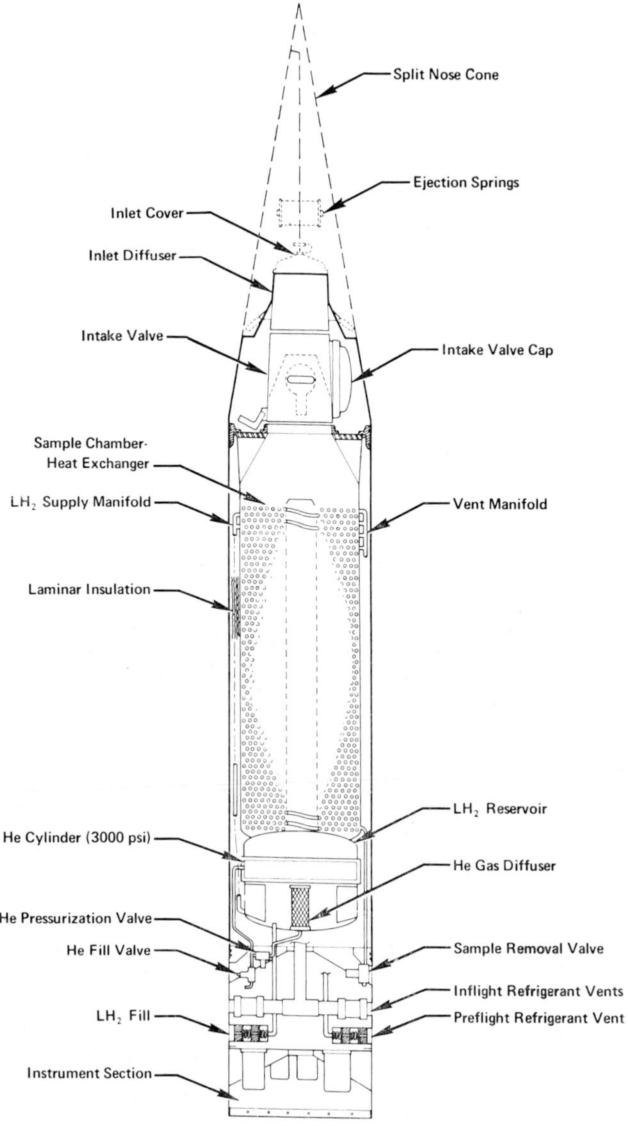

Fig. 2. Schematic diagram of rocket-borne cryogenic air sampler.

ascent. The sampled air condenses on stainless steel coils cooled with pressurized liquid H_2 or liquid Ne. After parachute recovery of the sampler, the collected air is allowed to warm to room temperature in the sample chamber and then is transferred by cryopumping into stainless steel pressure cylinders. Water vapor and CO_2 are quantitatively recovered in a liquid N_2 trap during air sample transfer. The remaining H_2O adsorbed on the sample chamber walls subsequently is recovered by high temperature bakeout of the sampler.

3.2. RESULTS

The concentrations of minor constituents present in the September 4, 1968 stratopause air sample are summarized in Tables I and II. Some of the results and the methods of analysis are discussed in the tabulated references (Table II).

TABLE I

Comparison of ^{20}Ne, ^{40}Ar, and ^{84}Kr in
air at the stratopause and in surface air

Isotope	Concentration ratio[a]
^{20}Ne	1.003 ± 0.004
^{40}Ar	1.002 ± 0.004
^{84}Kr	1.001 ± 0.004

[a] Ratio of the concentration of the given isotope in the high altitude air sample to its concentration in surface air (concentrations in atoms cm^{-3} STP)

TABLE II

The concentration of minor constituents at the stratopause, 31°N, September 1968

	Concentration (ppmv)	Reference
CH_4	0.25 ± 0.02	Ehhalt *et al.* (1972)
H_2	$0.40 \pm {}^{0.05}_{0.10}$	Scholz *et al.* (1970)
H_2O	~ 4.0	Figure 4 and text
CO_2	308 ± 3	See text
Ne	18	Bieri *et al.* (1970)
Ar	9.3×10^3	Bieri *et al.* (1970)
Kr	1.14	Bieri *et al.* (1970)
He	(5.24)	Bieri *et al.* (1970)
^{3}H (H_2O)	6.0×10^7 T.U.[a]	See text
^{3}H (H_2)	$\geq 1.7 \times 10^7$ T.U.[a]	See text
^{14}C (CO_2)	$(242 \pm 25)\%$ above recent natural[b]	

[a] 1 T.U. (tritium unit) = 1 tritium atom per 10^{18} H atoms.
[b] 95% of NBS Standard.

3.2.1. *Noble Gases*

Neon, Ar, and Kr were determined by isotope dilution mass spectrometry (Bieri *et al.*, 1970) and are compared with identical measurements for dry surface air (Table I). Errors given include the precision (3σ) of measurements and systematic errors. These inert gas results provide the first experimental confirmation that gravitational effects have negligible influence on the composition of the air in the upper stratosphere and lower mesosphere. The results also show that this sampling technique provides un-differentiated air samples and permits the quantitative recovery of the noncondensables (H_2, He and Ne) as well as the condensable gaseous constituents of air.

3.2.2. *Methane*

The concentration of CH_4 in the stratopause sample was measured by gas chromatography (Heidt and Ehhalt, 1972) and determined to be 0.25 ± 0.02 ppmv (Ehhalt *et al.*, 1972). This result, $\sim 20\%$ of the average concentration of CH_4 in the troposphere, is attributed to oxidation of CH_4 as it mixes up through the stratosphere, by reaction with O, OH, and HO_2. This result makes it evident that the destruction of CH_4 in the upper atmosphere takes place mainly by chemical reactions below 50 km rather than by photodissociation at higher levels (Nicolet, 1970). The measured CH_4 concentrations vs. altitude are given in Figure 3.

3.2.3. *Water Vapor*

Because of the presence of residual H_2O in the sample chamber wall in amounts comparable to the H_2O in the high altitude sample, the experimental determination of the stratopause water vapor concentration was difficult. Our published result, based on isotopic and tritium analysis of two recovered H_2O fractions, was 3 to 10 ppmv (Scholz *et al.*, 1970). A better estimate can be made using the measured H_2O content of the lower tropical stratosphere (Figure 4) and the CH_4 data (Figure 3). The 2 yr average concentration of H_2O vs. altitude based on balloon measurements over Trinidad (Mastenbrook, 1968) (Figure 4) shows a minimum mixing ratio of 2.1×10^{-6} near 20 km altitude and an increase at higher levels attributable to contamination by H_2O desorbed from detector surfaces. Assuming that the contaminant H_2O contribution varies inversely with atmospheric pressure above 22 km altitude, the corrected H_2O mixing ratio at 22 km is $\sim 1.5 \times 10^{-6}$. Due to the oxidation of CH_4 and H_2, the H_2O mixing ratio increases slowly with altitude above 22 km corresponding to $\sim 2.5 \times 10^{-6}$ at the stratopause and will reach $\sim 3.0 \times 10^{-6}$ in the lower mesosphere if the H_2 and CH_4 are completely oxidized (Figure 4). Based on the measured total tritium content of the water fraction (Scholz *et al.*, 1970) the corresponding tritium concentration of H_2O at the stratopause (Table II) was 6.0×10^7 T.U.

3.2.4. *Molecular Hydrogen*

The concentration of H_2 (Table II) also was determined by gas chromatography (Heidt and Ehhalt, 1972). Because pressurized liquid H_2 was used as cryogen in the

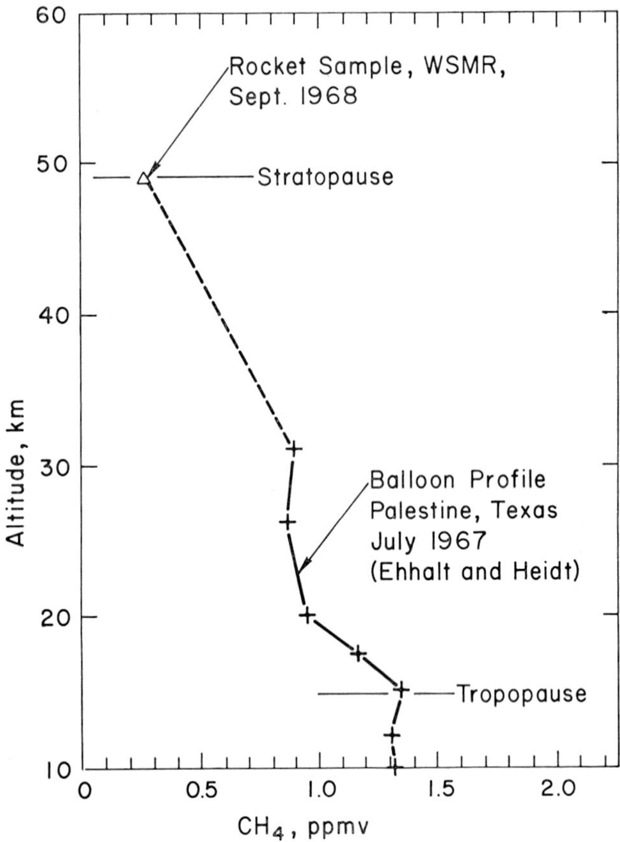

Fig. 3.   Methane concentration vs. altitude at $\sim 31\,^\circ$N based on balloon and rocket sampling results.

September 1968 rocket sampling flight, the extent of possible H_2 contamination of the sample remains somewhat uncertain. However further confidence in our result for H_2 is provided by recent determination that the tritium content of the H_2 fraction is 1.7×10^7 T.U. (Table II). For isotopic exchange equilibrium between H_2 and H_2O at -10°C (the mean temperature over the altitude interval of sampling), $K = = (HTO/H_2O) \times (H_2/HT) = 8.5$ (Bigeleisen, 1965). The tritium content of the H_2 and H_2O (Table II) corresponds to equilibrium at a higher temperature, possibly explained by the thermonuclear explosion origin of $\sim 99\%$ of the tritium (Scholz *et al.*, 1970). The tritium results indicate that an appreciable fraction of the H_2 must have been collected aloft and cannot be attributed to contamination by cryogen H_2 which had a substantially lower tritium content.

3.2.5. *Carbon Dioxide*

The concentration of CO_2 in the stratopause sample was determined by its quantitative separation and measurement. During air sample transfer, CO_2 and H_2O were collected in a liquid N_2 cooled trap. Subsequently, with the trap in dry ice, the CO_2 was transferred quantitatively to a vessel of calibrated volume and its temperature and pressure determined. A total of 67.27 ± 0.23 ml CO_2 was recovered from $218.1 \pm \pm 1.4$ l atmospheres of high altitude air, STP, corresponding to a volume concentration of 308 ± 3 ppm.

This result for CO_2 at the stratopause in September 1968 is significantly lower than the ~ 319 ppm CO_2 observed concurrently in surface air at Mauna Loa. The difference can be attributed to the long residence time of air at the stratopause and the lower

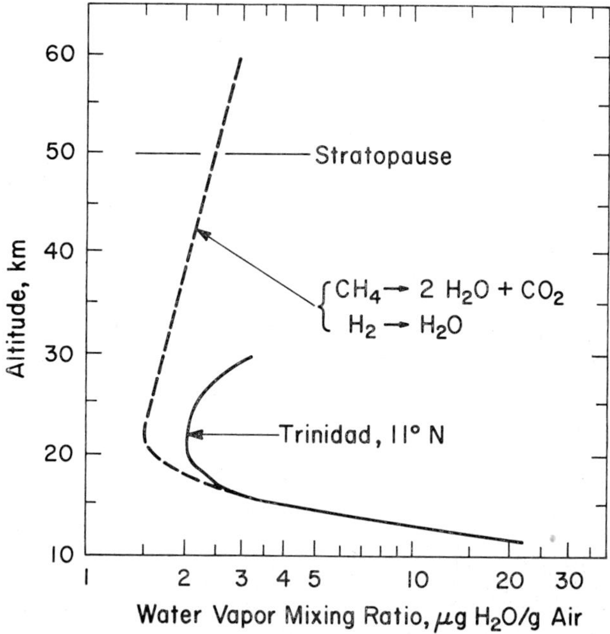

Fig. 4.   Water vapor concentration vs. altitude based on balloon and rocket measurements. Solid curve shows median H_2O mixing ratio over Trinidad during 1964 and 1965 (Mastenbrook, 1968). Dashed curve includes correction for contaminant H_2O and the increase in H_2O with altitude due to CH_4 and H_2 oxidation (see text).

concentrations of CO_2 in air which mixed up from the troposphere in the past. Based on observations at Mauna Loa since the IGY, the air concentration of CO_2 currently increases by ~ 0.7 ppm each year due to fossil fuel combustion (Keeling *et al.*, 1968). It also is possible that the lower CO_2 concentration at the stratopause is, in part, due to the photodissociation of CO_2 in the mesosphere. This possibility can be readily resolved by measurement of CO in high altitude air. Carbon-14 measurement of the CO_2 fraction (Table II) shows a ^{14}C excess ~ 4 times that concurrently observed in the lower atmosphere (Nydal and Lovseth, 1970), consistent with the longer residence time and excess tritium at the stratopause.

4. Future Plans

Following the successful rocket sampling flight of September 4, 1968, we have developed an improved sampler which employs liquid Ne as cryogen and is designed to allow bakeout of the sample chamber at 450°C. These two features should make possible the reliable measurement of H_2O and H_2 in high altitude air samples collected in the future. In addition, techniques for the fractional distillation of large air samples have been developed and tested to provide for the separation of H_2 for isotopic and tritium analysis and also to make possible quantitative enrichment of CH_4, CO, and other trace gases which may be present below their limits of detectability in the collected air samples. Laboratory techniques for the measurement of CO and N_2O also are being developed. It is planned to utilize the improved cryogenic sampler in a series of rocket flights at White Sands, beginning in late 1972. Air samples will be collected over the altitude intervals: 40 to 50, 50 to 60, and 60 to 75 km and subjected to detailed isotopic and trace gas analysis.

Acknowledgments

I am pleased to acknowledge the work of two close associates, D. H. Ehhalt and L. E. Heidt, who carried out most of the laboratory isotopic and trace gas measurements reported above. The cryogenic rocket sampler development and its research application were jointly supported by the National Aeronautics and Space Administration under Contract NASr-224 and by the National Center for Atmospheric Research, sponsored by the National Science Foundation.

References

Bhandari, N. and Lal and Rama, D. : 1966, *Tellus* **18**, 391.

Bigeleisen, J.: 1965, *Science* **147**, 463.

Bieri, R. H., Koide, M., Martell, E. A., and Scholz, T. G.: 1970, *J. Geophys. Res.* **75**, 6731.

Brewer, A. W.: 1949, *J. Roy. Meteorol. Soc.* **75**, 351.

Colegrove, F. D., Hanson, W. D. and Johnson, F. S.: 1965, *J. Geophys. Res.* **70**, 4931.

Dickinson, R. E.: 1972, 'Motions in the Stratosphere', in *Proceedings of Department of Transportation Conference on Climatic Impace Assessment, Cambridge, Mass.,* in press.

Ehhalt, D. H., Heidt, L. E., and Martell, E. A.: 1972, *J. Geophys. Res.* **77**, 2193.

Gault, W. A. and Hunten, D. M.: 1963, *Nature* **198**, 469.

Heidt, L. E. and Ehhalt, D. H.: 1972, *J. Chromatogr.* **69**, 103.

Johnson, F. S. and Wilkins, E. M.: 1965, *J. Geophys. Res.* **70**, 1281. Correction: 1965, *J. Geophys. Res.* **70**, 4063.

Kalkstein, M. I.: 1962, *Science* **137**, 645.

Keeling, C. D., Harris, T. B., and Wilkens, E. H.: 1968, *J. Geophys. Res.* **73**, 4511.

Kellogg, W. W.: 1964, *Space Sci. Rev.* **3**, 275.

Leipunskii, O. I., Konstantinov, J. E., Fedorov, G. A., and Scotnikova, O. G.: 1970, *J. Geophys. Res.* **75**, 3569.

Martell, E. A.: 1968, *J. Atmospheric Sci.* **25**, 113.

Martell, E. A.: 1970a, *Adv. Chem. Series* **93**, 138.

Martell, E. A.: 1970b, *J. Appl. Meteorol.* **9**, 170.

Mastenbrook, H. J.: 1968, *J. Atmospheric Sci.* **25**, 299.

Newell, R. E., Wallace, J. M., and Mahoney, J. R.: 1966, *Tellus* **18**, 363.

Nicolet, M.: 1970, *Ann. Geophys.* **26**, 531.

Nydal, R. and Lovseth, K.: 1970, *J. Geophys. Res.* **75**, 2271.
Reed, R. J. and Vlcek, C. L.: 1969, *J. Atmospheric Sci.* **26**, 163.
Salter, L. P.: 1964, in Fallout Program Quarterly Summary Report, HASL-142, United States Atomic Energy Commission Health and Safety Laboratory, p. 303.
Scholz, T. G., Ehhalt, D. H., Heidt, L. E., and Martell, E. A.: 1970, *J. Geophys. Res.* **75**, 3049.
Sullivan, H. M. and Hunten, D. M.: 1964, *Can. J. Phys.* **42**, 937.
Telegadas, K. and List, R. J.: 1964, *J. Geophys. Res.* **69**, 4741.
Wallace, J. M. and Newell, R. E.: 1966, *Quart. J. Roy. Meteorol. Soc.* **92**, 481.

AEROSOLS AND PARTICLES

F. LINK

Institut d'Astrophysique, Paris, France

1. Introduction

The existence of aerosols above 70 km in the atmosphere has been established by methods developed over two centuries. Only two optical methods are treated here, the extinction and the scattering of light by the aerosols. Long optical paths render the small effects readily observable and measurements are described. For non-absorbing particles extinction is caused by scattering alone, and a simple relation exists between extinction and scattering observations. If the particles are absorbing in addition, the extinction is augmented above the scattering value.

Various assumptions concerning the density and the vertical distribution of cosmic dust are possible depending on the distribution of particle size, law of fall that is applicable, etc....

All results are subject to variation by vertical air motion and to variations in the history of the dust particles in approaching the earth, i.e., at what level ablation may have occurred. In any case, the mixing ratio, $n(a)/n$ (aerosol density/air density) is a constant throughout the upper atmosphere >60 km altitude.

2. Extinction of Light

2.1. Bouguer's method

Bouguer introduced a method for determining atmospheric extinction which involves measuring the light intensity I of extraterrestrial sources, at different zenith angles z. In general light of intensity (I_0) outside the atmosphere may traverse a Rayleigh atmosphere, high absorbing layer, and troposphere haze layer with respective light paths $M(z)$, $G(z)$, and $H(z)$ and respective extinction coefficients A, B, and C. The intensity at the earth is then given by:

$$\log I = \log I_0 - AM(z) - BG(z) - CH(z). \tag{1}$$

Early observations by Müller (1883, 1893) were interpreted by Hausdorff (1895) as requiring the term $BG(z)$ to fit the data thus indicating the existence of a high absorbing layer. This method is subject to considerable ambiguity in the presence of low haze layers but subsequent observations by Bauer and Danjon (1923) and Link (1929) revealed at least sporadic indications of high absorbing layers. A more sophisticated data analysis method was developed by Linke (1932) and used by Link (1943) to analyze routine measurements of the solar constant for the possible existence of a high absorbing layer.

B. M. McCormac (ed.), Physics and Chemistry of Upper Atmospheres, 34–40. All Rights Reserved.
Copyright © 1973 by D. Reidel Publishing Company, Dordrecht-Holland.

2.2. Lunar eclipses

The extinction measurements discussed above are radial optical soundings whereas the occurence of lunar eclipse allows tangential soundings in the atmosphere. During an eclipse the shadow of the earth and its atmosphere is cast upon the moon. The measured angular value of the umbra projected on the moon is about 2% larger than that which would be predicted solely on the basis of parallaxes and the angular raduis of the sun. This correction varies somewhat throughout the year and shows a tendency towards enhancement during meteor showers (Link, 1969a).

An increase of 2% in the earth's shadow corresponds to an effective increase of 95 km in the radius of the earth. At this altitude the effect of a Rayleigh atmosphere is negligible and the shadow increase can be explained by the permanent presence of meteoric aerosols which are entrained during swarm activity. The mesopause, the altitude at the temperature minimum, is the boundary between the underlying turbulent region and the overlying more stable region. It is likely that the turbulent regime can check the relatively rapid fall of aerosols and create a concentration of dust at this level. The existence of noctilucent clouds in this region also indicates the presence of dust.

2.3. Luminosity of eclipses

Tangential optical sounding during eclipses provides long path lengths and hence small concentrations of dust can be detected. Also the rays lie mostly above the low altitude polluted tropospheric layers which degrade terrestrial measurements of extinction. Using several centuries of eclipse observations Švestka (1950) detected a variation in eclipse luminosity with meteoric showers. The luminosity declines for ~ 1 month after the peak of the shower and then returns to normal after 2 more months. This implies that the bulk of meteoric material needs at least 1 month to descend from the accretion level to the earth's surface. The same delay was found by Bowen (1953) in comparing meteoric shower and heavy rains or the number of condensation ice nuclei.

The *permanent* presence of aerosols in the upper atmosphere is more difficult to establish. Some eclipses, however, are darker than Rayleigh atmosphere calculations predict. This darkening cannot be reasonably explained by tropospheric pollution (since the shadow density does not diminish toward the edge of the umbra) or excess absorption by O_3 (Link, 1969b). The darkening can be quantitatively explained by horizontal rays passing through an absorbing layer (thick or thin).

3. Scattering of Light

3.1. General scope

An important domain of measurement is the scattering of sunlight during late twilight, at solar depression $> 7°$. The sky overhead is then illuminated only at altitudes above 50 km. The scattering from aerosols is then greatly enhanced and in fact, offers a much more sensitive method than extinction measurements. A typical aerosol volume that causes extinction of only 2% causes an increase of luminance 10^5 times larger than that of the twilight sky. The effect can also be obtained by rocket or satellite observations at 100 km. Studies at night with searchlight or laser methods are as yet limited.

For the present study, twilight measurements are assumed to contain only high component luminance due to direct illumination of the upper atmosphere. Multiple scattering is made small by use of balloon or high-mountain observations. Atmospheric emission and stellar emission are corrected for or removed by filters.

3.2. OPTICAL MANIFESTATIONS OF METEORITIC AEROSOLS

A key study has been the change in scattering associated with meteoric showers and their associated aerosols. Such a study was made from balloons at 30 km altitude during the maximum of the Orionids from October 19 to 26, 1970. The results on a narrow spectral region at 5100 Å are shown in Figure 1. The influence of meteoritic aerosols is clear. The data suggest that the meteoric effect should be detectable from mountain stations under excellent meteorological conditions with even greater solar depression angles (9.5 to 14.5°) with the effective region at 70 to 120 km. This was confirmed in tests run on several occasions since 1969 on the Orionids, the Quadrantids, and the Lyrids (Link and Robley, 1972).

Depending on the assumptions of rate the fall velocity of the aerosols and the distribution of particle sizes, it has been possible to compute the flux of meteoric dust from the observations of the 1970 Orionids as falling in the range from 7×10^{-13} to 2×10^{-14} g cm^{-2} s^{-1}; a high value compared with the accepted 10^{-16} value for sporadic flux. An assumption of local concentration of aerosols in a layer where the fall velocity is markedly reduced can minimize the apparent discrepancy.

3.3. DIRECTIONAL DISTRIBUTION OF AEROSOL SCATTERING

The presence of aerosols in the upper atmosphere can also be detected by the difference in the scattering directional pattern relative to the Raleighy form for air molecules. Such tests have been run at the 35 km level. A somewhat elaborate procedure is forced, for this analysis, by the increased relative importance of the multiple scattered component of the luminance. This component can no longer be neglected, and its influence has been eliminated by algebraic reduction of observations involving two unknowns, the direct and the multiple scattering components and two (or usually more) observations at various azimuthal angles. The chief result is the suggestion of a high-lying aerosol layer (Fehrenbach *et al.*, 1972).

3.4. ADDITIONAL MEASURING TECHNIQUES

Three other measurements on scattering by aerosols will now be discussed:

(1) Mikirov (1963) measured the diurnal brightness of the sky from rockets at 50 to 100 km altitude. By working in such a high range, he rather drastically eliminated multiple scattering. These observations showed the sky luminance to fall much more slowly with height than is predicted by Rayleigh scattering, and he attributed the

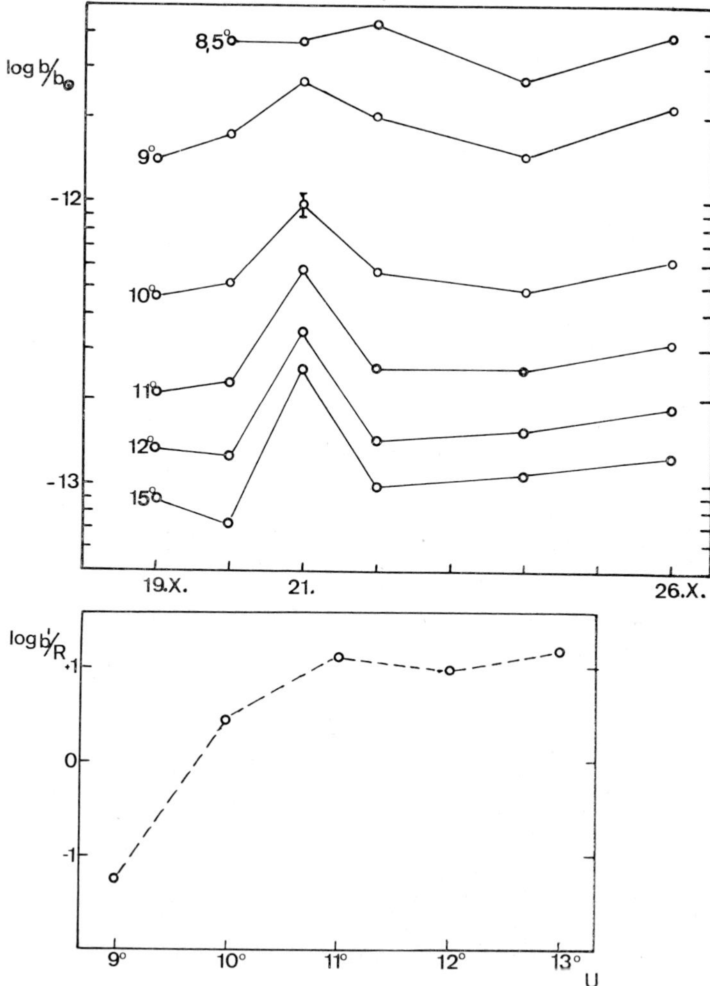

Fig. 1.   Daily variation of the luminance of twilight sky between 1970 October 19 and 26, at different solar depressions $U = 8.5°....15°$ obtained at 30 km (*top*). Ratio b'/R (meteoritic to Rayleigh scattering) on 1970 October 21 (*bottom*) (Fehrenbach *et al.*, 1971).

result to aerosols. He found an indication of a layer between 50 and 100 km whose position and importance depended on latitude. His results are shown in Figure 2. He found the ratio of aerosol concentration at 500 to 100 km to be 10^{-4} as contrasted with 5×10^{-6} in the U.S. Standard Atmosphere.

(2) Clark (1970) set up two 1.5 m diam. searchlights 45 km apart, one to emit a modulated light beam and the other to collect the scattered light from the atmosphere. By varying the height of the crossing point of the 'beams', heights from 20 to 50 km were explored. The findings were inconclusive in the respect that no results were found associated with one meteor shower whereas a strong density was found follow-

F. LINK

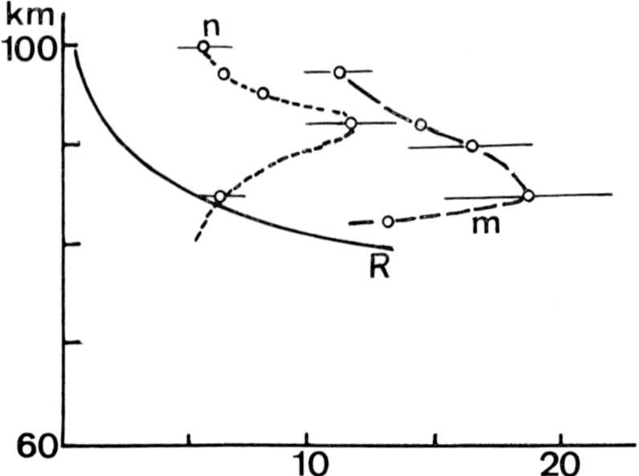

Fig. 2. *In situ* measurements of the diurnal sky luminance at different heights obtained at middle (*m*) and northern (*n*) geographical latitudes; *R* is the Rayleigh luminance (Mikirov, 1965).

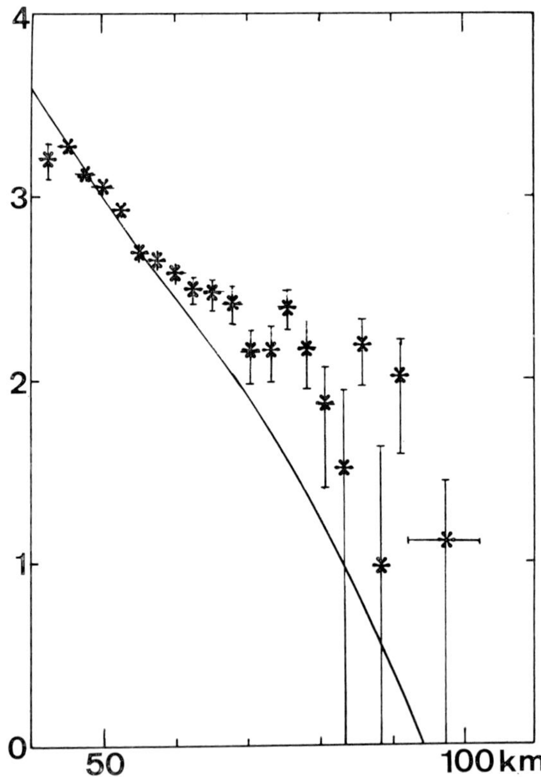

Fig. 3. Scattering diagram obtained between 1967 July 16 and August 1 by laser shots (Silverberg, 1970). Relative scattering power vs. altitude; the solid line corresponds to U.S. Standard Atmosphere.

ing a second shower but, at a delayed time that is inconclusive because the fall velocity was uncertain.

(3) Fiocco and Smulin (1963) used a pulsed light flash method with measurement of the back-scattered light intensity, essentially a RADAR type of measurement but with visible light. Beginning at 70 km altitude, the scattering exceeded that predicted by Rayleigh's law. The method suffers from the limiting sensitivity of the photometric method. Several investigations yielded results but without general agreement. Some of the results are shown in Figure 3.

In summary, the existence of aerosols above 70 km altitude arising from meteoritic events is not only derivable from the data cited but has been further reported by Kent *et al.* (1971), Zacharov (1957), and Fehrenbach *et al.* (1971).

4. Appearance of Cosmic Dust Particles

Hemenway (1963) developed a rocket-borne technique for collecting particles on films which at the same time distinguishes between cosmic and terrestrial ones. The recovered films are then examined under an electron microscope. The bulk of the particles was in the submicron range. The method and perhaps others like it yield the particle flux during the time of measurement, but this depends on both the mass and the velocity of the particles and hence limited information is obtained about the spatial density (Millman, 1970).

Hemenway (1972) noted that the chemical composition of the submicron particles (about 3/4 of the total) was heavily inclined to elements of high atomic weight, including La, Tu, Pr, Os, Yt, and Ta, and a deficiency of light atoms. Since this characteristic is more typical of stars than of the moon, meteors, asteroids, or comets, he proposed that the dust was formed in high temperature environments like sunspots and then ejected into space. The hypothesis requires further study.

5. Conclusion

An accumulation of data can be interpreted by assuming a permanent or transient layer of aerosols near the mesopause, or slightly above 80 km. The effects correlate with meteor showers leading to the belief that the aerosols are meteoric dust. The observations are the following:

(a) An excess of extinction over the Rayleigh atmosphere value as shown by illumination of the moon during lunar eclipses. (b) Enhanced brightness of the twilight sky in the upper atmosphere. (c) Deviation in directional distribution of scattered light from Rayleigh formula for air molecules.

References

Bauer, E. and Danjon, A.: 1923, *Compt. Rend. Acad. Sci. Paris* **176**, 761.
Bowen, E. G.: 1953, *Australian J. Phys.* **6**, 480.
Clark, D. D.: 1970, *Space Res.* **10**, 305.

Fehrenbach, M., Frimout, D., Link, F. and Lippens, C.: 1971, *Compt. Rend. Acad. Sci. Paris* **272B**, 913.

Fehrenbach, M., Frimout, D., Link, F., Lippens, C., and Weill, G.: 1972, *Compt. Rend. Acad. Sci. Paris* **275B**, 223.

Fiocco, G. and Smulin, L. D.: 1963, *Nature* **199**, 1275.

Hausdorff, F.: 1895, *Ber. Verh. Sächs. Akad. Wiss.* **47**, 401.

Hemenway, C. L.: 1963, *Smithsonian Contrib. Astrophys.* **7**, 93.

Hemenway, C. L.: 1972, Paper presented at XV Cospar Meeting, Madrid.

Kent, G. S., Sandford, M. C. W., and Keenliside, W.: 1971, *J. Atmospheric Terrest. Phys.* **33**, 1257.

Link, F.: 1929, *Bull. Obs. Lyon* **11**, 229.

Link, F.: 1943, *Gerl. Beitr. Geoph.* **60**, 139.

Link, F.: 1969a, *Eclipse Phenomena*, Springer-Verlag, Berlin, p. 113.

Link, F.: 1969b, *Eclipse Phenomena*, Springer-Verlag, Berlin, p. 67.

Link, F. and Robley, R.: 1972, *Space Res.* **12**, 813.

Linke, F.: 1932, *Gerl. Betr. Geoph.* **37**, 1.

Mikirov, A. E.: 1963, *Space Res.* **3**, 155.

Mikirov, A. E.: 1965, *Space Res.* **5**, 815.

Millman, M. P.: 1970, *Space Res.* **10**, 260.

Müller, G.: 1883, *Publ. Obs. Potsdam* **3**, 285.

Müller, G.: 1893, *Publ. Obs. Potsdam* **8**, 40.

Silverberg, E. C.: 1970, Ph.D. Thesis, University of Texas.

Švestka, Z.: 1950, *Bull. Astron. Inst. Czech.* **2**, 41.

Zacharov, I.: 1957, *Bull. Astron. Inst. Czech.* **8**, 139.

PART II

PHYSICAL PROCESSES

ELECTROMAGNETIC TRANSPORT PROCESSES IN THE IONOSPHERE

H. RISHBETH

S.R.C. Radio and Space Research Station, Slough SL3 9JX, U.K.

1. Motions of Ionospheric Ions and Electrons

The interesting electrical phenomena in the ionosphere largely depend on the complex way in which the charged particles move when acted upon by winds, E fields, and the geomagnetic field. This paper is concerned only with the thermal ions and electrons, not with energetic particles, and with large-scale motions with time-scales of the order of a day. To begin with, in the present section, the particle motions are described: the analysis is not rigorous, but seems adequate for understanding the physical situation at the heights with which this paper is concerned, viz. 70 to 300 km. Next, the sources of E fields and currents in the ionosphere are discussed: Section 2 deals with the 'atmospheric dynamo' and currents driven by magnetospheric processes. The way in which E fields affect the ionosphere is dealt with in Section 3, and the complementary topic of how electrical forces drive neutral-air winds in Section 4. Some useful formulas are given in Section 5.

1.1. Motions perpendicular to the geomagnetic field

Consider a particle of mass m and charge $+e$ (for an ion) or $-e$ (for an electron), acted on by a force $\mathbf{F}$ perpendicular to the geomagnetic induction $\mathbf{B}$. On a small scale, the particle gyrates around the magnetic field lines with an angular gyrofrequency $\omega = Be/m$, and collides with neutral particles with a collision frequency v. But if the motion is averaged over a period much longer than the gyroperiod $2\pi/\omega$ (which is a small fraction of a second for ionospheric particles) it approximates to a steady drift with velocity $\mathbf{V}$ given by

$$\mathbf{F} \pm e\mathbf{V} \times \mathbf{B} - mv\mathbf{V} = 0 \tag{1}$$

($+$ for ions, $-$ for electrons; the equation is easily adapted for multiply-charged or negatively-charged ions, but these probably play a negligible part in carrying ionospheric currents). If energetic particles were to be considered, further terms containing $\mathbf{B}$ would be required, but for thermal particles Equation (1) suffices. It should be mentioned that v can be defined in different ways, but the definition used here – which is in terms of the transfer of momentum between charged and neutral particles – seems the most useful for present purposes.

The terms in Equation (1) can be represented by the sides of a right-angled triangle of forces (Figure 1). The angle θ between $\mathbf{V}$ and $\mathbf{F}$ is given by

$$\tan \theta = \omega/v = Be/mv. \tag{2}$$

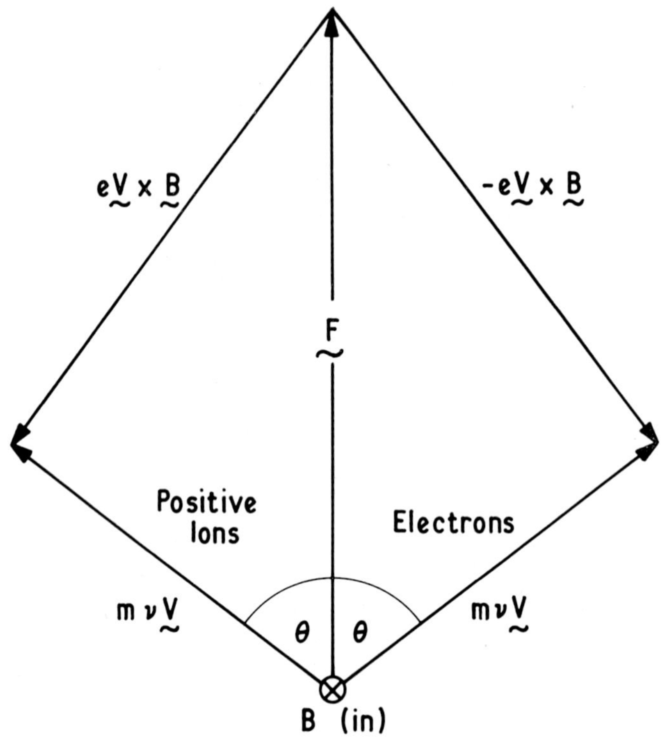

Fig. 1.   Triangle of forces for charged particles acted on by a force **F** perpendicular to the magnetic induction **B**, drawn for the case $Be/mv = \tan\theta = \frac{4}{3}$.

Since v is roughly proportional to the neutral gas density it decreases rapidly upwards so $\tan\theta$ increases with height h. There is a range of height, centered on the level where $\omega = v$, in which θ rapidly increases from near $0°$ to near $90°$.

Very roughly:

For electrons: $\theta \approx 0°$ at $h < 60$ km; $\theta = 45°$ at $h \approx 70$ km; $\theta \approx 90°$ at $h > 80$ km.

For ions: $\theta \approx 0°$ at $h < 105$ km; $\theta = 45°$ at $h \approx 125$ km; $\theta \approx 90°$ at $h > 150$ km.

Thus, between about 60 and 150 km a force **F** normal to **B** will move ions and electrons in different directions. An appreciable electric current will then flow if the electron and ion concentration N is sufficient, which normally is only the case by day and above about 100 km. Thus the 'dynamo region' in which currents flow is at 100 to 150 km, essentially the E layer of the ionosphere.

The force **F** may be due to gravity (though this is unimportant in the present context), a neutral-air wind, or an electrostatic field. A wind of velocity U produces a force mvU which can be combined with the term $-mv$**V** in Equation (1) to give a 'frictional force' $mv(\mathbf{U} - \mathbf{V})$ which balances e**V** × **B**. The drift speed at any height is

$$V = \frac{v}{\sqrt{(v^2 + \omega^2)}} U \tag{3}$$

with limiting cases $V = U$ low down $(v \gg \omega)$ and $V = Uv/\omega$ high up $(v \ll \omega)$. Assuming U independent of height, Figure 2 shows the form taken by Figure 1 at different heights. The right-hand half of the figure shows electron drifts $\mathbf{V}_e$ at 70 and 80 km, where the ions move with the wind $(\mathbf{V}_i = \mathbf{U})$ so there is an electric current in the direction of $(\mathbf{U} - \mathbf{V}_e)$, shown by dotted arrows. Normally, however, N is so small at these heights that the current is imperceptible. The left-hand half of the figure shows ion motions $\mathbf{V}_i$ above 100 km, where the electrons are virtually motionless.

An electric field $\mathbf{E}$ produces a force $\mathbf{F} = \pm e\mathbf{E}$, and the resulting drift has magnitude

$$V = \frac{\omega}{\sqrt{(v^2 + \omega^2)}} \frac{E}{B} = \frac{eE}{m\sqrt{(v^2 + \omega^2)}} \tag{4}$$

with limiting cases $V = eE/mv$ low down $(v \gg \omega)$ and $V = E/B$ high up $(v \ll \omega)$. The force $\mathbf{F}$ is approximately height-independent, unlike the 'wind' case in which the force depends on v. Partly for this reason, and partly because the electrons are driven in a direction opposite to $\mathbf{E}$, Figure 1 takes a rather different form for the 'electric field' case (see Figure 3). In this case the lower half of the diagram shows electron

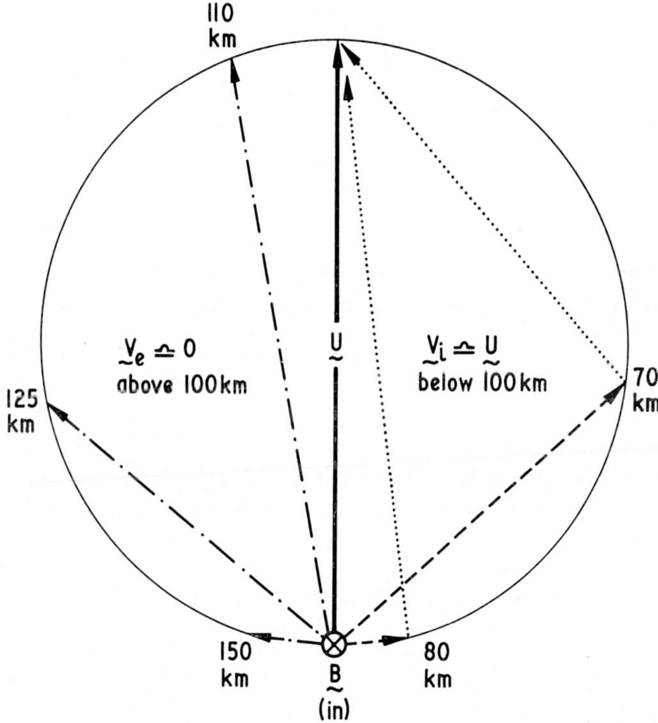

Fig. 2. Drifts of ions and electrons at various heights in the ionosphere produced by a wind $\mathbf{U}$, assumed independent of height. Ion drift $\mathbf{V}_i$ $-\cdot-\cdot\!\!>$; electron drift $\mathbf{V}_e$ $---\!\!>$; their difference $(\mathbf{V}_i - \mathbf{V}_e)$ $\cdots\cdots\!\!>$ is proportional to the electric current density (but is not shown at heights where $\mathbf{V}_e \approx 0$).

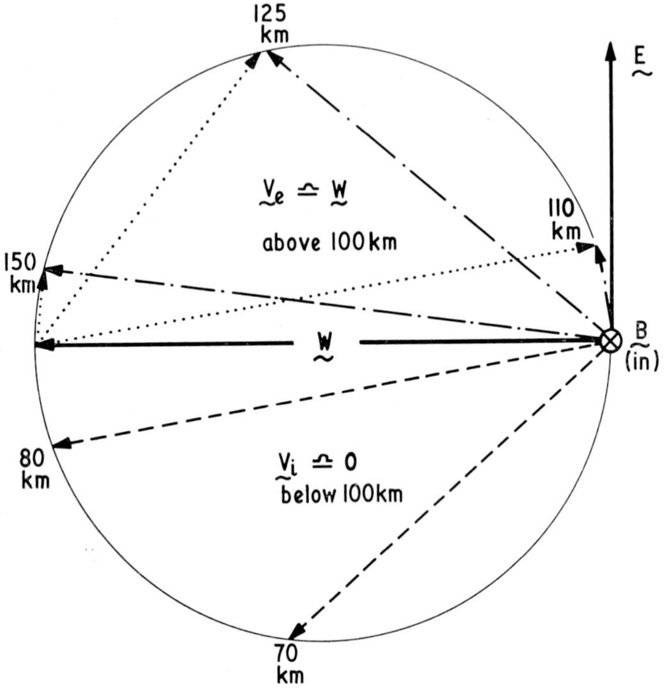

Fig. 3.   Drifts of ions and electrons at various heights in the ionosphere produced by an E field **E**, assumed independent of height. Ion drift V_i ─·─·─> ; electron drift V_e ───> ; their difference $(V_i - V_e)$ ········> is proportional to the electric current density (but is not shown at heights where $V_i \approx 0$). As in the text, $W = E \times B/B^2$.

drifts $\mathbf{V_e}$ at 70 and 80 km, where the ions hardly move, so the current is just opposite in direction to $\mathbf{V_e}$. Above 100 km, $\mathbf{V_e}$ is essentially equal to the 'electromagnetic drift velocity',

$$\mathbf{W} = \mathbf{E} \times \mathbf{B}/B^2 \tag{5}$$

i.e., normal to **E** and **B** with speed E/B. From 110 to 150 km the ion velocity $\mathbf{V}_i$ varies as shown, the electric current being in the direction of $(\mathbf{V}_i - \mathbf{W})$ as indicated by dotted arrows. Above 150 km both ions and electrons drift almost together with velocity **W** so little current flows.

1.2. Motions parallel to the geomagnetic field

If a force **F** is applied parallel to **B**, Equation (1) gives the resulting drift speed as simply $V = F/mv$. A field aligned wind U just takes the electrons and ions with it, because $F = mvU$ so that $V = U$. An E field drives ions and electrons in opposite directions with speed $V = \pm eE/mv$, producing an electric current: at E region heights this current is almost entirely carried by the electrons because

$$(m_i v_i)/(m_e v_e) = (v_i \omega_e)/(v_e \omega_i) \sim 4000. \tag{6}$$

Above about 80 km, electrons can move so readily along the geomagnetic field lines that it is difficult to maintain a steady E field parallel to **B**. Thus the geomagnetic field lines are approximately electric equipotentials, from which it follows that the electrostatic field normal to **B** varies very little with height along a given field line, at any rate within the ionosphere, as was assumed in Section 1.1 (Farley, 1960).

Since field aligned currents are almost entirely carried by electrons, and cross field currents are substantially carried by ions, it can be deduced that field aligned currents tend to cause depletion (accumulation) of ionization at places where they enter (leave) the ionosphere. Normally this effect is small, and easily compensated for by the production and loss processes for the ionization, but in certain circumstances it might have some significance (e.g. Boström, 1964).

2. The Atmospheric Dynamo

2.1. ATMOSPHERIC OSCILLATIONS

Although the idea that upper atmospheric winds generate electric current as they blow across the geomagnetic field dates back a century or more, the dynamo theory could not be put on a proper basis until the properties of the ionosphere were established. Many calculations pertaining to atmospheric tides have been carried out over the years (see Wilkes, 1949; Lindzen and Chapman, 1969), but the ionospheric side of the dynamo theory only took on its modern form with the theoretical studies of conductivity and E fields by Baker and Martyn (1952, 1953) and Fejer (1953).

The tidal oscillations are primarily driven by solar heating. Different solar radiations are absorbed in the troposphere, ozonosphere, and thermosphere and set up a series of oscillatory modes. In the idealized (but nevertheless complex) theory these modes correspond to eigenfunctions of the tidal equation of motion, which under certain assumptions are Hough functions characterized by indices $(m, \pm n)$: thus $m = 1, 2, \ldots$ for diurnal, semidiurnal, etc., modes and $|n|$ depends on the number of nodes in the latitude variation. There also exist small lunar semidiurnal, etc. 'L' modes that are excited gravitationally. The solar 'Sq' tides cause a fractional pressure oscillation , of the order of 10^{-3} at the ground and 10^{-1} at 100 km, which is accompanied by horizontal winds of the order of 5 cm s^{-1} at the ground and 50 m s^{-1} at 100 km.

For any particular mode of the tidal oscillation:

(1) The strength of the excitation depends on how its shape matches the distribution of heat input (or other driving force);

(2) in the absence of any trapping or dissipation, its amplitude increases upwards in such a way that the kinetic energy density remains constant as the gas density decreases;

(3) however, it may be reflected at certain heights – depending on the temperature profile – and it is dissipated in the ionosphere by viscosity and neutral-ion collisions, and also by nonlinear effects if its amplitude becomes large enough; and

(4) its efficiency in generating electric currents depends on how well it is transmitted

to the dynamo region (100 to 150 km) and on the pattern of winds it produces at that height.

2.2. ELECTRIC CURRENTS AND FIELDS

The tidal winds are nearly horizontal. Wherever they blow across the geomagnetic field they produce ion and electron drifts as shown in Figure 2. But only the ions are blown across the field at heights above 100 km where N is large enough for appreciable currents to flow. At any place the ion motion tends to pile up electric charge at a rate div $(Ne\mathbf{V}_i)$. Since there is no reason for this divergence to be zero everywhere, a polarization field $\mathbf{E}$ is set up. Moving freely along magnetic field lines (Section 1.2), electrons adjust the charge distribution in such a way that $\mathbf{E}$ becomes normal to $\mathbf{B}$ and nearly height-independent. The field $\mathbf{E}$ then causes ions and electrons to drift across field lines, as shown in Figure 3. A steady state is reached when $\mathbf{E}$ is so adjusted that the ion and electron drifts everywhere give a divergenceless current density $\mathbf{j}$, satisfying the equation

$$\operatorname{div}\mathbf{j} \equiv \operatorname{div} Ne\,(\mathbf{V}_i - \mathbf{V}_e) = 0. \tag{7}$$

Then no further charges accumulate. It should be mentioned that, even when the polarization charges are fully developed, the difference between the ion and electron concentrations is a minute fraction of N, and that the steady-state condition is set up very quickly (in <1 s). The current flow needed to satisfy Equations (7) may well include interhemispheric currents along field lines through the magnetosphere (Maeda and Murata, 1965; Van Sabben, 1966).

Customarily the relation between ionospheric winds, fields and currents is expressed in terms of a conductivity $[\boldsymbol{\sigma}]$ which is a tensor, the current $\mathbf{j}$ being not necessarily parallel to the applied field $\mathbf{E}'$. $\mathbf{E}'$ comprises an 'induced' component $\mathbf{U} \times \mathbf{B}$, representing the effect of the wind $\mathbf{U}$, and a 'polarization' component $\mathbf{E}$ which is derivable from a potential Φ. Then the Ohm's Law equation is

$$\mathbf{j} = [\boldsymbol{\sigma}]\,\mathbf{E}' \equiv [\boldsymbol{\sigma}]\,(\mathbf{U} \times \mathbf{B} - \nabla\Phi). \tag{8}$$

A major reason for the complexity of dynamo theory is that the charge distribution, and hence Φ, at any place is not simply related to the local wind vector $\mathbf{U}$, but depends on the global distribution of wind and conductivity. It can therefore only be determined, as a rule, by solving the equations globally. Very often it is assumed, for simplicity, that $\mathbf{U}$ is independent of height in the dynamo region; then $\mathbf{j}$ and $[\boldsymbol{\sigma}]$ can be replaced by height-integrated or 'layer' values. Unfortunately this assumption does not appear to be well justified by the available data on winds.

The tensor $[\boldsymbol{\sigma}]$ takes account of both ion and electron motions. Its components are functions of magnetic dip and of the three scalar conductivities, 'direct' σ_0, 'Pedersen' σ_1, and 'Hall' σ_2. (See for example Baker and Martyn (1953), Maeda and Kato (1966), Rishbeth and Garriott (1969).) Further details are given in Section 5.

Figure 4 shows schematically how the various elements of the dynamo theory are

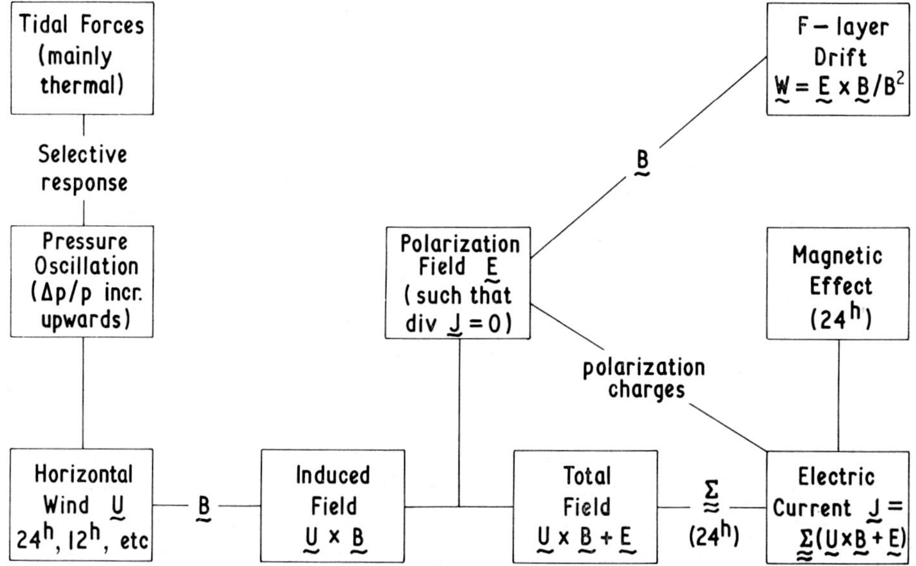

Fig. 4.   Block diagram of the dynamo theory. The dominant periods of some quantities are shown.

related. The ground level magnetic variations, which provide much of the observational data on the dynamo system, are not straightforward to interpret partly because
of induced ground currents, and partly because of contributions from magnetospheric currents.

2.3. MAGNETOSPHERICALLY-DRIVEN CURRENTS

In high latitudes, especially during magnetic disturbances, substantial ionospheric
currents (the auroral electrojets) are driven from magnetospheric sources, the circuits
being completed by field aligned currents in the magnetosphere (Boström, 1964) and
return currents in the ionosphere outside the auroral zone. These current systems are
primarily driven by electrostatic fields and not by the atmospheric dynamo, though
heat input in the auroral zone does modify the wind pattern. Even under quiet conditions, magnetospheric sources may possibly represent quite a large part of the ionospheric 'Sq' system (Matsushita, 1971); though this question is still controversial.

3. Electromagnetic Drifts in the Ionosphere

The electromagnetic (e.m.) drifts produced by ionospheric E fields are sometimes
pictured as being due to a 'motor' driven by the atmospheric 'dynamo'. They have
some influence on the structure of the ionosphere. In the continuity equation for the
ionization

$$\partial N/\partial t = q - l(N) - \mathrm{div}(N\mathbf{V} \tag{9}$$

(where q and l are the rates of production and loss), the 'transport term' contains a drift velocity $\mathbf{V}$ which may, by virtue of Equation (7), be taken as either $\mathbf{V}_e$ or $\mathbf{V}_i$. This term includes the effects of several different processes but, as far as the contribution of e.m. drifts is concerned, one may take $\mathbf{V}_e = \mathbf{W}$ at all heights in the E and F layers (Equation (5) and Figure 3). Then for e.m. drifts the transport term may be written

$$\operatorname{div}(N\mathbf{V}) = \operatorname{div}(N\mathbf{W}) = \mathbf{W}. \quad \operatorname{grad} N + N \operatorname{div} \mathbf{W}. \tag{10}$$

Normally div $\mathbf{W}$ is very small (Dougherty, 1959; Farley, 1960) so e.m. drifts are only effective in changing N if $\mathbf{W}$. grad N is appreciable, i.e., if $\mathbf{W}$ has an appreciable component parallel to grad N. In general N varies more rapidly vertically than horizontally, except at the peaks of the ionospheric layers; consequently e.m. drifts produce their most striking effects in the equatorial zone where $\mathbf{W}$ has a large vertical component.

The well-known 'fountain process' in the equatorial $F2$ layer, envisaged by Martyn (1947) and investigated in detail by Bramley and Peart (1965), and Hanson and Moffett (1966), depletes the ionization at the magnetic equator by day. The ionization is lifted upwards and subsequently diffuses outwards and downwards along field lines, forming 'crests' (maxima) of $NmF2$ in subtropical latitudes.

At mid-latitudes the e.m. drifts do not have much effect on $NmF2$, being in this respect much less effective than neutral air winds (Bramley and Rüster, 1971). This is partly because $\mathbf{W}$ is largely horizontal, owing to the inclination of the field lines; and partly because the vertical component of $\mathbf{W}$ tends to be nullified by the ion-drag effect (Section 4). At night ion-drag is less important and noticeable changes in the mid-latitude $F2$ layer can be produced by the E fields associated with magnetic bays (Rüster, 1969).

At high latitudes the e.m. drifts are mainly horizontal, and they are unlikely to produce large changes of N except where N has strong horizontal gradients, such as at the edges of the ionospheric 'troughs.' In such places the fields have the effect of moving the $F2$ layer structures bodily in a horizontal direction.

In the E layer e.m. drifts produce only minor effects, because the terms q and l in Equation (9) are relatively much more important than they are in the $F2$ layer; nevertheless, the perturbations due to e.m. drifts are readily detectable in E layer data (Beynon and Brown, 1959). In the equatorial E layer, very strong polarization fields develop because of the special geometrical situation, giving rise to the intense eastward current of the equatorial electrojet (Baker and Martyn, 1953; Sugiura and Cain, 1966).

4. Electrically-Driven Winds

The E and F layer ionization, drifting as described in Section 3, reacts on the neutral air and sets it in motion. This reaction is sometimes described as a $\mathbf{j} \times \mathbf{B}$ force acting on the conducting air. However, the $\mathbf{j} \times \mathbf{B}$ force acts in the first instance on the charged particles that carry the current (being represented by the $e\mathbf{V} \times \mathbf{B}$ term in Equation (1)) and is then transmitted to the neutral air by ion-neutral collisions;

electron-neutral collisions are of negligible importance in this respect. The resulting acceleration of the air is represented by the 'ion-drag' term, the first term on the right-hand side of the equation of motion of the air

$$\mathrm{d}\mathbf{U}/\mathrm{d}t = v'\,(\mathbf{V}_i - \mathbf{U}) + (\mu/\varrho)\,\nabla^2\mathbf{U} - (\nabla^* p)/\varrho\,. \tag{11}$$

Here ϱ is the neutral gas density, μ the coefficient of molecular viscosity, and $\nabla^* p$ is the horizontal gradient of gas pressure (the vertical gradient being balanced by gravity). v' is the collision frequency for neutral particles with ions, related to the ion-neutral collision frequency v_i by

$$Nm_i v_i = \varrho v'\,. \tag{12}$$

Suppose for the moment that ion-drag is the only force accelerating the air. Then Equation (11) shows that, if $\mathbf{V}_i$ is constant, the air takes up this velocity subject to a time constant $(1/v')$. At F layer heights $(1/v')$ is typically $\frac{1}{2}$ hr by day but 2 to 5 hr by night. Hence at mid-latitudes by day an E field, driving the ions with velocity $\mathbf{V}_i = \mathbf{W}$, rapidly sets the air into horizontal motion. It cannot move the air vertically, gravity being far too strong, and the air reacts in such a way that the ion motion ,too, is very nearly horizontal (Dougherty, 1961; Rishbeth and Garriott, 1969). As mentioned in Section 3, the influence of e.m. drifts on the F layer is thereby greatly diminished. At night when $(1/v')$ is longer, ion-drag has less effect on the e.m. drifts.

The electrostatic fields in the F layer therefore set up a system of horizontal winds, comparable in magnitude to the e.m. drift velocity, typically 50 m s^{-1} (Challinor, 1970). They are additional to the winds produced by the thermospheric pressure gradients $\nabla^* p$. However, the electrically-driven winds may influence the pressure distribution, should they cause air to pile up in, or be removed from, certain parts of the globe. The strong E fields and e.m. drifts in the auroral zone drive strong winds which circulate to lower latitudes (Rees, 1972), though the auroral zone heat input also modifies the atmospheric pressure distribution, producing further winds. The viscosity term in Equation (11) has the effect of smoothing out on the spatial variation of U, particularly the variation with height, so the wind velocity above 200 km is expected to be almost independent of height.

In the E layer $(1/v')$ is about 3 hr by day and much longer at night. Since $\mathbf{V}_i$ cannot be expected to remain constant for several hours, not much of the ion velocity is imparted to the neutral air. Figure 3 shows that below 150 km the direction in which the air is accelerated varies with height.

Another kind of interaction occurs in the F layer where winds, driven by a pressure gradient $\nabla^* p$, move ions across the geomagnetic field in accordance with Equation (3). This motion, though extremely slow because $(v/\omega)_i$ is only about 10^{-3} at 300 km, can establish a polarization field. Normally by day the polarization charges leak away along field lines and through the conducting E layer, but at night appreciable fields may develop, causing ion drift $\mathbf{V}_i$ which greatly modifies the ion-drag term in Equation (11). Rishbeth (1971) has suggested that a net W–E wind could result from this effect.

    H. RISHBETH

5. Collision Frequencies and Conductivities

This section gives some formulas for purposes of reference, all numerical values being m.k.s. According to the dipole approximation to the geomagnetic field, the value of B in the dynamo region varies with magnetic latitude ϕ in such a way that $B(0) = 2.9 \times 10^{-5}$ tesla and

$$B(\phi)/B(0) = f(\phi) \equiv \sqrt{(4 - 3 \cos^2 \phi)}. \tag{13}$$

According to Bennett *et al.* (1972) the monoenergetic collision frequency ν_M for ionospheric electrons is $9.9 \times 10^5 \, p$ at gas pressure p. For calculating the dc electrical conductivity the appropriate collision frequency is $1.5\nu_M$. Taking B and $f(\phi)$ to be given by Equation (13), it follows that

$$\nu_e/\omega_e = 0.28 \, p/f(\phi). \tag{14}$$

In the dynamo layer it may be assumed that the mean ion and neutral particle masses are equal. If n and ϱ are, respectively, the neutral gas concentration and density, then $\nu_i \approx 5 \times 10^{-16} \, n$ whence

$$\nu_i/\omega_i = 1.1 \times 10^8 \, \varrho/f(\phi). \tag{15}$$

As in Equation (1) let $\theta_e = \arctan(\omega_e/\nu_e)$, $\theta_i = \arctan(\omega_i/\nu_i)$. Then if an E field $\mathbf{E}$ is applied normal to $\mathbf{B}$, the current parallel to $\mathbf{E}$ is given by the Pedersen conductivity

$$\sigma_1 = (Ne/B)(\cos\theta_e \sin\theta_e + \cos\theta_i \sin\theta_i) \tag{16}$$

and the current normal to both $\mathbf{E}$ and $\mathbf{B}$ by the Hall conductivity

$$\sigma_2 = (Ne/B)(\sin^2\theta_e - \sin^2\theta_i). \tag{17}$$

For an electron density $N = 10^{11} \, \mathrm{m}^{-3}$, as might exist in the dynamo region by day,

$$Ne/B = 5.5 \times 10^{-4}/f(\phi). \tag{18}$$

In contrast the direct conductivity, which applies to a current along $\mathbf{B}$ produced by an E field along $\mathbf{B}$, is independent of B: it is

$$\sigma_0 = (Ne^2)\left[(m_e \nu_e)^{-1} + (m_i \nu_i)^{-1}\right]. \tag{19}$$

Its value is about 6×10^{-2} mho m^{-1} at $h = 100$ km; and as shown by Equation (6), virtually all of σ_0 is contributed by the electron term in Equation (19).

Acknowledgment

This paper is published by permission of the Director of the Radio and Space Research Station of the Science Research Council.

References

Baker, W. G. and Martyn, D. F.: 1952, *Nature* **170**, 1090.
Baker, W. G. and Martyn, D. F.: 1953, *Phil. Trans. Roy. Soc.* **A246**, 281.
Bennett, F. D. G., Hall, J. E., and Dickinson, P. H. G.: 1972, *J. Atmospheric Terrest. Phys.* 1321.
Beynon, W. J. G. and Brown, G. M.: 1959, *J. Atmospheric Terrest. Phys.* **14**, 138.
Boström, R.: 1964, *J. Geophys. Res.* **69**, 4983.
Bramley, E. N. and Peart, M.: 1965, *J. Atmospheric Terrest. Phys.* **27**, 1201.
Bramley, E. N. and Rüster, R.: 1971, *J. Atmospheric Terrest. Phys.* **33**, 269.
Challinor, R. A.: 1970, *Planetary Space Sci.* **18**, 1485.
Dougherty, J. P.: 1959, *J. Geophys. Res.* **64**, 2215.
Dougherty, J. P.: 1961, *J. Atmospheric Terrest. Phys.* **20**, 167.
Farley, D. T.: 1960, *J. Geophys. Res.* **65**, 869.
Fejer, J. A.: 1953, *J. Atmospheric Terrest. Phys.* **4**, 184.
Hanson, W. B. and Moffett, R. J.: 1966, *J. Geophys. Res.* **71**, 5559.
Lindzen, R. S. and Chapman, S.: 1969, *Space Sci. Rev.* **10**, 3.
Maeda, K. and Kato, S.: 1966, *Space Sci. Rev.* **5**, 57.
Maeda, K. and Murata, H.: 1965, *Rep. Ionospheric Space Res. Japan* **19**, 272.
Martyn, D. F.: 1947, *Proc. Roy. Soc.* **A189**, 241.
Matsushita, S.: 1971, *Radio Sci.* **6**, 279.
Rees, D.: 1972, *Phil. Trans. Roy. Soc.* **A271**, 563.
Rishbeth, H.: 1971, *Planetary Space Sci.* **19**, 357.
Rishbeth, H. and Garriott, O. K.: 1969, *Introduction to Ionospheric Physics*, Academic Press, New York.
Rüster, R.: 1969, *J. Atmospheric Terrest. Phys.* **31**, 765.
Sugiura, M. and Cain, J. C.: 1966, *J. Geophys. Res.* **71**, 1869.
Van Sabben, D.: 1966, *J. Atmospheric Terrest. Phys.* **28**, 965.
Wilkes, M. V.: 1949, *Oscillations of the Earth's Atmosphere*, Cambridge Univ. Press, England.

HEAT BALANCE AND THERMAL CONDUCTION

G. KOCKARTS

Institut d'Aéronomie Spatiale de Belgique, 3 Avenue Circulaire, B-1180 Bruxelles, Belgium

1. Introduction

A large amount of information regarding the atmospheric density in the heterosphere has been provided by satellite drag data analysis. The first results showed that the density decreases at a slower rate above 100 km altitude than in the homosphere. Such a decrease of the density gradient results from an increase of the atmospheric scale height, i.e., a diminution of the mean molecular mass and an increase of the temperature with altitude. The temperature increase is necessarily related to an energy absorption by the atmospheric constituents. Moreover, the hydrodynamical regime of the atmosphere above 100 km is such that conduction provides a very effective way for the energy transport (Spitzer, 1949; Bates, 1951). The first analysis of the time evolution of the atmosphere subject to heat conduction was made by Lowan (1955). The effect of a heat flow on the temperature gradient was demonstrated by Nicolet (1961), and the static diffusion models used for the interpretation of the satellite drag data (Jacchia, 1971) are a direct application of Nicolet's analysis. The time dependent solution given by Harris and Priester (1962) requires a hypothetical second heat source which is of the same order of magnitude as the solar UV heating. The second heat source of Harris and Priester (1962) is introduced in order to reduce the amplitude of the diurnal temperature variation and to shift the temperature maximum from 17 hr LT to 14 hr LT. Whereas the satellite drag data usually lead to a maximum of the density around 14 hr LT, it is now well established that the temperature obtained from incoherent backscatter measurements peaks around 17 hr LT. By introducing perturbations of the density and of the temperature at the base of the thermosphere, Rishbeth (1969) and Chandra and Stubbe (1970) attempted to explain this phase anomaly. It seems, however, that such an explanation is not fully satisfactory (Cummack and Butler, 1972). With a two dimensional time-dependent model, Mayr and Volland (1972) showed that a diurnal wind circulation shifts the maximum of the atomic oxygen concentration 1 or 2 hr before the temperature maximum. This controversial example indicates that some of the mechanisms responsible for the thermospheric structure are not completely understood.

Even if the numerical techniques were sufficiently developed to obtain a three dimensional solution of the conservation equations (mass, momentum, and energy), it would still be necessary to know exactly the heat sources and the heat sinks which can influence the behavior of the upper atmosphere. Some of these parameters are known qualitatively but there are still large uncertainties with regard to quantitative evaluation. Up to now, only solar UV heating and IR loss by O have been considered in the theoretical models. But it appears that other mechanisms such as tidal dissipa-

B. M. McCormac (ed.), Physics and Chemistry of Upper Atmospheres, 54–63. All Rights Reserved.
Copyright © 1973 by D. Reidel Publishing Company, Dordrecht-Holland.

tion (Lindzen and Blake, 1970) and Joule heating (Cole, 1971) cannot be neglected *a priori*.

2. Absorption Cross Sections

In a sunlit upper atmosphere, solar UV radiation is without any doubt an important energy source (Bates, 1951; Johnson, 1956; Nicolet, 1961). It is however necessary to investigate which wavelengths can be absorbed by the thermospheric constituents, i.e. by O, O_2, N_2, He and H. An analysis of the absorption cross sections combined with the total contents of each constituent shows that, above 100 km altitude, it is necessary to consider only the wavelength region from 1750 Å down to approximatively 80 Å. Ozone and O_2 are the most important absorbers above 1000 Å. Since we are dealing only with altitudes higher than 100 km, O_3 absorption is negligible and the absorption cross sections for O_2 can be taken from the analysis by Ackerman (1971) in the Schumann-Runge continuum. The O_2 absorption cross section varies from 2×10^{-19} cm^2 at 1750 Å, with a peak of 1.5×10^{-17} cm^2 at (1425 ± 25) Å, to 3×10^{-19} cm^2 around 1220 Å. Since the molecular oxygen total content reaches a value of 7×10^{16} molecules cm^{-2} around 115 km, it is clear that the maximum absorption in the Schumann-Runge continuum will occur below 120 km. It is therefor necessary to include this wavelength region in theoretical models for which the lower boundary conditions are adopted below 120 km.

The intense solar Ly-α radiation at 1216 Å can be neglected in the thermospheric thermal structure, since it penetrates into the mesosphere practically without any appreciable absorption. The much less intense Ly-β line at 1027.8 Å can, however, ionize O_2 in the E region since the absorption and ionization cross sections are respectively 1.6×10^{-18} cm^2 and 9.6×10^{-19} cm^2. The carbon C III line at 977 Å is absorbed by O_2 with a cross section of 4×10^{-18} cm^2.

Below 1000 Å, most of the atmospheric constituents are ionized and the experimental and theoretical determinations of the relevant cross sections do not always have the required accuracy for an aeronomical application. The available data have recently been reviewed by Hudson (1971). The measurements related to a continuum can usually be applied to the atmosphere. When a band spectrum is involved, it should, however, be necessary to take into account the rotational structure, particularly in the thermosphere where temperatures higher than 1000 K occur. An example of such a situation is given by the N_2 spectrum between 910 Å and 796 Å. In this wavelength region a complex band structure appears and the absorption cross section varies approximatively between 10^{-19} cm^2 and 10^{-16} cm^2 (Huffman *et al.*, 1963; Cook and Metzger, 1964; Samson and Cairns, 1964; Carter, 1972). This kind of difficulty was already pointed out by Nicolet and Swider (1963) for ionospheric problems.

Table I gives the adopted absorption cross sections between 910 Å and 80 Å (Nicolet, 1969). The wavelength intervals correspond respectively to the first ionization potential of O (910 Å), to the first ionization potential of N_2 (796 Å), to the second (732 Å) and to the third (665 Å) ionization potential of O. The other limits

TABLE I

Absorption cross sections (cm^2)

Wavelength (Å)	O	O_2	N_2	He	H
910–796	4.0×10^{-18}	7.5×10^{-18}	see text	–	5.5×10^{-18}
796–732	4.0×10^{-18}	2.0×10^{-17}	2.0×10^{-17}	–	4.0×10^{-18}
732–665	7.5×10^{-18}	2.0×10^{-17}	2.3×10^{-17}	–	3.2×10^{-18}
665–375	1.1×10^{-17}	2.5×10^{-17}	2.3×10^{-17}	6.0×10^{-18}	1.5×10^{-18}
375–275	8.0×10^{-18}	1.7×10^{-17}	1.4×10^{-17}	3.0×10^{-18}	3.0×10^{-19}
275–150	3.8×10^{-18}	7.5×10^{-18}	5.0×10^{-18}	1.0×10^{-18}	1.0×10^{-19}
150–80	1.8×10^{-18}	3.5×10^{-18}	2.2×10^{-18}	3.0×10^{-19}	3.0×10^{-20}

have been chosen to allow an extrapolation from the X-ray region. In Table I, no value is given for N_2 between 910 Å and 796 Å since the absorption cross section changes at least by two orders of magnitude in this wavelength region. In order to estimate the amount of energy absorbed by N_2, the 910 to 796 Å interval is subdivided by assuming that 5% of the available solar energy is absorbed with a cross section $\sigma = 1 \times 10^{-19}$ cm^2, 5% with $\sigma = 2.5 \times 10^{-19}$ cm^2, 10% with $\sigma = 5 \times 10^{-19}$ cm^2, 20% with $\sigma = 1 \times 10^{-18}$ cm^2, 20% with $\sigma = 5 \times 10^{-18}$ cm^2, 20% with $\sigma = 1 \times 10^{-17}$ cm^2, and 20% with $\sigma = 2 \times 10^{-17}$ cm^2. Using the absorption cross sections of Table I, it is possible to compute the transmission of the different wavelength regions through the

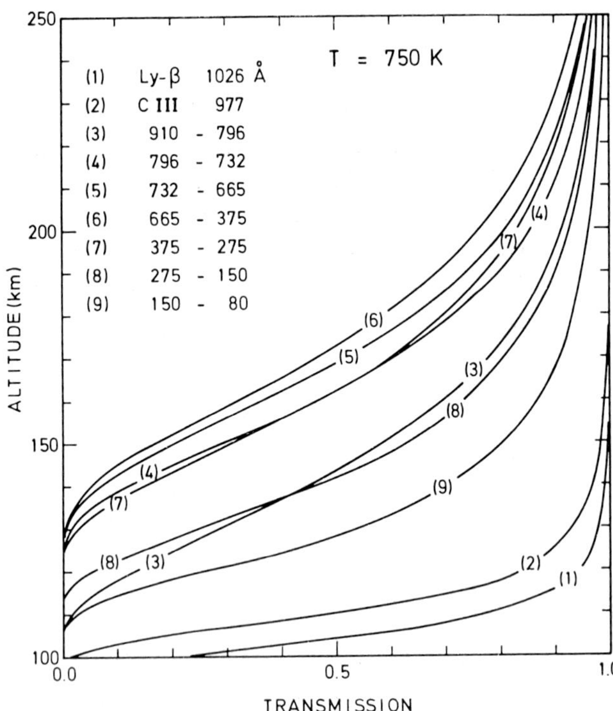

Fig. 1.   Transmission of solar radiation for several wavelength bands in a 750 K atmospheric model. Overhead sun conditions.

upper atmosphere. Figure 1 gives an example of such a computation for an atmospheric model characterized by a 750 K thermopause temperature. It can be seen from Figure 1 that the amount of energy deposited by the Ly-β and the CIII lines is almost negligible above 120 km altitude. For the other wavelength intervals, unit optical depth (transmission $= 0.369$) is reached between 120 and 160 km. It is also clear that the absorption between 910 Å and 796 Å can play an important role. This is unfortunately the wavelength region where the N_2 absorption cross section has a very complex structure.

3. Solar Ultraviolet Flux

A quantitative evaluation of the heat production also requires accurate knowledge of the solar UV flux available at the top of the atmosphere. From the recent discussion of the solar UV flux by Ackerman (1971), it appears that the present available data in the Schumann-Runge continuum are not necessarely accurate enough for an unambiguous aeronomic application. It is, for instance, not well established whether or not a variation with solar activity occurs in the Schumann-Runge continuum. Below 1000 Å, there is an important variation with the solar cycle and the amplitude of this variation increases with decreasing wavelength. However, due to the presence of numerous chromospheric emission lines, it is difficult to establish a general correlation with the radio electric fluxes which do not necessarily come from the same solar regions (Hall and Hinteregger, 1970). Moreover, the absolute calibration of the measured UV fluxes is so difficult that it is not always easy to distinguish between real variation and instrumental effects.

In the Schumann-Runge continuum, the available energy flux is of the order of 15 erg cm^{-2} s^{-1}, whereas between 910 and 80 Å the total UV flux ranges from 1.7 to 4.5 erg cm^{-2} s^{-1} depending on the solar activity. Table II gives the UV fluxes for the same intervals as in Table I. The quoted variations are intended to represent real variations as well as experimental uncertainties. It should be noted that the experimental values recommended by Hinteregger (1970) for medium solar activity fall approximately in the middle of the range given in Table II.

4. Heat Sources

The heat production $P(\lambda, z)$ resulting from the absorption of a solar radiation of wavelength λ at an altitude z can be written

$$P(\lambda, z) = \sum_i \varepsilon_i(\lambda, z)\, \sigma_i(\lambda)\, n_i(z)\, E_{\mathrm{UV}}(\lambda)\, e^{-\tau(\lambda, z)}, \tag{1}$$

where $\sigma_i(\lambda)$ is the absorption cross section of the constituent with concentration n_i. The summation is taken over all the components which can absorb the energy flux $E_{\mathrm{UV}}(\lambda)$ available at the top of the atmosphere. The heating efficiency $\varepsilon_i(\lambda)$ represents the fraction of the UV energy which is transformed into heat. The optical

58 G. KOCKARTS

TABLE II

Possible solar UV fluxes (erg cm^{-2} s^{-1})

λ (Å)	E_{UV} (erg cm^{-2} s^{-1})	λ (Å)	E_{UV} (erg cm^{-2} s^{-1})
910–796	$(2.6 \pm 0.9) \times 10^{-1}$	375–275	$(8.4 \pm 3.7) \times 10^{-1}$
796–732	$(3.9 \pm 1.3) \times 10^{-2}$	275–150	$(1.4 \pm 0.6) \times 10^{0}$
732–665	$(2.1 \pm 0.7) \times 10^{-2}$	150–80	$(2.4 \pm 1.4) \times 10^{-1}$
665–375	$(3.4 \pm 1.6) \times 10^{-1}$		

depth $\tau(\lambda, z)$ is defined by

$$\tau(\lambda, z) = \sum_i \int_z^\infty n_i(z)\, \sigma_i(\lambda)\, F \, dz, \tag{2}$$

where the optical depth factor F is simply $\sec\chi$, when the solar zenith distance χ is less than 75°. For greater solar zenith distances F is given by the so-called Chapman function which takes the earth's curvature into account. Several practical approximations of F are available in the literature (Fitzmaurice, 1964; Swider, 1964). The total heating production over a certain wavelength interval is obtained by integration of Equation (1) over λ.

When the heating efficiencies ε_i are unity, Equation (1) gives the amount of energy absorbed by the atmosphere. Figure 2 shows the results for overhead sun ($F = \sec\chi = 1$) in two atmospheric models characterized by thermopause temperatures of 750 and 2000 K, respectively. The solar UV flux is identical in the Schumann-Runge conti-

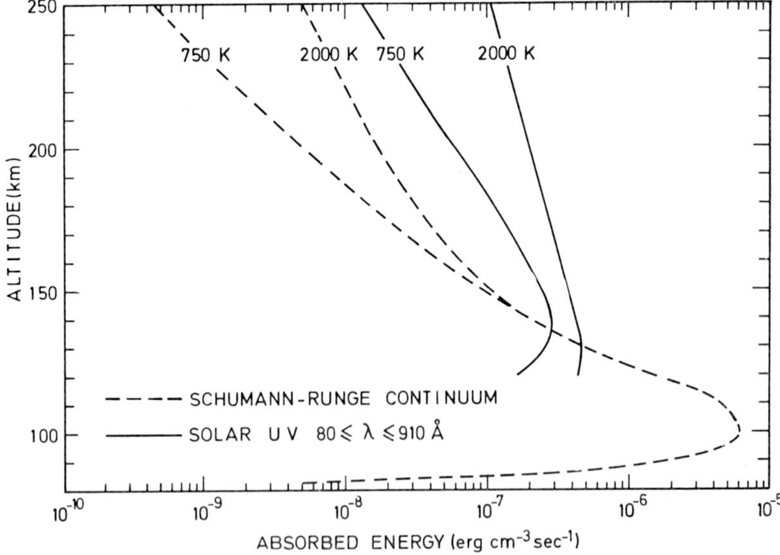

Fig. 2. Absorbed energy for overhead sun conditions in two atmospheric models with thermopause temperatures of 750 and 2000 K.

nuum for the two models. Between 80 and 910 Å, the minimum values of Table II are used for the 750 K model whereas the maximum values are adopted for the 2000 K model. Below 130 km altitude, the heating in the Schumann-Runge continuum is predominant. The heating efficiency, however, is less than one, since part of the available energy in the Schumann-Runge continuum is used for the photodissociation of molecular oxygen according to the process

$$O_2 + h\nu\,(\lambda < 1750\ \text{Å}) \rightarrow O\,(^3P) + O\,(^1D). \tag{3}$$

Any amount of energy above the dissociation threshold of 7.047 eV ($\lambda = 1750$ Å) should appear on the dissociation products. Furthermore, the metastable atom $O(^1D)$ can be deactivated to its ground state $O(^3P)$ by releasing an energy of 1.967 eV. Johnson and Gottlieb (1970) assume that any photon energy greater than 5.08 eV $=$ 7.047–1.967 eV is transformed into heat. The heating efficiency in the Schumann-Runge continuum is then simply

$$\varepsilon_{\text{S-R}} = 1 - 4.1 \times 10^{-4}\,\lambda, \tag{4}$$

where λ is the incident photon wavelength in Å. The efficiency increases linearly with decreasing wavelength from 0.28 at 1750 Å to 0.59 at 1026 Å. In a more detailed analysis (Izakov, 1970; Izakov and Morozov, 1970), collisional deactivation of $O(^1D)$ and of $O_2(b^1\Sigma_g^+)$ as well as radiative emissions from excited states are taken into account. Izakov and Morozov (1970) suggest an average efficiency of 0.3 between 1750 Å and 1350 Å. With this value, Figure 2 shows that a maximum heat production of 1.8×10^{-6} erg cm^{-3} s^{-1} occurs around 100 km due to the absorption in the Schumann-Runge continuum.

Below the first ionization potential of O_2 at 1026 Å, it is very difficult to estimate the efficiency since the UV radiation is simultaneously absorbed by O, O_2, and N_2. Photoelectrons can also produce secondary ionizations and excitations. Chemical reactions between ionized species and neutral particles lead to a very complex procedure for the energy transfer. Furthermore, thermal energy is not necessarily released at the height where absorption of the solar radiation occurs, since particle transport phenomena can be involved in an appreciable way. The heating efficiency could also depend on solar activity since the relative abundance of the atmospheric constituents is variable during a solar cycle. An order of magnitude of the amount of heat produced below 1000 Å can be obtained by multiplying the absorbed energies of Figure 2 by 0.3. It should also be realized that the daily solar UV heat production depends strongly on the geographic latitude and on the season. These factors are involved in Equation (2) through the optical depth factor F which changes as a function of latitude, solar declination, and local time.

Among the other heat sources which could play a role in the heterosphere, one should consider Joule heating (Cole, 1971) which is accompanied by viscous heat dissipation. Tidal dissipation from below has also been suggested as a non negligible heat source (Lindzen and Blake, 1970), but the excitation mechanism has not yet been fully explored (Kato, 1971).

Joule heating results from electric currents in the ionosphere. Convection E fields originate in the magnetosphere and lead then to complex movements of the ionospheric plasma in the earth's magnetic field. Fedder and Banks (1972) have shown that energies of the order of 10^{-6} erg cm^{-3} s^{-1} can be deposited over the polar cap in the E region. Such values are higher than the absorbed energy for $\lambda < 910$ Å given on Figure 2 and they are comparable to the energy absorbed in the Schumann-Runge continuum. It is still necessary, however, to determine which fraction of this energy can be converted into heat. Up to now Joule heating has not been introduced in theoretical models. It appears however that such a high latitude energy source could influence the global thermospheric structure through imporant horizontal movements.

5. Heat Sinks

Among the various atmospheric components which can play a role in the thermal structure of the heterosphere, only O has a permanent magnetic dipole moment. Therefore, IR emission of O can lead to an energy loss process in the heat balance. The excitation potentials of the two upper levels of the ground state $O(^3P)$ are of the same order of magnitude as the thermal energy of the atmospheric constituents. Infrared emissions are possible at 63 and at 147 μm (Bates, 1951). The intensity of the 147 μm emission is, however, negligible. For temperatures between 300 and 2000 K the IR flux at 63 μm is given approximately (Kockarts, 1971) for an optically thin atmosphere, by

$$L_{63\,\mu}(z) \simeq 7 \times 10^{-19} \int_z^\infty n(\mathrm{O})\,\mathrm{d}z \quad (\mathrm{erg\ cm^{-2}\ s^{-1}}) \tag{5}$$

which leads to a cooling flux of 1.3 erg cm^{-2} s^{-1} at 120 km. Equation (5) however overestimates the IR loss by several orders of magnitude at 100 km and by 10 to 20% at 150 km. Accurate results require the solution of a radiative transfer equation since O is not optically thin at 63 μm below 150 km. In this case the cooling flux ranges between 0.1 and 0.2 erg cm^{-2} s^{-1} for thermopause temperatures of 750 and 2000 K. A detailed discussion of the radiative transfer solution is given by Kockarts and Peetermans (1970). In any case, it appears that the IR loss is not sufficient to avoid the buildup of very large temperature gradients around 120 km.

6. Energy Balance

Swartz *et al.* (1971) studied the equatorial energy balance for an empirical model based on incoherent scatter temperature measurements and satellite drag density determinations. With a heating efficiency of the order of 0.4, Hinteregger's (1970) UV input fluxes were about 10% less than the total losses including the horizontal transport effects.

The most simple one-dimensional heat conduction equation is

$$\varrho c_v \frac{\partial T}{\partial t} + \frac{\partial E}{\partial z} = P - L \tag{6}$$

where the heat conduction flux E is given by

$$E = -\varkappa \frac{\partial T}{\partial z}. \tag{7}$$

In these equations ϱ is the total density, c_v the specific heat at constant volume, and $\varkappa$ is the heat conductivity. With the UV flux of Table II and the IR loss, Equation (6) leads to a diurnal amplitude of the temperature greater than a factor of 2, as shown by the dashed curve on Figure 3. It is possible, however, to *simulate* the effect of lateral compressional heating at night and of expansive cooling during the day. This can be done by introducing into the second member of Equation (6) an arbitrary production or loss

$$P_c = A \exp\left[-\frac{1}{2}\left(\frac{z - z_0}{s}\right)^2 \right], \tag{8}$$

where z_0 is the altitude of the maximum value for P_c. With $s = 50$ km and $A = 8 \times 10^{-10}$ erg cm^{-3} s^{-1}, the integrated flux obtained from Equation (8) is only 0.1 erg cm^{-2} s^{-1}. Using the time dependent distribution shown on Figure 3 for the integrated flux, the solution of Equation (6) leads to the diurnal thermopause temper-

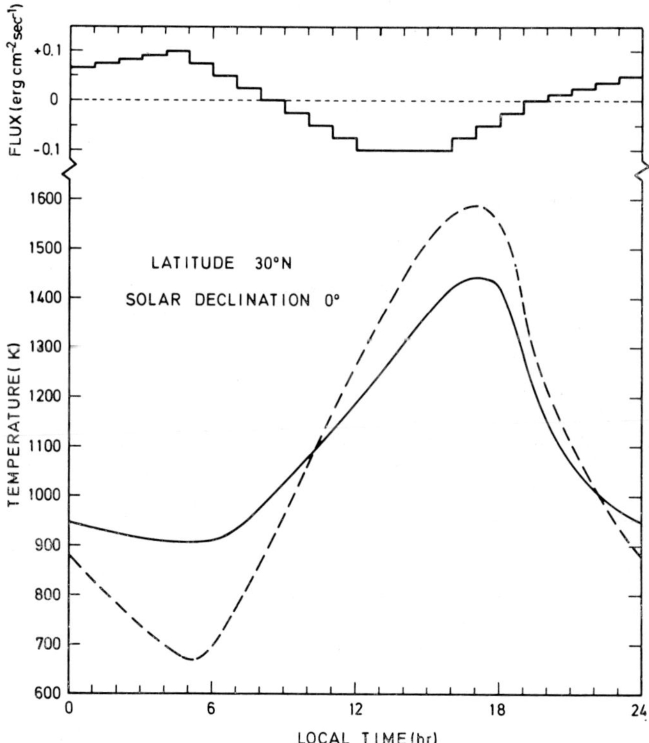

Fig. 3. Example of a diurnal thermopause temperature variation (full curve) with a small heat source or loss shown in the upper part of the figure (see Equation (8)). The dashed line corresponds to the case when only UV heating and IR loss are involved.

ature variation given by the full curve, when $z_0 = 350$ km in Equation (8). It is clear that the diurnal amplitude is strongly reduced. The total loss introduced through P_c during the day is approximatively equal to the total heat production during the night. The more or less arbitrary heat source (or loss) given on Figure 3 never exceeds 0.1 erg cm^{-2} s^{-1} whereas the maximum UV heating flux at 1200 LT is of the order of 2 erg cm^{-2} s^{-1} in this example. The additional source or loss introduced here is completely different in magnitude and in amplitude from the second heat source used by Harris and Priester (1962). It should be noted that even with P_c small compared to the UV heating below 910 Å, it is possible to build models in agreement with the incoherent scatter temperature measurements. The diurnal variation of P_c is a function of latitude and solar declination. No attempt has been made to fit any particular measurement and the Schumann-Runge continuum was not taken into account. When this wavelength region is included, turbulent downward heat transport must be included around 120 km to avoid too large temperature gradients below 150 km. This problem, however, is independent of the effect resulting from the term P_c.

At the present time, the general conservation equations have never been solved for a planetary upper atmosphere. It is almost impossible, therefore, to make a definite analysis of the heat balance. Furthermore, the heat sources and heat sinks are not known with enough accuracy. Continuous and accurate measurements of the solar UV spectrum are one of the major requirements for a good understanding of the mechanism responsible for the upper atmospheric thermal structure. The compressional heating at night and the expansive cooling during the day result from horizontal transports and it is difficult to make exact computations without solving the coupled three dimensional conservation equations. However, the use of a small nighttime production and daytime loss around 300 or 350 km can simulate horizontal energy transport.

References

Ackerman, M.: 1971, in G. Fiocco (ed.), *Mesospheric Models and Related Experiments*, D. Reidel Publishing Company, Dordrecht-Holland, p. 149.
Bates, D. R.: 1951, *Proc. Phys. Soc.* **64B**, 805.
Carter, V. L.: 1972, *J. Chem. Phys.* **56**, 4195.
Chandra, S. and Stubbe, P.: 1970, *Planetary Space Sci.* **18**, 1021.
Cole, K. D.: 1971, *Planetary Space Sci.* **19**, 59.
Cook, G. R. and Metzger, P. M.: 1964, *J. Chem. Phys.* **41**, 321.
Cummack, C. M. and Butler, P. M.: 1972, *Planetary Space Sci.* **20**, 289.
Fedder, J. A. and Banks, P. M.: 1972, *J. Geophys. Res.* **77**, 2328.
Fitzmaurice, J. A.: 1964, *Appl. Opt.* **3**, 640.
Hall, L. A. and Hinteregger, H. E.: 1970, *J. Geophys. Res.* **75**, 6959.
Harris, I. and Priester, W.: 1962, *J. Atmospheric Sci.* **19**, 286.
Hinteregger, H. E.: 1970, *Ann. Geophys.* **26**, 547.
Hudson, R. D.: 1971, *Rev. Geophys.* **9**, 305.
Huffman, R. E., Tanaka, Y., and Larrabee, J. C.: 1963, *J. Chem. Phys.* **39**, 910.
Izakov, M. N.: 1970, *Geomag. and Aeron.* **10**, 279.
Izakov, M. N. and Morozov, S. K.: 1970, *Geomag. Aeron.* **10**, 495.
Jacchia, L. G.: 1971, *Smithsonian Inst. Astrophys. Obs. Spec. Rep.* **332**.
Johnson F. S.: 1956, *J. Geophys. Res.* **61**, 71.
Johnson, F. S. and Gottlieb, B.: 1970, *Planetary Space Sci.* **18**, 1707.

Kato, S.: 1971, *Space Sci. Rev.* **12**, 421.

Kockarts, G.: 1971, in F. Verniani (ed.), *Physics of the Upper Atmosphere*, Editrice Compositori, Bologna, Italy, p. 389.

Kockarts, G. and Peetermans, W.: 1970, *Planetary Space Sci.* **18**, 271.

Lindzen, R. S. and Blake, D.: 1970, *J. Geophys. Res.* **75**, 6868.

Lowan, A. N.: 1955, *J. Geophys. Res.* **60**, 421.

Mayr, H. G. and Volland, H.: 1972, *J. Geophys. Res.* **77**, 2359.

Nicolet, M.: 1961, *Planetary Space Sci.* **5**, 1.

Nicolet, M.: 1969, private communication.

Nicolet, M. and Swider, W. Jr.: 1963, *Planetary Space Sci.* **11**, 1459.

Rishbeth, H.: 1969, *Ann. Geophys.* **25**, 495.

Samson, J. A. R. and Cairns, R. B.: 1964, *J. Geophys. Res.* **69**, 4583.

Spitzer, L., Jr.: 1949, in G. P. Kuiper (ed.), *The Atmospheres of the Earth and Planets*, The University of Chicago Press, Chicago, Illinois, p. 213.

Swartz, W. E., Rohrbaugh, J. L., and Nisbet, J. S.: 1971, *Ionospheric Research Sci. Rep. 383, Pennsylvania State University*.

Swider, W. Jr.: 1964, *Planetary Space Sci.* **12**, 761.

SOLAR RADIATION AND ITS ABSORPTION IN THE UPPER ATMOSPHERE

M. ACKERMAN

Institut d'Aéronomie Spatiale de Belgique, Brussels, Belgium

1. Introduction

As it is well known the solar radiation absorbed in the atmosphere initiates the photochemical reactions taking place in it and brings in energy at a rate almost equal to $2 \, \text{cal} \, \text{cm}^{-2} \, \text{min}^{-1}$. This value is the rather well known solar constant. One of the most complete reviews of its determinations has been published by Arvesen *et al.* (1969). Below 1000 Å, the photons are mostly absorbed in the thermosphere leading to photo-ionization processes as discussed by Kockarts (1973). Between 1000 and 10000 Å the solar radiation is of fundamental importance when the photochemistry of neutral species has to be considered. Most of our knowledge in this range was summarized in 1970 by Ackerman (1971). Since then, some progress has been made concerning the intensity of the solar UV radiation and the absorption cross sections of atmospheric constituents. The purpose of this short note is to point out the few new data relevant to these questions.

2. Absorption Cross Sections

Recent data on absorption cross sections of aeronomic interest have been reviewed by Ackerman (1972). The most important new data concern CO_2 and are mostly of interest for the study of Mars and Venus. Data obtained by Shemansky (1972) lead to a much lower absorption cross section, dominated by Rayleigh scattering in the region longward of 2035 Å, than reported previously by Ogawa (1971).

3. Solar UV Radiation

At wavelengths shorter than 3000 Å only absolute measurements performed from balloon gondolas or from rockets have provided data. Spectra of the 2000 Å window obtained by Ackerman *et al.* (1971) lead to the following conclusions: the radiation in the vicinity of 2200 Å had been underestimated in the earlier work of Detwiler *et al.* (1961) and the 2085 Å discontinuity is sharper than previously mentioned. The first conclusion may be explained by Pitz (1971) who obtained a higher spectral irradiance of the carbon arc than was determined by Johnson (1956) at wavelengths shorter than 2300 Å.

The measurements of Parkinson and Reeves (1969), Widing *et al.* (1970), and Brueckner and Moe (1972), have shown that the solar intensity below 1800 Å was much lower than previously reported (Detwiler *et al.*, 1961) giving some support to the second conclusion. A discrepancy however still exists between the values of Widing *et al.* (1970) and of Parkinson and Reeves (1969) reaching more than a factor of two. As shown in Figure 1 the values obtained at low resolution by means of ESRO rockets on February 28, 1972, by Ackerman and Simon (1973) are closer to the latter ones.

Does the solar intensity decreases smoothly below 1950 Å or is there another discontinuity? A rather abrupt decrease could take place in the 1900 to 1930 Å range.

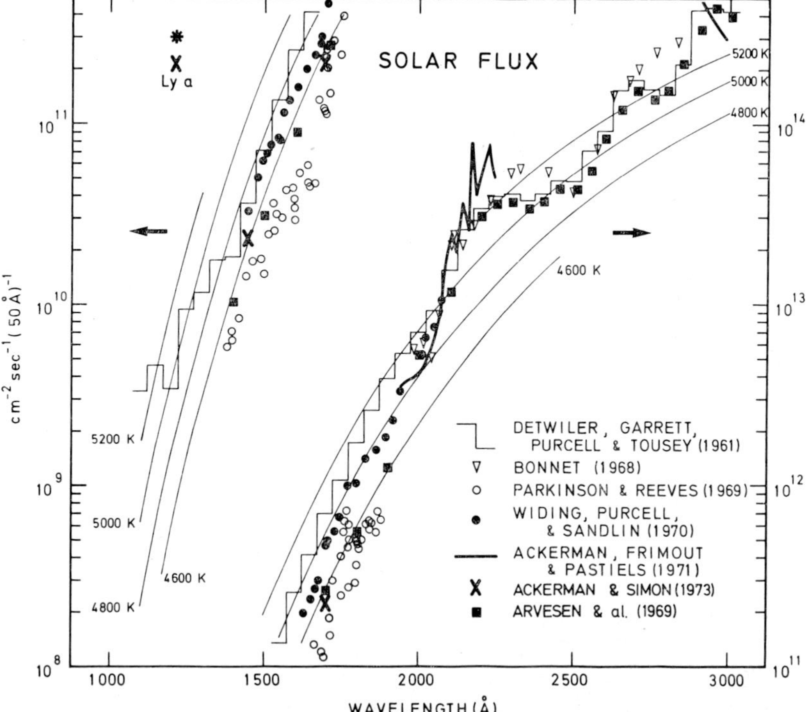

Fig. 1. UV solar flux values vs. wavelength obtained by various authors. Values corresponding to various black-body temperatures in degrees Kelvin are also represented.

The variability of solar UV radiation is also being studied (Heath, 1969; Hall and Hintereger, 1970). More determination of its dependence on wavelength is needed at long wavelengths.

4. Conclusion

It is now certain that the effective black-body temperature of the global sun can decrease by some 800 K from 2100 to 1600 Å. The steps are possibly as follows: 5200 K at 2100 Å, 4800 K at 2000 Å, 4650 K at 1900 Å, 4550 K at 1700 Å and 4600 K at 1450 Å with a minimum at 1600 Å of the order of 4400 K.

References

Ackerman, M.: 1971, in G. Fiocco (ed.), *Mesospheric Models and Related Experiments*, D. Reidel Publishing Company, Dordrecht-Holland, p. 149.

Ackerman, M.: 1972, *Ann. Geophys.* **28**, 79.
Ackerman, M., Frimout, D., and Pastiels, R.: 1971, in F. Labuhn and R. L. Lüst (eds.), *New Techniques in Space Astronomy*, D. Reidel Publishing Company, Dordrecht-Holland, p. 251.
Ackerman, M. and Simon, P.: 1973, to be published.
Arvesen, J. C., Griffin, R. N., and Pearson, B. D.: 1969, *Appl. Opt.* **8**, 2215.
Bonnet, R. M.: 1968, *Space Res.* **7**, 458.
Brueckner, G. E. and Moe, O. K.: Paper, Cospar, Seattle 1971.
Detwiler, C. R., Garrett, D. L., Purcell, J. D., and Tousey, R.: 1961, *Ann. Geophys.* **17**, 9.
Hall, L. A. and Hinteregger, H. E.: 1970, *J. Geophys. Res.* **75**, 6959.
Heath, D. F.: 1969, *J. Atmospheric Sci.* **26**, 1157.
Kockarts, G.: 1973, this volume, p. 54.
Ogawa, M.: 1971, *J. Chem. Phys.* **54**, 2550.
Parkinson, W. H. and Reeves, E. M.: 1969, *Solar Phys.* **10**, 342.
Pitz, E.: 1971, *Appl. Opt.* **10**, 813.
Shemansky, D. E.: 1972, *J. Chem. Phys.* **56**, 1582.
Widing, K. G., Purcell, J. D., and Sandlin, G. D.: 1970, *Solar Phys.* **12** 52.

PRECIPITATING ENERGETIC ELECTRONS IN THE MID-LATITUDE LOWER IONOSPHERE

T. A. POTEMRA

Applied Physics Laboratory, The Johns Hopkins University, Silver Spring, Md., U.S.A.

Abstract. Satellite and rocket measurements of precipitating electrons with energies $\geqslant 10$ keV are reviewed for Mid-latitudes. The ionization rates due to these electrons are compared with the most recent estimates of other ionization sources which are expected to be important in the day and night ionosphere below 100 km. The results show that precipitating electrons provide an important ionization source in the mid-latitude D and lower E regions during undisturbed and disturbed conditions at night and during the day for large solar zenith angles or high magnetic activity.

1. Introduction

Energetic electrons precipitating on the ionosphere from the trapped radiation belts have been suggested as the cause of numerous ionospheric disturbances, especially those associated with magnetic activity (see for example the review in Potemra and Zmuda, 1970). Many satellite and rocket measurements of precipitating electron fluxes at mid-latitude have been made, but direct measurements of electron fluxes only recently have been directly correlated with ionospheric measurements of disturbances with ground based techniques. Ivanov-Kholodny and Kazatchevskaya (1971), using rockets at mid-latitudes in the U.S.S.R. measured greater electron fluxes with energies in the 1 to 40 keV range when a higher critical frequency of the E-region was observed. Potemra and Rosenberg (1972) have reported disturbances to the phase of several long distance mid-latitude VLF transmissions, reflected from the D region, which occurred nearly simultaneously with balloon observations of Bremsstrahlung X-rays (due to precipitated energetic electrons with energies >30 keV) at Siple Station ($\sim 60°$ geomagnetic latitude). Preliminary analysis of data from the polar orbiting satellite STP 71-2 by Johnson *et al.* (1972) and Imhof (1972) show that large fluxes of electrons with energies >130 keV were precipitated at mid-latitudes during the large magnetic storm which began on December 17, 1971. During this same period, ionospheric absorption was measured by Johnson *et al.* (1972) with the 'Earth-Reflection Ionospheric Sounder' experiment on the same satellite and perturbations to the phase of several mid-latitude VLF transmissions received at the Applied Physics Laboratory (near Washington, D.C.) were observed by the author.

The purpose of this paper is to supplement Potemra and Zmuda's (1970) review of satellite and rocket measurements of precipitating electrons >40 keV at mid-latitudes with more recent measurements of electrons >10 keV. Comparison is made of the ionization due to these electrons with the most up-to-date estimates of all other sources of ionization below 100 km, such as galactic X-rays and cosmic rays, the ionization of NO by direct and scattered HLy-α, and solar X-ray and UV radiation. This comparison strengthens the view that constantly precipitating energetic electrons provide an important ionization source in the mid-latitude D and E regions during undisturbed and disturbed conditions at night (Potemra and Zmuda, 1970; Manson, 1971) and during the day for large solar zenith angles or high magnetic activity (Ivanov-Kholodny, 1970).

2. Particle Data

Potemra and Zmuda (1970) reviewed the satellite measurements of precipitated electrons with $E \geqslant 40$ keV during quiet magnetic conditions. These data showed large variability of the precipitating electrons at a given latitude and a general increase in the precipitated flux with increasing latitude (see their Figure 1). For example, this flux ranges between 300 to 2000 electrons cm^{-2} s^{-1} sr^{-1} at a 40° IN Lat but at 50° they range between 500 and 10^4 electrons cm^{-2} s^{-1} sr^{-1} and at 60° the range is from 2000 to 10^5 cm^{-2} s^{-1} sr^{-1}. Figure 1 shows the energy spectrums determined

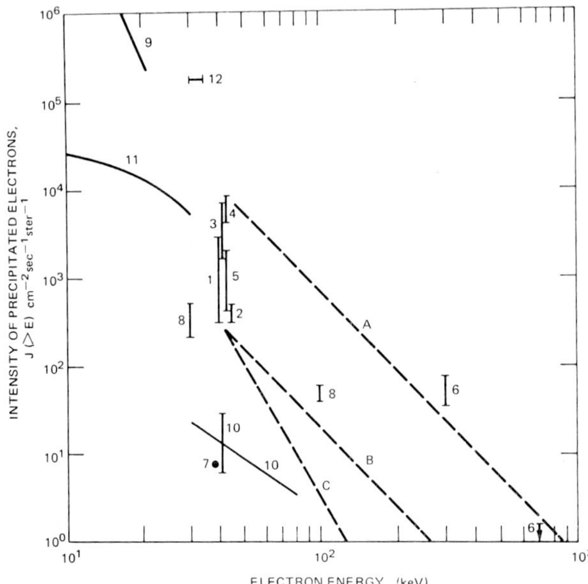

Fig. 1. Intensity of precipitated electrons at mid-latitudes (35° to 55° geomagnetic) from the following satellites: (1) Injun 1 during June–September 1961, O'Brien (1962); (2) Injun 3, January 1963, O'Brien (1964); (3) Injun 3, February–October 1963, Fritz (1967, 1968, 1970); (4) Explorer 12, August–September 1961, O'Brien and Laughlin (1963); (5) Alouette, October 1962 to January 1963, McDiarmid *et al.* (1963); (6) 1964–45A at ∼ 58° magnetic latitude in August 1964 in the southern hemisphere Paulikas *et al.* (1966) – the value at 700 keV represents an upper limit. The intensities from the following rocket flights are included: (7) From Wallops Island (∼ 50°) in July 1964, O'Brien *et al.* (1965); (8) 'Mid-latitude' in U.S.S.R. in summer, Tulinov (1967); (9) Mean integral spectrum Ivanov-Kholodny (1968), see also Antonova and Kazatchevskaya (1970); (10) Summary of 7 rocket flights, 0° to 64° geomagnetic latitude in the Indian and Pacific Oceans, Tulinov *et al.* (1969); (11) Night of July 31, 1968, mid-latitudes U.S.S.R., Kazatchevskaya and Koryagin (1969); (12) 'Mid-latitude' in U.S.S.R., October 18, 1962, solar angle $= 68°$, $K_p = 4$, Antonova *et al.* (1971).

from these satellite fluxes in the mid-latitude range $35°$ to $55°$. This figure is taken from Potemra and Zmuda (1970) with the rocket determinations of Ivanov-Kholodny (1968), Kazatchevskaya and Koryagin (1969), Tulinov *et al.* (1969), Antonova and Kazatchevskaya (1970) and Antonova *et al.* (1971) added. This figure shows model power law spectra as dashed lines labeled 'A, B, C' that were adopted by Potemra and Zmuda (1970) for the calculation of ionization rate profiles. These are of the form $J(>E) \sim E^{-\gamma}$ with $\gamma = 3$ and $J(>40 \text{ keV}) = 8000 \text{ cm}^{-2} \text{ s}^{-1} \text{ sr}^{-1}$ for model A, $\gamma = 3$ and $J(>40 \text{ keV}) = 300 \text{ cm}^{-2} \text{ s}^{-1} \text{ sr}^{-1}$ for B and $\gamma = 5$ and $J(>40 \text{ keV}) = 300$ for C.

It is difficult to compare rocket measurements with satellite observations of precipitating electrons because of their variation with geomagnetic conditions and location. An additional complication exists in the comparison of satellite measured electron spectra at ~ 1000 km altitude with rocket measured spectra at ~ 100 km altitude because the atmospheric column between these altitudes significantly alters the spectrum. For example, the 10 keV electrons measured by a satellite at 1000 km altitude cannot be detected below a 100 km altitude because these particles will be stopped at this altitude. Tulinov *et al.* (1969) have computed the relative particle flux as a function of total atmospheric mass penetrated, and its variation with different electron spectrums. A greater reduction of particle flux occurs with decreasing altitude for softer electron spectrums (a greater proportion of lower to higher energy electrons). The net effect is to produce a smaller total electron flux at rocket altitudes than at satellite altitudes.

The rocket measurements do have the advantage that they are *in situ* and all the available measurements have been combined in Figure 1 to estimate the importance of precipitating electrons to the lower ionosphere. The spectrum for electrons with energies <20 keV labeled '9' in Figure 1 from Ivanov-Kholodny (1968) was 'constructed on the basis of all the available experimental data on fluxes of precipitated electrons at mid-latitudes,' but is noticeably inconsistent with all the other measurements. These large fluxes are more representative of auroral latitudes than of mid-latitudes. The range of electron fluxes with energies > 40 keV measured by Tulinov *et al.* (1969) and their spectrum $dJ/dE \sim E^{-3}$ are both labeled '10' in this figure. The latter electron spectrum was determined by virtue of the differential particle fluxes measured at different atmospheric altitudes combined with the variation of the electron range (or stopping power) with height. Kazatchevskaya and Koryagin (1969) determined the spectrum $dJ/dE = 4.4 \times 10^3 \exp(-E/13 \text{ keV}) \, (\text{cm}^{-2} \text{ s}^{-1} \text{ sr}^{-1} \text{ keV}^{-1})$ which is labeled '11'. The latter two rocket observations are more consistent with the satellite measurements. The measurements of Antonova *et al.* (1971) (labeled '12') were made during relatively high magnetic activity ($K_p = 4$) which could account for this large flux in comparison to most of the others in Figure 1.

3. Ionization Rates

Figures 2 and 3 show the height profiles of ion-pair production rates computed from some of the fluxes of precipitating electrons shown in Figure 1. Figure 2 also shows

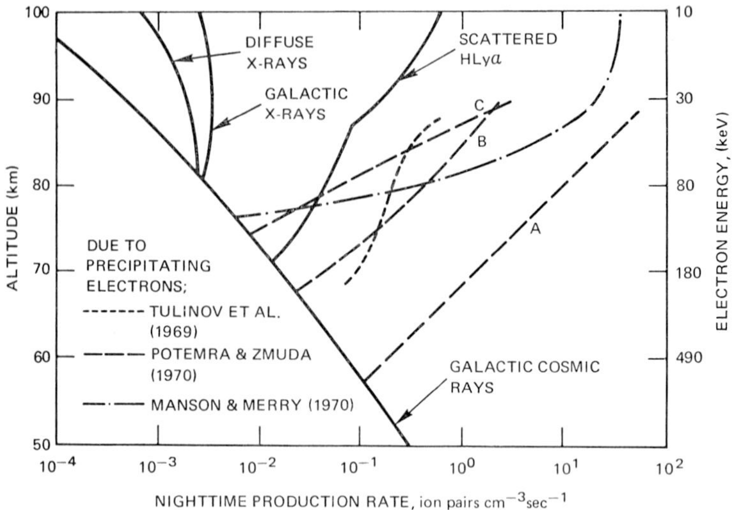

Fig. 2.   Ionization rates in the nighttime lower ionosphere.

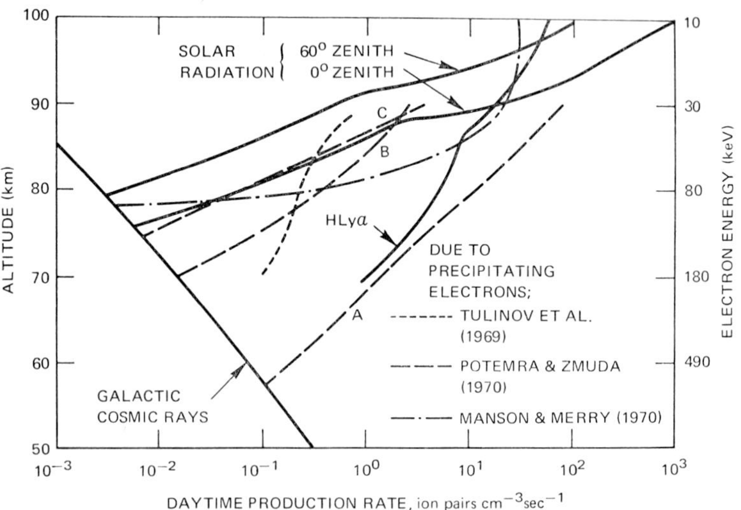

Fig. 3.   Ionization rates in the daytime lower ionosphere.

the production rates due to the following sources which are expected to be important at night: galactic cosmic rays for solar minimum conditions at a $50°$ geomagnetic latitude from Webber (1962); the ionization of NO by scattered HLy-α (determined from Meira's (1971) NO profile with an energy flux of scattered HLy-α equal to 5.5×10^{-2} erg cm^{-2} s^{-1} outside the atmosphere); the diffuse X-ray background computed by Francey (1970) with the spectrum measurement of Henry *et al.* (1968);

and galactic X-rays summed over 50 sources for a $30°\,N$ geographic latitude in July computed by Mitra and Ramanamutry (1972). The dashed curves labeled 'A, B, and C' in these figures were computed by Potemra and Zmuda (1970) from their model electron spectrums, shown in Figure 1, using the specific ionization rates from Rees (1963, 1965). The production rates computed by Tulinov *et al.* (1969) from their electron spectrum shown in Figure 1 are also shown in Figure 2. And the ionization rates computed by Manson and Merry (1970) with $J(>40\,\mathrm{keV})=10^3$ electrons $\mathrm{cm}^{-2}\,\mathrm{s}^{-1}\,\mathrm{sr}^{-1}$ and a spectrum of the form $J(>E)\sim\exp(-E/15\,\mathrm{keV})$ is also shown in this figure.

Figure 3 shows the same ionization rate profiles due to precipitating electrons as in Figure 2 and with the following sources which are expected to be important in the daytime ionosphere: the same galactic cosmic ray source shown in Figure 2; solar X-rays for quiet conditions for a $0°$ and $60°$ solar zenith angle from Swider (1969), and ionization of NO by direct HLyα (with Meira's (1971) NO profile and a HLy-α intensity of 5.5 erg $\mathrm{cm}^{-2}\,\mathrm{s}^{-1}$ outside the atmosphere). Figures 2 and 3 also show on the right hand scale the energy an electron must have to penetrate down to the altitude on the corresponding left scale.

Figure 2 shows that all computed ionization rates due to energetic electrons exceed the other nighttime source above 75 km, and Potemra and Zmuda's (1970) 'curve A' dominates above 60 km. This would strengthen Potemra and Zmuda's (1972) suggestion that N_2^+ emissions at 3914 Å and 4278 Å caused by precipitated energetic electrons are a permanent feature of the night sky at mid-latitudes.

For the daytime condition shown in Figure 3 only the model A curve and possibly Manson and Merry's (1970) curve in the altitude range between 85 and 95 km (depending on the solar zenith angle) are competitive with the daytime ionization sources due to solar X-rays and galactic cosmic rays. Ionization due to the energetic electrons is competitive with the daytime sources only during disturbed conditions when larger fluxes would be expected.

Simultaneous measurements of precipitating electron fluxes and ionospheric disturbances are difficult and until recently few have been made. This is partly due to the fact that very small fluxes of electrons are required to disturb the D region. For example, the model spectrums labeled 'B' and 'C' in Figures 2 and 3 were computed for fluxes of only $J(>40\,\mathrm{keV})=300$ electrons $\mathrm{cm}^{-2}\,\mathrm{s}^{-1}\,\mathrm{sr}^{-1}$ (in both cases equivalent to an energy flux of less than 2×10^{-4} erg $\mathrm{cm}^{-2}\,\mathrm{s}^{-1}$). This flux is considerably smaller than the flux of mirroring particles at satellite altitude (Fritz, 1967, 1968) and is close or below the sensitivity of most existing satellites and rocket particle detectors. But variations in this flux could produce detectable disturbances in the nighttime D and lower E regions. Satellite measurements of larger electron fluxes such as during the severely disturbed period beginning December 17, 1971, have already been made (Johnson *et al.*, 1972), but the sensitivity of these detectors will have to be improved (for example to detect precipitated fluxes of $J(>40\,\mathrm{keV})=100\,\mathrm{cm}^{-2}\,\mathrm{s}^{-1}\,\mathrm{sr}^{-1}$) to positively identify constantly precipitating electrons as an important source of the undisturbed lower ionosphere.

Acknowledgments

I thank Dr A. J. Zmuda for many profitable discussions. This work was supported by the Naval Ordnance Systems Command, Department of the Navy, under contract N00017-72-C-4401.

References

Antonova, L. A. and Kazatchevskaya, T. V.: 1970, *Space Res.* **10**, 757.
Antonova, L. A., Ivanov-Kholodny, G. S., and Kazatchevskaya, T. V.: 1971, Proceedings of the COSPAR Symposium on *D*- and *E*-Region Ion Chemistry Univ. of Illinois, Urbana, Ill.
Francey, R. J.: 1970, *J. Geophys. Res.* **75**, 4849.
Fritz, T. A.: 1967, Ph. D. Thesis, Univ. of Iowa, Iowa City.
Fritz, T. A.: 1968, *J. Geophys. Res.* **72**, 7245.
Fritz, T. A.: 1970, private communication.
Henry, R. C., Fritz, G., Meekins, J. F., Friedman, H., and Byram, E. T.: 1968, *Astrophys. J.* **153**, L11.
Imhof, W. L.: 1972, private communication.
Ivanov-Kholodny, G. S.: 1968, *Transactions of the Summer School on Space Physics*, Irkutsk.
Ivanov-Kholodny, G. S.: 1970, *Ann. Geophys.* **26**, 575.
Ivanov-Kholodny, G. S. and Kazatchevskaya, T. V.: 1971, *J. Atmospheric Terrest. Phys.* **33**, 285.
Johnson, R. G., Reagan, J. B., and Bradbury, J. N.: 1972, *Trans. Am. Geophys. Union* **53**, 455.
Kazatchevskaya, T. V. and Koryagin, A. I.: 1969, *Komich. Issled.* **7**, 950.
Manson, A. H.: 1971, *Planetary Space Sci.* **19**, 270.
Manson, A. H. and Merry, M. W. J.: 1970, *J. Atmospheric Terrest. Phys.* **32**, 1169.
McDiarmid, I. B. J., Burrows, J. R., and Budzinski, E. E.: 1963, *Can. J. Phys.* **41**, 2064.
Meira, L. G.: 1971, *J. Geophys. Res.* **76**, 202.
Mitra, A. P. and Ramanamutry, Y. V.: 1972, *Radio Sci.* **7**, 67.
O'Brien, B. J.: 1962, *J. Geophys. Res.* **67**, 3687.
O'Brien, B. J.: 1964, *J. Geophys. Res.* **69**, 13.
O'Brien, B. J. and Laughlin, C. D.: 1963, *Space Res.* **3**, 399.
O'Brien, B. J., Allum, F. R., and Goldwire, H. C.: 1965, *J. Geophys. Res.* **70**, 161.
Paulikas, G. A., Blake, J. B., and Freden, S. C.: 1966, *J. Geophys. Res.* **71**, 3165.
Potemra, T. A. and Zmuda, A. J.: 1970, *J. Geophys. Res.* **75**, 7161.
Potemra, T. A. and Rosenberg, T. J.: 1972, *Trans. Am. Geophys. Union* **53**, 455.
Potemra, T. A. and Zmuda, A. J.: 1972, *Radio Sci.* **7**, 63.
Rees, M. H.: 1963, *Planetary Space Sci.* **11**, 1209.
Rees, M. H.: 1965, private communication.
Swider, W., Jr.: 1969, *Rev. Geophys.* **7**, 573.
Tulinov, V. F.: 1967, *Space Res.* **7**, 368.
Tulinov, V. F., Shibaeva, L. V., and Jakovlev, S. G.: 1969, *Space Res.* **9**, 231.
Webber, W.: 1962, *J. Geophys. Res.* **67**, 5091.

TRANSAURORAL IONOSPHERE, MAGNETOSPHERIC IMPLICATIONS

J. J. BERTHELIER and M. SYLVAIN

Groupe de Recherches Ionosphériques, 4, Avenue de Neptune, 94100-Saint-Maur, France

1. Introduction

Recent low energy particle measurements (Frank, 1971; Frank and Ackerson, 1972) and photometric observations (Eather and Mende, 1971) have shown that in a ring extending (for $K_p \lesssim 3$) from $\sim 78°$ to $85°$ IN Lat in the noon sector and $\sim 72°$ to $80°$ in the midnight sector most of the corpuscular energy input is due to soft particles: by transuroral regions we shall refer to this ring-shaped zone. Its existence was disclosed a few years ago on the basis of ionospheric observations (e.g., Nelms and Chapman, 1970) and it is clear that the behavior of the transauroral ionosphere is essentially dominated by magnetospheric processes and magnetosphere-ionosphere coupling.

Electric fields and particle precipitations are the major parameters of interest and in the first part of this paper a review will be given of the recent results. In the second part we shall present some observations made at an IN Lat of $81°$ in the Antarctic related to particle precipitations and simultaneous ionospheric perturbations.

2. Electric Field Observations

Initial measurements of convection E fields at latitudes higher than $70°$ (Wescott *et al.*, 1969; Heppner *et al.*, 1971) showed the E field to be directed roughly from dawn to dusk and of the order of 20 to 40 mV m^{-1}. Considerable progress has been made during the last 2 yr due to several double probe experiments flown onboard satellites, particularly polar orbiting satellites, i.e., Injun 5 (Cauffman and Gurnett, 1971; Frank and Gurnett, 1971) and OGO 6 (Heppner, 1972). Most of the results published so far have been obtained near the dawn-dusk meridian approximately between 0300 to 0900 and 1500 to 2100 LT confirming the model proposed by Axford and Hines (1961). The most striking feature is the existence of a "convection auroral belt" between approximately $65°$ and $75°$ IN lat, where the plasma motion is sunward, and a 'polar cap', poleward of this auroral zone, where the convection is antisunward (Figure 1). The boundary between these two regions often has the aspect of sharp reversals or field shears with possible multiple zero crossings of the E field. In fact, this general scheme represents an average and simplified picture and analysis of individual OGO 6 passes (Heppner, 1972) shows a more complex and variable structure in transauroral regions. Contrary to what was deduced from Injun 5 data (Frank and Gurnett, 1971) OGO 6 results do not seem to indicate any systematic existence of a polar cavity where the convective motion of plasma would be very slow compared to surrounding regions; a possible reason for this disagreement is the relatively low

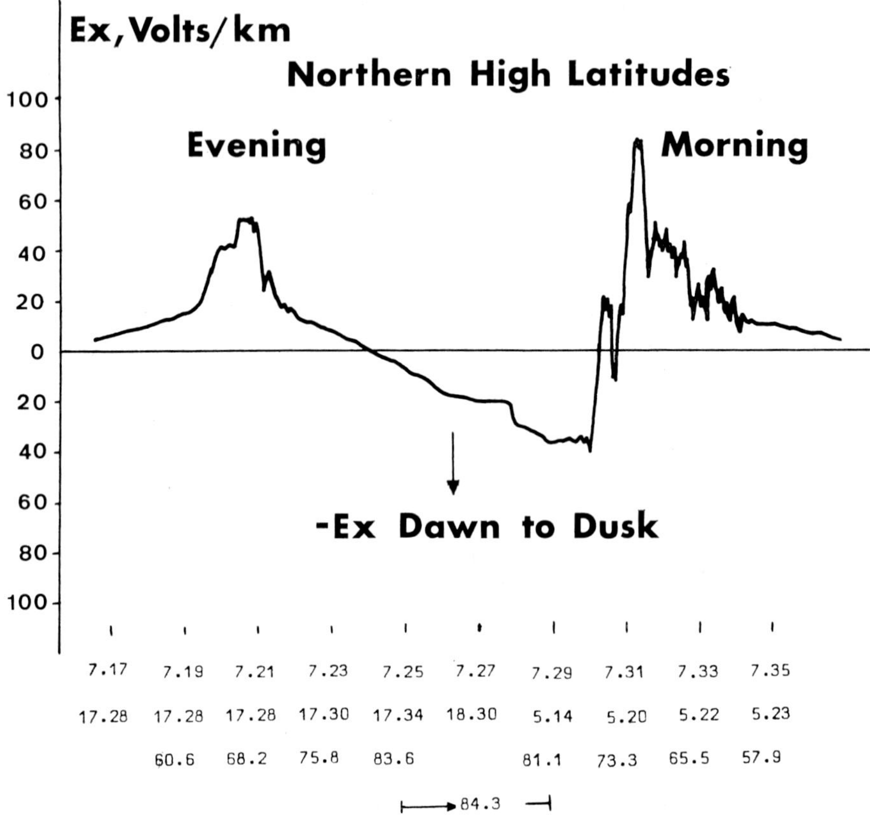

Fig. 1.   Typical convection E field for a OGO 6 northern polar pass (after Heppner, 1972).

sensitivity of the Injun 5 experiment (from $\pm$ 10 to $\pm$ 30 mV m^{-1}) due to wake and shadowing from the satellite. Preliminary results from a comparison between the interplanetary magnetic field (IMF) and the magnetospheric convection E field have disclosed the importance of the direction of the IMF in the ecliptic plane. When the IMF is directed away from the sun the convection E field is higher on the northern hemisphere morning side and on the southern hemisphere evening side where are rooted those geomagnetic field lines which are parallel to the IMF near the magneto-pause. The effect of the polarity of the IMF – and more specifically of the Y component of the IMF in G.S.M. coordinates – has already been observed in analyzing the pertubations of the vertical component of the geomagnetic field at high latitudes (Friis-Christensen *et al.*, 1972) and these observations may be accounted for by the corresponding changes in E field configuration (Heppner, 1972). A study of the dependence of the horizontal component of the geomagnetic field variations on the IMF direction at very high latitudes (Berthelier, 1972) seems to confirm this result since it can be interpreted as the consequence of a higher E field on the northern hemisphere morning side when the Y component of the IMF is positive (Figure 2).

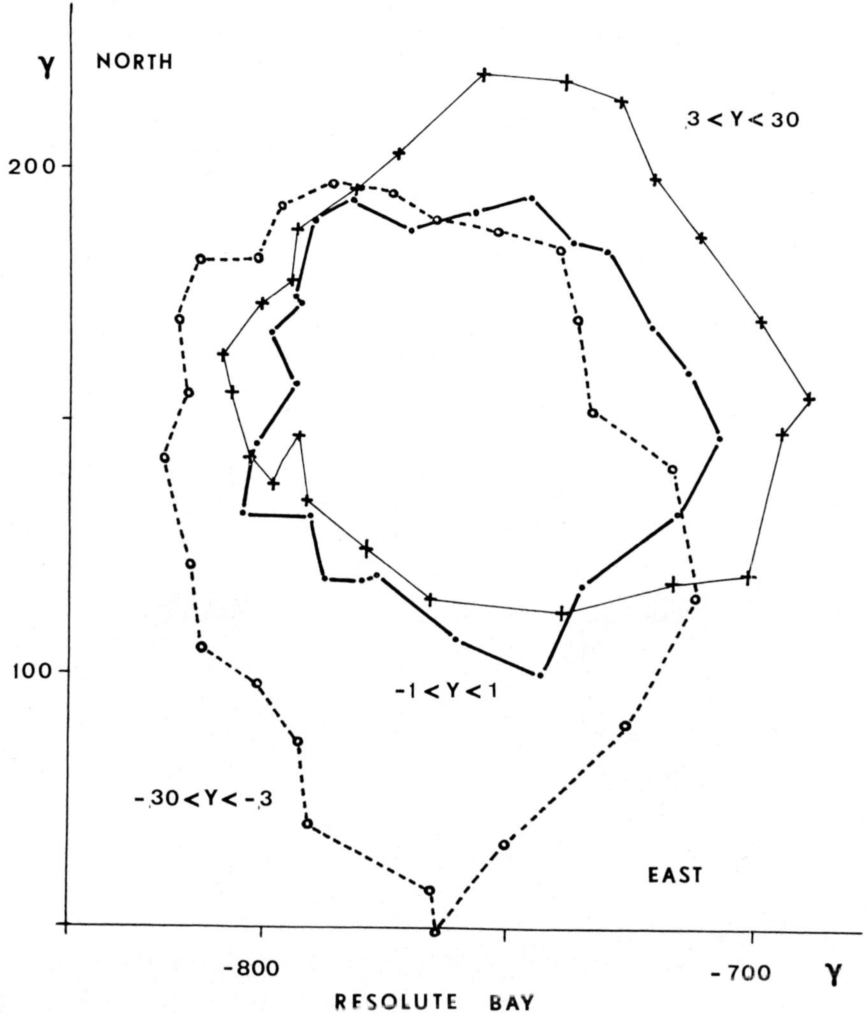

Fig. 2. Influence of the Y component of the IMF on the horizontal component of magnetic field at high latitude.

It must be pointed out that the major limitation of the above mentioned satellite experiments was that only one component of the convection E field was measured. The apparent differences between OGO 6 and INJUN 5 results, especially near the reversal zones, may be in part due to the different orientations of the probe axis at time of measurements. Preliminary results from a balloon experiment (Mozer, 1972) measuring the true vector E field do indeed indicate that, at least during morning hours, convection velocity may be oriented along the 1000 to 2200 hr meridian.

A symmetry of the electrical properties of the magnetospheric cavity along this direction has been theoretically predicted (Forbes and Speiser, 1971) on the basis of the average direction of the IMF at the orbit of the earth.

3. Particle Precipitations in Transauroral Regions

Data on low energy particle precipitations at transauroral latitudes were obtained several years ago through photometric measurements and more recently by means of direct measurements onboard satellites. Extensive studies of atmospheric emissions led Sandford (1964, 1968, 1970) to the distinction of a nightside peak aurora coincident with the nighttime auroral zone and a dayside peak aurora, near the location of the midday auroral oval. Recently several air-borne expeditions were made through the auroral and transauroral zones (Eather, 1969; Eather and Mende, 1971, 1972; Heikkila *et al.*, 1972; Romick and Brown, 1971) and results confirming the general features previously observed are summarized in Figure 3.

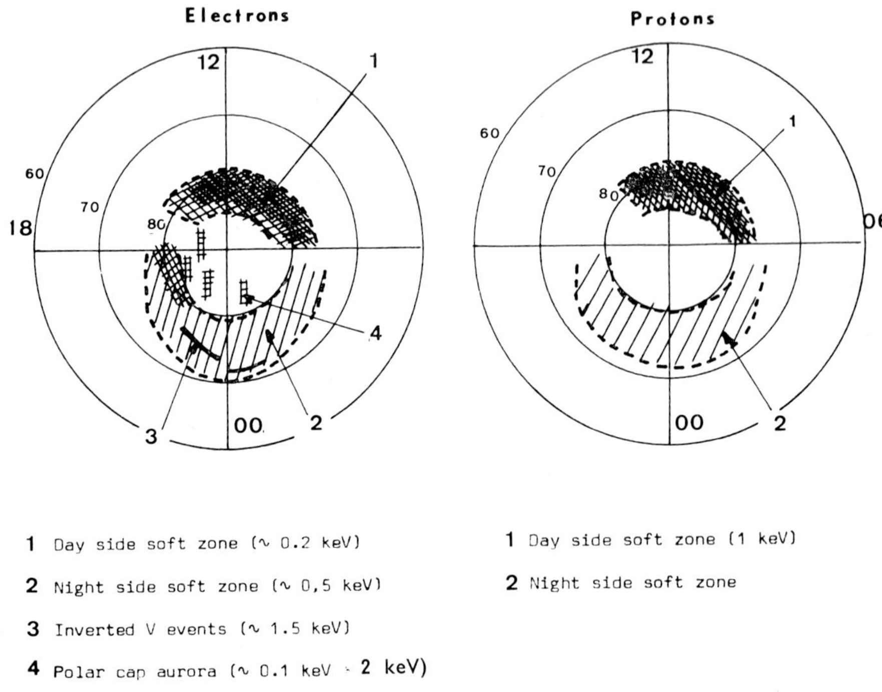

Fig. 3.   Low energy particle precipitation pattern at high latitudes.

(a) In the central part of the polar cap there is a polar cavity where no diffuse emissions at $\lambda\,4278\mathrm{N}_2^+$ are observed which sets an upper limit of $\sim 10^{-3}$ erg cm^{-2} s^{-1} for the corpuscular energy input. Discrete and rapidly variable emissions are observed between 80 to 85° IN Lat with energy input of ~ 0.15 erg cm^{-2} s^{-1} corresponding to polar cap discrete auroraes due to electrons of ~ 100 eV with a 2 keV superimposed population (Romick and Brown, 1971).

(b) In the transauroral regions there are two main zones of low energy particle precipitations:

(i) A nighttime soft zone extending from ~ 72 to $80°$ IN Lat and from dusk to dawn where the electron characteristic energy, decreasing with increasing latitude, averages about 0.5 keV. Protons seem to be absent and the origin of precipitating electrons is thought to be the plasma sheet (Eather and Mende, 1971) although previous magnetic models indicate that the plasma sheet should not map above $\sim 74°$ IN Lat The average energy input is $\lesssim 0.1$ erg cm^{-2} s^{-1}.

(ii) A dayside soft zone where both protons (~ 1 keV) and electrons (0.2 keV) precipitate between 77 and $82°$ IN Lat, where the spectral ratio $I(6300)/I(4278)$ is of the order of 20 to 30 and indicates a very soft electron spectrum. Short lived arcs or patches are superimposed on this continuum; the total amount of energy deposited averages about 0.15 erg cm^{-2} s^{-1} and the particles are thought to originate in magnetosheath plasma.

These low energy particle fluxes were recently identified by satellite experimenters (Burch, 1968; Maehlum, 1969; Heikkila and Winningham, 1971). The dayside soft zone has been shown to extend from aproximately 06.00 to 16.00 LT and to be unambiguoulsy the signature of the direct entry of magnetosheath plasma in the polar ionosphere. The latitudinal width of this zone is of the order of 2 to $3°$ and its position depends upon the tilt angle β between the geomagnetic axis and the sun-earth line (Maehlum, 1969) and also upon magnetic activity (Winningham, 1971). A statistical study of the 0.7 keV electron bursts detected onboard OGO 4 (Hoffman and Berko, 1971) is in very good agreement with above mentioned results and seems to indicate a symmetry with respect to the 1000 to 2200 hr meridian.

In contrast with photometric observations, particle measurements have disclosed the presence of protons with measurable intensities in the late evening and midnight soft zone which have been tentatively identified as having their origins within the downstream magnetosheath (Frank and Ackerson, 1972). There is also disagreement on the position of the boundaries which, according to these authors, are significantly lower in the midnight sector. A major difference, which may be due in part to the averaging tendency of photometric measurements, is the disclosure of the importance of highei energy (~ 1 to 5 keV) 'inverted V electron events' which have been identified only in satellite results and carry a significant part of the energy.

The last point to be noted is the excellent agreement between low energy electron measurements obtained by Whalen *et al.* (1971) during a rocket flight through a polar cap aurora and the previously mentioned photometric results.

4. Ionospheric Absorption Events and Related Phenomena in Transauroral Regions

Using data from a 30.1 MHz riometer located at Dumont d'Urville (Antarctica), $81°$ IN Lat, an analysis of ionospheric absorption events has been made. Events are mainly characterized by a long duration (~ 1 to 3 hr) and a weak (~ 0.3 dB) slowly varying absorption (typical rise and fall time ~ 5 to 15 min). Details of data analysis may be found in Lavergnat (1970), Vassal (1971) and Sylvain (1972). Results of this analysis disclose the existence of 3 different groups of events distinguishable by their different diurnal and seasonal variations:

(a) Events of the first group (type I) only occurring under summer conditions (September to April). with a peak occurrence between November and January, are associated with the disappearance of the echos from F_1 and/or F_2 layers on ionograms ('F lacunae') and are described in details in Section 4.1.

(b) Events of the second group (type II) occur predominantly during winter time with a peak occurrence in June and almost completely disappear between November and January. The diurnal curve of occurrence peaks near 1900 to 2100 LT (Vassal, 1971) (Figure 4).

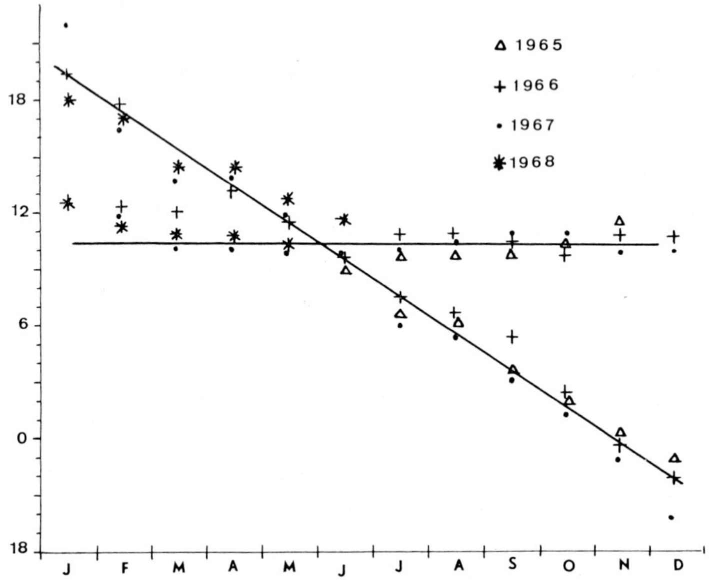

Fig. 4.   Hours of maximum of occurrence of type II and III absorption events at Dumont d'Urville (81° IN Lat, Antarctica).

(c) Events of the third group (type III) occur mainly near the equinoxes and their hour H_m of maximum of occurrence shows a sideral like variation all along the year following the approximate relation

$$H_m(\mathrm{UT}) = 2M + 22, \quad M = \text{number of month}$$
(Figure 4).

4.1. TYPES II AND III EVENTS

Type II events display a strong correlation with local magnetic activity (K index) and substorm activity (AE index). On the contrary type III events do not correlate with either K or AE, and this striking difference certainly implies different origins and precipitation mechanisms. For both types of events there is a clear correlation between absorption as measured on a riometer and variations of f_{min} which indicates that energetic ($\geqslant 10$ keV) electrons precipitate since protons of ~ 100 keV which could have the same effect are not likely to be present under non-PCA conditions. A comparison between events of both types and zenithal aurora recorded on an all-sky camera (Weill, 1972) gave a null correlation thus indicating again that the bulk of energy is carried by relatively hard electrons (Eather and Mende, 1971).

The time of occurrence of type II events fits very well with the existence of a secondary peak in 0.7 keV electron precipitation found by Hoffman and Berko (1971) near 1800 to 2100 LT. However these authors did not give any indication on fluxes or energy spectra and the origin of these electrons remains highly speculative at the time. In spite of the high IN Lat the close association with substorm activity favors a plasma sheet origin where intense fluxes of energetic electrons sufficient to produce a few tenths of a dB absorption are known to be present during substorms (Akasofu *et al.*, 1971). Seasonal variation in occurrence could be due to seasonal variation of plasma sheet position (Montgomery, 1968); however, the evening occurrence is opposite to

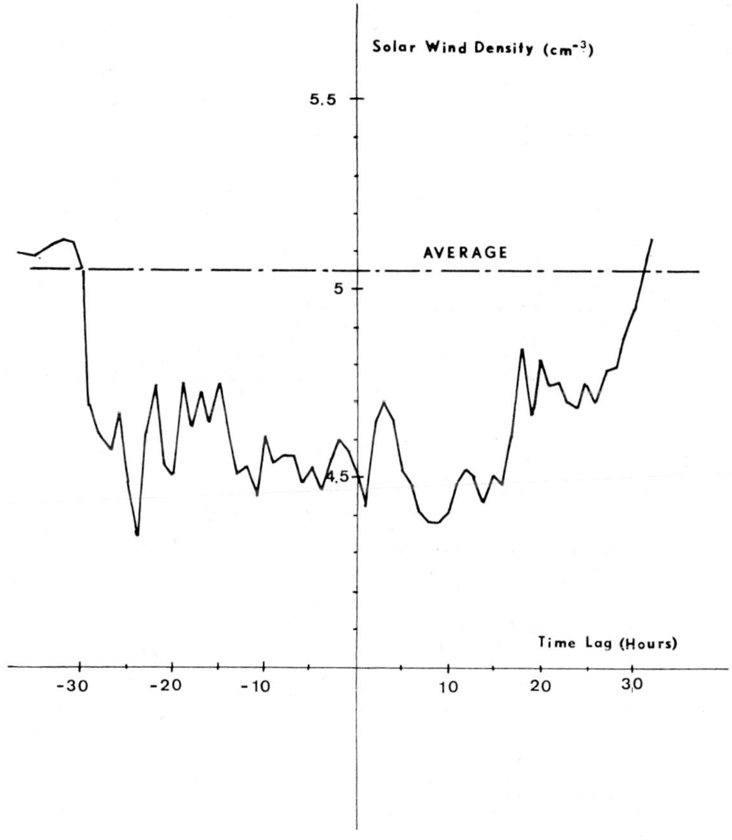

Fig. 5.   Time-lag analysis of solar wind density variations during type III events.

the dawn-dusk asymmetry of the plasma sheat and this observation has not been interpreted so far.

The most striking feature of type III events is their sideral like time variation of H_m which leads to the assumption that these events are a result of a favorable position of the magnetosphere with respect to the ecliptic plane but the dipole axis does not show any remarkable position at the time of maximum occurrence H_m. A preliminary analysis of interplanetary medium parameters does however, indicate that events occur preferably when the solar wind density is lower than normal (Figure 5).

4.2. F Lacunae events

The disappearance of echoes from F_1 and/or F_2 regions (Figure 6) is often observed on ionograms from transauroral stations (Cartron, 1962; Lebeau, 1965; Olesen, 1971). Durations are extremely variable ranging from a few minutes to a few hours and in this latter case a succession of different types of lacunae is generally observed with intervals showing normal ionograms. Changes in the receiver gain may affect the observations because with a 25 dB increase in the gain $\sim 20\%$ F lacunae ionograms recover a normal aspect (Sylvain, 1972). The F_2 lacunae cannot be ascribed to a decrease of f_0F_2 below f_0F_1 (G. Condition) since in many cases the pre- and post-lacuna ionograms display a well-behaved F layer with $f_0F_2 - f_0F_1 \geqslant 1$ MHz and also because

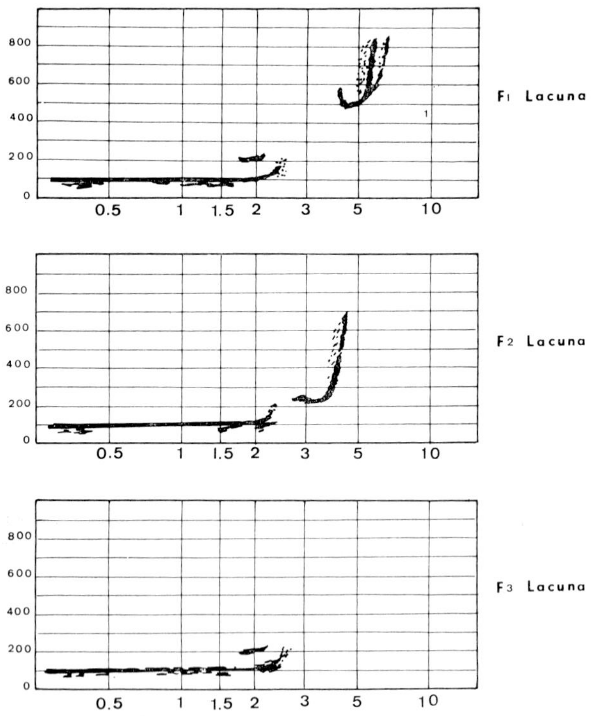

Fig. 6.   Sketch of typical ionograms during F_1, F_2 and F_3 lacunae events.

it is sometimes possible to see the F_2 layer reappearing when the gain is increased. The lacuna events show a pronounced maximum of occurrence in December-January in Antarctica and are never seen between May and August; they are daytime events observed between 0600 to 1500 LT with a peak occurrence at 1100 LT for F_1 lacunae and two peaks at 0800 to 13.00 LT for F_2 lacunae. We have looked for a possible explanation in three different ways: collisional absorption of radio waves emitted by the sounder, enhanced diffusion in the F layer, and nonhorizontal isodensity contours in the F region preventing reflected radio-waves to be returned to the receiver. The third hypothesis seems to the most likely (Sylvain, 1972) but the physical processes responsible for these large scale disturbances (gravity waves, field aligned currents, etc.) await further more precise measurements to be understood.

Acknowledgments

We acknowledge fruitfull discussions with our colleagues A. Berthelier, and J. Lavergnat. Interplanetary magnetic field data were kindly provided by Dr D. S. Colburn and solar wind parameters by Dr H. S. Bridge and A. J. Lazarus.

References

Axford, W. I. and Hines, C. O.: 1961, *Can. J. Phys.* **39**, 1433.
Akasofu, S. I., Hones, E. W., Montgomery, M. D., Bame, S. J., and Singer, S.: 1971, *J. Geophys. Res.* **76**, 5985.
Berthelier, A.: 1972, to be published.
Burch, J. L.: 1968, *J. Geophys. Res.* **73**, 3585.
Cartron, S.: 1962, in *Monographie sur les observations ionosphériques à la station Dumont d'Urville* (ed. by C.N.R.S.).
Cauffman, D. P. and Gurnett, D. A.: 1971, *J. Geophys. Res.* **76**, 6014.
Eather, R. H.: 1969, *J. Geophys. Res.* **74**, 153.
Eather, R. H. and Mende, S. B.: 1971, *J. Geophys. Res.* **76**, 1746.
Eather, R. H. and Mende, S. B.: 1972, *J. Geophys. Res.* **77**, 660.
Forbes, T. G. and Speiser, T. W.: 1971, *J. Geophys. Res.* **76**, 7542.
Frank, L. A.: 1971, *J. Geophys. Res.* **76**, 5202.
Frank, L. A. and Ackerson, K. L.: 1972, *J. Geophys. Res.* **77**, 4116.
Frank, L. A. and Gurnett, D. A.: 1971, *J. Geophys. Res.* **76**, 6829.
Friis-Christensen, E., Lassen, K., Wilhjem, J., Wilcox, S. M., Gonzalez, W., and Colburn, D. S.: 1972, *J. Geophys. Res.* **77**, 3371.
Heikkila, W. J. and Winningham, J. D.: 1971, *J. Geophys. Res.* **76**, 883.
Heikkila, W. J., Winningham, J. D., Eather, R. H., and Akasofu, S. I.: 1972, *J. Geophys. Res.* **77**, 4100.
Heppner, J. P., Stolarik, J. C., and Wescott E. M.: 1971, *J. Geophys. Res.* **76**, 6028.
Heppner, J. P.: 1972, *J. Geophys. Res.* **77**, 4877.
Hoffman, R. A. and Berko, F. N.: 1971, *J. Geophys. Res.* **76**, 2967
Lavergnat, J.: 1970, *Thèse de 3e Cycle*, Université de Paris.
Lebeau, A.: 1965, *Ann. Geophys.* **21**, 167.
Maehlum, B. N.: 1969, *J. Atmospheric Terrest. Phys.* **31**, 531.
Montgomery, M. D.: 1968, *J. Geophys. Res.* **73**, 871.
Mozer, F. S.: 1972, private communication.
Nelms, G. L. and Chapman, J. H.: 1970, in G. Skovli (ed.), *The Polar Ionosphere and Magnetospheric Processes*, Gordon and Breach, New York.

Olesen, J. K.: 1971, Danish Meteorological Institute, Ion. Lab. Tech. Report.
Romick, G. J. and Brown, N. B.: 1971, *J. Geophys. Res.* **76**, 8420.
Sandford, B. P.: 1964, *J. Atmospheric Terrest. Phys.* **26**, 749.
Sandford, B. P.: 1968, *J. Atmospheric Terrest. Phys.* **30**, 1921.
Sandford, B. P.: 1970, in G. Skovli (ed.), *The Polar Ionosphere and Magnetospheric Processes*, Gordon and Breach, New York.
Sylvain, M.: 1972, *Thèse de 3e Cycle*, Université de Paris.
Vassal, J.: 1971, *Thèse de 3e Cycle*, Université de Paris.
Weill, G.: 1972, private communication.
Wescott, E. M., Stolarik, J. D., and Heppner, J. P.: 1969, *J. Geophys. Res.* **74**, 3469.
Whalen, B. A., Miller, J. R., and McDiarmid, I. B.: 1971, *J. Geophys. Res.* **76**, 6847.
Winningham, J. D.: 1971, in B. M. McCormac (ed.), *Earth's Particles and Fields*, D. Reidel Publishing Company, Dordrecht, Holland.

PART III

CHEMICAL PROCESSES AND MODELS

NEUTRAL ATMOSPHERIC CHEMISTRY –
INTRODUCTION AND REVIEW

HAROLD I. SCHIFF

Centre for Research in Experimental Space Science,
York University, Downsview, Ont., Canada

1. Introduction

The explanation for the surprising observation of a temperature minimum near
11 km was made by Sydney Chapman, the recognized father of aeronomy. Chapman
(1930, 1943) pointed out that most of the UV radiation incident on this planet is
absorbed in the upper regions of the atmosphere where it is converted into chemical
energy in the form of dissociated and ionized gases, and finally deposited in the form
of heat. The principal atmospheric constituent which participates in this energy conversion is O_2.

Oxygen dissociates in two wavelength regions. Absorption in the Schumann-Runge
continuum at wavelengths less than 1759 Å results in the production of two O atoms,
one in the ground state, the other in the first, excited state:

$$O_2 + h\nu \rightarrow O(^3P) + O(^1D). \tag{1}$$

So strong is this absorption that radiation of these wavelengths cannot penetrate
below about 70 km, while above about 120 km most of the oxygen is present in the
atomic form.

Absorption in the relatively weak Herzberg continuum, at wavelengths less than
2420 Å produces two ground state (^3P) atoms. This process can occur at altitudes
down to 30 km.

Recombination of the atoms can occur either directly

$$O + O + M \rightarrow O_2 + M + 118 \text{ kcal} \tag{2}$$

or via the intermediate formation of O_3:

$$O + O_2 + M \rightarrow O_3 + M + 24 \text{ kcal} \tag{3}$$
$$O + O_3 \rightarrow O_2 + O_2 + 94 \text{ kcal}. \tag{4}$$

The net result of reactions (3)+(4) is identical to that of reaction (2). The relative
importance of these two mechanisms depends on the relative concentrations of O and
O_2. The second mechanism is predominant in the stratosphere. These reactions, along
with the photolysis of O_3:

$$O_3 + h\nu \rightarrow O_2 + O \tag{5}$$

were essentially those proposed by Chapman to explain the 'ozone layer'. This relatively simple set of chemical reactions led to a lively, and fruitful interaction among
atmospheric observers, atmospheric modellers and laboratory kineticists.

B. M. McCormac (ed.), Physics and Chemistry of Upper Atmospheres, 85–98. All Rights Reserved.
Copyright © 1973 by D. Reidel Publishing Company, Dordrecht-Holland.

Reliable reaction rate data for these reactions have been surprisingly difficult to attain for several reasons. Ozone has a relatively weak bond strength (24 kcal mole^{-1}) and is, therefore, subject to catalytic destruction by impurities in the gases or on the surfaces of reaction vessels. Secondly, the electrical discharges used in many of the earlier investigations to generate O atoms also produce electronically and vibrationally excited species capable of reacting with O_3. Only recently have thermal and photolytic sources of O atoms yielded reliable data for the rate constants* for N_2 as the third body in Equations (2) and (3).

$$k_2 = 7 \times 10^{-31} \, T^{-1.0} \qquad \text{(Kondratiev, 1970)}$$
$$k_3 = 1.1 \times 10^{-34} \exp(0.52 \times 10^3 / T) \qquad \text{(Huie and Davis, 1972)}$$
$$k_4 = 1.0 \times 10^{-11} \exp(-2.15 \times 10^3 / T) \qquad \text{(McCrumb and Kaufman, 1972;}$$
$$\text{Krezenski } et \, al., \text{ 1971)}.$$

Insertion of these rate constants into the Chapman mechanism predicts a total global atmospheric O_3 content some 50% larger than observed. This discrepancy, in itself, would not be too disturbing in view of the combined uncertainties in the observations, rate data, solar flux and in the absorption coefficients and quantum yields of the O_2 and O_3 photolysis. But the mechanism is incapable of accounting for the seasonal and latitudinal dependence of the O_3 distribution, the most glaring example of which is the occurrence of the maximum O_3 densities near the poles in late winter. The O_3 must have been produced at lower latitudes and transported polewards. The entire atmosphere is obviously not in a photochemical steady state. In fact, the Herzberg continuum is so weak that virtually no O_3 is formed below 30 km, where the O_3 density is near its maximum. Transport processes, therefore, determine the O_3 densities at low altitudes while photochemistry determines the O_3/O ratio.

Above 50 km, where the reaction times (half-lives) are all less than one day, photochemical steady state should exist. But even at these altitudes the simple Chapman mechanism consistently predicts too much O_3 production. Other chemical reactions must be occurring which destroy O_3.

2. Reactions Involving Hydrogen Oxides

Hampson (1964) was the first to point out that, in the mesosphere, O_3 densities may be controlled by H containing, minor constituents. The most effective catalytic chain appears to be:

$$H + O_3 \rightarrow OH + O_2 + 78 \text{ kcal} \qquad (6)$$
$$OH + O \rightarrow O_2 + H + 16 \text{ kcal} \qquad (7)$$

the net effect of which is identical to reaction (4), but with an effective rate constant about 10^3 times larger.

---

* Throughout this article rate constants will be given with concentration units of particles cm^{-3}, time in seconds and temperature T in degrees Kelvin. Original articles and critical reviews should be consulted for estimates of uncertainties in the rate constants.

To assess the importance of this chain it is necessary to consider the sources and sinks of the chain carriers, H and OH. The obvious sources and H_2O, CH_4 and H_2 which originate at the earth's surface. But how are these compounds dissociated? Hydrogen molecules and CH_4 photodissociate only in the vacuum UV, but H_2O can be dissociated in roughly equal proportion by Ly-α or Schumann-Runge radiation which can penetrate into the mesosphere. In the stratosphere, however, photodissociation is relatively unimportant. Although ground state O atoms react only very slowly, O atoms in their first excited, (^1D) state, react rapidly with H_2O, CH_4 and H_2 to produce OH. These reactions are undoubtedly the main sources of the chain carriers in the stratosphere.

We must, however, also consider other reactions which these chain carriers can undergo. Figure 1 illustrates most of the reactions which may occur in a H-O system. In this figure the reaction rate constants are given at 300 K as negative exponents to the base 10. Measured temperature coefficients, B, defined by the expression $k = A \exp(-B \times 10^3/T)$ are given in square brackets.

2.1. REACTIONS OF OH

Reactions of OH which terminate the chain by producing H_2O occur with OH and

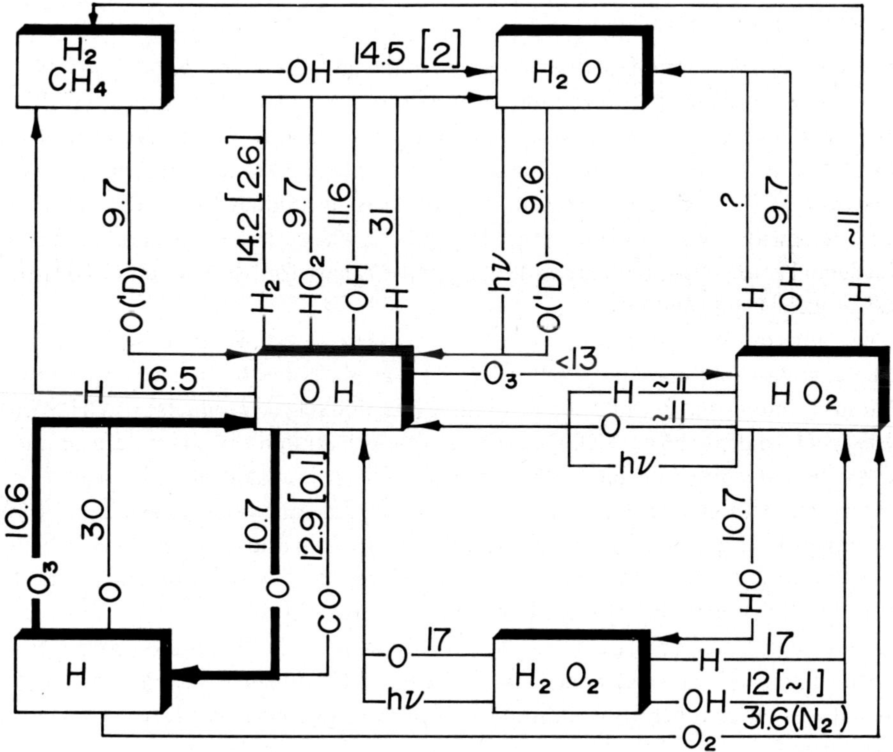

Fig. 1. Reaction paths in the O-H system. Reaction rates given as negative powers of ten, temperature coefficients in square brackets. Values are taken from the National Bureau of Standards, Chemical Kinetics Data Survey, or given in the text.

HO_2, both of which are fast. The three-body recombination with H is too slow to be significant. The reaction with O_3, at one time believed to provide an important alternate chain for O_3 destruction, is now believed to be too slow to be important.

The reaction of OH with CO is responsible for oxidizing CO to CO_2 in the earth's atmosphere and probably in the atmospheres of Mars and Venus. It does not, however, stop the chain decomposition of O_3 since the other chain carrier, H, is a product of the reaction. It simply takes the place of reaction (7) in the chain mechanism.

The reactions of OH with CH_4 and H_2 are not important chain termination steps at stratospheric temperatures since both have activation energies of about 4 kcal mole^{-1} (Greiner, 1969, 1970).

The other chain carrier, H, not only reacts with O_3 but, in the stratosphere, reacts even more rapidly with O_2 in a three-body process to form HO_2:

$$H + O_2 + M \rightarrow HO_2 + M. \tag{8}$$

This reaction does not terminate the chain since HO_2 is rapidly attacked by O:

$$HO_2 + O \rightarrow OH + O_2 \tag{9}$$

thereby producing the other chain carrier, OH.

2.2. REACTIONS OF HO_2

The occurrence of reaction (8) does however, make HO_2 an important minor constituent in the atmosphere and other loss processes for this radical must be considered. Considerable uncertainty still exists about the rates of many of the reactions of this radical (Lloyd, 1970). Much of the data derives from indirect, albeit ingeneous deductions from studies of flames. Recently, photolysis techniques have been combined successfully with UV absorption of the HO_2 radical. The best established rate constants (Hochendal *et al.*, 1972; Paukert and Johnston, 1972) are for the reaction of HO_2 with itself to yield H_2O_2 and O_2.

The rate constant for reaction (9) has not been measured directly but indirect evidence suggests (Kaufman, 1969) it has a value of about 10^{-11}. The reaction with H atoms is interesting because there are three possible sets of products: (i) OH + OH; (ii) H_2 + O; and (iii) H_2O + O. Channel (i) would maintain the chain destruction of O_3, (ii) would constitute chain termination while (iii) would actually lead to O_3 production. Although the rate for this reaction has not been measured directly, McConnell (1972) has recently analyzed existing data and suggests that the rate constants for the overall reaction may be as high as 10^{-10} with a branching ratio for channels (i), (ii), and (iii) of 0.35, 0.39 and 0.26 respectively. It is rather difficult to accept so high a probability for channel (iii) which involves the simultaneous rupture and formation of chemical bonds. Nevertheless, McConnell is probably correct in concluding that channel (ii) is an important process in converting H_2O to H_2 in the mesosphere.

H_2O_2 is relatively inert to attack by H and O atoms and is lost in the atmosphere mainly by photolysis and by reaction with OH.

2.3. MODEL CALCULATIONS FOR AN OXYGEN-HYDROGEN ATMOSPHERE

There now appears to be general agreement among various atmospheric models on the effect of H compounds on atmospheric chemistry.

The lifetime for achieving photochemical steady states, τ_{chem}, for H_2 and H_2O is greater than the mixing lifetime, τ_{mix}, so that the distributions of these compounds in the stratosphere and mesosphere are controlled by transport. Since CH_4 is not regenerated by any atmospheric process it is consumed in the stratosphere.

The radicals H, OH and HO_2 have $\tau_{\text{chem}} < \tau_{\text{mix}}$ in the mesosphere and so should be in a photochemical steady state. But their ratios cannot be unequivocally predicted because of their (largely unknown) interactions with the oxides of N (Nicolet, 1970).

At altitudes less than about 40 km the effect of the hydrogenous constituents on the O_3 distribution is smaller than the effect of transport, while above this altitude they play the controlling role. The H chain is, therefore, incapable of explaining the discrepancies between the Chapman mechanism and observations in the stratosphere.

3. Reactions Involving Nitrogen Oxides

There is another chain mechanism for O_3 decomposition involving oxides of N which has received a great deal of attention recently in connection with the SST problem (Crutzen, 1970, 1972; Johnston, 1971):

$$O_3 + NO \rightarrow NO_2 + O_2 + 47 \text{ kcal} \tag{10}$$

$$O + NO_2 \rightarrow NO + O_2 + 47 \text{ kcal.} \tag{11}$$

Again, the net effect of this sequence is identical to that of reaction (4) but with a much larger effective rate constant. Either NO or NO_2 can initiate the chain and the two together are often referred to as NO_x.

NO_2 also undergoes photodecomposition at wavelengths less than 3980 Å

$$NO_2 + h\nu \rightarrow NO + O. \tag{12}$$

Reaction (12) sidetracks the chain since the O atoms will recombine by reaction (4) to form O_3. Because reaction (12) can occur with visible light, any process that leads to NO_2 formation can give rise to O_3 down to the earth's surface. Such processes play important roles in O_3 smog formations in the lower troposphere.

There is only one chemical reaction which produces NO_x in the stratosphere, and that is the rapid reaction of $O(^1D)$ with N_2O which has been transported from the troposphere where it probably has a biological origin. This reaction has several possible channels

$$O(^1D) + N_2O \rightarrow NO + NO \tag{13a}$$

$$\rightarrow N_2 + O_2 \tag{13b}$$

$$\rightarrow NO_2 + N. \tag{13c}$$

Laboratory measurements show that channel (c) has a very low probability while channels (a) and (b) have roughly equal probabilities (Goldman *et al.*, 1971; Scott *et al.*, 1971).

In addition to upward transport of N_2O, there is the possibility of direct transport of NO_x, from the troposphere as well as of NH_3, which will be rapidly oxidized.

There are several processes for NO_x production in the mesosphere. These include dissociation of N_2 by UV light, by X-rays and by energetic particles. Because of its low ionization potential, NO^+ is a major product of ion-molecule reactions in the *D* and *E* regions of the ionsphere. Rapid dissociative recombination of NO^+

$$NO^+ + e \rightarrow N + O \tag{14}$$

can produce N atoms in the first, excited (^2D) state which reacts rapidly with O_2 to form NO (Black *et al.*, 1969; Lin and Kaufman, 1971). Ground state N atoms react only slowly with O_2 but rapidly with NO to form N_2 and O. Thus NO is both produced and destroyed chemically in the mesosphere. Moreover, it has been recently shown that NO can be predissociated in the Schumann-Runge wavelength region. It is now believed (Brasseur and Cieslik, 1972) that there is no net production of NO in the mesosphere and, therefore, little, if any, downward transport into the stratosphere.

The chemistry of NO_x is represented in Figure 2. Oxidation of NO by O_2 is a slow three body process important only near ground level. The three body reaction with O atoms is accompanied by the familiar 'air afterglow' reaction, responsible for most

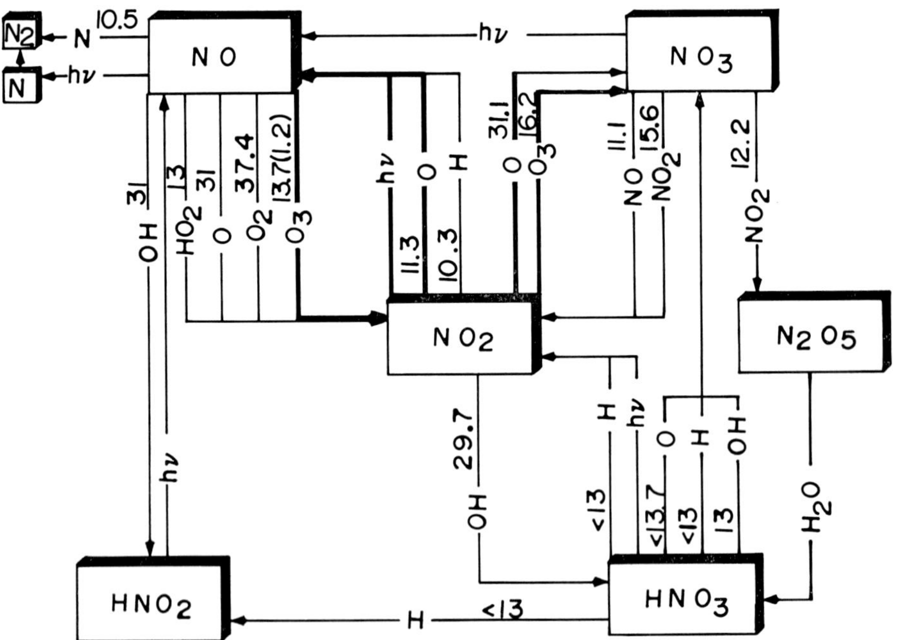

Fig. 2.   Reaction paths involving the oxides of N.

of the airglow continuum, but relatively unimportant in determining the NO_x distribution.

The reaction:

$$NO + HO_2 \rightarrow NO_2 + OH \tag{15}$$

may be important in coupling the N and H systems but its rate has not yet been established.

3.1. REACTIONS OF NO_2

In addition to its reaction with O atoms, NO_2 can also be reduced to NO by H atoms. Although this is a very rapid reaction it does not have aeronomic significance because these reactants reach their maximum concentrations at very different altitudes.

NO_2 can be oxidized by O in a three body reaction or by O_3 in a two body reaction to form NO_3. Although NO_3 can be reduced by further reaction with either NO_2 or NO, it can also react with NO_2 to form N_2O_5. The rates of these reactions are not well established due to complications of surface effects in laboratory studies, but they are probably too slow to be significant in the stratosphere. The most likely fate of NO_3 is photodissociation which, surprisingly enough produces NO and O_2 (Johnston, 1972).

The major sinks for NO_x in the stratosphere appears to be the formation of HNO_3 and HNO_2 by the reactions

$$NO_2 + OH + M \rightarrow HNO_3 + M \tag{16}$$
$$NO + OH + M \rightarrow HNO_2 + M \tag{17}$$

Nitric acid has indeed been observed in appreciable quantities in the stratosphere (Murcray *et al.*, 1969).

An important question is whether these sinks leak. Despite earlier indications (Berces and Forgeteg, 1970) it now appears that the reactions of HNO_3 with O, H, and OH are slow (Morris and Niki, 1971) and that the main loss processes are photolysis. Still needed are quantum yields and the mode of photodecomposition as a function of wavelength. At the long wavelength end of the absorption spectrum of HNO_3 the major photoproducts appear to be OH and NO_2 (Johnston, 1972) thus reversing reaction (16). HNO_3 thus simply holds up the chain process, but this may be sufficient to reduce the catalytic destruction of O_3 if the HNO_3 can be removed by downward transport into the troposphere where it will be rapidly removed in rain.

3.2. SUMMARY OF THE OXYGEN-HYDROGEN-NITROGEN CHEMISTRY

Although the Chapman mechanism provides the basic model for the composition of the stratosphere and mesosphere it is inadequate for quantative predictions. Above 40 km reactions involving H compounds and, below this altitude, reactions involving N compounds play important roles. In the lower stratosphere transport processes are more important than chemistry. Improvement in current models will require better reaction rate data, detailed knowledge of the solar flux, absorption coefficients, and quantum yield data, Above all, they require better understanding of the large and small scale transport processes, including the exchange between the troposphere and stratosphere.

3.3. CO_x CHEMISTRY

The chemistry of CO_2 and CO, while of only secondary interest in the earth's atmosphere, is fundamental in the understanding of the atmospheres of Mars and Venus.

In the earth's atmosphere, the CO_2 mixing ratio is constant up to the turbopause, while the CO concentration is controlled by chemical reaction, primarily with OH. Carbon dioxide is involved in the chemistry of both positive and negative ions in the D region ionosphere (Reid, 1973).

Carbon dioxide is the major component in the atmospheres of Mars and Venus where it should readily undergo photolysis to form CO and O. The puzzle is why there is so little O in the upper atmosphere, and so little O_2 in the lower atmosphere of Mars. Recent laboratory studies (Slanger *et al.*, 1972) have shown that, at the pressure existing on the Martian surface, the reaction

$$CO + O + M \rightarrow CO_2 + M \tag{18}$$

will be a three-body process with a rate constant given by the expression $k_{18} = 3.2 \times \times 10^{-33} \exp(-2.2 \times 10^3/T)$ which is too small to provide effective reoxidation of CO. At higher altitudes, where the temperature and pressure are both lower, the rate of reaction (18) will be even slower.

It is interesting to note that reaction (18) is one of the few three-body reactions known to have a positive temperature coefficient. This is related to the fact that the reaction involves a change in spin. It has been proposed (Clyne and Thrush, 1962) that it proceeds in several stages. The initial process

$$CO(^1\Sigma^+) + O(^3P) \rightarrow CO_2(^3B_2)$$

which conserves spin by producing electronically excited CO_2 is slower than the curve-crossing processes which lead to ground state CO_2.

Since the UV photolysis of CO_2 produces $O(^1D)$, it was originally suggested that CO is rapidly oxidized by the spin-allowed recombination

$$O(^1D) + CO \overset{M}{\rightarrow} CO_2. \tag{19}$$

However, laboratory results (Slanger and Black, 1970) have shown that the rapid quenching of $O(^1D)$ by CO results in physical quenching

$$O(^1D) + CO \rightarrow O(^3P) + CO \tag{20}$$

rather than the chemical reaction (19).

Suggestions that $O(^1D)$ might react with CO_2 to provide a reactive CO_3 intermediate

$$CO_2 + O(^1D) \rightarrow CO_3^* \tag{21}$$

$$CO_3^* + CO \rightarrow CO_2 + CO_2 \tag{22}$$

have also been largely discredited by laboratory evidence (Loucks and Cvetanovic, 1972; Slanger and Black, 1971).

The remaining explanation for the planetary observations would be the oxidation of CO by OH. The problem has shifted, therefore, to finding a satisfactory explanation for the presence of this radical on the atmospheres of Mars and Venus.

4. Reactions Involving Excited States

Observations of emissions from airglow and auroras have shown that there are appreciable quantities of excited atoms and molecules in the earth's atmosphere. Kineticists have long been aware that the rates and even the reaction mechanisms of chemical reaction are strongly dependent on the internal energy of the reagents. But, until recently, little quantitative data were available, due mainly to the difficulty of producing sufficient concentrations of these species for kinetic studies using existing detectors. The development of flash and laser sources and pulse counting detection techniques is now providing reliable data for the rates of deactivation and, in many cases, for identification of the reaction products. What follows is a brief review of some of the more recent laboratory results for those species of aeronomic interest.

4.1. $O(^1D)$ [1.97 eV, 110 S]*

This species is produced in the atmosphere by the UV photodissociation of O_2 and O_3, by electron excitation of O, by dissociative recombination of O_2^+ and NO^+, and by relaxation from $O(^1S)$.

The quenching rate constants at 300 K with atmospheric gases are: N_2, $5(-11)$; O_2, $6(-11)$; O_3, $2.5(-10)$; N_2O, $2(-10)$; H_2O, $3(-10)$; CH_4, $2(-10)$; CO_2, $1(-10)$; and H_2, $2(-10)$ where the numbers in parenthesis refer to powers of 10. The relative values of these quantities are much more reliable than are the absolute values.

Molecular nitrogen is the major atmospheric quencher, and, at atmospheric pressures, the quenching is entirely physical, with the possibility of energy transfer to produce vibrationally excited N_2. Quenching by O_2 and O_3 will be discussed in some detail in a later section. It has now been established that the quenching by H_2O leads predominantly to the formation of OH. Chemical reaction also occurs with N_2O, leading, in approximately equal proportion to the two sets of products $N_2 + O_2$ and $NO + NO$. Chemical reaction is the favored process with CH_4 with $CH_3 + OH$ being the dominant products (Greenberg and Heicklen, 1972) although CH_3OH, CH_2O, $CH_2 + H_2O$ have also been postulated (DeMore and Raper, 1967; Groth, 1964).

4.2. $O(^1S)$ [4.19 eV, 0.745]

The maximum intensity of the $(^1D \leftarrow {}^1S)$ airglow emission at 5577 Å occurs near 100 km. $O(^1S)$ has been shown (Young, 1969) by laboratory studies, to be formed by

* the quantities in square brackets represent the energy, in electron volts of the species above the ground electronic state, and the radiative lifetime.

the mechanism originally proposed by Chapman (1931):

$$O + O + O \rightarrow O_2 + O(^1S). \tag{23}$$

At higher altitudes it is also produced by electron impact excitation of $O(^3P)$ and by dissociative recombination of O_2^+.

The quenching rate constants at 300 K are: N_2, $< 5(-17)$; O_2, $3.5(-13)$; H_2O, $4(-10)$; H_2, $1(-15)$; NO, $5(-10)$; $O, 8(-12)$; and O_3, $6(-10)$.

The very low quenching rate with N_2 explains why the 5577 Å emission is seen at 100 km whereas the subsequent $(^3P \leftarrow {}^1D)$ emission at 6300 Å is not. Another interesting comparison of the relative quenching rates of these two species is the fact that H_2O rapidly quenches both $O(^1D)$ and $O(^1S)$ while H_2 is a good quencher only for $O(^1D)$. This difference may have some aeronomic significance (Schiff, 1969).

Several recent laboratory measurenemts have interesting consequences in connection with the Chapman reaction (23). Young (1969) originally pointed out that his determination of $k_{23} \simeq 10^{-35}$, when applied to the observed airglow intensity, implied an O atom density at 100 km much higher than those given in model atmospheres. The determination of k_{23} was, however, based on a quenching rate constant of $O(^1S)$ by $O(^3P)$ of 1.3×10^{-13}. New measurements (Felder and Young, 1972) increase this value to 7.5×10^{-12} which results in a corresponding higher value for k_{23}, which now makes the O atom densities required to explain the airglow intensities consistent with atmospheric models.

Molecular oxygen was previously believed to be the main atmospheric quencher for $O(^1S)$. However, Atkinson and Welge (1972) have found that quenching by O_2 is temperature dependent with an activation energy of about 1.5 kcal. This means that O_2 is much less important and $O(^3P)$ much more important in quenching $O(^1S)$ in the green line layer than previously believed. Moreover, quenching and radiative lifetimes are roughly comparable. These results lead to a quite different dependence of the $O(^1S)$ lifetime on altitude than was previously believed.

The large quenching rate constant with NO also suggests that there should be a large attenuation of the $O(^1S)$ emission intensities in those auroral arcs for which large concentrations of NO have been reported by Zipf *et al.* (1970).

4.3. $O_2(^1\Delta_g)$ [0.98 eV 45 m]

Emission from the atmospheric IR bands at 1.27 and 1.58 μ is the strongest features in the dayglow. Its main source is photolysis of O_3 in the UV. Other mechanisms must also occur to account for the persistence of the emission during the night but have not been unequivocally established. This will be discussed in detail in one of the following papers (Llewellyn and Evans, 1973).

The quenching rate constants at 300 K are: N_2, $<1(-20)$; O_2, $2.2(-18)$; N, $<1(-15)$; and H, $3(-14)$. O zone is decomposed by $O_2(^1\Delta_g)$ with a rate constant given by the expression $k = 4.6 \times 10^{-11} \exp(-2.8 \times 10^3/T)$ (Findlay and Snelling, 1971; Becker *et al.*, 1972).

All these quenching rates are small, with O_2 being the principal atmospheric quencher. The reaction with N was, at one time, suggested as a possible source of NO in the mesosphere (Hunten and McElroy, 1968) but its rate is too small to be significant. In fact, recent studies (Schiff, 1972a) have indicated that the the rate constant may actually be slower than 10^{-15} since the observed decreases in $O_2(^1\Delta_g)$ used to calculate the rate may have been caused by an energy exchange process with excited N_2 molecules also present in the laboratory systems. Energy exchange have also been observed (Bader and Ogryzlo, 1964) in systems containing only $O_2(^1\Delta_g)$, e.g.

$$O_2(^1\Delta_g) + O_2(^1\Delta_g) \rightarrow O_2 + O_2 + h\nu\,(\lambda = 6340\ \text{Å}) \tag{24}$$
$$\rightarrow O_2(^1\Sigma_g^+) + O_2(^3\Sigma_g^-). \tag{25}$$

$O_2(^1\Delta_g)$ plays a role in ionospheric chemistry by the utilization of its electronic energy to detach electrons from O_2^- and O_3^- (Fehsenfeld *et al.*, 1969):

$$O_2(^1\Delta_g) + O_2^- \rightarrow O_2 + O_2 + e \tag{26}$$
$$O_2(^1\Delta_g) + O^- \rightarrow O_3 + e. \tag{27}$$

Its role as a source of O_2^+ by photoionization is now believed to be of minor importance due to the absorption of solar radiation in that spectral region by CO_2 (Huffman *et al.*, 1971).

4.4. $O_2(b\,^1\Sigma_g^+)$ [1.64 eV, 12 s]

This species is formed in the atmosphere by resonance absorption and by the reaction of $O(^1D)$ with $O_2(^3\Sigma_g^-)$. O atom recombination can produce $O_2(^1\Sigma_g^+)$ as well as $O_2(^1\Delta_g)$ and $O_2(X\,^3\Sigma_g^-)$ but the relative efficiencies of these processes have not been established.

The quenching rate constants at 300 K are: N_2, $2(-15)$; $O_2(-16)$; H_2O, $4(-12)$; CO_2, $3(-13)$; H_2, $6(-13)$; and O_3, $2.5(-11)$.

4.5. $N_2(^2D^\circ)$ [2.38 eV, 26 h]

This species may be formed in the atmosphere by photon or electron impact on N_2 and N and by dissociative recombination of NO^+ and N_2^+, the latter process being identified in the atmosphere by Weill (1969).

The quenching rate constants are: O_2, $6(-12)$; N_2, $2(-14)$; NO, $1(-10)$; CO_2, $5(-13)$; and N, $6(-12)$.

The reaction with O_2 is probably an important source of NO in the mesosphere and thermosphere.

5. Ozone Photochemistry

Because of its leading role in atmospheric chemistry, the final section of this review will be concerned with some recent results on O_3 photochemistry. Its importance stems from several unique properties of this molecule. It absorbs light over a wide wavelength region, particularly strongly in the UV Hartley band. Due to its low bond strength it can be dissociated at wavelengths less than 11 800 Å. Therefore, in the visible and UV absorption products may be found with large amounts of internal energy. Both photoproducts, O and O_2 have long-lived, low-lying electronic states.

Thus a variety of excited species can be formed at low altitudes in the atmosphere. Not all combinations are permitted by the spin selection rules but these rules may be rigorously obeyed only where the absorption is strong.

In the Chappuis band there is strong laboratory evidence that both products are formed in their ground electronic states. Near the center of the Hartley band at 2500 Å the evidence is for the production of $O(^1D)$ with near unit efficiency. It is energetically possible to form $O(^1S)$ in a spin allowed process at wavelengths less than 1995 Å, but this possibility has yet to be investigated. If it occurs it could give rise to this species in the lower atmosphere because of windows in the Schumann-Runge absorption in this general region.

Of great importance is the quantum yield for $O(^1D)$ production, φ at the long wavelength limit of the Hartley band. This is the only spectral region where solar UV radiation can penetrate into the lower stratosphere, while the increase in solar flux with wavelength compensates for the decrease in O_3 absorption coefficient. We have seen earlier that $O(^1D)$ is required to initiate most of the N and H chemistry in the stratosphere. $O(^1D)$ can be formed in a spin-allowed process only at wavelengths less than 3100 Å but, as mentioned above, this does not rule out its occurrence in this spectral region where the O_3 absorption is weak.

The experimental measurements are far from conclusive. Atmospheric modellers frequently use the value of $\varphi(3130$ Å$)=0.25$, obtained by DeMore and Raper (1966) from experiments in liquid O_3/Ar solutions at 87 K, conditions somewhat removed from those in the atmosphere. Moreover, the analysis of their results has been challenged (Schiff, 1972b). Jones and Wayne (1970) report $\varphi(3130$ Å$)=0.10$ and $\varphi(3340$ Å$)=0.02$. This work is also suspect since the same experiments gave overall quantum yields for O_3 decomposition as high as 16. Castellano and Schumacher (1972) have recently reported $\varphi(3130$ Å$)\geqslant 0.9$. But, in addition, they report an overall quantum yield of 6, which requires both photoproducts to be in singlet states. This is energetically prohibited at this wavelength. A definitive measurement is yet to be made.

At 2500 Å the molecular product has now been shown to be $O_2(^1\Delta_g)$. Nothing is yet known about the molecular states formed at shorter wavelengths where, in addition to the two singlets $O_2(^1\Delta_g)$ and $O_2(^1\Sigma_g^+)$, several of the higher triplet states are energetically possible. There is evidence (Wayne, 1972) that singlet O_2 is also formed in a spin-forbidden process in the short wavelength end of the Huggins band. The quantum yield of $O_2(^1\Delta_g)$ production as a function of wavelength is of considerable atmospheric interest since relatively simple measurements of the 1.27 μ emission intensity provide an attractive method for monitoring O_3 densities in the mesosphere.

The secondary reactions of the intial photoproducts are also of interest and are still in a state of controversy. The fast reaction of $O(^1D)$ with O_3 can proceed by several routes:

$$O(^1D) + O_3 \rightarrow O_2 + O + O \tag{28}$$

$$\rightarrow O_2 + O_2 \tag{29}$$

$$\rightarrow O_2^* + O_2, \tag{30}$$

where O_2^* is an energy-rich molecule capable of decomposing an additional O_3 molecule. Webster and Bair (1970) believe reaction (28) is the predominant channel while Giachardi and Wayne (1972) believe it has a relative probability of 30%. Davenport *et al.* (1972) find that $k_{28} \simeq k_{29}$ which rules out long, energy chains.

Two possibilities have been proposed for the reaction of $O(^1D)$ with O_2:

$$O(^1D) + O_2 \rightarrow O_2(^1\Sigma_g^+) + O \tag{31}$$
$$\rightarrow O_2^\dagger + O \tag{32}$$

where $O_2^\dagger$ represents vibrationally excited O_2 in the ground state. Reported values of the ratio k_{32} range from 0 to 0.8.

The reader may feel that too much stress had been placed on O_3 chemistry for an article on atmospheric chemistry. This may reflect, in part, the author's long-time interest in the subject and, in part, the recent public interest stimulated by the SST controversy. But the author believes that the more detailed articles which follow will show that this molecule does indeed play a central role in the chemistry of the earth's atmosphere below 100 km.

References

Atkinson, R. and Welge, K. H.: 1972, *J. Chem. Phys.*, to be published.
Bader, L. and Ogryzlo, E. A.: 1964, *Disc. Faraday Soc.* **37**, 46.
Becker, K. H., Groth, W., and Schurath, U.: 1972, *Chem. Phys. Letters* **14**, 489.
Berćes, T. and Forgeteg, S.: 1970, *Trans. Farad. Soc.* **66**, 640
Black, G., Slanger, T. G., St. John, G. A., and Young, R. A.: 1969, *J. Chem. Phys.* **51**, 116.
Brassseur, G. and Cieslik, S.: 1972, Symposium on Atmospheric Ozone, Arosa, Switzerland.
Castellano, E. and Schumacher, H. J.: 1972, Tenth Informal Conference on Photochemistry.
Chapman, S.: 1930, *Memoirs Roy. Meteorol. Soc.* **3**, 103.
Chapman, S.: 1931, *Proc. Roy. Soc. (London)* **A132**, 353.
Chapman, S.: 1943, *Reports Prog. Phys.* **9**, 92.
Clyne, M. A. A. and Thrush, B. A.: 1962, *Proc. Roy. Soc. (London)* **A269**, 404.
Crutzen, P. J.: 1970, *Quart. J. Roy. Meteorol. Soc.* **96**, 320.
Crutzen, P. J.: 1972, *Ambio* **1**, 41.
Davenport, J., Ridley, B., Schiff, H. I., and Welge, K. H.: 1972, *Disc. Farad. Soc.*, to be published.
DeMore, W. B. and Raper, O. F.: 1966, *J. Chem. Phys.* **44**, 1780.
DeMore, W. B. and Raper, O. F.: 1967, *J. Chem. Phys.* **46**, 2500.
Fehsenfeld, F. C., Albritton, D. L., Burt, J. A., and Schiff, H. I.: 1969, *Can. J. Chem.* **47**, 1793.
Felder, W. and Young, R. A.: 1972, *J. Chem. Phys.* **56**, 6028.
Findlay, F. D. and Snelling, D. R.: 1971, *J. Chem. Phys.* **54**, 2750.
Giachardi, D. J. and Wayne, R. P.: 1972, Tenth Informal Conference on Photochemistry, Oklahoma.
Goldman, C. S., Greenberg, R. I., and Heicklen, J.: 1971, *Int. J. Chem. Kinetics* **3**, 501.
Greenberg, R. I. and Heicklen, J.: 1972, Pennsylvania State University Ionospheric Research Scientific Report, 383.
Greiner, N. R.: 1969, *J. Chem. Phys.* **51**, 5049.
Greiner, N. R.: 1970, *J. Chem. Phys.* **53**, 1070.
Groth, W.: 1964, *Disc. Farad. Soc.* **37**, 210.
Hampson, J.: 1964, Canadian Armament Research and Development Establishment, Technical Note 1627.
Hochandel, C. J., Ghormley, J. A., and Ogren, P. J.: 1972, *J. Chem. Phys.* **56**, 4426.
Huffman, R. E., Paulsen, D. E., Larrabee, J. C., and Cairns, R. B.: 1971, *J. Geophys. Res.* **76**, 1028.
Huie, R. and Davis, D. D.: 1972, private communication.
Hunten, D. M. and McElroy, M. B.: 1968, *J. Geophys. Res.* **73**, 2421.
Johnston, H. S.: 1971, *Science* **173**, 517.

Johnston, H. S.: 1972, private communication.

Jones, I. T. N. and Wayne, R. P.: 1970, *Proc. Roy. Soc. (London)* **A319**, 273.

Kaufman, F.: 1969, *Can. J. Chem.* **47**, 1917.

Kondratiev, V. N.: 1970, in *Rate Constant of Gas Phase Reactions*, U.S.S.R. Academy of Science, Moscow.

Krezenski, D. C., Simonaitis, R., and Heicklen, J.: 1971, *Int. J. Chem. Kinetics* **3**, 467.

Lin, C. and Kaufman, F.: 1971, *J. Chem. Phys.* **55**, 3760.

Llewellyn, E. J., Evans, W. F. J., and Wood, H. C.: 1973, this volume, p. 193.

Lloyd, A. C.: 1970, 'Evaluated and Estimated Kinetic Data for the Gas Phase Reactions of the Hydroperoxyl Radical', Nat. Bur. Stand. (U.S.) Report 10447.

Loucks, L. F. and Cvetanovic, R. J.: 1972, *J. Chem. Phys.* **56**, 321.

McConnell, J. C.: 1972, *Planetary Space Sci.*, to be published.

McCrumb, L. M. and Kaufman, F.: 1972, *J. Chem. Phys.* **57**, 1270.

Morris, E. D. and Niki, H.: 1971, *J. Phys. Chem.* **75**, 3193.

Murcray, D. R., Kyle, T. G., Murcray, F. H., and Williams, W. J.: 1969, *J. Opt. Soc. Amer.* **59**, 1131.

Nicolet, M.: 1970, *Ann. Geophys.* **26**, 531.

Paukert, T. T. and Johnston, H. S.: 1972, *J. Chem. Phys.* **56**, 2824.

Reid, G. C.: 1973, this volume, p. 99.

Schiff, H. I.: 1969, *Ann. Geophys.* **25**, 815.

Schiff, H. I.: 1972a, *Ann. Geophys.* **28**, 67.

Schiff, H. I.: 1972b, unpublished work.

Scott, P. M., Preston, K. F., Anderson, R. J., and Quick, L. M.: 1971, *Can. J. Chem.* **49**, 1808.

Slanger, T. G. and Black, G.: 1970, *J. Chem. Phys.* **53**, 3722.

Slanger, T. G. and Black, G.: 1971, *J. Chem. Phys.* **54**, 1889.

Slanger, T. G., Wood, B. J., and Black, G.: 1972, *J. Chem. Phys.* **57**, 233.

Wayne, R. P.: 1972, *Disc. Farad. Soc.*, to be published.

Webster, H. and Bair, E. J.: 1970, *J. Chem. Phys.* **53**, 4532.

Weill, G. M.: 1969, in B. M. McCormac and A. Omholt (eds.), *Atmospheric Emissions*, Van Nostrand Reinhold Company, New York, p. 449.

Young, R. A.: 1969, *Can. J. Chem.* **47**, 1927.

Zipf, E. C., Borst, W. L., and Donahue, T. M.: 1970, *J. Geophys. Res.* **75**, 6371.

ION CHEMISTRY OF THE D AND E REGIONS

GEORGE C. REID

*Aeronomy Laboratory, National Oceanic and Atmospheric Administration,
Boulder, Colo., U.S.A.*

1. Introduction

The ion chemistry of the lower ionosphere is characterized by a high degree of complexity, caused by a combination of relatively high pressures, leading to fast chemical reactions, and strong sunlight rich in photochemically active UV radiation. This presentation will be concerned mainly with the results of laboratory investigations of the relevant reactions, and their applicability to the lower ionosphere; the mass-spectrometric observations of ion composition will be reviewed in a later paper (Narcisi, 1973).

In broad outline, we are concerned with three distinct species: electrons, positive ions, and negative ions. The primary sources of ionization produce electrons and positive ions, while negative ions are formed by attachment, particularly at the lower altitudes. Many secondary species of both positive and negative ions are produced by ion-neutral reactions, and the ultimate sinks are a recombination of electrons and positive ions, and mutual neutralization (or ion-ion recombination) of positive and negative ions. In the D region, we are probably safe in neglecting transport processes in general, since lifetimes of most species are short in comparison to the time required to transport them appreciable distances. This may not be a valid assumption, however, in the neighborhood of steep gradients, where the shape of the gradient may be influenced by transport. In the E region, transport processes cannot be neglected, and become of major importance in the case of long-lived metallic ions.

2. Negative Ions in the D Region

For many years it was suspected that the principal negative-ion species in the D region was probably O_2^-, since O_2 is a major neutral constituent, and N_2 does not form a stable negative ion. Laboratory experiments (Chanin *et al.*, 1959) also showed that three-body attachment to O_2 was a fairly efficient process at D region temperatures, with a rate coefficient of about 10^{-30} cm^6 s^{-1}. O_2^- was known, however, to undergo photodetachment by visible sunlight, since its electron affinity is only about 0.45 eV, but early observations of the twilight variation of PCA events showed that the primary response of the D region was to UV, rather than visible, light (Reid, 1961), leading to the suspicion that other negative ions were involved, having higher electron affinities than that of O_2^-.

The situation was greatly clarified by the laboratory work of Fehsenfeld *et al.* (1967, 1969), who used a flowing afterglow system to study a wide range of negative-ion reactions of potential ionospheric importance. Their results showed that an O_2^- ion in the D region has a choice of two fates: either it can react with O through

$$O_2^- + O \rightarrow O_3 + e \tag{1}$$

which releases a free electron, or it can react with O_3 in a charge-exchange reaction

$$O_2^- + O_3 \rightarrow O_3^- + O_2. \tag{2}$$

The O_3^- ion can initiate a chain of reactions with such other atmospheric constituents as CO_2 and NO, leading ultimately to NO_3^-, which is presently thought to be the 'terminal' species of the chain. It has an electron affinity of about 3.9 eV (Ferguson *et al.*, 1972), which would account for the observed dependence of the D region on UV, rather than visible light.

Our present understanding of D region negative-ion chemistry is summarized in the block diagram of Figure 1, which shows the three distinct types of negative ions

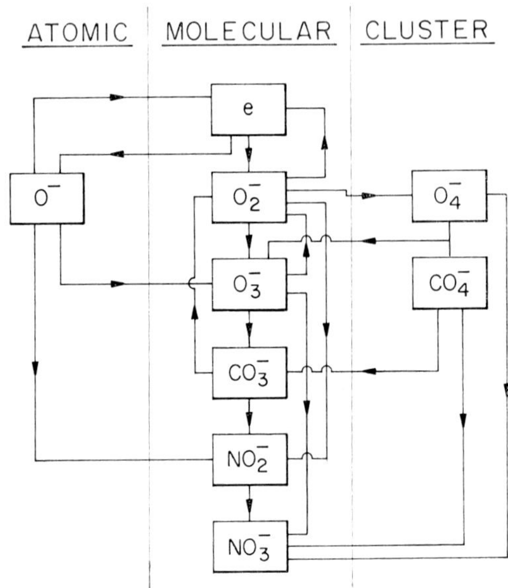

Fig. 1.   Schematic diagram of proposed negative-ion reactions in the D region. The end products of reactions of electrons and negative ions with neutrals are labeled in each block.

that are expected to be present: atomic, molecular, and cluster. The only atomic ion shown is O^-, whose role is uncertain, though it is known to exist from mass-spectrometer observations. It can form from the dissociative-attachment reaction

$$O_3 + e \rightarrow O^- + O_2 \tag{3}$$

but this is a slow reaction whose rate is uncertain to within an order of magnitude (Phelps, 1969). Cluster ions are species containing a weak bond between a molecular ion and a neutral constituent – the two species shown here are O_4^- ($O_2^- \cdot O_2$) and CO_4^- ($O_2^- \cdot CO_2$) – and are likely to be of particular importance when water vapor is present, since the large dipole moment of the water molecule leads to the formation of relatively strong electrostatic bonds and large clusters. All hydrogen chemistry has been omitted from the scheme shown in Figure 1, but is not expected to lead to major changes, apart from the formation of water clusters, though Fehsenfeld and Ferguson (1972) have recently pointed out that the reactions

$$NO_2^- + H \rightarrow OH^- + NO \tag{4}$$

$$NO_2^- + H \rightarrow HNO_2 + e \tag{5}$$

$$OH^- + O \rightarrow HO_2 + e \tag{6}$$

may be significant additional source of free electrons in the *D* region.

The rates of the various reactions entering into Figure 1 will not be discussed here, and the reader is referred to the reviews and tabulations that exist in the literature (e.g., Ferguson, 1969; Phelps, 1969; Reid, 1970). Using these reaction rates, it is possible to compute the concentration of the various negative-ion species that should exist in the *D* region under steady-state conditions, and the most important result of these calculations is that the concentration of negative ions should be very small compared to that of electrons at altitudes above 75 km in daytime. At night, they should continue to be important up to altitudes of over 80 km.

The results of a typical calculation of ambient negative-ion composition are shown in Figure 2, where it is obvious that there is a marked change in the composition at an altitude near 77 km. Below this level, where negative ions occur in significant concentrations, the species are primarily the more complex ions, with NO_3^- becoming

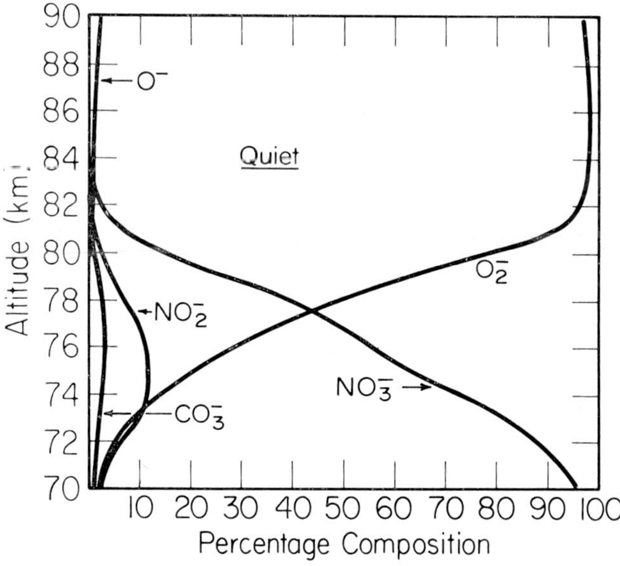

Fig. 2.   Results of a model calculation of the steady state negative-ion composition of the *D* region for quiet conditions. Only those species that reach at least 1 % of the total are shown.

more dominant with decreasing altitude. Above 77 km, O_2^- is the dominant species, but the actual concentrations are insignificant in comparison to the electron concentration.

The basic reason for this behavior is the competition between the two reactions in Equations (1) and (2) suffered by an O_2^- ion. At the lower altitudes, O_3 is more abundant than O, and the primary O_2^- ions are rapidly converted into the more tightly-bound species; at the higher levels, O is more abundant than O_3, so that the primary O_2^- ions are quickly destroyed. This leads to low concentrations, but the negative ions that are present are almost entirely O_2^-, since there is very little conversion to the more complex species. The ratio of [O] to [O_3] is thus a very important parameter in D region negative-ion chemistry, and exerts a controlling influence both on the concentration of negative ions and on the distribution among the various species. The marked differences that occur in this ratio between daytime and night-time conditions in the mesosphere are probably largely responsible for the corresponding differences in negative-ion concentrations.

The two principal deficiences in the negative-ion model are the lack of knowledge of photodetachment cross sections for most of the species involved, and the lack of detailed information on ion-ion mutual neutralization rates. In the calculations described above, all species were assumed to have an identical mutual-neutralization coefficient of 10^{-7} cm^3 s^{-1}, and photodetachment was neglected for all but the species for which its rate is known (O^-, O_2^-, and O_3^-). Since these two processes represent the ultimate sink for negative ions, it is obviously vitally important that their rate be determined, particularly for the 'terminal' species NO_3^-. The importance of mutual

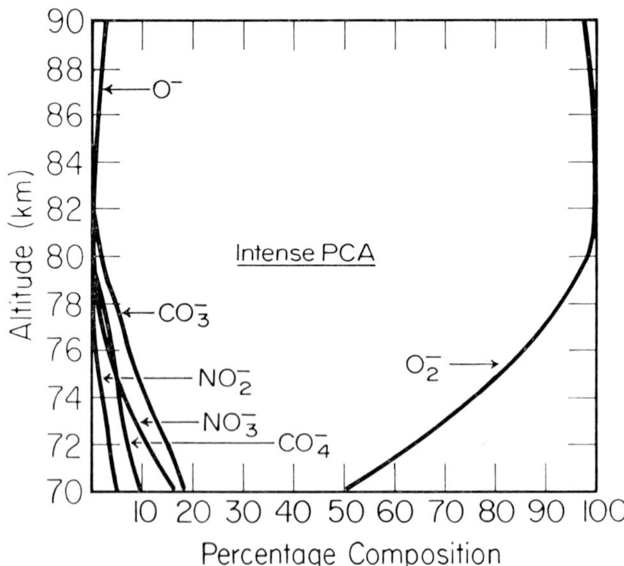

Fig. 3.   Results of an identical model calculation for an intense *PCA* event.

neutralization is illustrated by Figure 3, which shows the negative-ion composition computed for an intense PCA event, in which ion production rates are much larger than in the quiet conditions illustrated in Figure 2. There are obvious differences, the chief one being the dominance of O_2^- down to altitudes below 70 km in the PCA case, as well as the appearance of significant concentrations of CO_4^-. The reason for this is the increased rate of mutual neutralization caused by the large concentration of positive ions in the PCA case. Each of the chemical reactions that form the higher members of the negative-ion series must compete against mutual neutralization, so that when positive-ion concentrations are high the lower members of the series are favored in the ambient composition.

In the case of negative ions, the theory (based on laboratory measurements) has out-stripped observation, and there is a real need for definitive measurements of negative-ion composition. This is a much more difficult task than that of measuring the positive-ion composition, and the few observations that have been made with mass spectrometers have produced conflicting results. It is important that these efforts continue, however, since the results of model calculations such as those presented above can hardly be regarded as more than educated guesses without some direct observational testing.

3. Positive Ions in the D Region

The primary positive ions produced in the D region under quiet conditions are NO^+ (from Ly-α ionization of NO), O_2^+, and N_2^+ (from ionization of O_2 and N_2 by UV and X-radiation, energetic electrons, and cosmic rays). N_2^+ is removed very rapidly by charge exchange with O_2 to form O_2^+, which in turn charge-exchanges with NO to form NO^+. Thus for many years it was suspected that the dominant positive-ion species in the D region should be NO^+, with smaller amounts of O_2^+. The first indication that this picture was oversimplified came from the early mass spectrometer observations of Narcisi and Bailey (1965), who found that below about 80 km the dominant species were water derived ions of mass 19 (H_3O^+), 37 ($H_5O_2^+$), and 55 ($H_7O_3^+$). Above about 85 km the main species were indeed NO^+ and O_2^+, as had been expected.

The question of the origin of the water cluster ion sequence (each member of which is formed by adding a water molecule to the preceding member) immediately became a serious challenge. H_3O^+ had long been known to form in the laboratory through the reaction

$$H_2O^+ + H_2O \rightarrow H_3O^+ + OH \tag{7}$$

but this seems an unlikely reaction for the ionosphere, since even if H_2O^+ were formed it would charge-exchange rapidly with O_2 instead of reacting through Equation (7). A chain of reactions leading rapidly to the water clusters was soon found by Fehsenfeld and Ferguson (1969), and by Good $et\ al.$ (1970), starting with the primary ion O_2^+.

The principal reactions are:

$$O_2^+ + O_2 + M \rightarrow O_4^+ + M \tag{8}$$

$$O_4^+ + H_2O \rightarrow O_2^+ \cdot H_2O + O_2 \tag{9}$$

$$O_2^+ \cdot H_2O + H_2O \rightarrow H_3O^+ + OH + O_2 \tag{10a}$$

producing H_3O^+. Reaction (10a) can follow an alternative path, leading to the higher water clusters:

$$O_2^+ \cdot H_2O + H_2O \rightarrow H_3O^+ \cdot OH + O_2 \tag{10b}$$

$$H_3O^+ \cdot OH + H_2O \rightarrow H_3O^+ \cdot H_2O + OH. \tag{11}$$

Successive three-body reactions of the type

$$H_3O^+ \cdot (H_2O)_n + H_2O + M \rightarrow H_3O^+ \cdot (H_2O)_{n+1} + M \tag{12}$$

then lead to more complex clusters, the chain ending only when thermal decomposition (the inverse of Equation (12)) becomes more rapid than formation for the weakly bound complex clusters.

The key to the rapidity of this reaction chain is the existence of a fast clustering reaction in Equation (8) to a major atmospheric constituent (O_2), followed by a fast two-body reaction (Equation (10)) that produces the water clusters. By assuming that the water vapor concentration in the mesosphere was only a few parts per million, Ferguson and Fehsenfeld (1969) were able to predict water cluster ion concentrations that turned out to be in reasonable qualitative agreement with the mass-spectrometer observations. They assumed, however, that the rate of production of the primary ion O_2^+ was that predicted by Hunten and McElroy (1968) to arise from UV ionization of metastable $O_2(^1\Delta_g)$. This estimate neglected strong absorption of the UV radiation by CO_2, and more recent estimates by Huffman *et al.* (1971) have greatly reduced the strength of this source, especially below 80 km and at large solar zenith angles.

At present, it appears that the dominant *primary* positive ion throughout most of the D region is NO^+, rather than O_2^+, and thus there is a pressing need to find a reaction chain analogous to that above, but starting with NO^+. The only such chain that appears to be firmly substantiated so far begins with three successive clustering reactions to form $NO^+ \cdot (H_2O)_3$, which can then undergo a fast two-body reaction:

$$NO^+ \cdot (H_2O)_3 + H_2O \rightarrow H_3O^+ \cdot (H_2O)_2 + HNO_2 \tag{13}$$

(Fehsenfeld and Ferguson, 1969; Lineberger and Puckett, 1969). Unfortunately, this chain enters the water cluster sequence at $H_3O^+ \cdot (H_2O)_2$, leaving the lower members to be produced by thermal decomposition, and present indications are that it is too slow to account for the observed water cluster dominance below 80 km (Reid, 1971). The three initial clustering reactions were originally thought to be of the form

$$NO^+ \cdot (H_2O)_n + H_2O + M \rightarrow NO^+ \cdot (H_2O)_{n+1} + M \tag{14}$$

with $n = 0, 1, 2$, but Dunkin *et al.* (1971a) have shown that intermediate steps involving CO_2 may be involved, in which

$$NO^+ \cdot (H_2O)_n + CO_2 + M \rightarrow NO^+ \cdot (H_2O)_n \cdot CO_2 + M \qquad (15)$$

is followed by the 'switching' reaction

$$NO^+ \cdot (H_2O)_n \cdot CO_2 + H_2O \rightarrow NO^+ \cdot (H_2O)_{n+1} + CO_2 \cdot \qquad (16)$$

These reactions are fast, but the clustering is with such minor constituents as H_2O and CO_2, so that the sequence is much slower than the corresponding O_2^+ sequence in which clustering takes place to a major constituent. Heimerl and Vanderhoff (1971) have recently reported rapid clustering of NO^+ to N_2, in contradiction to the results of Dunkin $et\ al.$ (1971a). If borne out, this should help to solve the problem, though the requirement for three such three-body clustering reactions before the fast two-body reaction (Equation (13)) will still make the sequence considerably slower than the O_2^+ sequence.

One marked feature of the water cluster ion distribution that was not explained by Ferguson and Fehsenfeld (1969) is the sudden disappearance of the ions above some level that usually lies near the mesopause at 80 to 85 km. Subsequently, it was found that the reaction

$$O_4^+ + O \rightarrow O_2^+ + O_3 \qquad (17)$$

was fast, and could effectively short-circuit the O_2^+ sequence leading to water cluster formation. When this reaction is included in the sequence, it does indeed produce a sharp cutoff in the water cluster distribution near the observed level, where the O concentration increases steeply with increasing altitude. If the reaction chain is to start with NO^+, rather than O_2^+, presumably a corresponding fast reaction with O will have to be identified.

This sharp transition from water cluster ions below to molecular ions above has a marked effect on the electron concentration, since the dissociative recombination coefficients of the water cluster ions are much larger than those of the molecular ions. It appears to explain, qualitatively at least, the steep gradient in electron density that is almost invariably observed at some altitude in the vicinity of 85 km (Reid, 1970). Above the 'ledge', electron densities in the quiet daytime ionosphere are typically a few thousand per cm^3, leading to inferred recombination coefficients of a few times 10^{-7} cm^3 s^{-1}, while below the ledge they are typically a few hundred per cm^3, suggesting recombination coefficients of a few times 10^{-5} cm^3 s^{-1}. The former number is close to the known recombination coefficients of O_2^+ and NO^+, but recent laboratory measurements of the recombination coefficients of water cluster ions (Biondi $et\ al.$, 1971) suggest that 10^{-5} cm^3 s^{-1} is about the maximum that can be expected. There thus remains a discrepancy in the region below the ledge that can perhaps be removed if our current estimates of the rate of ion production are too high by a factor of 3 to 5, or if some unknown factor is maintaining negative ions in the region between 70 km and the ledge with concentrations comparable to that of electrons.

The outstanding problem in D region positive-ion chemistry thus appears to be that of idetifying a reaction chain that will convert NO^+ into water cluster ions in a time comparable with that needed to convert O_2^+. Without such a mechanism, it is difficult to understand the observed dominance of the water cluster ions below 80 km, and the low electron densities in the same region.

4. Positive Ions in the E Region

Negative ions form a negligible component of the E region, except possibly at night, and we shall concern ourselves solely with the positive-ion chemistry. In general, this is much simpler than the positive-ion chemistry of the D region, since cluster formation is not of major importance. The situation is complicated, however, by the presence of metallic ions of meteoric origin whose chemistry is relatively unclear, largely due to the major difficulties involved in measuring their reaction rates in the laboratory. These metallic ions appear throughout the E region (and also to some extent in the D region, where they form a relatively minor ionic constituent, however), but are particularly important in sporadic E layers, where dynamical effects cause them to be preferentially concentrated because of their low recombination coefficient.

Leaving the metallic ions aside for the moment, the major ionization sources in the E region produce initially N_2^+ and O_2^+, with smaller quantities of NO^+, N^+, and O^+ (though near twilight the major primary ion is probably NO^+, created by Ly-α ionization of NO (Swider and Keneshea, 1968), just as in the case of the daytime D region). The principal reactions undergone by these species are listed in Table I, together with the best current estimates of their rates.

TABLE I

Ion-molecule reactions in the E region

$N^+_2+O_2\rightarrow O^+_2+N_2$	(18)	$1\ \ \times 10^{-10}\,\mathrm{cm^3\,s^{-1}}$	Ferguson (1967)
$N^+_2+O\rightarrow NO^++N$	(19)	1.4×10^{-10}	Fehsenfeld *et al.* (1970)
$O^+_2+NO\rightarrow NO^++O_2$	(20)	6.3×10^{-10}	Fehsenfeld *et al.* (1970)
$O^+_2+e\rightarrow O+O$	(21)	2.0×10^{-7}	Biondi (1968)
$NO^++e\rightarrow N+O$	(22)	4.0×10^{-7}	Biondi (1968)
$O^++O_2\rightarrow O^+_2+O$	(23)	2.0×10^{-11}	Dunkin *et al.* (1968)
$O^++N_2\rightarrow NO^++N$	(24)	1.2×10^{-12}	Ferguson (1967)
$O^++NO\rightarrow NO^++O$	(25)	$<1.3\times 10^{-12}$	Dunkin *et al.* (1971b)

The primary N_2^+ rapidly charge-transfers with O_2 to produce O_2^+; the O_2^+ disappears either through direct dissociative recombination with electrons, or through charge transfer with NO, producing NO^+. NO^+ can also be produced directly from N_2^+ and O^+, and its ultimate loss is through dissociative recombination. The general scheme is thus a cascade toward NO^+, the ion with the lowest ionization potential, and leads to the prediction that NO^+ should be an important, if not dominant, positive-ion species in the E region, although it is not a major primary ion except under special conditions. In general, these conclusions appear to be borne out by observation in the quiet E region outside sporadic E layers (Keneshea *et al.*, 1970). During auroral conditions in the E region, however, extremely large NO^+/O_2^+ ratios have been observed by mass spectrometers (Donahue *et al.*, 1970; Zipf *et al.*, 1970), apparently in association with very large amounts of neutral NO. These observations cannot be explained on the basis of the known ion chemistry.

As mentioned above, metallic ions form an important component of the E region, generally appearing in thin layers that are probably related to the mid-latitude sporadic E layers seen by ionosondes. Several metals have been identified by mass spectrometers, including Na^+, Mg^+, Ca^+, Fe^+, and Si^+, as well as the oxide SiO^+. The atomic metallic ions are characterized by extremely low recombination coefficients (by comparison with molecular ions), and thus can survive long enough to be redistributed by winds. Their source is almost certainly ionization of metallic atoms arising from ablation of meteors (Gadsden, 1970), and their ultimate sink is probably in the lower atmosphere, where they can be transported by dynamical effects (Chimonas and Axford, 1968). An alternative sink, through oxidation followed by dissociative recombination of the molecular oxide ion, is not very effective, since the reaction

$$MO^+ + O \rightarrow M^+ + O_2 \tag{18}$$

is likely to be rapid, and to overwhelm dissociative recombination for most metals (Ferguson and Fehsenfeld, 1968). Clustering reactions might also contribute to the loss of metallic ions, as suggested by Keller and Beyer (1971), who measured the rate of clustering of Na^+ to atmospheric constituents in the laboratory; the dissociative recombination rate of the resultant cluster ions is likely to be high.

Lack of space does not allow an adequate review of the problems of atmospheric metal-ion chemistry here, and the reader is referred to several pertinent discussions that exist in the literature (Swider, 1969; Gadsden, 1970; Ferguson, 1972). The interrelationship between metallic ions, metallic-oxide ions, and the corresponding neutral species is chemically complex, and theoretical explanations of their altitude distribution is made even more difficult by the uncertain role of atmospheric dynamics, and by the complicated ablation process that creates them. This field is likely to present a significant challenge for several years to come.

5. Summary

In summary, the ion chemistry of the lower ionosphere is fraught with complexity. Our knowledge has made major advances in recent years, through a combination of laboratory measurements of reaction rates and direct observations, particularly with mass spectrometers. Several baffling problems remain, however, among which the following can be singled out as being especially important:

(1) The loss mechanism for NO^+ in the D region needs to be identified; known mechanisms fall far short of being able to explain the low electron densities, and the dominance of water cluster ions, that are consistently observed.

(2) Laboratory measurements of negative-ion reaction rates have progressed to the point of allowing quite detailed predictions of the negative-ion composition of the D region. Direct observations, on the other hand, have been very few, and have not been entirely consistent either with the laboratory predictions or with each other. More definitive observations are badly needed.

(3) There is a need for definitive laboratory measurements of the rates of photodetachment and mutual neutralization for most of the relevant negative-ion species. In spite of all the recent advances in our knowledge of D region negative-ion chemistry, the role of photodetachment is still almost unknown.

(4) The difficult field of metal-ion chemistry in the atmosphere needs more extensive investigation, both *in situ* and in the laboratory.

References

Biondi, M. A.: 1968, *Can. J. Chem.* **47**, 1711.

Biondi, M. A., Leu, M. T., and Johnsen, R.: 1971, Paper, COSPAR Symposium on D- and E-Region Ion Chemistry.

Chanin, L. M., Phelps, A. V., and Biondi, M. A.: 1959, *Phys. Rev. Letters* **2**, 344.

Chimonas, G. and Axford, W. I.: 1968, *J. Geophys. Res.* **73**, 111.

Donahue, T. M., Zipf, E. C., and Parkinson, T. D.: 1970, *Planetary Space Sci.* **18**, 171.

Dunkin, D. B., Fehsenfeld, F. C., Schmeltekopf, A. L., and Ferguson, E. E.: 1968, *J. Chem. Phys.* **49**, 1365.

Dunkin, D. B., Fehsenfeld, F. C., Schmeltekopf, A. L., and Ferguson, E. E.: 1971a, *J. Chem. Phys.* **54**, 3817.

Dunkin, D. B., McFarland, M., Fehsenfeld, F. C., and Ferguson, E. E.: 1971b, *J. Geophys. Res.* **76**, 3820.

Fehsenfeld, F. C. and Ferguson, E. E.: 1969, *J. Geophys. Res.* **74**, 2217.

Fehsenfeld, F. C. and Ferguson, E. E.: 1972, *Planetary Space Sci.* **20**, 295.

Fehsenfeld, F. C., Schmeltekopf, A. L., Schiff, H. I., and Ferguson, E. E.: 1967, *Planetary Space Sci.* **15**, 373.

Fehsenfeld, F. C., Ferguson, E. E., and Bohme, D. K.: 1969, *Planetary Space Sci.* **17**, 1759.

Fehsenfeld, F. C., Dunkin, D. B., and Ferguson, E. E.: 1970, *Planetary Space Sci.* **18**, 1267.

Ferguson, E. E.: 1967, *Rev. Geophys.* **5**, 305.

Ferguson, E. E.: 1969, *Can. J. Chem.* **47**, 1815.

Ferguson, E. E.: 1972, *Radio Sci.* **7**, 397.

Ferguson, E. E. and Fehsenfeld, F. C.: 1968, *J. Geophys. Res.* **73**, 6215.

Ferguson, E. E. and Fehsenfeld, F. C.: 1969, *J. Geophys. Res.* **74**, 5743.

Ferguson, E. E., Dunkin, D. B., and Fehsenfeld, F. C.: 1972, *J. Chem. Phys.*, **57**, 1459.

Gadsden, M.: 1970, *Ann. Geophys.* **26**, 141.

Good, A., Durden, D. A., and Kebarle, P : 1970, *J. Chem. Phys.* **52**, 222.

Heimerl, J. M. and Vanderhoff, J. A.: 1971, Paper, American Geophysical Union Annual Fall Meeting.

Huffman, R. E., Paulsen, D. E., Larrabee, J. C., and Cairns, R. B.: 1971, *J. Geophys. Res.* **76**, 1028.

Hunten, D. M. and McElroy, M. B.: 1968, *J. Geophys. Res.* **73**, 2421.

Keller, G. E. and Beyer, R. A : 1971, *J. Geophys. Res.* **76**, 289.

Keneshea, T. J., Narcisi, R. S., and Swider, W.: 1970, *J. Geophys. Res.* **75**, 845.

Lineberger, W. C. and Puckett, L. J.: 1969, *Phys. Rev.* **87**, 286.

Narcisi, R. S.: 1973, this volume, p. 171.

Narcisi, R. S. and Bailey, A. D.: 1965, *J. Geophys. Res.* **70**, 3687.

Phelps, A. V.: 1969, *Can. J. Chem.* **47**, 1783.

Reid, G. C.: 1961, *J. Geophys. Res.* **66**, 4071.

Reid, G. C.: 1970, *J. Geophys. Res.* **75**, 2551.

Reid, G. C.: 1971, in G. Fiocco (ed.), *Mesospheric Models and Related Experiments*, D. Reidel

Publishing Company, Dordrecht-Holland, p. 198.
Swider, W.: 1969, *Planetary. Space Sci.* **17**, 1233.
Swider, W. and Keneshea, T. J.: 1968, in *Space Res.* **8** (ed. by A. P. Mitra, L. G. Jacchia, and W. S. Newman), North-Holland Publishing Company, Amsterdam, p. 370.
Zipf, E. C., Borst, W. L., and Donahue, T. M.: 1970, *J. Geophys. Res.* **75**, 6371.

GAS-PHASE NITROGEN AND METHANE CHEMISTRY
IN THE ATMOSPHERE

PAUL J. CRUTZEN

University of Stockholm, Sweden

1. Introduction

The chemistry (and meteorology) of the stratosphere has become a very important and controversial topic, since it was proposed that large scale SST operation could cause a reduction in atmospheric O_3 due to the catalytic action of nitrogen oxides emitted from the exhaust (Crutzen, 1971; Johnston, 1971.) There are strong indications that oxides of nitrogen play the dominant role in controlling the natural atmospheric O_3 content (Crutzen, 1970), but too little is known about the details of the chemistry of nitrogen oxides in the stratosphere, a subject which is treated by Nicolet (1971). Important new ideas have emerged in recent years about the main chemical processes in the stratosphere. The main purpose of this paper is to outline some of these ideas and to indicate the main problems. Some possible consequences for the gas phase chemistry of the troposphere have been indicated on the basis of recent models by Levy (1971, 1972) and McConnell *et al.* (1971).

2. Production and Destruction of Ozone

Production of atmospheric O_3 takes place by the reactions (Chapman, 1930)

R1 $O_2 + h\nu \rightarrow 2O$ $\lambda < 242$ nm, Ackerman (1971)

R2 $O + O_2 + M \rightarrow O_3 + M$ $k_2 = 1.1 \times 10^{-34} \exp(500/T)$,

Davis *et al.* (1972).

The total production of O_3 molecules due to this pair of reactions in a vertical column of the atmosphere amounts to about 3×10^{13} molecules cm^{-2} s^{-1}. The mean flux of O_3 molecules downwards through the surface boundary layer is estimated to be only about 5×10^{10} molecules cm^{-2} s^{-1} (Fabian and Junge, 1970). Thus, only a very minute part of all O_3 molecules produced in the upper atmosphere can escape chemical destruction within the atmosphere. The reactions (Chapman, 1930)

R3a $O_3 + h\nu \rightarrow O(^1D) + O_2$ $\lambda < 310$ nm

R3b $O_3 + h\nu \rightarrow O + O_2$ 310 nm $< \lambda < 1140$ nm

R4 $O + O_3 \rightarrow 2O_2$ $k_4 = 2 \times 10^{-11} \exp(-2395/T)$ Johnston (1971)

are not fast enough to counter balance the total O_3 production rate by reactions R1 and R2. The reactions between odd oxygen (O, O_3) and some of the odd hydrogen compounds (H, OH, HO_2, H_2O_2) are more efficient as was first shown by Bates and

Nicolet (1950) for the mesosphere and lower thermosphere

R5	$O + OH \rightarrow H + O_2$	$k_5 = 2.5 \times 10^{-11}$, Kaufman (1969)
R6	$H + O_3 \rightarrow OH + O_2$	$k_6 = 2.6 \times 10^{-11}$
R7	$HO_2 + O \rightarrow OH + O_2$	$k_7 \geqslant 10^{-11} \, (7 \times 10^{-11})$
R8	$OH + O_3 \rightarrow HO_2 + O_2$	$k_8 < 10^{-16}$ (room temperature;

Langley and McGrath, 1971).

However, despite the fact that reaction R7 may be very fast $k_7 \approx 7 \times 10^{-11}$ (Hochanadel *et al.*, 1972) it may still be shown that below 45 km the amount of H, OH, and HO_2 is too small to play a major role in O_3 destruction (Crutzen, 1970). Since the total production by processes R1 and R2 in this region is still of the order of 10^{13} molecules $cm^{-2} \, s^{-1}$ and since this region contains above 99% of all atmospheric O_3 it is of the utmost importance to find out what additional reactions lead to a loss of O_3.

It is likely that catalytic reactions between odd oxygen and NO_x-particles (x = 1, 2, 3,) are of decisive importance in the stratosphere (Crutzen, 1970). The chain of reactions

R9	$NO + O_3 \rightarrow NO_2 + O_2$	$k_9 = 10^{-12} \exp(-1250/T)$,
		Clyne *et al.* (1964)
R10	$NO_2 + O \rightarrow NO + O_2$	$k_{10} = 9.2 \times 10^{-12}$, Herron *et al.* (1972)

$$O + O_3 \rightarrow 2O_2$$

effectively replaces the Chapman reaction R4.

An additional chain which is important below 20 km, and therefore at SST levels, is as follows (Johnston, 1971 and Crutzen, 1971)

R9	$NO + O_3 \rightarrow NO_2 + O_2$	
R11	$NO_2 + O_3 \rightarrow NO_3 + O_2$	$k_{11} = 10^{-11} \exp(-3000/T)$,
		Johnston and Yost (1949)
R12a	$NO_3 + h\nu \rightarrow NO + O_2$	$J_{12} \approx 5 \times 10^{-2} \, s^{-1}$ (?).

$$2O_3 \rightarrow 3O_2$$

Experiments indicate that the photodissociation process R12a does indeed occur, producing $O_2 (^1\Sigma_g^+)$ in red light (Johnston, 1972c). Furthermore, the following sequence of reactions involving the formation of N_2O_5, in combination with reactions R9 and R12a, may destroy O_3, especially below 20 km

R13	$NO_3 + NO_2 \rightleftarrows N_2O_5^*$	$\begin{cases} k_{13} = 7 \times 10^{-12}, \quad \text{Johnston (1951)} \\ k_{-13} = 2 \times 10^{8} \, s^{-1} \end{cases}$
R14	$N_2O_5^* + M \rightleftarrows N_2O_5 + M$	$\begin{cases} k_{14} = 1.7 \times 10^{-10} \\ k_{-14} = 2 \times 10^{-5} \exp(-9650/T) \end{cases}$
R15a	$N_2O_5 + h\nu \rightarrow NO_3 + NO_2$ (?)	
R15b	$N_2O_5 + h\nu \rightarrow NO + NO_2 + O_2$ (?).	

The O_3 molecule is restored however, if

R15c $N_2O_5 + hv \rightarrow 2NO_2 + O$

which is followed by reaction R2. It is interesting to note that the cycle involving N_2O_5 formation probably leads to catalytic destruction of O_3; however, the loss of NO_x into N_2O_5 reduces the catalytic destruction of O_3 by odd nitrogen particles above 20 km, since the chain R9–R10 is more efficient than the chain R9–R11 – R13– R14–R15a, b in this altitude region.

Additional reactions which must be considered, but which do not participate in chains of reactions leading to net O_3 loss, are

R12b $NO_3 + hv \rightarrow NO_2 + O$
R16 $NO_3 + NO \rightarrow 2NO_2$ $k_{16} \approx 10^{-11}$
R17a $NO_2 + hv \rightarrow NO + O$.

Unfortunately, many rate coefficients are not sufficiently well known, especially those for reactions R9–R17. The fast and temperature independent rate for reaction R10 recently reported by Davis *et al.* (1972) agrees with the experimental finding of $k_{10} = 8.3 \times 10^{-12}$ (at room temperature) by Clyne and Cruse (1971). Previous laboratory work had indicated a lower rate (Schofield, 1967). The dissociation rates J_{12a} and J_{12b} are not well known, but J_{12a} may be large, may be even as large as 5×10^{-1} s^{-1}. Different values for k_{16} exist (Heicklen and Cohen, 1968). This reaction rate is of great importance for calculations of the rates of O_3 destruction below 20 km. Also the extrapolation of laboratory results of k_{13}–k_{14} (Johnston, 1951) to stratospheric conditions may cause an error. A very important problem is that of the quantum yield for reaction R17a near 400 nm, including a pronounced temperature dependence (Calvert and Pitts, 1967) between 365 nm and 410 nm. The value of J_{17a} lies between 3×10^{-3} and 8×10^{-3} s^{-1}. Absorption of solar radiation near and above 400 nm leads to the production of electronically excited NO_2 which is quickly quenched by air molecules (M)

R17b $NO_2 + hv \rightarrow NO_2{}^*$
R18a $NO_2{}^* + M \rightarrow NO_2 + M$ $k_{18a} \approx 3 \times 10^{-11}$.

In about 7% of the cases, the quenching by O_2 leads to the formation of excited molecular oxygen $O_2(^1\Delta_g)$, mainly in the first vibrational level (Frankiewicz and Berry, 1972; Berry, 1972)

R18b $NO_2{}^* + O_2 \rightarrow NO_2 + O_2(^1\Delta_g)$.

The rate of production of $O_2(^1\Delta_g)$ by this mechanism is given by

$$P(^1\Delta_g) \approx 4 \times 10^{-4}(NO_2) \tag{1a}$$

and the resulting $O_2(^1\Delta_g)$ concentration is

$$(^1\Delta_g) \approx 10^{15}\frac{(NO_2)}{(M)} \tag{1b}$$

in the stratosphere and in a cloudless troposphere. This process is an important source for $O_2(^1\Delta_g)$ below 30 km, as was indicated by Frankiewicz and Berry (1972).

Taking into consideration, as first approximation, only the catalytic reactions R9–R10 and R9–R11–R12a, the equation for the rate of change of O_3 concentrations may be written

$$\frac{d}{dt}(O_3) = 2J_1(O_2) - 2k_{10}(O)(NO_2) -$$

$$- 2k_{11}(NO_2)(O_3)\frac{J_{12a}}{J_{12} + k_{16}(NO)} + F \tag{2}$$

with

$$(O) = \frac{J_3(O_3) + J_{17a}(NO_2)}{k_2(O_2)(M)} \tag{3}$$

$$\frac{(NO)}{(NO_2)} = \frac{J_{17a} + k_{10}(O)}{k_9(O_3)} \tag{4}$$

F denotes convergence of the O_3 flux, caused by atmospheric motions, which in this study was approximated by

$$F = K(M)\frac{d}{dz}\left(\frac{(O_3)}{(M)}\right)$$

where K, the vertical eddy diffusion coefficient was assumed to be equal to $10^4\ \mathrm{cm}^2\ \mathrm{s}^{-1}$.

3. Sources and Sinks for NO_x

Applying the most recent data on k_2 and k_{11} from Davis *et al.* (1972), the mixing ratio of $NO_x(NO + NO_2)$, which is required to give a satisfactory O_3 balance in the stratosphere, can be estimated to be about 10^{-8} at 30 km; it is somewhat larger above 30 km and smaller below 30 km. Nitric oxide and NO_2 may indeed be prevalent in such quantities in the stratosphere, although measurements to confirm this have not been performed or reported.

There are two important sources for stratospheric NO_x:

(a) Downward diffusion from the ionospheric production region (Nicolet, 1970; Norton and Barth, 1970). Nitric oxide may be present at 70 km with a mixing ratio of 4×10^{-8} (Meira, 1971). On its way down into the stratosphere, dissociation of NO and conversion to N_2 occurs (Callear and Smith, 1964). Although an attempt has been made to estimate theoretically the stratospheric abundance of NO_x due solely to this ionospheric source (Strobel, 1971), reliable modeling is not possible as this requires good knowledge of the global transport processes in the lower thermosphere and mesosphere.

(b) The oxidation of N_2O which is produced in the soil by microbiological action (Goody and Walshaw, 1953).

R19a $N_2O + O(^1D) \rightarrow 2NO$ $k_{19a} \approx 10^{-10}$ (Scott *et al.*, 1971;

Greenberg and Heicklen, 1970).

This reaction is important, despite the fact that $O(^1D)$ is efficiently deactivated by air molecules with a rate coefficient of 5×10^{-11} cm^3 molecule^{-1} s^{-1} (Noxon, 1970). Additional reactions affecting N_2O are

R19b $N_2O + O(^1D) \rightarrow N_2 + O_2$ $k_{19b} \approx 10^{-10}$
R20 $N_2O + hv \rightarrow N_2 + O(^1D)$ $\lambda < 337$ nm.

The rate of the latter reaction above 20 km, and thereby the production of NO_b in the stratosphere, depends strongly on the intensity of penetrating solar radiation in the Schumann-Runge bands of O_2 between 175 nm and 200 nm (Ackerman, 1971; Kockarts, 1971) and knowledge of the vertical eddy diffusion coefficient. The estimates of the concentrations of $O(^1D)$ atoms in the lower stratosphere are also indefinite due to serious uncertainties in the quantum yield of reaction R3a.

Loss of NO and NO_2 occurs mainly by their storage as HNO_3 and N_2O_5. The formation of HNO_3 occurs via

R21a $OH + NO_2 \rightarrow HNO_3{}^*$
R21b $HNO_3{}^* + M \rightarrow HNO_3 + M$
R22 $HNO_3 + hv \rightarrow OH + NO_2$.

From the study by Simonaitis and Heicklen (1972) the following rate coefficient

$$k_{21}(M = Ar) = 1.05 \times 10^{-11} \exp\left(-170/T\right) \frac{2 \times 10^{-11}(M)}{2 \times 10^{-11}(M) + \dfrac{10^{13}}{(1 + 5000/T)^4}}$$

can be deduced for the combined reaction

R21 $OH + NO_2 + (Ar) \rightarrow HNO_3 + (Ar)$.

Additional recent experimental work on this reaction has been reported by Morley and Smith (1972) over a pressure range of 20 to 300 Torr of He, at $T = 300$ K and $T = 416$ K. Their experimental values are close to those which may be calculated from the expression for k_{21} given above. If for M = air the rate coefficient k_{21b} is twice as large as for M = Ar we may write

$$k_{21}(M = air) = 1.05 \times 10^{-11} \exp\left(-170/T\right) \frac{4 \times 10^{-11}(M)}{4 \times 10^{-11}(M) + \dfrac{10^{13}}{(1 + 5000/T)^4}}$$

The photodissociation reaction R22 may be very slow below about 30 km (Johnston, 1972b) if the results of the experimental study by Bérces and Förgeteg (1970) are correct. These workers deduced a small quantum yield of only 0.1 at 265 nm and of 0.3 at 253.7 nm. They also state that quenching of the excited triplet state by O_2 may even further reduce the rate of R22. There is an urgent need to check these findings, which should not be accepted without further proof.

The presence of OH-radicals in the stratosphere, which are not numerous enough to significantly affect the O_3 destruction rates directly, leads therefore to an important removal of catalytic NO_x molecules. Nitric acid forms a bridge between odd nitrogen

(NO, NO_2, NO_3, $2N_2O_5$, HNO_3) and odd hydrogen compounds (H, OH, HO_2, $2H_2O_2$, HNO_3) and the chemistry of odd hydrogen particles is closely interrelated to that of the odd nitrogen particles.

4. Odd Hydrogen Reactions

Production of odd hydrogen occurs by the following reactions

R23a $\quad O(^1D) + H_2O \rightarrow 2OH$
R23b $\quad O(^1D) + CH_4 \rightarrow CH_3 + OH$
R23c $\quad O(^1D) + H_2 \rightarrow H + OH$

with rate coefficients between 10^{-10} and 5×10^{-10} (Greenberg and Heicklen, 1972; Paraskevopoulos and Cvetanovic, 1971). The rate coefficients $k_{23a,b,c}$ were assumed to be equal to 3×10^{-10} cm^3 molecule^{-1} s^{-1} in this study.

Removal of odd hydrogen occurs mainly by the reactions

R24 $\quad OH + HO_2 \rightarrow H_2O + O_2 \quad k_{24} = 2 \times 10^{-10}$, $\quad$ Hochanadel *et al.* (1972)
R25 $\quad OH + HNO_3 \rightarrow H_2O + NO_3$.

The rate for reaction R25 is not well-known, but its stratospheric value may be about 10^{-13} cm^3 molecule^{-1} s^{-1} (Johnston 1972b). If reaction R22 is slow, then reaction R25 could represent the main loss process for HNO_3 below 25 km. Besides reactions R5-R8, R21, and R22 the following additional reactions lead to relatively fast redistributions among the odd hydrogen compounds

R26 $\quad H + O_2 + M \rightarrow HO_2 + M \quad k_{26} = 3 \times 10^{-32}$, $\quad$ Kaufman (1969)
R27 $\quad OH + OH + M \rightarrow H_2O_2 + M \quad k_{27} \approx 10^{-30}$

R28 $\quad HO_2 + HO_2 \rightarrow H_2O_2 + O_2 \quad k_{28} = 8 \times 10^{-11} \exp\left(\dfrac{-1000}{T}\right),$

$$\text{Nicolet (1971)}$$

R29 $\quad H_2O_2 + h\nu \rightarrow 2\,OH \quad \lambda < 565 \text{ nm}$

R30 $\quad OH + CO \rightarrow H + CO_2 \quad k_{30} = 2.1 \times 10^{-13} \exp\left(\dfrac{-115}{T}\right),$

$$\text{Greiner (1969)}$$

R31 $\quad OH + H_2 \rightarrow H + H_2O \quad k_{31} = 3.7 \times 10^{-11} \exp\left(\dfrac{-2575}{T}\right),$

$$\text{Drysdale-Lloyd (1970)}$$

R32 $\quad HO_2 + NO \rightarrow OH + NO_2 \quad k_{32} \approx 10^{-12}$, $\quad$ Davis (1972).

A potentially important loss process for OH radicals is further provided by the reaction

R33 $\quad CH_4 + OH \rightarrow CH_3 + H_2O \quad k_{33} = 5.5 \times 10^{-12} \exp\left(\dfrac{-1900}{T}\right),$

$$\text{Greiner (1970).}$$

However, as three hydrogen atoms are contained in CH_3 it is necessary to consider subsequent reactions.

5. Oxidation of Methane

A good understanding about the way in which CH_4 is oxidized in the atmosphere is important. For example, CO may be an intermediate species and it has been postulated by McConnell *et al.* (1971) that the CH_4 oxidation process provides the dominant production mechanism for atmospheric CO. The oxidation routes are not known with certainty at present, but some postulated chains for the atmospheric region below 30 km are indicated. We consider the following main steps

A: *From CH_4 to CH_3O_2*

R33 $CH_4 + OH \rightarrow CH_3 + H_2O$

R34 $CH_3 + O_2 + M \rightarrow CH_3O_2 + M$ $k_{34} = 8 \times 10^{-32}$,   Heicklen (1968)

A: $CH_4 + O_2 + OH \rightarrow CH_3O_2 + H_2O$.

B: *From CH_3O_2 to CH_2O*

The route followed in the atmosphere is not well established and considerable uncertainty exists.

The set of reactions preferred by Levy (1972) and McConnell *et al.* (1971) is as follows

R35 $CH_3O_2 + NO \rightarrow CH_3O + NO_2$ $k_{35} = 2.7 \times 10^{-12} \exp\left(\dfrac{-500}{T}\right)$,

Levy (1972)

$\dfrac{\text{R17a}}{+ \text{R2}}$ $NO_2 + hv + O_2 \rightarrow NO + O_3$

R36 $CH_3O + O_2 \rightarrow CH_2O + HO_2$ $k_{36} = 10^{-12} \exp\left(\dfrac{-3300}{T}\right)$,

Levy (1972)

R32 $HO_2 + NO \rightarrow OH + NO_2$

$\dfrac{\text{R17a}}{+ \text{R2}}$ $NO_2 + hv + O_2 \rightarrow NO + O_3$

B1: $CH_3O_2 + 3O_2 \rightarrow CH_2O + OH + 2O_3$.

The value for the reaction coefficient k_{36}, especially its activation energy, is uncertain. Furthermore, Spicer *et al.* (1972) claim not to find any experimental evidence for reaction R35 to occur. These workers and Wiebe *et al.* (1972) instead advocate (with $x = 1,2$) the reactions:

R37 $CH_3O_2 + NO_x \rightarrow CH_3O_2NO_x$ $k_{37} \approx 5 \times 10^{-14}$ (room temperature)

R38 $CH_3O + NO_x \rightarrow CH_3O\,NO_x$ $k_{38} \approx 5 \times 10^{-14}$ (room temperature)

i.e., a potential loss for NO_x. The further fate of these species, which may become mainly photodissociated is therefore important. Additional reactions proposed by Spicer *et al.* (1972) and Wiebe *et al.* (1972) are ($x=1,2$):

R39 $CH_3O_2 + NO_x \rightarrow CH_2O + HNO_{x+1}$ $k_{39} = 1\text{–}2 \times 10^{-14}$

R40 $\underline{HNO_{x+1} + h\nu \rightarrow OH + NO_x}$ $J_{HNO_2} \approx 5 \times 10^{-4}\,s^{-1}$

B2: $CH_3O_2 \rightarrow CH_2O + OH$.

C: *From CH_2O to CO*

The following competing routes (C1–C3) exist:

C1: $CH_2O + h\nu \rightarrow CO + H_2$ $J_{C1} \approx 1.1 \times 10^{-4}\,s^{-1}$, Calvert *et al.* (1972)

R41 $CH_2O + h\nu \rightarrow CHO + H$ $J_{41} \approx 3.3 \times 10^{-5}\,s^{-1}$, Calvert *et al.* (1972)

R42 $CHO + O_2 \rightarrow CO + HO_2$ $k_{42} = 10^{-13}$, McMillan and Calvert (1965)

R26 $H + O_2 + M \rightarrow HO_2 + M$

R32 $HO_2 + NO \rightarrow OH + NO_2\ (2^x)$

   $\underline{NO_2 + h\nu + O_2 \rightarrow NO + O_3\ (2^x)}$

C2: $CH_2O + 4\,O_2 \rightarrow CO + 2\,OH + 2\,O_3$

R43 $CH_2O + OH \rightarrow CHO + H_2O$ $k_{43} = 1.4 \times 10^{-11}$, Morris and Niki (1972)

R44 $CHO + O_2 \rightarrow CO + HO_2$

R32 $HO_2 + NO \rightarrow OH + NO_2$

   $\underline{NO_2 + h\nu + O_2 \rightarrow NO + O_3}$

C3: $CH_2O + 2\,O_2 \rightarrow H_2O + CO + O_2$

D: *From CO to CO_2*

R30 $CO + OH \rightarrow H + CO_2$

R26 $H + O_2 + M \rightarrow HO_2 + M$

R32 $HO_2 + NO \rightarrow OH + NO_2$

   $\underline{NO_2 + h\nu + O_2 \rightarrow NO + O_3}$

D: $CO + 2\,O_2 \rightarrow CO_2 + O_3$.

E: *From H_2 to H_2O*

R31 $H_2 + OH \rightarrow H + H_2O$

R26 $H + O_2 + M \rightarrow HO_2 + M$

R32 $HO_2 + NO \rightarrow OH + NO_2$

   $\underline{NO_2 + h\nu + O_2 \rightarrow NO + O_3}$

E: $H_2 + 2\,O_2 \rightarrow H_2O + O_3$.

Summing up some of the possible net results of the CH_4 oxidation up to the formation of CO, we find

(a)	A + B1 + C1:	$CH_4 + 4O_2 \rightarrow CO + H_2 + H_2O + 2O_3$
(b)	A + B1 + C2:	$CH_4 + 8O_2 \rightarrow CO + H_2O + 2\,OH + 4O_3$
(c)	A + B1 + C3:	$CH_4 + 6O_2 \rightarrow CO + 2\,H_2O + 3O_3$
(d)	A + B2 + C1:	$CH_4 + O_2 \rightarrow CO + H_2 + H_2O$
(e)	A + B2 + C2:	$CH_4 + 5O_2 \rightarrow CO + H_2O + 2OH + 2O_3$
(f)	A + B2 + C3:	$CH_4 + 3O_2 \rightarrow CO + 2\,H_2O + O_3$.

It follows that the oxidation of CH_4 is an important potential source for atmospheric CO as was first proposed by McConnell *et al.* (1971). It may furthermore lead to enhanced production of OH (Cases (b) and (e): the initial investment in reaction R33 is thus repaid with high interest) and to production of H_2 (Cases (a) and (d)). It should not be excluded, however, that additional oxidation routes exist which, for example, lead to net removal of odd hydrogen or odd nitrogen (Crutzen, 1972b). An interesting possibility is O_3 formation in the course of CH_4 oxidation, even in the unpolluted lower atmosphere (the formation of O_3 is well known to take place when hydrocarbons are oxidized in polluted air in the presence of NO_x; for a review see Berry and Lehmann (1971)). This becomes especially clear if we add steps, D and E:

(a)	A + B1 + C1 + D + E:	$CH_4 + 8O_2 \rightarrow CO_2 + 2H_2O + 4O_3$
(b)	A + B1 + C2 + D:	$CH_4 + 10O_2 \rightarrow CO_2 + H_2O + 2OH + 5O_3$
(c)	A + B1 + C3 + D:	$CH_4 + 8O_2 \rightarrow CO_2 + 2H_2O + 4O_3$
(d)	A + B2 + C1 + D + E:	$CH_4 + 5O_2 \rightarrow CO_2 + 2H_2O + 2O_3$
(e)	A + B2 + C2 + D:	$CH_4 + 7O_2 \rightarrow CO_2 + H_2O + 2OH + 3O_3$
(f)	A + B2 + C3 + D:	$CH_4 + 5O_2 \rightarrow CO_2 + 2H_2O + 2O_3$

Net O_3 formation would occur even considering the fact that O_3 is initially destroyed by reactions R3a + R23a.

Adopting the concentration of OH (about $3 \times 10^6 \, cm^{-3}$) as estimated by Levy (1972), and a tropospheric mixing ratio of methane of 1.5×10^{-6} (Bainbridge and Heidt, 1966) the number of reactions between CH_4 and OH in a 2 km layer near the earth's surface is of the order of 2×10^{11} molecules $cm^{-2} s^{-1}$. In good agreement, Koyama (1963) estimated the production rate of CH_4 in the soil to be about 1.3×10^{11} molecules $cm^{-2} s^{-1}$. Robinson and Robbins (1968) predicted larger production rate of CH_4 of about 5×10^{11} molecules $cm^{-2} s^{-1}$. If any of the oxidation routes given above are correct then a production rate of O_3 within the troposphere of the order of 10^{12} molecules $cm^{-2} s^{-1}$ would result. This value is large in comparison with the estimated flux of O_3 through the boundary layer of 5×10^{10} molecules $cm^{-2} s^{-1}$ according to Fabian and Junge (1970), who assume the production of O_3 in the stratosphere to be the main source for tropospheric O_3. This comparison indicates that our present knowledge in this area is inadequate. The one month lifetime of tropospheric CO (Weinstock and Niki, 1971) agrees with daytime OH concentrations of about 5×10^6 molecules cm^{-3}, (Levy, 1972), provided CH_4 oxidation is indeed the main atmospheric CO source. In the presence of OH and about 1 ppb. of NO the oxidation of CO to CO_2 leads to O_3 production via the reaction sequence D (Westberg *et al.*, 1971). The prevailing OH concentrations are smaller than estimated by Levy (1972) if the photodissociation process R22 is indeed slow and if additional reactions play a role such as

$$\text{R45} \qquad NH_3 + OH \rightarrow NH_2 + H_2O \qquad k_{45} = 1.3 \times 10^{-12} \exp\left(\frac{-550}{T}\right),$$

Drysdale and Lloyd (1970),

if further oxidation reactions of NH_2 do not release odd hydrogen. This reaction, the rate coefficient of which is only theoretically estimated, may initiate important gas phase conversion from NH_3 to NO_x in the troposphere. With $(OH) = 5 \times 10^6$ the estimated lifetime of NH_3 due to reaction R45 alone is about 3 weeks to be compared with an estimated lifetime of about 1 week, including all loss processes (Robinson and Robbins, 1970).

To return to the problem of the tropospheric O_3 budget, it must be remembered that the set of reactions $R9 + R11 + R12a$ leads to O_3 destruction especially in the lowest two kilometers of the troposphere. Assuming mixing ratios of NO and NO_2 of 2×10^{-9} this destruction rate could reach a value of

$$2 \int_0^{2\text{ km}} k_{11}(NO_2)(O_3) \frac{J_{12a}}{J_{12} + k_{16}(NO)} \, dz \approx 10^{12} \text{ molecules cm}^{-2}\text{ s}^{-1},$$

adopting $(O_3) \approx 10^{12}$ molecules cm^{-3} (Hering and Borden, 1964).

Also this rate of destruction is very large in comparison to the estimated flux of O_3 through the boundary layer of 5×10^{10} molecules cm^{-2} s^{-1}. There is thus need to make detailed studies of the tropospheric O_3 budget and $CH_4 - NO_x$ chemistry.

6. Model Calculations

In Tables I and II are presented two calculated distributions of some of the trace gases in the stratosphere. For these calculations the tropopause level was assumed to be at an altitude of 10 km and adopted lower boundary molar mixing ratios were as follows: $\mu(CH_4) = 1.5 \times 10^{-6}$, $\mu(N_2O) = 2.5 \times 10^{-7}$, $\mu(H_2O) = 3 \times 10^{-6}$, $\mu(H_2) = 5 \times 10^{-7}$, $\mu(HNO_3) = 10^{-10}$, $\mu(NO_x) = 3 \times 10^{-9}$. The solar zenith angle was kept constant at 45°. The photodissociation rates adopted in this work are calculated from the sources given in a previous paper (Crutzen, 1971). The low mixing ratio of HNO_3 at the tropopause level was chosen merely in order to study the degree of conversion of NO_x to HNO_3 solely by the chemical processes in the stratosphere. The routes of CH_4 oxidation adopted in these calculations all go via alternative B1. In calculating the results presented in Table I it was assumed that reaction R8 could be neglected, while for Table II the value $k_8 = 3 \times 10^{-11} \exp(-1500/T)$ was adopted. In the former case the model calculations indicate, due to the prevalence of much more OH, a more pronounced conversion of NO_x to HNO_3, thereby permitting larger O_3 concentra-

TABLE I

Distribution of some minor constituents in the stratosphere. Reaction R8 neglected.

Altitude km	(O_3)	$\mu(NO_x)$	$\mu(HNO_3)$	$\mu(N_2O)$	$\mu(CH_4)$	$\mu(H_2)$	$\mu(CO)$	(OH)	(CH_2O)
40	0.6 (12)	2.0 ($-$8)	3.4 ($-$11)	7.0 ($-$10)	0.2 ($-$6)	2.4 ($-$7)	7.0 ($-$8)	6.9 (6)	2.3 (6)
35	1.6 (12)	1.9 ($-$8)	3.3 ($-$10)	1.0 ($-$8)	0.3 ($-$6)	4.0 ($-$7)	4.0 ($-$8)	6.9 (6)	6.5 (6)
30	3.8 (12)	1.4 ($-$8)	2.2 ($-$9)	5.3 ($-$8)	0.5 ($-$6)	5.6 ($-$7)	2.6 ($-$8)	6.1 (6)	1.6 (7)
25	4.6 (12)	5.9 ($-$9)	4.9 ($-$9)	1.3 ($-$7)	0.9 ($-$6)	6.3 ($-$7)	2.0 ($-$8)	4.4 (6)	2.8 (7)
20	3.7 (12)	2.8 ($-$9)	4.1 ($-$9)	1.9 ($-$7)	1.1 ($-$6)	6.2 ($-$7)	2.0 ($-$8)	3.1 (6)	4.8 (7)
15	2.4 (12)	1.9 ($-$9)	1.4 ($-$9)	2.2 ($-$7)	1.4 ($-$6)	5.6 ($-$7)	3.2 ($-$8)	2.7 (6)	7.9 (7)
10	5.0 (11)	3.0 ($-$9)	1.0 ($-$10)	2.5 ($-$7)	1.5 ($-$6)	5.0 ($-$7)	1.0 ($-$7)	1.5 (6)	7.9 (7)

TABLE II

Distribution of some minor constituents in the stratosphere. The reaction coefficient for reaction R8: $k_8 = 3 \times 10^{-11} \exp(-1500/T)$.

Altitude km	(O_3)	$\mu(NO_x)$	$\mu(HNO_3)$	$\mu(N_2O)$	$\mu(CH_4)$	$\mu(CO)$	(OH)	(CH_2O)
40	0.6 (12)	2.0 (−8)	1.5 (−11)	0.7 (−9)	4.1 (−7)	1.7 (−7)	3.0 (6)	4.2 (6)
35	1.5 (12)	1.8 (−8)	5.0 (−11)	1.0 (−8)	0.7 (−6)	2.1 (−7)	1.0 (6)	6.8 (6)
30	3.5 (12)	1.5 (−8)	1.7 (−10)	5.2 (−8)	1.0 (−6)	1.4 (−7)	4.7 (5)	7.1 (6)
25	3.9 (12)	9.4 (−9)	1.0 (−9)	1.3 (−7)	1.2 (−6)	6.2 (−8)	5.7 (5)	8.6 (6)
20	3.0 (12)	5.0 (−9)	1.7 (−9)	1.9 (−7)	1.3 (−6)	3.9 (−8)	7.6 (5)	1.7 (7)
15	1.9 (12)	3.0 (−9)	1.5 (−9)	2.3 (−7)	1.4 (−6)	4.1 (−8)	1.2 (6)	4.3 (7)
10	5.0 (11)	3.0 (−9)	1.0 (−10)	2.5 (−7)	1.5 (−6)	1.0 (−7)	1.3 (6)	6.6 (7)

122   PAUL J. CRUTZEN

tions. For the same reason the calculated CO concentrations are smaller in Table I
than in Table II. The rate coefficient k_8 therefore remains an important parameter. In
calculating the concentrations of CO the photodissociation process

R46 $CO_2 + hv \rightarrow CO + O,$

which is important above 25 km, was included.

Interesting features revealed by the results in Table I and II are:

(a) The pronounced conversion of NO_x molecules to HNO_3 below 30 km. The
major odd nitrogen species in this altitude region may be HNO_3 (Johnston, 1972b;
Harries *et al.*, 1972; Murcray, 1973).

(b) The calculated slight increase of the mixing ratio of H_2 with altitude is due to
this gas being an intermediate in the methane oxidation, and shows that this process
may contribute significantly to atmospheric H_2. There is a sink for H_2 in the upper
stratosphere due to reaction R23c.

(c) The sharp decline of CO mixing ratios above the tropopause is due to fast
chemical action in comparison with transport processes in the stratosphere and reflects the fact that the activation energy of react on R33 is larger than that of reaction
R30. It may be expected that changes in the tropospheric CO content can not penetrate high enough to influence the O_3 layer.

(d) The increase of odd nitrogen mixing ratios with altitude due to the oxidation
of N_2O and the decrease of the mixing ratios of N_2O and CH_4. The computed profiles
of CH_4 and N_2O agree roughly with measured profiles (Ackerman *et al.*, 1972;
Schütz *et al.*, 1970).

7. The SST Problem

It has been stated before (Crutzen, 1971; Johnston, 1971), and there is no reason now
to change this statement, that the emissions of NO_x gases from a large fleet of SSTs
may lead to a decrease in the atmospheric O_3 level. This problem has recently been
reviewed by Johnston (1972a) and Crutzen (1972a,b). One additional aspect needs to
be discussed in some detail. In the natural stratosphere a significant proportion of the
catalytic NO_x may be converted to HNO_3. In connection with the SST problem it is
essential to know the time in which NO_x is stored into HNO_3. If this time is long
then appreciable destruction of O_3 may take place before any appreciable proportion
of the injected NO_x is converted to HNO_3. In a parcel of air with initial mixing ratios
$\mu(NO_2)$ and $\mu(H_2O)$ the minimum time of conversion τ_c may roughly be estimated
from

$$\tau_c = \frac{\mu(NO_2)}{2k_{23}(O^*)\,\mu(H_2O)}.$$

Near 17 km $(O^*) < 0.5\ \text{cm}^{-3}$ and therefore $\tau_c > 2$ weeks if $\mu(NO_2) = 10^{-9}$ and
$\mu(H_2O) = 3 \times 10^{-6}$, and $\tau_c > 0.5$ yr if $\mu(NO_2) = 10^{-8}$ and $\mu(H_2O) = 3 \times 10^{-6}$. Therefore, areas with high traffic intensities may always be characterized by large NO and
NO_2 concentrations relative to those of HNO_3 and faster O_3 destruction. It is there-

fore important to study in detail the spreading of the injected material for different locations and seasons together with the chemistry. It must be emphasized that in making these estimates all loss processes of odd hydrogen molecules as well as the possible effect of reaction R22 were neglected. Thus the actual conversion times from NO_x to HNO_3 may be much larger than indicated above. On the other hand some enhanced production of odd hydrogen via the CH_4 oxidation cycle may take place.

8. Conclusions

Many problems of both chemical and meteorological nature must be resolved before any reliable predictions can be made of O_3 reductions due to the effects of large scale SST operation. They have been indicated in this paper. These problems are also problems of tropospheric chemistry. For example, the oxidation of CH_4 to H_2O and CO_2 may imply a sink or a source for atmospheric NO_x and HO_x. In this process tropospheric O_3 may be formed. This paper has strongly concentrated on gas phase chemistry, but it is conceivable that heterogeneous reactions may also play an essential role in the lower troposphere (Levy, 1972).

References

Ackerman, M.: 1971, in G. Fiocco (ed.), *Mesospheric Models and Related Experiments*, D. Reidel Publishing Company, Dordrecht-Holland, p. 149.

Ackerman, M., Frimout, D., Lippes, C., and Muller, C.: 1972, *Aeronomica Acta* **A97**.

Bainbridge, A. E. and Heidt, L. E.: 1966, *Tellus* **18**, 221.

Bates, D. R. and Nicolet, M.: 1950, *J. Geophys. Res.* **55**, 189.

Bérces, T. and Förgeteg, S.: 1970, *Trans. Far. Soc.* **66**, 633.

Berry, R. S.: 1972, private communication.

Berry, R. S. and Lehman, P. A.: 1971, *Ann. Rev. Phys. Chem.* **47**, 84.

Brooks, J. N., Goldman, A., Kosters, J. J., Murcray, D. G., Murcray, F. H., and Williams, W. J.: 1973, this volume, p. 278.

Callear, A. B. and Smith, I. W. M.: 1964, *Disc. Far. Soc.*, p. 96.

Calvert, J. G. and Pitts, J. N. Jr.: 1967, *Photochemistry*, John Wiley and Sons, New York, p. 218.

Calvert, J. G., Kerr, J. A., Demerjian, K. L., and McQuigg, R. D.: 1972, *Science* **175**, 751.

Chapman, S.: 1930, *Quart. J. Roy. Metcorol. Soc.* **3**, 103.

Clyne, M. A. A. and Cruse, H. W.: 1971, *Trans. Far. Soc.* **67**, 2869.

Clyne, M. A. A., Thrush, B. A., and Wayne, R. P.: 1964, *Trans. Far. Soc.* **60**, 359.

Crutzen, P. J.: 1970, *Quart. J. Roy. Meteorol. Soc.* **96**, 320.

Crutzen, P. J.: 1971, *J. Geophys. Res.* **30**, 7311.

Crutzen, P. J.: 1972a, *Ambio* **1**, 41.

Crutzen, P. J.: 1972b, Proceedings Climatic Impact Assessment Program, in press.

Davis, D. D.: 1972, private communication.

Davis, D. D., Herron, J. T., and Huie, R. E.: 1972, *J. Chem. Phys.*, submitted.

Drysdale, D. D. and Lloyd, A. C.: 1970, in C. F. H. Tipper (ed.), *Oxidation and Combustion Reviews*, Elsevier **4**, 157.

Fabian, P. and Junge, C. E.: 1970, *Arch. Meteorol. Geophys. Biokl.* **A19**, 161.

Frankiewicz, T. and Berry, R. S.: 1972, *Envir. Sci. Techn.* **6**, 365.

Goody, R. M. and Walshaw, C. D.: 1953, *Quart. J. Roy. Meteorol. Soc.* **79**, 496.

Greenberg, R. I. and Heicklen, J.: 1970, *Int. J. Chem. Kin.* **2**, 185.

Greenberg, R. I. and Heicklen, J.: 1972, Report 383, Ionosphere Research Laboratory, The Pennsylvania State University.

Greiner, N. R.: 1969, *J. Chem. Phys.* **51**, 5049.

Greiner, N. R.: 1970, *J. Chem. Phys.* **53**, 1070.

Harries, J. E., Swann, N. R. W., Beckman, J. E., and Ade, P. A. R.: 1972, *Nature* **236**, 159.

Heicklen, J.: 1968, *Advances in Chemistry Series* **76**, 23.

Heicklen, J. and Cohen, N.: 1968, in W. A. Noyes, G. S. Hammond, and J. N. Pitts, Jr. (eds.), *Advances in Photochemistry*, Interscience Publishers, **5**, p. 157.

Hering, W. S. and Borden, Jr, T. R.: 1964, Report AFCRL-64-30(11), Air Force Cambridge Research Laboratories.

Hochanadel, C. J., Ghormley, J. A., and Orgren, P. J.: 1972, *J. Chem. Phys.* **56**, 4426.

Johnston, H. S.: 1951, *J. Amer. Chem. Soc.* **73**, 4542.

Johnston, H. S.: 1971, *Science* **173**, 517.

Johnston, H. S.: 1972a, Proceedings Climatic Impact Assessment Program, in press.

Johnston, H. S.: 1972b, *J. Atmospheric Sci.*, to be published.

Johnston, H. S.: 1972c, private communication.

Johnston, H. S. and Yost, D. M.: 1949, *J. Chem. Phys.* **17**, 386.

Kaufman, F.: 1969, *Can. J. Chem.* **47**, 1917.

Kockarts, G.: 1971, in G. Fiocco (ed.), *Mesospheric Models and Related Experiments*, D. Reidel Publishing Company, Dordrecht-Holland, p. 160.

Koyama, T.: 1963, *J. Geophys. Res.* **68**, 3971.

Langley, K. F. and McGrath: 1971, *Planetary Space Sci.* **19**, 413.

Levy, H.: 1971, *Science* **173**, 141.

Levy, H.: 1972, *J. Atmospheric Sci.*, in press.

McConnell, J. C., McElroy, M. B., and Wofsy, S. C.: 1971, *Nature*, **233**, 187.

McMillan, G. R. and Calvert, J. G.: 1965, *Oxidation Comb. Rev.* **1**, 83.

Meira, L. G., Jr.: 1971, *J. Geophys. Res.* **76**, 202.

Morley, C. and Smith, I. W. M.: 1972, *J. Chem. Soc. Far. Trans. II.* **68**, 1016.

Morris, E. D. and Niki, H.: 1972, *J. Chem. Phys.*, to be published.

Nicolet, M.: 1970, *Planetary Space Sci.* **18**, 1111.

Nicolet, M.: 1971, in G. Fiocco (ed.), *Mesospheric Models and Related Experiments*, D. Reidel Publishing Company, Dordrecht-Holland, p. 1.

Norton, R. B. and Barth, C. A.: 1970, *J. Geophys. Res.* **75**, 3903.

Noxon, J. F.: 1970, *J. Chem. Phys.* **52**, 1852.

Paraskevopoulos, G. and Cvetanovic, R. J.: 1971, *Chem. Phys. Letters* **9**, 603.

Robinson, E. and Robbins, R. C.: 1968, Final Report, SRI Project, PR-6755, Stanford Research Institute.

Robinson, E. and Robbins, R. C.: 1970, *APCA J.* **5**, 303.

Schofield, K.: 1967, *Planetary Space Sci.* **15**, 643.

Schütz, K., Junge, C., Beck, R., and Albrecht, B.: 1970, *J. Geophys. Res.* **75**, 2230.

Scott, P. M., Preston, K. F., Andersen, R. J. and Quick, L. M.: 1971, *Can. J. Chem.* **49**, 1808.

Simonaitis, R. and Heicklen, J.: 1972, Report 380, Ionosphere Research Laboratory, The Pennsylvania State University, 1972.

Spicer, C. W., Villa, A., Wiebe, H. A., and Heicklen, J.: 1972, CAES Report No. 223-71, Dept. of Chemistry, The Pennsylvania State University.

Strobel, D. R.: 1971, *J. Geophys. Res.* **76**, 8384.

Weinstock, B. and Niki, H.: 1971, Ford Preprint Publication.

Westberg, K., Cohen, N., and Wilson, K. W.: 1971, *Science* **174**, 1013.

Wiebe, H. A., Villa, A., Hellman, T., and Heicklen, J.: 1972, CAES Publication No. 170-70, Dept of Chemistry, The Pennsylvania State University.

REACTIONS INVOLVING EXCITED STATES OF O AND O_2

R. P. WAYNE

Physical Chemistry Laboratory, Oxford, England

1. Introduction

Excited species are of interest in atmospheric chemistry since they may undergo reactions that are slow, or impossible, for the ground state species. Further, the excited species can undergo radiative energy loss thus possibly contributing to the energy balance and providing emission from the atmosphere. Excited O and O_2 are certainly present in the earth's atmosphere since features in the day and night airglow derive from optical transition from these species. The excited species of oxygen to be discussed in this paper are all singlets $(O(^1D), O(^1S), O_2(^1\Delta_g), O_2(^1\Sigma_g^+))$. The electric dipole selection rule, $\Delta S = 0$, therefore formally 'forbids' optical transitions to the ground, triplet, states, and the species are therefore 'metastable': that is, they have long natural lifetimes in the absence of reactive or quenching collisions. In addition, these states of O and O_2 have low excitation energies, so that they can become excited in relatively low energy chemical or photochemical processes. Table I shows the excitation energies, optical transitions and lifetimes for the species under consideration.

TABLE I

The low lying singlet states of O and O_2

Species	Excitation energy (eV)	Transition	Wavelength (nm)	Lifetime (s)
$O(^1D)$	1.97	$^1D \rightarrow ^3P_2$	630.0 (Chamberlain, 1961)	145
		$^1D \rightarrow ^3P_1$	636.4 (Chamberlain, 1961)	455
$O(^1S)$	4.17	$^1S \rightarrow ^1D$	557.7 (Corney and Williams, 1972)	0.8
		$^1S \rightarrow ^3P_1$	297.2 (Chamberlain, 1961)	13
$O_2(^1\Delta_g)$ ($v'=0$)	0.98	$^1\Delta_g \rightarrow ^3\Sigma^-_g$	1269 (Badger *et al.*, 1965)	3876
$O_2(^1\Sigma^+_g)$ ($v'=0$)	1.63	$^1\Sigma^+_g \rightarrow ^1\Delta_g$	1908 (Noxon, 1961)	400
		$^1\Sigma^+_g \rightarrow ^3\Sigma^-_g$	761.9 (Wallace and Hunten, 1968)	12

Evaluation of the importance of atmospheric processes should, where possible, be supported by reliable laboratory data. Development of a hypothetical model for an atmosphere and subsequent manipulation of rate coefficients and other parameters to fit observables is not satisfactory unless the accuracy of the observations permits choice of a unique mechanism with unique rate coefficients and concentrations. The

B. M. McCormac (ed.), Physics and Chemistry of Upper Atmospheres, 125–132. All Rights Reserved.
Copyright © 1973 by D. Reidel Publishing Company, Dordrecht-Holland.

present paper describes laboratory investigation of excited O and O_2. Unfortunately, conflicting data have often been presented by different research groups: the conflict has frequently resulted from unexpected effects which invalidate presumptions about the system, rather than from inaccurate measurements. An attempt is made to show the problems arising in the laboratory work.

There are dangers inherent in assuming an atmosphere to be a giant laboratory. First, laboratory systems have 'walls,' and surface to volume ratios are incomparably higher than those found in the atmosphere (although the presence of particulate matter should always be – but usually is not – considered in atmospheric studies). Some progress has been made in recent laboratory studies to reduce surface-to-volume ratios, notably by the research group at Bonn, under Professors Groth and Becker: these workers currently use a reaction chamber of volume $220\ \mathrm{m}^3$. Secondly, in laboratory kinetic experiments, reaction pressures and reactant concentrations are generally higher than in the atmosphere. As a result, multiple step processes may be mistaken for elementary processes in the laboratory; processes which are predominantly third-order in the laboratory may have a second-order pathway which is important in the atmosphere. Thirdly, atmospheric temperatures may differ considerably from the 'room' temperature at which many laboratory experiments are performed. It is thus important, for reactions thought to be temperature sensitive, to investigate the temperature dependence. In this connection, it is worth remembering that third-order reactions frequently have a small *negative* activation energy, that reactions with a rate constant approaching the gas kinetic value must have a near-zero activation energy, and that collisional quenching reactions also often have very small activation energies; wherever a reaction is endothermic the activation energy must be equal to or greater than the endothermicity.

2. Atmospheric Production of Excited Species

There are three possible routes to electronic excitation in the neutral atmosphere: (a) direct absorption of light; (b) chemical reaction or (c) energy transfer. Direct absorption is bound to be inefficient for the singlet species discussed here, since optical transitions from the ground state are forbidden. However, Wallace and Hunten (1968) have, for example, shown that solar absorption in the $\lambda = 761.9$ nm band may make an appreciable contribution to the total $[O_2(^1\Sigma_g^+)]$ at some altitudes. The remainder of this section is nevertheless restricted to consideration of routes (b) and (c).

2.1. Photochemical production

The likely photochemical precursors of excited O and O_2 in the earth's atmosphere are obviously O_2 and O_3, with possible minor contributions from CO_2, H_2O, and the oxides of nitrogen. We now consider primary photochemical processes in these species.

It is essential that the radiation absorbed possesses sufficient energy for both photodissociation and excitation. Table II shows the limiting wavelengths for formation of various products (data from Herzberg, 1950, 1966).

TABLE II

Excited products of primary photochemical step

Ground state precursor	Products	Approximate short wavelength limit (nm)	Remarks
$O_2(^3\Sigma^-_g)$	$O(^3P) + O(^1D, ^1S)$	176, 133	$\phi_{O(^1D)} \sim 1$ at $\lambda < 176$ nm [a]
$O_3(^1A_1)$	$O(^3P) + O_2(^1\Delta_g, ^1\Sigma^+_g)$	611, 463	$\phi_{O_2(^1\Delta_g)} \sim 1$ a) $\lambda = 334$ nm [b]
	$O(^1D) + O_2(^3\Sigma^-_g)$	411	Unimportant at $\lambda = 334$ nm [b]
	$O(^1D) + O_2(^1\Delta_g, ^1\Sigma^+_g)$	310, 266	$\phi_{O(^1D)}, o_2(^1\Delta_g) \sim 1$ at $\lambda = 254$ nm [c]
	$O(^1S) + O_2(^3\Sigma^-_g)$	234	$O_2(^1\Sigma^+_g)$ *not* formed at $\lambda = 254$ nm [d]
	$O(^1S) + O_2(^1\Delta_g, ^1\Sigma^+_g)$	196, 179	
$N_2O(^1\Sigma^+)$	$N_2(^1\Sigma^+_g) + O(^1D, ^1S)$	340, 237	$\phi_{O(^1S)} = 0.55$; $\phi_{O(^1D)} = 0.5$ at $\lambda = 147$ nm [d]
$N_2(^2A_1)$	$NO(^2\Pi) + O(^1D, ^1S)$	244, 170	$O(^1D)$ produced at $\lambda = 229$ nm [e]
	$N(^4S) + O_2(^1\Delta_g, ^1\Sigma^+_g)$	230, 202	
$NO(^2\Pi)$	$N(^4S) + O(^1D, ^1S)$	146(?), 117(?)	
$CO_2(^1\Sigma^+_g)$	$CO(^1\Sigma^+) + O(^1D, ^1S)$	167, 129	$O(^1S)$ produced at $\lambda = 105, 107$ nm [f, g]
$H_2O(^1A_1)$	$H_2(^1\Sigma^+_g) + O(^1D, ^1S)$	175, 133	

[a] Noxon (1970)
[b] Jones and Wayne (1970)
[c] Jones and Wayne (1971)
[d] Gilpin *et al.* (1971)
[d] Young *et al.* (1968a)
[e] Preston and Cvetanovic (1966)
[f] Young *et al.* (1969)
[g] Slanger and Black (1971)

Reference is made to recent laboratory studies concerning the photolytic pathways In most cases it is not established that the *possible* products are, in fact, formed. It is sometimes argued that the overall dissociation process must be 'spin-allowed': that is, $\Delta S = 0$ for an allowed optical transition, and the products must correlate with the upper state, so that spin is 'conserved' in the photolysis. However, where an absorption to an upper molecular state is weak, or a 'forbidden' predissociation is still the major loss process for the excited state, then this conservation may no longer be expected to hold rigorously. For example, the photolysis of O_3 (a singlet in the ground state) yields a triplet atomic and a singlet molecular fragment in the weak tail of the UV absorption band (Jones and Wayne, 1970). The importance at any wavelength of a photolytic process in the atmosphere depends on a combination of the quantum yield for the process, the absorption cross section, and the solar intensity at that wavelength. Relatively weak absorption combined with high solar intensity can still lead to significant production of excited species, as in the case of $O_2(^1\Delta_g)$ formation from O_3 at $\lambda > 310$ nm for low altitudes (Crutzen *et al.* 1971).

2.2. CHEMICAL REACTION

Chapman (1937) proposed that the three body process involving ground state O atoms

$$O + O + O \rightarrow O(^1S) + O_2 \tag{1}$$

could lead to excitation of $O(^1S)$ in the atmosphere. Young and Sharpless (1963) have demonstrated in the laboratory that this reaction does, in fact, occur, and further suggest that the reaction

$$O(^1S) + O(^3P) \rightarrow 2O(^1D) \tag{2}$$

accounts for $O(^1D)$ production. Felder and Young (1972) have recently measured the rate constant for process (1) to be 4.8×10^{-33} cm^6 molecule s^{-1}.

Excited molecular oxygen may also be formed in atom recombination reactions. The production of $O_2(^1\Sigma_g^+)$ by the process

$$O + O + M \rightarrow O_2(^1\Sigma_g^+) + M \tag{3}$$

has been observed in laboratory systems, and Young and Black (1966) give a rate constant of 1.7×10^{-37} cm^6 molecule^{-2} s^{-1} for $M = N_2$. No experimental evidence exists to indicate that $O_2(^1\Delta_g)$ can be excited by recombination involving two atoms. However, Evans et al. (1972) have shown that atmospheric observations suggest that the process

$$O + O_2 + O_2 \rightarrow O_3 + O_2(^1\Delta_g) \tag{3a}$$

excites $O_2(^1\Delta_g)$ on every recombination event.

Energy transfer can lead to production of excitation in species other than in which the excess energy originally resides, and, indeed, the process in Equation (2) may be an example of such energy transfer. Transfer of energy from $O(^1D)$ to ground state $O_2(^3\Sigma_g^-)$ is known to excite $O_2(^1\Sigma_g^+)$ (Young and Black, 1967; Izod and Wayne, 1968)

$$O(^1D) + O_2(^3\Sigma_g^-) \rightarrow O_2(^1\Sigma_g^+) + O(^3P). \tag{4}$$

The efficiency α of this energy transfer relative to the overall quenching of $O(^1D)$ by O_2 has been the subject of some controversy, although Noxon (1970) has shown $\alpha \nless 0.3$, and recent estimates (Giachardi and Wayne, 1972) give values of $\alpha = 0.5$ to 0.6. The value of α cannot be unity, since vibrational excitation of $O_2(^3\Sigma_g^-)$ occurs in the quenching process (McCullough and McGrath, 1971).

In the laboratory, $O_2(^1\Sigma_g^+)$ may be excited by the 'energy pooling' process

$$O_2(^1\Delta_g) + O_2(^1\Delta_g) \rightarrow O_2(^1\Sigma_g^+) + O_2 \tag{5}$$

but reliable determinations of the rate constant $(2 \times 10^{-17}$ cm^3 molecule^{-1} s^{-1}; Derwent and Thrush, 1971) show that the reaction cannot be important for the concentrations of $O_2(^1\Delta_g)$ in the earth's atmosphere. Energy transfer from carbonyl compounds or hydrocarbons to O_2 may be an important source of $O_2(^1\Delta_g)$ in polluted atmospheres (see Wayne (1970) p. 224 for a discussion). Excited NO_2, produced by light absorption at $\lambda > 400$ nm, can also transfer energy to O_2 to produce $O_2(^1\Delta_g)$ (Jones and Bayes, 1971):

$$NO_2^* + O_2 \rightarrow O_2(^1\Delta_g) + NO_2. \tag{6}$$

3. Loss Processes

Certain reactions of excited O and O_2 may be of atmospheric significance; these reactions together with physical deactivation compete with the production processes to establish the atmospheric concentrations. In some cases the products of reaction are open to doubt, and attention is drawn below to important examples. Table III then lists those laboratory measurements of rate constants for *loss* which the author regards as reliable. All rate constants are in units of $cm^3 \, molecule^{-1} \, s^{-1}$ and refer to 'room temperature' (ca 295 K). Temperature dependent studies seem to have been made only for reaction of $O_2(^1\Delta_g)$ with O_3 (see below), quenching of $O_2(^1\Delta_g)$ by O_2 $(k = 2.22 \pm 0.03 \times 10^{-18} (T/300)^{0.78 \pm 0.32}$; Findlay and Snelling, 1971b) and quenching of $O(^1S)$ by $CO_2 (k = 5 \pm 2 \times 10^{-11} \exp(-2.7 \pm 0.4 \, kcal \, mol^{-1}/kT)$; Welge *et al.*, 1971).

3.1. OZONE REACTIONS

The reaction of $O(^1D)$ with O_3 may proceed *via* several routes, one of which yields some excited O_2 molecules capable of propagating chains at high concentrations. There is still some doubt about the chain process (Wayne, 1972), but it is certain that chains are not propagated at $[O_3] < 10^{16}$ molecule cm^{-3} (i.e., at atmospheric concentrations). Webster and Bair (1970) have suggested that another route might be

$$O(^1D) + O_3 \rightarrow 2O + O_2 \tag{7}$$

and the most recent studies (Giachardi and Wayne, 1972) indicate that this route is followed for about one third of reactive collisions. Singlet molecular oxygen seems *not* to be formed in the reaction (Wayne, 1969) although there is sufficient energy for its excitation.

Singlet O_2 reacts with O_3 according to

$$O_2(^1\Delta_g, {}^1\Sigma_g^+) + O_3 \rightarrow 2O_2 + O. \tag{8}$$

For $O_2(^1\Delta_g)$ there is good agreement between the results of different workers for room temperature measurements, although estimates of the activation energy range from 3.1 kcal mol^{-1} (Clark *et al.*, 1970) to 5.6 kcal mol^{-1} (Findlay and Snelling, 1971a; Becker *et al.*, 1972).

3.2. REACTION OF $O(^1D, {}^1S)$ WITH H_2O

The reactions of excited O with H_2O (or H_2) are of interest in that OH is almost certainly a product (e.g., McGrath and Norrish, 1960)

$$O(^1D, {}^1S) + H_2O \rightarrow 2 \, OH. \tag{9}$$

The problem is whether the OH radicals formed can lead to a catalytic chain for O_3 decomposition. A chain decomposition certainly occurs in wet O_3 in the laboratory (Norrish and Wayne, 1965; Jones and Wayne, 1970; Lissi and Heicklen, 1972), but it is not clear if OH (and HO_2) could propagate this chain in the atmosphere. If the OH chain is effective, then it can be a critical feature in determining atmospheric O_3 profiles. Because of the greater stability of $O(^1S)$ to quenching by N_2 or O_2 compared

TABLE III
Rate constants for loss of excited O and O_2

Excited species	O₂	N₂	CO₂	O₃	H₂O	Reference
O(^{1}D)		$2 \pm 1 \cdot 10^{-11}$		$3 \cdot 10^{-10}$		Snelling and Bair (1967)
	$4 \cdot 10^{-11}$	$5 \cdot 10^{-11}$	$2.3 \cdot 10^{-10}$			Young et al. (1968b)
					$\sim 8 \cdot 10^{-12}$ [b]	Katakis (1967)
					$1.4 \pm 0.6 \cdot 10^{-11}$ [b]	Biedenkapp et al. (1970)
	$6 \pm 3 \cdot 10^{-11}$	$9 \pm 4 \cdot 10^{-11}$	$3 \pm 1 \cdot 10^{-12}$			Noxon (1970)
	$5 \pm 5 \cdot 10^{-11}$			$2.5 \pm 1.0 \cdot 10^{-10}$		Gilpin et al. (1971)
			$3 \pm 1 \cdot 10^{-10}$			Slanger and Black (1971)
					$1.5 \cdot 10^{-11}$ [b]	Langley and McGrath (1971)
		$4 \pm 2 \cdot 10^{-11}$ [a]	$1.1 \pm 0.6 \cdot 10^{-10}$ [a]			Lowenstein (1971)
		$4 \,\text{or}\, 7 \cdot 10^{-11}$ [a]	$1.6 \,\text{or}\, 2.1 \cdot 10^{-10}$ [a]			Paraskevopoulos et al. (1971)
			$3 \cdot 10^{-10}$			Clark and Noxon (1972)
O(^{1}S)	$1 \cdot 10^{-13}$	$< 10^{-17}$	$2.5 \cdot 10^{-14}$			Young and Black (1966)
	$5 \cdot 10^{-13}$	$3 \cdot 10^{-15}$	$3 \cdot 10^{-13}$			Black et al. (1969)
	$2.1 \pm 0.4 \cdot 10^{-13}$					Zipf (1969)
	$3.6 \cdot 10^{-13}$	$< 2 \cdot 10^{-16}$	$3.6 \cdot 10^{-13}$		$7 \cdot 10^{-11}$	Filseth et al. (1970)
				$5.8 \pm 1 \cdot 10^{-10}$		London et al. (1971)
O₂($^1\Delta_g$)	$2.4 \cdot 10^{-18}$	$< 10^{-19}$	$4 \cdot 10^{-18}$		$1.5 \cdot 10^{-17}$	Clark and Wayne (1969)
	$2.2 \cdot 10^{-18}$	$< 3 \cdot 10^{-21}$				Findlay et al. (1969)
			$< 1.5 \cdot 10^{-20}$		$4.5 \cdot 10^{-18}$	Findlay and Snelling (1971b)
				$3.0 \cdot 10^{-15}$		Clark et al. (1970)
				$3.5 \cdot 10^{-15}$		Findlay and Snelling (1971a)
				$5 \cdot 10^{-15}$		Becker et al. (1972)
	$1.7 \cdot 10^{-18}$	$< 10^{-20}$	$< 8 \cdot 10^{-20}$		$4.4 \cdot 10^{-18}$	Becker et al. (1971)
O₂($^1\Sigma^+_g$)	$< 10^{-16}$	$2.3 \cdot 10^{-15}$		$7 \cdot 10^{-12}$		Izod and Wayne (1968)
		$2.5 \cdot 10^{-15}$	$5 \cdot 10^{-14}$		$1 \cdot 10^{-13}$	Ogryzlo (1969)
	$1.5 \cdot 10^{-16}$	$2.0 \cdot 10^{-15}$	$3 \cdot 10^{-13}$			Noxon (1970)
	$1 \cdot 10^{-16}$	$3 \cdot 10^{-15}$	$1.5 \cdot 10^{-13}$		$4 \cdot 10^{-12}$	O'Brien and Myers (1970)
	$4.5 \cdot 10^{-16}$	$1.8 \cdot 10^{-15}$	$4.4 \cdot 10^{-13}$		$3.3 \cdot 10^{-12}$	Filseth et al. (1970)
				$2.5 \pm 1.0 \cdot 10^{-11}$		Gilpin et al. (1971)
	$1.5 \cdot 10^{-16}$	$2.2 \cdot 10^{-15}$	$3 \cdot 10^{-13}$	$7 \cdot 10^{-12}$	$5 \cdot 10^{-12}$	Becker et al. (1971)

[a] Based on $k_{O_2} = 6 \cdot 10^{-11}$.　　[b] Based on $k_{O_3} = 3 \cdot 10^{-10}$.

with $O(^1D)$ (see Table III), small sources of $O(^1S)$ (e.g. O_3 photolysis or the reaction in Equation (1)) might be important to the O_3 balance.

References

Badger, R. M., Wright, A. C., and Whitlock, R. F: 1965, *J. Chem. Phys.* **43**, 4345.

Becker, C. H., Groth, W., and Schurath, U.: 1971, *Chem. Phys. Letters* **8**, 259.

Becker, C. H., Groth, W., and Schurath, U.: 1972, *Chem. Phys. Letters*, in press.

Biedenkapp, D., Hartshorn, L. G., and Bair, E. J.: 1970, *Chem. Phys. Letters* **5**, 379

Black, G., Slanger, T. G., St. John, G. A., and Young, R. A.: 1969, *Can. J. Chem.* **47**, 1872.

Chamberlain, J. W.: 1961, in *Physics of the Aurora and Airglow*, Academic Press, New York.

Chapman, S.: 1937, *Phil. Mag*, **23**, 657.

Clark, I. D. and Noxon, J. F.: 1972, *J. Chem. Phys.*, in press.

Clark, I. D. and Wayne, R. P.: 1969, *Proc. Roy. Soc.* **A314**, 111.

Clark, I. D., Jones, I. T. N., and Wayne, R. P.: 1970, *Proc. Roy. Soc.* **A317**, 407.

Corney, A. and Williams, O. M.: 1972, *J. Phys. B.: Atom. Molec. Phys.* **5**, 686

Crutzen, P. J., Jones, I. T. N., and Wayne, R. P.: 1971, *J. Geophys. Res.* **76**, 1490.

Derwent, R. G. and Thrush, B. A.: 1971, *Trans. Far. Soc.* **67**, 2036.

Evans, W. F. J., Llewellyn, E. J., and Vallance-Jones, A.: 1972, *J. Geophys. Res.*, in press.

Felder, W. and Young, R. A.: 1972, *J. Chem. Phys.*, **56**, 6028.

Filseth, S. V., Stuhl, F., and Welge, K. H.: 1970a, *J. Chem. Phys.* **52**, 239.

Filseth, S. V., Zia, A., and Welge, K. H.: 1970b, *J. Chem. Phys.* **52**, 5502.

Findlay, F. D. and Snelling, D. R.: 1971a, *J. Chem. Phys.* **54**, 2750.

Findlay, F. D. and Snelling, D. R.: 1971b, *J. Chem. Phys.* **55**, 545.

Findlay, F. D., Fortin, C. J., and Snelling, D. R.: 1969, *Chem. Phys. Letters* **3**, 204.

Giachardi, D. J. and Wayne, R. P.: 1972, *Proc. Roy. Soc.* **A330,** 131.

Gilpin, R., Schiff, H. I., and Welge, K. H.: 1971, *J. Chem. Phys.* **55**, 1087.

Herzberg, G.: 1950, *Spectra of Diatomic Molecules*, Van Nostrand, Princeton, N.Y.

Herzberg, G.: 1966, *Electronic Spectra of Polyatomic Molecules*, Van Nostrand Reinhold, New York.

Izod, T. P. J. and Wayne, R. P.: 1968, *Proc. Roy. Soc.* **A308**, 81.

Jones, I. T. N. and Bayes, K. D.: 1971, *Chem. Phys. Letters* **11**, 163.

Jones, I. T. N. and Wayne, R. P.: 1970, *Proc. Roy. Soc.* **A319**, 273.

Jones, I. T. N. and Wayne, R. P.: 1971, *Proc. Roy. Soc.* **A321**, 409.

Katakis, D.: 1967, *J. Chem. Phys.* **47**, 541.

Langley, K. F. and McGrath, W. D.: 1971, *Planetary Space Sci.* **19**, 413.

Lissi, E. and Heicklen, J.: 1972, *J. Photochem.*, **1,** 39.

London, G., Gilpin, R., Schiff, H. I., and Welge, K. H.: 1971, *J. Chem. Phys.* **54**, 4512.

Lowenstein, M.: 1971, *J. Chem. Phys.* **54**, 2282.

McCullough, D. W. and McGrath, W. D.: 1971, *Chem. Phys. Letters* **8**, 353.

McGrath, W. D. and Norrish, R. G. W. 1960, *Proc. Roy. Soc.* **A254**, 317.

Norrish, R. G. W. and Wayne, R. P.: 1965, *Proc. Roy. Soc.* **A288**, 361.

Noxon, J. F.: 1961, *Can. J. Phys.* **39**, 1110.

Noxon, J. F.: 1970, *J. Chem. Phys.* **52**, 1852.

O'Brien, R. J. and Myers, G. H.: 1970, *J. Chem. Phys.* **53**, 3832.

Ogryzlo, E. A.: 1969, *Can. J. Chem.* **47**, 1871.

Paraskevopoulos, G., Preston, K. J., and Cvetanovic, R. J.: 1971, *J. Chem. Phys.* **54**, 3907.

Preston, K. J. and Cvetanovic, R. J.: 1966, *J. Chem. Phys.* **45**, 2888.

Slanger, T. G. and Black, G.: 1971, *J. Chem. Phys.* **54**, 1889.

Snelling, D. R. and Bair, E. J., 1967: *J. Chem. Phys.* **47**, 228.

Wallace, L. and Hunten, D. M.: 1968, *J. Geophys. Res.* **73**, 4813.

Wayne, R. P.: 1969, *Advan. Photochem.* **7**, 311.

Wayne, R. P.: 1970, *Photochemistry*, Butterworth, London.

Wayne, R. P.: 1972, *Disc. Far. Soc.* **53**, in press.

Webster, H. and Bair, E. J.: 1970, *J. Chem. Phys.* **53**, 4532.

Welge, K. H., Zia, A., Vietske, E., and Filseth, S. V.: 1971, *Chem. Phys. Letters* **10**, 13.

Young, R. A. and Black, G.: 1966, *J. Chem. Phys.* **44**, 3741.

Young, R. A. and Black, G.: 1967, *J. Chem. Phys.* **47**, 2311.
Young, R. A. and Sharpless, R. L.: 1963, *J. Chem. Phys.* **39**, 1071.
Young, R. A., Black, G., and Slanger, T. G.: 1968a, *J. Chem. Phys.* **49**, 4769.
Young, R. A., Black, G., and Slanger, T. G.: 1968b, *J. Chem. Phys.* **49**, 4758.
Young, R. A., Black, G., and Slanger, T. G.: 1969, *J. Chem. Phys.* **50**, 309.
Zipf, E. C.: 1969, *Can. J. Chem.* **47**, 1863.

THE OXYGEN-HYDROGEN ATMOSPHERE

L. THOMAS

S.R.C., Radio and Space Research Station, Ditton Park, Slough, Bucks, U.K.

1. Introduction

The O-H constituents of the atmosphere have attracted considerable attention during the last two decades. It is believed that these represent the products of chemical and photochemical processes involving H_2O, CH_4 and H_2 transported from the earth's surface and also O_2 and O_3. Thus in the stratosphere these H compounds undergo chemical reactions with ground-state or excited O atoms, these atoms arising from the photodissociation of O_3. At still greater heights, in the lower mesosphere, the photodissociation of H_2O leads to the production of the constituents H, OH, HO_2, H_2O_2, and H_2, and the photodissociation of O_2 also becomes important. At thermospheric heights the dissociation of H_2O is more rapid and recombination of the dissociation products is negligible; it appears that in the upper thermosphere H atoms represent the only significant form of hydrogen content.

The O-H constituents in the height region 60 to 100 km are of particular interest because of their involvement in the airglow, the ion chemistry of the lower ionosphere, the thermal balance of this part of the atmosphere, and other aspects of aeronomy. For instance, the importance of H_2O has been emphasized by the observation of water cluster ions $(H^+ \cdot (H_2O)_n)$ at heights below 85 km (Narcisi and Bailey, 1965), by its possible role in the quenching of CO_2 vibrationally excited during the absorption of solar radiations (Houghton, 1969), and by its part in the formation of noctilucent clouds (Humphreys, 1933).

The photochemistry of an O-H atmosphere was treated originally by Bates and Nicolet (1950) and subsequently by Hunt (1966). Hesstvedt (1968) extended the model to take account of the departure from photochemical equilibrium of certain constituents by incorporating vertical transport by eddy diffusion, and more recent treatments have been carried out by Bowman *et al.* (1970), Shimazaki and Laird (1970), Hunt (1971), and Thomas and Bowman, (1972). Unfortunately, there have been very few measurements of H-O constituents at heights above the stratopause with which to compare these theoretical models.

The purpose of the present paper is to indicate our present understanding and knowledge of the H-O atmosphere in the height range 60 to 140 km. At lower heights account needs to be taken of reactions involving nitrogen oxides, as pointed out by Nicolet (1970), and at greater heights the main interest is in the escape of H atoms. Recent developments in theoretical concepts and in the input parameters for models are indicated in Section 2. Section 3 describes the temporal variations of these constituents as predicted by theoretical models, and Section 4 outlines the chemical processes which determine these variations in the mesosphere. In Section 5 the theore-

B. M. McCormac (ed.), Physics and Chemistry of Upper Atmospheres, 133–142. All Rights Reserved.
Copyright © 1973 by D. Reidel Publishing Company, Dordrecht-Holland.

tical results are compared with the available experimental data, and finally in Section 6 attention is drawn to particular problems and requirements to be considered to ensure a better understanding of the O-H atmosphere.

2. Recent Developments in Theoretical Models

The most significant recent advance in the photochemistry of the O-H atmosphere has been the realization that predissociation in the Schumann-Runge bands (1750 to 2000 Å) leads to a considerable increase in the O_2 dissociation rate in the mesosphere (Hudson *et al.*, 1969; Nicolet, 1970). Ackerman *et al.* (1970) have determined the relevant cross sections and have pointed out that for a complete treatment account needs to be taken of the temperature-dependence of these cross sections. The Schumann-Runge bands are also important for the photodissociation of H_2O in the lower parts of the mesosphere, Ly-α being more effective in the upper parts. A further development concerned with photodissociation processes has been the demonstration that $O_2(^1\Delta_g)$ is produced by O_3 photolysis (Gauthier and Snelling, 1970; Jones and Wayne, 1970).

In calculations of photodissociation rates the solar radiation intensities in the range 1400 to 1900 Å available until recently were those of Detwiler *et al.* (1961). However, observations by Ackerman *et al.* (1968) and Parkinson and Reeves (1969) have indicated substantially smaller intensities, and Hinteregger (1970) has recommended that the early values be reduced by a factor of about 3.

It has been customary in theoretical models to include the effects of vertical transport by turbulent processes by means of an eddy diffusion coefficient analogous to that representing molecular diffusion (Colegrove *et al.*, 1965). Both height-independent and varying values of eddy diffusion coefficient have been adopted in theoretical models, the magnitude of the parameter being adjusted in some cases to give agreement with particular measurements, such as the ratio of O and O_2 concentrations at 120 km. However, thermal data have been used to obtain estimates of the height variation of the coefficient (Johnson and Wilkins, 1965; Johnson and Gottlieb, 1970), since the thermal balance of the mesosphere and lower thermosphere depends on the downward transport of heat by eddy conduction. Estimates of eddy diffusion coefficients can also be made from observations of chemical releases in the atmosphere, and Keneshea and Zimmerman (1970) have drawn attention to the reduction in turbulence near 100 km indicated by such observations.

A major difficulty in formulating a theoretical model of atmospheric composition is the need to incorporate realistic upper and lower boundary conditions. For the O-H atmosphere, account needs to be taken of the downward flux of $O(^3P)$ from the thermosphere and the corresponding upward flux of O_2, and also the escape of hydrogen from the top of the atmosphere. These vertical fluxes render the assumption of diffusive equilibrium at the upper boundary invalid, and this and other boundary conditions adopted in previous models have been critically reviewed by Strobel (1972).

Theoretical models developed to date have not fully incorporated these recent developments. Furthermore, the restriction to a single dimension in these models, and the neglect of horizontal and vertical motions associated with large-scale circulation, represent very serious limitations. It seems evident, therefore, that the results presently available do not provide an adequate description of the behavior of the oxygen-hydrogen atmosphere. However, it is possible that some of the major features illustrated by these models are representative of the atmosphere, and it is particularly useful to consider the importance of photochemistry in the temporal changes expected in the mesosphere.

3. General Features of Theoretical Models

3.1. HEIGHT DISTRIBUTIONS IN DAYTIME

A feature common to all theoretical models is a fairly rapid increase of $O(^3P)$ concentration with increasing height above about 80 km in daytime, the height of maximum concentration depending on the height variation of eddy diffusion coefficient adopted. The distribution computed for noon by Thomas and Bowman (1972) is shown by the full line for (O^3P) in Figure 1a. Associated with this increase is a more gradual increase of H concentration and rapid decreases in concentrations of OH, HO_2 and H_2O_2; these features for H and OH are shown in the full curves of Figure 1b. The form of the O_3 distribution shows a general reduction with increasing height and a small maximum near 85 km (Figure 1a.) Calculations have shown that the distinctness and magnitude of this O_3 layer near 85 km and also the rate of decrease in H_2O concentration with height are also dependent on the assumed values of eddy diffusion coefficient.

3.2. DIURNAL VARIATIONS

The large day-to-night reduction in $O(^3P)$ concentrations below about 80 km first demonstrated for an O atmosphere by Hunt (1965a) has been confirmed for an O-H atmosphere in which account is taken of eddy and molecular diffusion (Shimazaki and Laird, 1970; Hunt, 1971; Thomas and Bowman, 1972). These models indicate that a similar reduction in concentrations of H occurs below 80 km at nighttime. However, the O_3 concentrations are increased at most heights during nighttime and no significant diurnal variations occur for H_2 and H_2O. For the other O-H constituents the diurnal changes are rather more complicated. Thus for OH the concentrations below about 80 km are reduced during the nighttime whereas above this level concentrations are increased. These diurnal changes for $O(^3P)$, O_3, H and OH are illustrated by the distributions shown in Figure 1a and 1b; the full and broken curves for each constituent correspond to noon and pre-sunrise times as computed by Thomas and Bowman (1972).

3.3. VARIATIONS DURING AN ECLIPSE

A photochemical treatment of an O atmosphere by Hunt (1965b) indicated a change by only 0.6% in total O_3 content during an eclipse, the effects being confined to

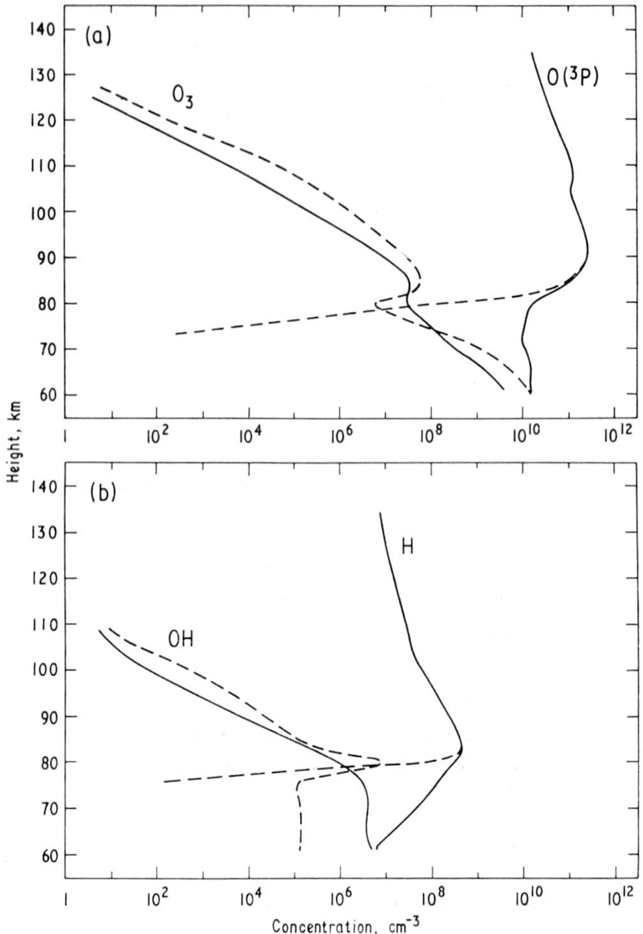

Fig. 1. The height distributions of O(³P), O₃, H and OH computed for noon (———) and pre-sunrise (–––––) (after Thomas and Bowman, 1972).

heights above about 45 km. The treatment was extended to an O-H atmosphere by Keneshea (1969), and more recently Thomas and Bowman (1973) have examined the changes in height distributions of a number of constituents in a model in which account is taken of eddy and molecular diffusion and predissociation of O_2. The results show that the maximum effect occurs for $O(^3P)$ near 75 km, the concentrations decreasing by about an order of magnitude. The changes for O_3 are rather more complicated; the concentration at each height increases up to the time of maximum obscuration and subsequently decreases to a minimum value, the magnitude of the change depending on the height. Hydrogen concentrations at heights below 80 km decrease during the eclipse, but both OH and HO_2 show increased concentrations up to about the time of maximum obscuration with the form of the return to post-eclipse values depending on height.

3.4. Seasonal variations

An attempt to predict the seasonal variations to be expected for an O-H atmosphere has been made by Shimazaki and Laird (1972). The seasonal differences encountered arose from the large summer-to-winter change in solar zenith angle for daytime hours, the change in the duration of the sunlit period, and an assumed change in the height variation of the eddy diffusion coefficient. The greater photodissociation in summer was found to have a major influence on the seasonal changes. Thus because of the increase in O_2 photodissociation the $O(^3P)$ and $O(^1D)$ concentrations above 70 km were an order of magnitude larger in summer than winter, and the increased photo-dissociation of H_2O and H_2O_2 also led to increased concentrations of OH, H, HO_2 and H_2.

4. Chemical and Photochemical Processes and Their Control of Temporal Variations in an Oxygen-Hydrogen Atmosphere

The O-H reactions operating throughout the whole atmosphere have been de-scribed in some detail by Nicolet (1970, 1971). It has been seen from Section 3 that the mesosphere is a region of particular interest since it shows major diurnal and other temporal variations. This is because it corresponds to the lowest heights at which photodissociations of O_2 and H_2O are important and the constituents resulting from these dissociations have chemical lifetimes which are significant but less than several hours. By contrast, at stratospheric heights H_2O is destroyed by reactions with $O(^1D)$ but the molecules are rapidly reformed by reactions involving H, OH and HO_2, and in the thermosphere the rapid photodissociation of H_2O results in H, and perhaps H_2, being the only significant form of hydrogen.

The important production and loss processes controlling the densities of $O(^3P)$, O_3 and H_2O in the mesosphere have been outlined by Bowman *et al.* (1970). Production of $O(^3P)$ results from the photodissociation of O_2 and O_3, either directly or via the $O(^1D)$ state after quenching. In the middle of the mesosphere reactions with OH and HO_2 represent the major loss processes for $O(^3P)$, and these account for the rapid decrease in $O(^3P)$ concentrations below about 80 km associated with the increases in OH and HO_2. The three-body reaction involving $O(^3P)$ and O_2 to form O_3 is also important, especially at the lower heights, and this reaction also represents the main production mechanism for O_3. Photodissociation of O_3 by the Hartley continuum is important over a range of heights but in the middle mesosphere the reaction with H to form OH is the dominant loss process for O_3. For H_2O, production is chiefly by the reaction of HO_2 and OH, and the main loss is by photodissociation; to compensate for this loss in the middle mesosphere and above an upward flux of H_2O is required.

It is evident from the above that the concentrations of $O(^3P)$, O_3, and H_2O in the middle mesosphere are dependent upon the concentrations of H, OH, and HO_2; the distributions of these latter constituents are also of interest in their own right, as in the interpretation of positive-ion composition measurements in the lower ionosphere (Burke, 1970; Niles *et al.*, 1972). It is therefore useful to consider in some detail the major reactions controlling the concentrations of H, OH, and HO_2 in the middle mesosphere. These reactions are indicated for a height of 70 km in Figure 2 in which (a) and (b) represent conditions at noon and at the end of the nighttime period, just prior to sunrise, respectively. The line representing each reaction is numbered in accordance with the list given in Table I and the reactant involved is also shown. The significance of each reaction is indicated: a broken line represents a production process; a dotted line a loss process; and a chain line represents a process important simultaneously for production and loss. It can be seen that the photodissociation of H_2O represents the primary source of the O-H constituents, directly for H and OH,

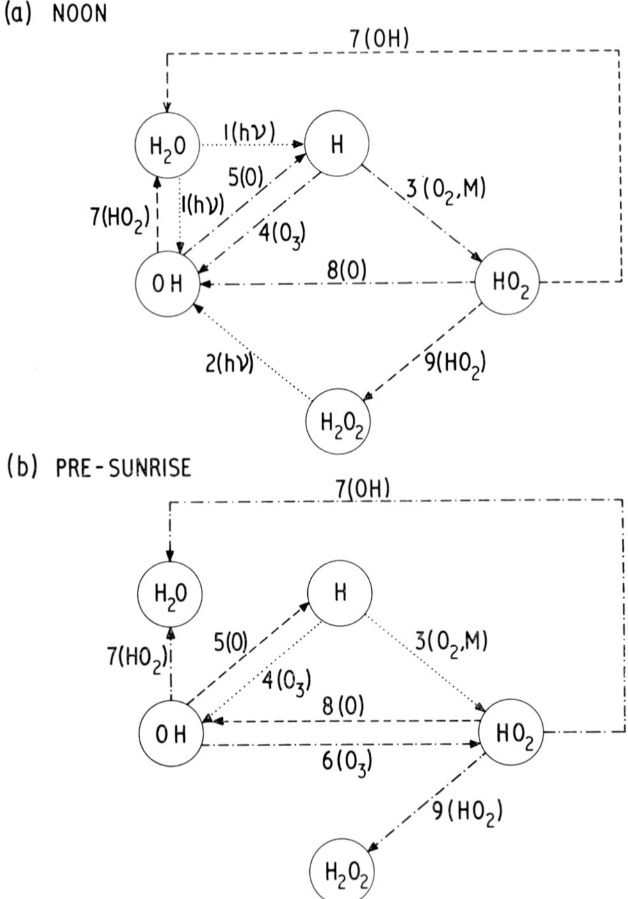

Fig. 2.   Schematic representation of changes controlling the concentrations of H, OH, HO_2, H_2O, and H_2O_2 in the mesosphere at noon and pre-sunrise. The individual processes of Table I which are important at a height of 70 km are shown together with the ractants involved. The significance of each reaction is also indicated: ––––– production; ..... loss; –·–·– production and loss.

TABLE I

Chemical and photochemical processes controlling the concentrations of H, OH,
HO_2, H_2O, and H_2O_2 in the middle mesosphere

Reaction	Rate coefficient c.g.s. units	Reference
1. $H_2O + hv \rightarrow H + OH$	4.0×10^{-7} [a]	Thomas and Bowman (1972)
2. $H_2O_2 + hv \rightarrow 2OH$	10^{-4} [a]	Thomas and Bowman (1972)
3. $H + O_2 + M \rightarrow HO_2 + M$	5.0×10^{-32}	Kaufman (1969)
4. $H + O_3 \rightarrow O_2 + OH$	2.6×10^{-11}	Kaufman (1964)
5. $OH + O \rightarrow H + O_2$	5.0×10^{-11}	Kaufman (1964)
6. $OH + O_3 \rightarrow HO_2 + O_2$	5.0×10^{-13}	Kaufman (1964)
7. $OH + HO_2 \rightarrow H_2O + O_2$	1.0×10^{-11}	Kaufman (1964)
8. $HO_2 + O \rightarrow O_2 + OH$	1.0×10^{-11}	Kaufman (1964)
9. $HO_2 + HO_2 \rightarrow H_2O_2 + O_2$	1.5×10^{-12}	Kaufman (1964)

[a] At 70 km.

and via additional reactions for HO_2 and H_2O_2. The main losses of OH and HO_2 during the day are by recombination with $O(^3P)$ atoms, in reactions 5 and 8, respectively; the corresponding loss for H_2O_2 in daytime is about an order of magnitude slower than that due to photodissociation, and at nighttime this and other loss processes for the molecule are very slow. It can be seen from Figure 2a and 2b that the major difference between day and night is in the production and loss processes for HO_2. The large decrease in H concentration below 80 km at nighttime (Section 3.2) renders production by reaction 3 unimportant compared with reaction 6. The corresponding reduction in $O(^3P)$ results in the major loss process for HO_2, reaction 8, being superseded at nighttime by reactions 7 and 9. The production of OH from H atoms by reaction 4 is also unimportant at nighttime but this reaction still represents a major loss power for H atoms.

It is of interest to note that the hydrogen constituents H, OH and HO_2 have chemical time constants in the mesosphere which are short in comparison with the mixing time constant H_{av}^2/K, where H_{av} represents the average scale height and K the eddy diffusion coefficient. These three constituents are, therefore, in photochemical equilibrium with one another. In temporal changes at mesospheric heights, although the individual concentration of H, OH, and HO_2 can vary rapidly, the total odd H content changes relatively slowly as determined by the reactions producing and destroying water, represented by 1 and 7 in Table I and Figure 2a and 2b.

For H_2, which is formed by a reaction involving H and HO_2, the main loss is by a reaction with $O(^1D)$. This and a corresponding reaction for H_2O become increasingly important in the stratosphere. Furthermore, at these lower heights reactions with nitrogen oxides become important as loss processes for HO_2 and H_2O_2.

5. Comparisons with Measurements

For the constituents considered in the theoretical models of the O-H atmosphere the greatest experimental information is available for $O(^3P)$ and O_2. Mass spectrometer measurements extending down to about 120 km have been carried out by a number of groups (Hedin *et al.*, 1964; Kasprzak *et al.*, 1968; Krankowsky *et al.*, 1968; Mauersberger *et al.*, 1968; Schaeffer, 1969; Von Zahn and Gross, 1969) and preliminary values down to about 86 km have also been reported by Philbrick *et al.* (1971). Of particular interest in these measurements is the ratio of $O(^3P)$ to O_2 concentrations at 120 km. The measurements show values varying from about 0.4 to 1.5 and the theoretical values predicted by Shimazaki and Laird (1970), Hunt (1971), and Thomas and Bowman (1972) are near unity. It is to be noted, however, that Von Zahn (1967) has argued that mass spectrometer measurements have provided lower limits for the $O(^3P)$ concentrations because of the reactions of atoms at the source walls. The ratio of 3.5 found with the use of an improved ion source (Offermann and Von Zahn, 1971) is larger than any value derived for an O-H atmosphere. The information available for $O(^3P)$ concentrations at lower heights has been obtained from the mass spectrometer observation of Philbrick *et al.* (1971) and from rocket-borne observations of the oxidation of silver strips (Henderson, 1971). The two sets of measurements show $O(^3P)$ concentrations of about 10^{11} near 90 km, the results of Philbrick *et al.* showing little change with increasing height whereas Henderson reported an increase in concentration to about 8×10^{11} at 94 km. This rapid increase in concentration found by Henderson is analogous to that predicted theoretically but is located almost 10 km higher.

Estimates of O_3 concentrations up to about 70 km have been made from chemiluminescent detectors released from rockets (Hilsenrath, 1971) and from rocket and satellite observations of the attenuation of solar UV radiations at sunrise and sunset (Johnson *et al.*, 1952; Rawcliffe *et al.*, 1963). A comparison of theoretical values of O_3 concentration with the experimental estimates has shown reasonable agreement. For heights near 80 to 85 km information on O_3 concentrations can be obtained from the height distributions of $O_2(^1\Delta_g)$ derived from rocket measurements of the IR atmospheric emission in the dayglow, on the basis that the excited molecule is produced by the photolysis of O_3 (Evans and Llewellyn, 1970). It is interesting to note that the height distribution of $O_2(^1\Delta_g)$ provided by the theoretical studies has similar forms to those deduced from measurements but a quantitative comparison indicates that the theoretical values of O_3 concentrations in the 80 to 85 km region are too small, especially in the morning hours (Shimazaki and Laird, 1972; Thomas and Bowman, 1972).

The only attempt to measure an O-H constituent in the mesosphere has been made by Anderson (1971a, b). Rocket-borne observations of resonantly scattered sunlight in an electronic transition of the OH radical at 3064 Å have been used to estimate OH concentrations at evening twilight during June 1969 and April 1971. These observations indicated upper limits of 6×10^{12}, 3×10^{12}, and 7×10^{11} cm^{-2} at 65, 75, and 85 km, respectively, for the OH column densities during the first flight, and local concentrations of 4.4×10^6, 5.5×10^6, and 3.5×10^6 cm^{-3} at 50, 60, and 70 km, respectively, on the second flight. These latter values show reasonable agreement with the daytime results of theoretical models (Shimazaki and Laird 1970; Hunt, 1971; Thomas and Bowman, 1972) but are rather larger than the predicted twilight or nighttime values.

6. Conclusions and Future Progress

A major problem in theoretical models of the O-H atmosphere is the need to apply specific boundary conditions which take full account of vertical fluxes of constituents. An improved description of turbulence is also required for these models, based on experimental observations such as those employing chemical tracers. Furthermore, a better physical understanding of the eddy diffusion coefficient is essential in order to help predict its temporal and spatial variability from the available measurements. A more general aim for the dynamical aspects of these models should be to include the effects of large-scale circulation as well as the small-scale mechanisms considered to date.

One of the principal requirements in the photochemistry of theoretical models is the improved measurement of solar flux intensities in the wavelength range 1300 to 2000 Å. In addition, the realization of the importance of predissociation in the Schumann-Runge bands in the dissociation of O_2 in the mesosphere has represented a significant advance in theoretical models and attempts should also be made to examine in more detail the absorption of solar radiations by other constituents.

It is evident that an improved formulation of the continuity equations, the incorporation of more realistic transport processes, and the adoption of improved input parameters will lead to more representative models of the O-H atmosphere. However, it will be essential that measurements of individual constituents be made in order to check and apply constraints to these theoretical models. It is particularly desirable that measurements of $O(^3P)$ concentrations be extended down to the lower mesosphere and that further measurements of O-H constituents at these heights be initiated.

Acknowledgment

This paper is published by permission of the Director of the Radio and Space Research Station of the Science Research Council.

References

Ackerman, M., Frimout, D., and Pastiels, R.: 1968, *Ciel Terre* **84**, 408.
Ackerman, M., Biaume, F., and Kockarts, G.: 1970, *Planetary Space Sci.* **18**, 1639.
Anderson, J. G.: 1971a, *J. Geophys. Res.* **76**, 4634.
Anderson, J. G.: 1971b, *J. Geophys. Res.* **76**, 782.
Bates, D. R. and Nicolet, M.: 1950, *J. Geophys. Res.* **55**, 301.
Bowman, M. R., Thomas, L., and Geisler, J. E.: 1970, *J. Atmospheric Terrest. Phys.* **32**, 1661.
Burke, R. R.: 1970, *J. Geophys. Res.* **75**, 1345.
Colegrove, F. D., Hanson, W. B., and Johnson, F. S.: 1965, *J. Geophys. Res.* **70**, 4931.
Detwiler, C. R., Garrett, D. L., Purcell, J. D., and Tousey, R.: 1961, *Ann. Geophys.* **17**, 265.
Evans, W. F. J. and Llewellyn, E. J.: 1970, *Ann. Geophys.* **26**, 167.
Gauthier, M. and Snelling, D. R.: 1970, *Chem. Phys. Letters* **5**, 93.

Hedin, A. E., Avery, P., and Tschetter, C. D.: 1964, *J. Geophys. Res.* **69**, 4637.
Henderson, W. R.: 1971, *J. Geophys. Res.* **76**, 3166.
Hesstvedt, E.: 1968, *Geophys. Pub.* **27**, 1.
Hilsenrath, E.: 1971, *J. Atmospheric Sci.* **28**, 295.
Hinteregger, H. E.: 1970, *Ann. Geophys.* **26**, 547.
Houghton, J. T.: 1969, *Quart. J. R. Meteorol. Soc.* **95**, 1.
Hudson, R. D., Carter, V. L., and Breig, E. L.: 1969, *J. Geophys. Res.* **74**, 4079.
Humphreys, W. J.: 1933, *Mon. Weather Rev.* **61**, 228.
Hunt, B. G.: 1965a, *J. Atmospheric Terrest. Phys.* **27**, 133.
Hunt, B. G.: 1965b, *Tellus* **17**, 516.
Hunt, B. G.: 1966, *J. Geophys. Res.* **71**, 1385.
Hunt, B. G.: 1971, *J. Atmospheric Terrest. Phys.* **33**, 1869.
Johnson, F. S and Wilkins, E. M.: 1965, *J. Geophys. Res.* **70**, 1281.
Johnson, F. S and Gottlieb, B.: 1970, *Planetary Space Sci.* **18**, 1707.
Johnson, F. S., Purcell, J. D., Tousey, R., and Watanabe, K.: 1952, *J. Geophys. Res.* **57**, 157.
Jones, I. T. N. and Wayne, R. P.: 1970, *Proc. Roy. Soc.* **A319**, 273.
Kasprzak, W. T., Krankowsky, D., and Nier, A. O.: 1968, *J. Geophys. Res.* **73**, 6765.
Kaufman, F.: 1964, *Ann. Geophys.* **20**, 106.
Kaufman, F.: 1969, *Can. J. Chem.* **47**, 1917.
Keneshea, T. J.: 1969, Aeronomy Report No. 32, University of Illinois, Urbana, Illinois, p. 400.
Keneshea, T. J. and Zimmerman, S. R.: 1970, *J. Atmospheric Sci.* **27**, 831.
Krankowsky, D., Kasprzak, W. T., and Nier, A. O.: 1968, *J. Geophys. Res.* **73**, 7291.
Mauersberger, K., Müller, D., Offermann D., and Von Zahn, U.: 1968, *J. Geophys. Res.* **73**, 1071.
Narcisi, R. S. and Bailey, A. D.: 1965, *J. Geophys. Res.* **70**, 3687.
Nicolet, M.: 1970, *Ann. Geophys.* **26**, 531.
Nicolet, M.: 1971, in G. Fiocco (ed.), *Mesospheric Models and Related Experiments*, D. Reidel,
    Publishing Company, Dordrecht-Holland, p.1.
Niles, F. E., Heimerl, J. M., and Keller, G. E.: 1972, *Trans. Amer Geophys. Union* **53**, 456.
Offerman, D. and Von Zahn, U.: 1971, *J. Geophys. Res.* **76**, 2520.
Parkinson, W. H. and Reeves, E. M.: 1969, *Solar Phys.* **10**, 342.
Philbrick, C. R., Faucher, G. A., and Wlodyka, R. A.: 1971, A.F.C.R.L. Report-71-0602, L. G.
    Hanscom Field, Bedford, Mass.
Rawcliffe, R. D., Meloy, G. E., Friedman, R. M., and Rogers, E. H.: 1963, *J. Geophys. Res.* **68**, 6425.
Schaeffer, E. J.: 1969, *J. Geophys. Res.* **74**, 3488.
Shimazaki, T. and Laird, A. R.: 1970, *J. Geophys. Res.* **75**, 3221.
Shimazaki, T. and Laird, A. R.: 1972, *Radio Sci.* **1**, 23.
Strobel, D. F.: 1972, *Radio Sci.* **1**, 1.
Thomas, L. and Bowman, M. R.: 1972, *J. Atmospheric Terrest. Phys.* **34**, 1843.
Thomas, L. and Bowman, M. R.: 1973, *J. Atmospheric Terrest. Phys.*, to be submitted.
Von Zahn, U.: 1967, *J. Geophys. Res.* **72**, 5933.
Von Zahn, U. and Gross, J.: 1969, *J. Geophys. Res.* **74**, 4055.

AEROSOL CHEMISTRY

A. W. CASTLEMAN, JR.

Brookhaven National Laboratory, Upton, New York, U.S.A.

1. Introduction

Upper atmospheric aerosols contribute to many observable optical phenomena and their existence was speculated as early as the late 1800's. Nevertheless, systematic quantitative measurements have only been made during the last decade and a half, and there is still a paucity of information concerning their origin and composition.

Recent reviews by Reiter (1971) and Rosen (1969) have given adequate historical surveys of the early work, and for this reason detailed reviews will not be given here. Furthermore, Link (1973) has surveyed the topic of aerosols and particles as they pertain to the structure and composition of the neutral atmosphere at elevations $\leqslant 1000$ km. We consider here only those aspects which are directly germane to the subject of aerosol chemistry.

2. Classification and Nature

The definition of an aerosol is rather arbitrary. In order to specify the upper size limit of an aerosol particle, we adopt the criterion that the ratio of particle settling velocity to average molecular velocity of the gas be less than unity. The lower limit is specified only by the requirement that the particle represents a distinct separate phase from the distribution of gaseous molecules. In practice, the range covers from approximately 10^{-7} to 10^{-2} cm.

Aerosol particles may be broadly classified as Aitken, $\approx 10^{-3}$ to 10^{-1} μm in diameter, large particles, 10^{-1} to 1 μm, and giant particles, > 1 μm. Those either formed by chemical reaction or participating as chemical reaction centers are most importantly the large particles, followed by the Aitken particles, and to a much less extent, the giant ones. Consequently, most of the particles dealt with in this paper may be considered to be in the free molecular regime in order to calculate their dynamics. This limiting case applies where the ratio of radius to mean free path of the gas molecules ≤ 10, i.e., all particles $\lesssim 10^{-4}$ cm, at altitudes above ~ 35 km. In this regime, the settling velocity is proportional to the radius, r, rather than r^2 as in the case where Stokes settling predominates.

One important property of aerosols is their tendency to coagulate upon collision. Although most formation processes do not produce aerosols of uniform size, subsequent second-order collision processes will always result in a polydispersed size distribution (Lindauer and Castleman, 1971). The size distribution can often be approximated by the Junge distribution

$$\mathrm{d}N/\mathrm{d}r = cr^{-\mathrm{p}},$$

B. M. McCormac (ed.), Physics and Chemistry of Upper Atmospheres, 143–157. All Rights Reserved.
Copyright © 1973 by D. Reidel Publishing Company Dordrecht-Holland.

where N is the total number of particles, and c and p are constants (Junge, 1963). For lower atmospheres, $p \simeq 4$, but its value is usually 3 or $3\frac{1}{2}$ at the upper altitudes (Junge, 1963; Reiter, 1971). Another important property of aerosols is their tendency to charge-exchange with ions in the air, tending to a Boltzmann form of charge-distribution. As a result, atmospheric aerosols influence the ion distribution in their local vicinity.

The presence of aerosols results in electromagnetic wave scattering and has a direct influence on the earth's albedo. Furthermore, they present potential sites for hetero-geneous nucleation. Finally, those which are formed by gas-to-particle conversion reactions often play a role in the global cycle of some of the minor constituents of the atmosphere. Consequently, interest in aerosols stems not only from a scientific interest in the composition of the upper atmosphere, but also from the fact that they have an influence on such diverse things as atmospheric electricity, long distance communication, and potential global climate modifications.

3. Sources of Aerosols in the Upper Atmosphere

Clearly, there are three main sources of aerosols in the upper atmosphere: (a) tropo-spheric, (b) extraterrestrial, and (c) those originating from *in situ* formation processes.

3.1. TROPOSPHERIC

Tropospheric aerosols considerably diminish in concentration with increasing altitude, approaching a first minimum at the tropopause. Those originating over the oceans, such as sea salt particles, exist in significant concentrations only up to ~ 3 km, while continental aerosols reach similar low concentration levels at ~ 5 km, the general region of natural haze layers (Junge, 1963). Although Hunten and Godson (1967) have raised the possibility of some minor sea salt particle transport to the 90 km level, it is generally conceded that the transport of tropospheric aerosols having radii >0.1 μm, to altitudes above the tropopause, is minimal. As a result of Hadley cells and tropical upwelling air masses, as well as the occurrence of high altitude thunder storms with attendant large vertical velocity profiles, the possibility cannot be totally dismissed.

In populated areas, aerosols are composed of as much as 40 to 60% sulfates, $(NH_4)_2SO_4$ presumably being the most abundant single sulfur compound. Other major inorganic components of tropospheric aerosols are silicates, oxides of Al, Fe, Mn and Mg, $CaCO_3$, and components of sea salt including K, Na, their chlorides, and to a lesser extent carbonates, and bromides. Organic compounds comprised of the photochemical reaction products of α and β terpenes, which emanate from large forest areas, are also other constituents of tropospheric aerosols.

3.2. EXTRATERRESTRIAL SOURCES

Aerosols resulting from extraterrestrial sources include cosmic dust and meteorites, meteoritic ablation and fragmentation products, zodiacal cloud dust and possibly particles of solar origin. Giant particles may reach a sufficiently high temperature to melt as they enter and pass through the earth's atmosphere. These give rise to the spherules collected by balloons in the upper atmosphere and also in some ground level sampling (Kuiz, 1962; Zacharov, 1962; Rosen, 1969). Rosen (1969) estimates that only a few percent of the Junge layer is composed of particles which can be classified as spherules. Gadsden (1968) has shown that ablation occurs as meteors enter the earth's atmosphere and that this may give rise to the Na and other metal ion layer at ~ 93 km. These metals undoubtedly produce oxide aerosols and may be the source of Fe in those aerosols which cannot be identified as spherules.

Depending on the type of the meteor, we expect to obtain material with varying amounts of Fe, Al, and Ni compounds, as well as silicates. Except for that fraction of the vaporized metals which is converted to ions, we expect the metals to react rapidly to form the stable (solid) oxides Fe_2O_3, FeO, Al_2O_3, and NiO. Alkali metal aerosols would be expected to undergo additional chemical transformations, a topic to be discussed in a later section of this paper.

3.3. AEROSOLS – 'in situ' FORMATION

Another source of aerosols in the upper atmosphere is production via *in situ* chemical reactions. Reactions that potentially occur on aerosol surfaces may lead to growth but generally not to the formation of new particles. Although these growth processes may increase particle size to a point where electromagnetic wave scattering by the aerosols is appreciable, this aspect of the reaction chemistry will be deferred to Section 4.3. In this section, we consider the reactions leading to the formation of new particles and therefore concentrate on gas-phase reactions (or ones leading to a vaporized product) followed by nucleation; depending on the conditions, the latter may be either homogeneous or heterogeneous. There are five broad classes of reactions which are known, on the basis of laboratory studies, to lead to particle formation. We consider as examples only those of potential importance in the atmosphere above the tropopause.

3.3.1. *Reactions Leading to Products Which Undergo Homogeneous Nucleation*

Consider a metal compound MX reacting with oxygen to form a vapor oxide product which is relatively nonvolatile and readily undergoes homogeneous nucleation

$$MX + \tfrac{n}{2}O_2 \rightarrow MO_n(g) + X(\cdots) \tag{1}$$
$$MO_n(g) \rightarrow MO_n(s). \tag{2}$$

We may write an overall rate expression for the formation and nucleation of MO_n. If the rate of formation is rapid enough so that supersaturation ratios $\geqslant 6$ (with respect to the condensed phase) develop faster than the product gas dissipates by convective and molecular diffusion, an aerosol will be generated. In the case of meteors containing sulfides and carbides, the reaction may occur at the surface of the incoming body, leading to the formation of an oxide. At high temperatures, the oxides may vaporize and recondense to form aerosols. In other cases, the metals directly vaporize upon heating and oxidation reactions occur in the vapor phase. For those constituents of the meteorites which are already in the highest oxidation state, the oxide may evaporate by ablation and only Step 2 is operative.

The metal atoms which form following ion neutralization at 93 km may also undergo oxidation reactions followed by nucleation, the important oxidizing species being O_3 (see Rundle, 1971). Other potential reactions leading to products which readily nucleate include

$$NH_3 + HNO_2 + M \rightarrow NH_4NO_2 + M \tag{3}$$

and

$$NH_3 + HNO_3 + M \rightarrow NH_4NO_3 + M; \tag{4}$$

M designates a third body.

3.3.2. *Chemical Reaction Followed by Heteromolecular Nucleation*

Sulfuric acid is an important component of atmospheric aerosols and is a major component of the stratospheric Junge layer (Lazrus *et al.*, 1971). The first step in the formation of this aerosol is the oxidation of SO_2 (or perhaps H_2S) to SO_3, followed by a process which we term 'heteromolecular' nucleation. Several mechanisms have been proposed as important for the oxidation of SO_2 to SO_3 (see Berry and Lehman, 1971; Bufallini, 1971).

It has generally been considered that the three body oxidation of SO_2 by O would be of most importance in the upper atmosphere (Cadle and Powers, 1966):

$$SO_2 + O + M \rightarrow SO_3 + M. \tag{5}$$

The reaction rate constant is 7.4×10^{-33} cm^6 molecule^{-2} s^{-1} (Mulcahy *et al.*, 1967). Other proposed mechanisms include

$$SO_2 + h\nu \left(^{3840}_{2940} \text{ Å}\right) \rightarrow {}^1SO_2^* + [O_2] \rightarrow {}^3SO_2^* \text{ (longlived)}$$
$$+ O_2 \left(^3\Sigma, {}^1\Delta, {}^1\Sigma\right) \tag{6}$$
$${}^3SO_2^* + O_2 + M \rightarrow SO_4 + M \tag{7}$$
$$SO_4 + O_2 \rightarrow SO_3 + O_3. \tag{8}$$

Another potentially important mechanism is

$$SO_2 + HO_2 \cdot \rightarrow SO_3 + OH. \tag{9}$$

Irrespective of the SO_2 oxidation mechanism, the next stage involves heteromolecular nucleation via the formation of sulfuric acid and associated cluster species:

$$SO_3 + H_2O + M \rightarrow H_2SO_4 + M \tag{10}$$

and

$$nH_2SO_4 + mH_2O + M \rightarrow (H_2SO_4)_n(H_2O)_m \text{(aerosol)} + M. \tag{11}$$

It has also been suggested that heteromolecular nucleation may occur as a result of water vapor clustering about free radicals such as OH, H, or about H_2O_2 (Allen and Kassner, 1969; Clark and Noxon, 1971; Burke, 1972). Experiments performed by

Clark and Noxon (1971) showed that aerosols form upon exposing He, containing very low partial pressures of water vapor, to UV radiation. Presumably similar conditions might exist in the upper atmosphere.

3.3.3. *Ion Cluster-Switching Reactions*

Ferguson and Libby (1971) have suggested that, under some conditions, atmospheric nitrogen fixation may occur as a result of ion-cluster switching reactions. Mohnen (see Coffey and Mohnen, 1972) has identified at least two rapid switching reactions which are of potential importance in aerosol formation processes. These include

$$H_3O^+ (H_2O)_n + NH_3 + M \rightarrow NH_4^+ (H_2O)_{n+1} + M \tag{12}$$

followed by

$$NH_4^+ (H_2O)_m + HNO_3 \rightarrow NH_4NO_3 (H_2O)_n + H_3O^+ (H_2O)_{m-n} \tag{13}$$

$$+ HCl \rightarrow NH_4Cl(H_2O)_n + H_3O^+ (H_2O)_{m-n}. \tag{14}$$

It is uncertain whether these aerosols are formed by growth with H_2O vapor or directly by homogeneous nucleation, but these potential mechanisms clearly warrant further attention.

3.3.4. *Nucleation About Ions*

Strictly speaking, this heterogeneous nucleation process is physical rather than chemical. Nevertheless, since the formation of the prenucleation ion clusters involves chemical reaction, and because of its potential importance in some upper atmospheric aerosol formation mechanisms (Witt, 1969), we will discuss the phenomena in this section. The role of small clusters in nucleation about ions has long remained obscure and the evidence was conflicting as to whether nucleation proceeds from the ion-cluster stage or ions merely stabilize pre-existing neutral clusters. Recently, Castleman and Tang (1972) resolved the conflict and established the fact that the small ion cluster distribution is a segment of the overall nucleation size spectrum.

On the basis of earlier observations, it had been conjectured that the difference in the magnitude of the barrier height to nucleation was due to the sign of the ionic charge. The recent work of Castleman and Tang (1972) has shown that the barrier height is instead governed by central ion-ligand interactions and may differ even for ions of like sign. Experimental results show that hydrated protons (oxonium ions) have one of the lowest nucleation barriers and should lead to nucleation at lower supersaturation ratios than required for most other ion-cluster complexes. Tang and Castleman (1972) found that monovalent lead ions, which are progeny of radon in the atmosphere, are more stable for the case of the first two clusters than those for the closed shell alkali metal ions of comparable size. A study of the cluster stability of other non-closed shell metal ions such as Fe and Al is in progress in our laboratory.

The potential role of ion clusters in upper atmospheric nucleation processes will be discussed in Section 4.2.

3.3.5. *Addition and Polymeric Reactions*

Several authors (see Mohnen, 1972) have observed the occurrence of a direct reaction between NH_3 and SO_2. Work is currently in progress (Mohnen, 1972) to measure the rate constants and conditions under which the reaction products are stable. Preliminary evidence suggests that the product is polymeric species of the form $[(NH_3)_2SO_2]_n$.

Other polymeric species can result from photochemical reactions involving organic compounds. CH_4 will react with OH radicals or $O(^1D)$ to form $CH_3\cdot$ and water. Tang and Castleman (1970) have shown that the relative rate constants for the further oxidation of $CH_3\cdot$ to $CH_3O_2\cdot$ and to H_2CO, are approximately comparable at third body concentrations of $\gtrsim 10^{18}$ molecules cm^{-3}; at lower number densities the reaction may favor the formation of H_2CO. Subsequently, this product molecule will largely react or decompose to form HCO, $H\cdot$, CO, CO_2, OH, and HO_2. In regions of high concentration, polymeric species could form and may in fact be the products observed in some laboratory experiments involving aerosol formation in the presence of UV and ionizing radiation (Castleman *et al.*, 1971). Although these reactions are unlikely to be of major importance in the atmosphere, they should be considered in interpreting the results of simulation laboratory experiments where high concentrations of $CH_3\cdot$ radicals can form.

4. Aerosol Layers Above the Tropopause

Aerosols exist at ground level in concentrations varying from $\sim 10^3$ in 'clean' air to $\sim 10^5$ in city air, the concentration being a balance between formation, coagulation, and removal. The first systematic study of change in aerosol concentration with altitude was reported by Junge *et al.* (1961) on the basis of direct measurement. Their results showed the expected continuous decrease in aerorsol concentration up to and through the tropopause.

Unexpected, however, was their finding of a globally distributed persistent stratospheric aerosol layer having a rather broad maximum around 20 km. Although the concentration of this layer exhibits some spatial and time variations, it is known to have existed as a relatively well defined layer since at least 1959. This first layer above the tropopause is referred to as the 'Junge layer.'

As shown by Bigg *et al.* (1970), the influence of this layer on stratospheric aerosol concentration persists up to ~ 37 km. On the basis of twilight measurements, Volz (1970) concluded the layer reaches highest altitudes in the tropics, to elevations of at least 25 km. Using optical radar techniques (Lidar), Schuster (1970) showed the existence of significant quasi-stable layers between 25 and 40 km, all presumably associated with the Junge layer.

There is a dearth of accurate information on higher altitude layers; recent data show considerably less aerosol concentration at all altitudes than had been inferred on the basis of earlier measurements. The early twilight observations of Volz and

Goody (1962) indicated a more or less uniform mixing ratio (gram of aerosol per gram of air) at about 25% of its 15 km value for altitudes above 30 km. There was some evidence of a weak secondary maximum near 50 km, the vicinity of the stratopause. Although the existence of this 'layer' is questionable, the data do show the presence of aerosols at all elevations up to 65 km. The particle concentration at 60 km was found to be 4×10^{-4} smaller than at 15 km, assuming size distribution to be relatively independent of height.

Using Lidar, Fiocco and Smullins (1963) obtained echoes localized at heights below 100 km and again between 110 and 140 km. It is well known that noctilucent clouds appear in extreme northern latitudes in summer months. These have been observed at heights varying from ~ 76 to 85 km but usually occur between 80 and 85 km (Reiter, 1971). It is generally agreed that their location is associated with the mesopause where the lowest atmospheric temperature of 140 K occurs during the summer months.

In the summer, during noctilucent cloud displays, Fiocco and Grams (1969) found appreciable quantities of particulate matter in the 60 to 70 km region. The measurements, which were made using Lidar, indicated a maximum in the interval 66 to 68 km. The concentration was calculated to be $\sim 0.2 \, \text{cm}^{-3}$ assuming the aerosols to be of 0.15 μm radius with a refractive index of 1.5. No return signal was obtained at altitudes above 90 km. Fechtig *et al.* (1968) made direct rocket collections of material at altitudes from 65 to 145 km. The data show that the flux of cosmic microparticles is smaller by three orders of magnitude than previous satellite and rocket measurements had indicated.

The use of in-flight shadowing techniques has been of considerable value in establishing the origin of collected particles. As a result of improved measurement techniques, it has now been established that earlier reported concentrations were several orders of magnitude (high) in error. Recent measurements by Hallgren *et al.* (1972) showed the particle concentrations to be $\leqslant 6 \times 10^{-3} \, \text{cm}^{-3}$ when directly sampling a noctilucent cloud and only $\sim \frac{1}{2} \times 10^{-3} \, \text{cm}^{-3}$ at other times. It should be noted that these concentrations were based on rather poor statistics, total numbers of particles being only 8 to 89 on the collection surfaces. Using similar direct rocket sampling techniques, Skrivanek (1972) (also see Chrest *et al.*, 1972) has failed to detect any particles directly attributable to collections at these altitudes, either in the presence or absence of noctilucent cloud displays. The sampling procedure is valid for collecting particles at least as small as 0.1 μm and should be applicable for particles down to 0.05 μm in size. Photometric instrumentation, also onboard the rockets, clearly indicated the presence of particles in the cloud displays of sufficient concentration for easy collection.

As discussed in a later section, one of the models for explaining the diurnal and seasonal variations in the Na airglow phenomena is based on the existence of a Na containing aerosol layer at ~ 93 km. There appears to be no direct evidence for such a layer at this altitude, but some aerosols have been detected in this region. This is also the approximate region where meteors undergo oxidation, undoubtedly giving rise to some oxide aerosols and probably fragmentation products. Twilight measurements by Volz and Goody (1962) provide strong evidence against the presence of aerosol layers between 100 and 150 km that have been speculated by others. Likewise, the rocket sampling measurements of Fechtig *et al.* (1968) contradict earlier high flux values and also do not indicate a dust belt around the earth.

Particles from extraterrestrial sources such as zodiacal clouds and other interplanetary dust may be captured by the earth's atmosphere. On the other hand, depending on their injection velocity and size, the particles may continue to orbit and eventually spiral into the sun. Schmidt and Elsässer (1967) have indicated that the earth's gravitational and magnetic field may trap solar-orbiting particles, thus leading to an increase in particle concentration within the earth's magnetosphere. Several authors (see Reiter, 1971) have suggested a geocentric condensation of interplanetary dust, which should increase with decreasing particle size. Although this was indicated by early rocket measurements, recent rocket and satellite observations did not confirm the existence of a geocentric 'dust belt.' Bandermann and Singer (1969) have suggested that the 'dust belt' reported on the basis of early measurements might have been a consequence of the measurement techniques employed. The giant particles which do pass through these upper altitudes will be considered a transient phenomenon and not be considered in discussing aerosol chemistry.

Next, we consider the implications of the foregoing reactions in establishing the chemistry of the aerosols known to exist in the upper atmosphere. We confine ourselves to the three regions of primary interest.

4.1. JUNGE LAYER

The stratifications in the region of the Junge layer exist between 17 and 40 km, their relative concentration varying with time but not in a regular seasonal fashion (Reiter, 1971). The particle sizes appear to be made up of three distinct distributions being those <0.1 μm, those with sizes between 0.1 and 5 μm, and those with particle sizes above 5 μm (Rosen, 1969). It has been implied that particles $\leqslant 0.1$ μm in size may be of terrestrial origin. The particles with diameters greater than 5 μm are of less importance and may represent meteoritic material, their concentration being only a few percent.

Measurements by Junge *et al.* (1961) showed the effective average particle size of stratospheric aerosols to be ~ 0.3 μm diameter, and the most abundant single chemical component to be sulfate. This fact was subsequently confirmed by Friend (1966) and Shedlovsky and Paisley (1966). Although early measurements indicated that the sulfate ions were chemically combined with ammonium ions, subsequent studies by Manson *et al.* (1961) showed that this conclusion must be viewed with caution since handling procedures often introduce ammonia as contamination, leading to the production of ammonium sulfate from sulfuric acid droplets.

More recent measurements by Cadle *et al.* (1970) have shown that particles collected in both the tropical and mid-latitude stratosphere contain sulfate, Si, Na, Cl, nitrate, Mn, and Br, but no K or nitrite; sulfate ions being the major single component.

Of the stratospheric particles collected in the upper mid-latitudes, a maximum average of 13.4% of the sulfate was chemically combined as ammonium sulfate. Samples from the tropic and lower mid-latitudes failed to show any ammonium content. Additional measurements were made by Lazrus *et al.* (1971) to determine the percentage of stratospheric sulfate present as ammonium sulfate at different locations and altitudes. The data showed that the percentage at 18 ± 1 km elevation varied between 0 and 38.5%, but no trend with global latitude was readily apparent.

On the basis of morphologic observations, Bigg *et al.* (1970) concluded that the percentage of the sulfate existing as ammonium sulfate increased somewhat with altitude. The data of Lazrus *et al.* (1971) averaged and recalculated by the present author, showed the respective percentages at 21 km to vary from 20 to $>100\%$. At 24 km the percentage was 18.7% while at 27 km it was $>100\%$. Presumably, where the percentage exceeded 100%, the remainder was bound with the chloride anion.

Sulfur compounds, such as SO_2 and H_2S, may enter the stratosphere by three major processes. These include transport in tropical upwellings and major storms, eddy diffusional processes, and direct injection by pyroclastic volcanoes. It is a well known fact that occasional volcanic eruptions input material into the stratosphere. Cronin (1971) has suggested that this may be the major source of sulfate in the Junge layer region. The concentration of aerosols in the Junge layer is known to have increased following the Mt. Agung eruption, the major volcanic event during the last decade. Nevertheless, until the recent work of Castleman *et al.* (1972) a direct correlation between volcanic input and changes in stratospheric sulfur concentration was not available.

A clear correlation between sulfate concentration and the Mt. Agung eruption is seen in Figure 1. Particles have an approximate stratospheric half-life of 1 to $1\frac{1}{2}$ yr and the semi-logarithmic plot of concentration vs. time shows the expected rate of decrease in concentration.

A study of isotopic ratios for the S^{32} to S^{34} isotopes provides additional evidence for the sources of sulfate. The symbol del, δ, is defined by

$$\delta = \left(\frac{[\text{ratio } S^{32}/S^{34}] \text{ standard}}{[\text{ratio } S^{32}/S^{34}] \text{ sample}} - 1 \right) \times 10^3 .$$

Values are indicative of both source and chemical changes, the latter generally due to the dependence of reaction rate on mass. Isotopic ratios for sea salt sulfates invariably average approximately $+15$, while those from volcanoes were found to average approximately $+2$.

A mass and isotopic balance for data obtained from samples collected in both the northern and southern hemispheres from 1962 to 1970, suggest that the major contribution to the sulfate layer is via volcanic activity, but some contribution from upwelling air is apparent. The SO_2 and H_2S evolved during explosive events are injected directly into the stratosphere and subsequently react, probably via the reaction in Equation (5). The possibility that direct reactions between NH_3 and SO_2 play some

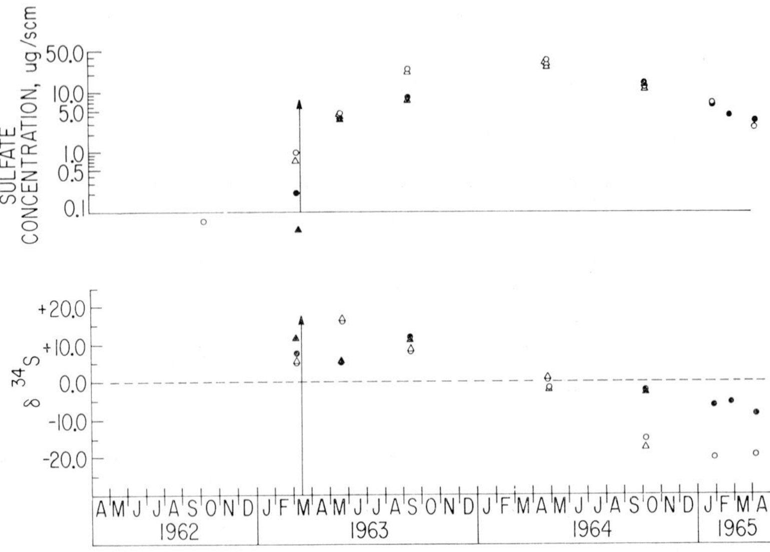

Fig. 1. Plot of stratospheric sulfate concentration (μg m^{-3}) and δ values for the southern hemisphere, 17–43° south latitude, showing changes resulting from the eruption of Mt. Agung. The eruption date is indicated by the vertical arrow. Closed data points represent samples taken at altitudes between 15.2–16.7 km, open data points 16.7–19.8 km.

role at the low temperature of the stratosphere cannot be totally dismissed and awaits further research. In any event, as the reactants diffuse upward, the reaction proceeds, leading to a product with an ever-decreasing δ value with altitude. Since the upper layer particles slowly settle, the del value of the sampled material would be expected to accordingly decrease for some period of time long after the eruption. This trend was observed in the data collected after each major volcanic eruption. In the absence of major volcanic events, the δ value of the Junge layer returns to an average of approximately $2\frac{1}{2}$. The corresponding stratospheric sulfate concentrations return to a low value, the latter presumably being formed from S compounds introduced by upwelling air masses and by diffusion.

Following conversion to SO_3, heteromolecular nucleation results in particle formation, the product being largely sulfuric acid. The H_2SO_4 subsequently reacts with the NH_3 diffusing into the stratosphere, thereby leading the to partial conversion of the acid molecules to $(NH_4)_2SO_4$. In the case of the volcanic inputs, NH_3, H_2S, and SO_2 are injected together, probably leading to a larger percentage of the sulfate combined as ammonium sulfate. It had originally been assumed that SO_2 would not survive in upwelling air as it approached the tropopause, but would convert to SO_3 and subsequently nucleate. The resulting particles would be retained in the lower atmosphere. Reassessing the extent of the reaction in Equation (5) in terms of recently determined rate constants and re-evaluated time estimates for upward transport by eddy diffusion, we conclude that substantial fractions of the SO_2 may reach altitudes near the Junge layer before being completely converted to SO_3.

4.2. NOCTILUCENT CLOUD LAYER

The recent findings of low particle concentrations in the vicinity of 85 km, even during noctilucent displays, suggest that the phenomena may not be totally due to pre-existing particles. Two possibilities must, therefore, be considered: nucleation on particles arising from extraterrestrial sources (as they pass downward through this region of minimum temperature) and nucleation about ions or free radicals.

Hemenway *et al.* (1972a, b) have attempted electron microprobe analyses of some of the particles collected during noctilucent displays. Since the collected particle number is far below the background contamination on the collection surfaces, exact particle concentrations and composition are very difficult to deduce. Morphologically, the particles appear to be composed of two components, a dense core surrounded by a deformable coating. In no case has a definite identification been possible, but certain spectral lines have been detected which allow a list of possible elements to be selected. In each case the possible elements comprised largely the heavy elements from Groups II–VIII. Since these elements have intensities comparable in magnitude to the usual meteoritic components (Al, Si, and Fe) it has been suggested by the authors that these particles are of solar origin. If the heavy elements had arisen from meteoritic origin, their concentration should be orders of magnitude below that of the usual components.

Hemenway *et al.* (1972a, b) and Fechtig (1972) have suggested that the constant influx of particles from the sun may lead to water nucleation as they pass through the region of the mesopause during times when its temperature is at the minimum value. On the other hand, photometric devices onboard the rockets showed the particle concentrations in noctilucent displays to exceed measured particle concentrations by 2 to 3 orders of magnitude (Fechtig *et al.*, 1968; Chrest *et al.*, 1972; Hallgren *et al.*, 1972). This has led Witt (1969) to suggest the possibility that noctilucent clouds occur as a result of nucleation on ions.

Our results (Castleman and Tang, 1972) show that the barrier for nucleation about H_3O^+ is the lowest of any of the positive ion clusters reported to date. Oxonium ions and associated hydrated clusters are known to exist in the vicinity of the mesopause and nucleation about these must be considered as a possible mechanism for noctilucent cloud formation. A precise evaluation awaits the resolution of the classical-non-classical theory of nucleation and a theory for calculating the values of supersaturation required for nucleation about specific ions.

Witt (1969) has raised the interesting question of whether stable ligand structures involving $Fe(H_2O)_6^+$ may play a role in these processes. Preliminary results from our laboratory indicate that hydrated Fe^+ clusters are comparable in stability to those of Pb^+; however, Al^+ clusters have unusual stability. Another possibility is that negative ions result in nucleation, perhaps as a result of their formation following electron showers. Far less is known about the properties of negative ion clusters, but Mohnen (1971) has calculated that hydrates of NO_3^- should predominate up to altitudes at least as high as 60 km.

A further possibility is that water nucleation is induced by a free radical such as OH. This possibility awaits further research to clarify the importance of these later processes. At this time these processes appear to be far less likely than nucleation about ions or condensation about incoming particles.

4.3. THE Na LAYER IN TERMS OF AN AEROSOL MODEL

Here we consider only those aspects of Na chemistry which have a potential bearing on the aerosol model as an explanation of the Na layer. Other aspects of alkali metal chemistry are covered elsewhere in this volume.

TABLE I

Free energies for reactions involving Na aerosols

Reaction	$-\Delta G$, kcal/mole				
	150K	190K	200K	250K	300K
$Na_2O(s) + H_2O(g) = 2NaOH(s)$	44.30	42.66	42.21	40.78	37.24
$2NaO(g) + H_2O(g) = NaOH(s) + \frac{1}{2}O_2(g)$	157.47	153.89	152.96	149.02	143.03
$2NaO_2(s) + H_2O(g) = 2NaOH(s) + \frac{3}{2}O_2(g)$	21.00	22.23	22.43	22.49	22.64
$2Na_2O_2(s) + H_2O(g) = 2NaOH(s) + \frac{1}{2}O_2(g)$	21.80	21.68	21.54	20.90	20.26
$2NaOH(s) + CO_2(g) = Na_2CO_3(s) + H_2O(g)$	25.71	25.27	25.27	25.27	25.26
$Na_2SO_4(s) + CO_2(g) + H_2O(g) = 2NaCO_3(s) + H_2SO_4(g)$	-44.21	-45.69	-45.31	-46.68	-48.04
$Na_2CO_3(s) + 2HNO_3(g) = 2NaNO_3(s) + CO_2(g) + H_2O(g)$	44.06	43.21	43.01	43.83	44.65
$Na_2SO_4(s) + 2HNO_3(g) = 2NaNO_3(s) + H_2SO_4(g)$	$-$	-2.48	$-$	$-$	-3.39

Calculated on basis of data taken from (1) JANAF Tables, and (2) NBS Circular 500.

In order to understand the possible interconversions of Na compounds on the surfaces of aerosols, we compiled the thermodynamic data given in Table I. One of the major gas-phase reactions in the aerosol model for the Na layer involves the conversion of Na to NaO via the reaction

$$Na + O_3 \rightarrow NaO + O_2 \cdot \tag{15}$$

Consider first the formation of sodium oxide aerosols and their stability in the presence of other constituents of the upper atmosphere. For the temperatures and oxygen partial pressures existing at 93 km, the stable oxide (solid) phases are Na_2O_2, NaO_2, and Na_2O. However, a consideration of the H_2O partial pressures existing at 93 km shows that at least the surface layers of these oxides would readily convert to NaOH. Further chemical conversion of the bulk particle would proceed, but at a slow rate due to the resistance to oxygen diffusion presented by the outer oxide layers. At the expected CO_2 concentration level, NaOH would further convert to

Na_2CO_3, the more stable solid phase. Additional calculations show that for HNO_3 partial pressures as low as 1.20×10^{-38} atm, conversion to $NaNO_3$ results. The carbonates and nitrates are quite stable and it is unlikely that a reversible chemical cycle involving these compounds can be found, which will produce the observed gas phase Na concentration.

Although Junge *et al.* (1962) have discounted the possibility that NaCl aerosols can be carried to high altitude, Hunten and Godson (1967) have considered additional transport mechanisms which may alter the earlier conclusions. Hunten and Wallace (1967) have suggested that the decomposition of these aerosols may give rise to the Na atoms. Employing appropriate thermodynamic values, we have considered the temperature to which NaCl particles would have to be heated in order to produce $\sim 10^3$ atoms of Na cm^{-3} via combined dissociation and oxidation mechanisms. The result (Figure 2) shows that the aerosol particles would have to be heated to greater than 700 K. Reasonable energy balance calculations indicate that the particles may attain a temperature of 400 to 500 K. It may be possible to invoke some additional photodissociation mechanism, but it seems unlikely that it will give rise to the observed Na concentration.

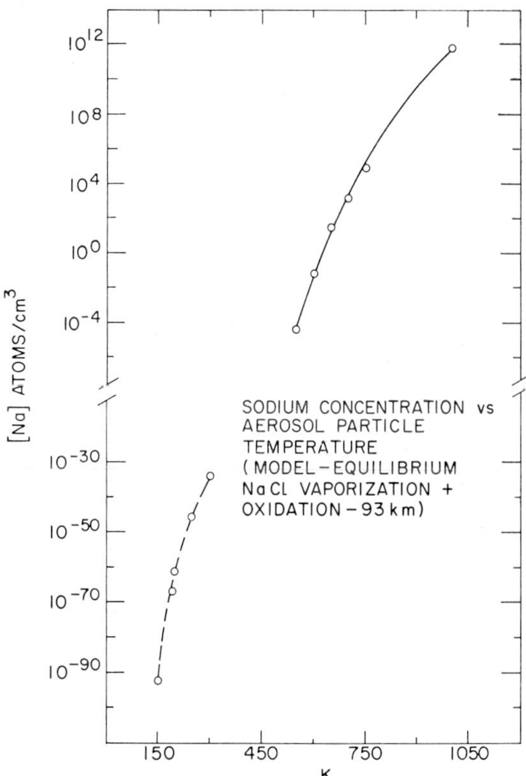

Fig. 2.   Plot of gas phase Na concentration (atoms cm^{-3}) as a function of temperature, K. Calculated on basis of an equilibrium model accounting for NaCl dissociation and Na oxidation.

One additional consideration also indicates the unlikely possibility that aerosols provide the reversible sink for the Na layer which has been suggested in the literature. Since at ~ 93 km, the Knudsen number is $\geqslant 10$ for all aerosol sizes, the maximum removal rate of sodium atoms may be calculated from molecular theory. Employing the measured aerosol number concentration, our calculations indicate a Na atom half-life of $\sim 5 \times 10^7$ s, even assuming a sticking coefficient of unity. This is a rather long half-life to expect that the aerosols will provide both a source and sink for Na atoms. Apparently, other models will have to be proposed to explain the observed Na concentrations.

Acknowledgments

The author acknowledges the invaluable assistance of his colleagues H. R. Munkelwitz, Dr Wm. Wood, and I. N. Tang. He also thanks Bernard Manowitz for his encouragement in undertaking the Junge layer studies, and Dr L. Newman and J. Forrest for performing the del value analyses. The assistance of Philip Krey, AEC-HASL, and individuals in the DOD in obtaining appropriate high altitude samples, is also gratefully acknowledged. This work was performed under the auspices of the U.S. Atomic Energy Commission and was partially financed by the Environmental Protection Agency under an interagency agreement.

References

Allen, L. B. and Kassner, J. L. Jr.: 1969, *J. Colloid Interface Sci.* **30**, 81.
Bandermann, L. W. and Singer, S. F.: 1969, *Rev. Geophys.* **7**, 759.
Berry, R. S. and Lehman, P. A.: 1971, in *Ann. Rev. Phys. Chem.* **22** (ed. by H. Eyring, C. J. Christensen and H. S. Johnston) Ann. Reviews Inc., p. 47.
Bigg, E. K., Ono, A., and Thompson, W. J.: 1970, *Tellus* **22**, 550.
Bufallini, M.: 1971, *Env. Sci Tech.* **5**, 685.
Burke, R. R.: 1972, *J. Colloid Interface Sci.* **38**, 660.
Cadle, R. D. and Powers, J. W.: 1966, *Tellus* **18**, 176.
Cadle, R. D., Lazrus, A. L., Pollock, W. H., and Shedlovsky, J. P.: 1970, *Proc. Symp. Tropical Meteorol.* **KIV**, 1-7.
Castleman, A. W. Jr., Munkelwitz, H. R., and Manowitz, B.: 1972, in preparation.
Castleman, A. W. Jr. and Tang, I. N.: 1972, *J. Chem. Phys.* **57**, 3629.
Castleman, A. W. Jr., Tang, I. N., and Munkelwitz, H. R.: 1971, unpublished results.
Chrest, S. A., Carnevale, R. F., Ryan, T. G., Skrivanek, R. A., Wilhelm, N., and Witt, G.: 1972, Rocket-Borne Particle Collection and Photometric Measurement of Noctilucent Clouds, to be published.
Clark, I. D. and Noxon, J. F.: 1971, *Science* **174**, 941.
Coffey, P. and Mohnen, V. A.: 1972, *Bull. Am. Phys. Soc.* **17**, 392.
Cronin, J. F.: 1971, *Science* **172**, 847.
Fechtig, H.: 1972, *Space Res.* **12**, in press.
Fechtig, H., Gerloff, U., and Weinrauch, J. H.: 1968, *J Geophys. Res.* **73**, 5029.
Ferguson, E. E. and Libby, W. F.: 1971, *Nature* **229**, 37.
Fiocco, G. and Smullins, L. D.: 1963, *Nature* **199**, 1275.
Fiocco, G. and Grams, G.: 1969, *J. Geophys. Res.* **74**, 2453.
Friend, J. P.: 1966, *Tellus* **18**, 465.
Gadsden, M.: 1968, *J. Atmospheric Terrest. Phys.* **30**, 151.
Hallgren, D. S., Schmalberger, D. C., and Hemenway, C. L.: 1972, Paper e-14 COSPAR, Madrid, Spain.

Hemenway, C. L., Hallgren, D. S., and Schmalberger, D. C.: 1972a, *Nature*, in press.

Hemenway, C. L., Erkes, J. W., Greenberg, J. M., Hallgren, D. S., and Schmalberger, D. C.: 1972b, Paper e-16 COSPAR, Madrid, Spain.

Hunten, D. M. and Godson, W. L.: 1967, *J. Atmospheric Sci.* **24**, 80.

Hunten, D. M. and Wallace, L.: 1967, *J. Geophys. Res.* **72**, 69.

Junge, C. E.: 1963, *Air Chemistry and Radioactivity*, Academic Press, New York and London.

Junge, C. E., Chagnon, C. W., and Manson, J. E.: 1961, *J. Meteorol.* **18**, 81.

Junge, C. E., Oldenberg, O., and Wasson, J. T.: 1962, *J. Geophys. Res.* **67**, 1027.

Kuiz, Z.: 1962, in K. Spurny (ed.), *Aerosols, Physical Chemistry and Applications*, Gordon and Breach, New York, p. 621.

Lazrus, A. L., Gandrud, B., and Cadle, R. D.: 1971, *J. Geophys. Res.* **76**, 8083.

Lindauer, G. C. and Castleman, A. W. Jr.: 1971, *J. Aerosol Sci.* **2**, 85.

Link, F.: 1973, this volume, p. 34.

Manson, J. E., Junge, C. E., and Chagnon, C. W.: 1961, *J. Meteorol.* **18**, 139

Mohnen, V. A.: 1971, Pageoph **84**, 141.

Mohnen, V. A.: 1972, *Atmospheric Sciences Research Center, Pub. No. 165*, (Kushnir, J. M., Malkin, H. I., Mohnen, V. A., Yencha, A. J., and McLaren, E. H.), State Univ. of New York at Albany.

Mulcahy, M. F. R., Steven, J. R., and Ward, J. C.: 1967, *J. Phys. Chem.* **71**, 2124.

Reiter, E. R.: 1971, *Atmospheric Transport Processes, Part 2, Chemical Tracers*, AEC Critical Review Series, U.S. Atomic Energy Commission, Division of Technical Information, Washington, D.C.

Rosen, J. M.: 1969, *Space Sci. Rev.* **9**, 58.

Rundle, H. N.: 1971, in B. M. McCormac (ed.), *The Radiating Atmosphere*, D. Reidel Publishing Company, Dordrecht-Holland, p. 90.

Schmidt, T. and Elsässer, H.: 1967, in Report NASA-SP-150, National Aeronautics and Space Administration, p. 287.

Schuster, B. G.: 1970, *J. Geophys. Res.* **75**, 3123.

Shedlovsky, J. P. and Paisley, S.: 1966, *Tellus* **18**, 499.

Skrivanek, R. A.: 1972, private communication.

Tang, I. N. and Castleman, A. W. Jr.: 1970, *J. Phys. Chem.* **74**, 3933.

Tang, I. N. and Castleman, A. W. Jr.: 1972, *J. Chem Phys.*, **57**, 3638.

Volz, F. E.: 1970, *J. Geophys. Res.* **75**, 1641.

Volz, F. E. and Goody, R. M.: 1962, *J. Atmospheric Sci.* **19**, 385.

Witt, G.: 1969, *Space Res.* **9**, 157.

Zacharov, I.: 1962, in K. Spurny (ed.), *Aerosols, Physical Chemistry and Applications*, Gordon and Breach, New York, p. 613.

ALKALI CHEMISTRY PROBLEMS OF
THE UPPER ATMOSPHERE

G. KVIFTE

Dept. of Physics, Agricultural University of Norway, Aas, Norway

1. Introduction

A vast amount of experimental data has been collected about the alkali metals in the upper atmosphere since the first observation of a yellow radiation in the nightglow was reported by Slipher in 1929, a radiation which turned out to originate from Na and to be located within the atmosphere of the earth. Summaries of the early history of Na are given by Kvifte (1953) and Chamberlain (1961). Lithium was reported observed in twilight almost simultaneously by Delannoy and Weill (1958), and Gadsden and Salmon (1958). Sullivan and Hunten (1964) were the first to furnish proof of the existence of K in the earth's upper atmosphere.

The first serious attempt to treat qualitatively the observational results of Na as a chemical problem was done by Chapman (1939). Later authors have supplemented and quantified his work on the basis of new experimental evidence.

Reviews of experimental status and theoretical interpretations have been given by Chamberlain (1961), Hunten (1967, 1971), Gadsden (1967), Noxon (1967), Llewellyn and Evans (1971), and Kvifte (1972).

Before discussing the chemistry of the alkali metals, a theoretical summary of the experimental results will be given.

2. Observational Results

2.1. Sodium

The radiation from Na in the upper atmosphere is by far the most studied of the alkali metals, and the techniques for evaluating atomic contents and height distribution of neutral atoms from the resonance radiations from twilight and dayglow are well established thanks to the pioneering work of the groups of Chamberlain and Hunten (see Chamberlain, 1961) and around Blamont (Gadsden *et al.*, 1966).

There are several significant observations on which, it seems, everybody agrees.

2.1.1. The column number density is around 5×10^9 neutral atoms cm^{-2}, varying up and down by a factor of 10 with time and location. Table I gives monthly means of the Na column density measured from twilight and dayglow observations by ground based instruments and rocket flights. The recent laser radar measurements of nightglow content (Sandford and Gibson, 1970; Gibson and Sandford, 1971) are also included. A seasonal variation is evident in the twilight observations from both hemispheres, especially at higher latitudes, with a main maximum about 1 mo before

TABLE I

Monthly means of Na abundances (in 10^9 cm^{-2}) from Dayglow (D) and Twilight (T)

Station	Lat.		Jan.	Feb.	Mar.	April	May	June	July	Aug.	Sept.	Oct.	Nov.	Dec.	Years	Ref.	Comments
Tromsö	70°	D	10.0	11.7	6.4	6.9									63.70	1	
		T	2.9	5.0	4.0	5.9											
College	65°	D															
		T	5.7		2.2								5.0	3.7	59–60	2	
Aas	60°	D	8.1	10.2								10.1	6.3	8.0	66–70	1	
		T	5.2	4.2								4.9	6.1	4.2			
Ft. Churchill	59°	D							5.1	1.5					64/66	3	Rocket
		T															
Saskatoon	52°	D															
		T	7.7	7.0	4.8	3.3	2.1	1.8	1.8	3.5	4.0	5.7	7.0	6.4	58–63	4	
Winkfield	51°	N	10.8	7.8	4.2	3.9	(2)	(1)	2.0	3.0	3.5	4.1	7.3		69–70	5	Night
Moscow U.S.A.	47°	D															
		T	6.7	7.4	9.7				0.9	2.6	4.3	6.7	8.4	11.0	67–68	6	
Haute Provence	44°	D	13.6	16.2	14.2	11.7	10.8	13.5	14.7	18.7	14.7	16.9	14.4	12.4	61–65	7	
		T	5.8	5.4	5.5	4.0	3.2	2.8	3.1	4.8	4.5	5.5	7.4	6.3	61–65		
Madison	43°	D	3.8	4.4	2.0	2.0			2.5	2.3	5.2	4.3	5.3	5.4	61–62	8	
		T															
Abastuman	42°	D															
		T	2.1	2.4	1.7	1.9	1.1	0.8	1.3	1.3	1.7	1.9	2.5	2.4	62–66	9	
Fritz Peak	40°	D	(2.8)	(3.0)	(3.2)	(3.0)	(1.8)	(1.8)		(1.7)	(1.6)	(1.7)	(2.6)	(2.8)	67–69	10	Rel. values
		T															
Wallops Isl. Virginia	37°	D									16.0				64	11	Rocket
		T															
White Sands	33°	D				10		1.7					3.8		63/64	12, 3	Rocket
		T															
Kitt Peak	32°	D															
		T	5.1	5.0	4.2	4.8	4.5	3.9	4.0	4.6	5.1	5.3	4.0	3.5	64–66	4	
Christchurch	−43°	D															
		T	2.7	4.5	6.4	11.4	14.2	10.7	8.6	7.0	7.6	3.5	2.7	2.5	61–62	13	
Lauder	−45°	D															
		T	1.5	2.4	2.3	2.5	2.6	2.2	2.1	2.9	2.5	1.8	1.5	1.4	62–63	14	

1. Kvifte, Kvifte, Hansen, Holm (unpublished). 2. Rees and Deehr (1962). 3. Ahmed *et al.* (1970) 4. Hunten (1967). 5. Sandford and Gibson (1970); Gibson and Sandford (1971). 6. Graham *et al.* (1971). 7. Gadsden *et al.* (1966). 8. McNutt and Mack (1963). 9. Toroshelidze (1968). 10. Gadsden and Purdy (1970) 11. Donahue and Meyer (1967). 12. Hunten and Wallace (1967). 13. Hunten *et al.* (1964). 14. Gadsden (1964a).

the winter solstice, perhaps a secondary minor one near the spring equinox (at the height of radiation), and a minimum near the summer solstice. The mean abundances vary from maximum to minimum by about 3:1 at mid-latitudes, somewhat less at lower latitudes and somewhat more at higher latitudes. Roughly the same variation is indicated at night through the series from Sandford and Gibson (1970), Gibson and Sandford (1971), and through the mean monthly intensities given in Table II. Here, however, the relative seasonal amplitude is not smaller at lower latitudes as for the twilight number densities. It may even be considerably greater, though the absolute intensity seems to be predominantly greater at higher latitudes. The years over which the different measurement series have been taken do not always overlap, and admittedly other effects (solar cycle, local or global irregularities) may lie masked in the data.

There are some differences of opinion as to the variation with season of the daytime content (Table I). Gadsden *et al.* (1966) find no seasonal variation at Haute Provence, whereas Gadsden and Purdy (1970) give evidence for one at Fritz Peak (in relative measure) of a magnitude of about 2 to 1 from winter to summer. Later observations at Haute Provence (Albano *et al.*, 1970) support this result even if their winter/summer ratio of about 3:1 will have to be modified downwards by taking into account the Ring effect (Noxon, 1968; Hunten, 1970).

It seems to have been generally accepted that the Na content shows a diurnal effect. Hunten (1970), referring to the results of Gadsden and Purdy (1970) ,Albano *et al.* (1970), and Donahue (1969), states that the values for low sun could be less by a factor between 2 and 5 of the maximum near noon. These numbers may well be reduced by a factor of 1.5 or more (Llewellyn and Evans, 1971) by a plausible reduction for the Ring effect. This would bring the magnitude of the diurnal variation more in correspondance with the (almost) simultaneous measurements of dayglow and twilight contents (Hunten and Wallace, 1967) and to the monthly means of daytime and twilight contents from Tromsö, Aas, and Haute Provence. The Tromsö and Aas results have been obtained by a method from Kvifte and Wallace (1970).

Quite recently, Gibson and Sandford (1971) have gotten their radar-laser instrument working in daytime. They find no variation from night to day of the Na content (Thomas, 1972). A diurnal variation of the Na content, therefore, seems questionable.

Few seem to have paid attention to the intensity curves given by Blamont and Donahue (1964, Figure 1 to 3) from Tromsö (1962) indicating a lower Na content just after sunset than both in daytime and in twilight. Similar results have been obtained in Tromsö (1970) with the Kvifte-Wallace (1970) method. It may have some relevance for the discussion of the Na chemistry in connection with O_3.

There is an indication of a latitude variation of the twilight Na content, with a small maximum around 45 to 50° N and a minimum probably around 40° N. This situation seems not to be mirrored on the southern hemisphere. With regard to the daytime content, too few observation series at different latitude have been obtained. The sporadic measurements from rockets give rather divergent results.

Davis and Smith (1965) have reported an observational series of Na nightglow along the 70° W meridian from 40 N to 60° S latitude. They find a small maximum of

TABLE II

Monthly means of Na nightglow intensities (in R)

Station	Lat.	Jan.	Febr.	Mar.	April	May	June	July	Aug.	Sept.	Oct.	Nov.	Dec.	Years	Ref.
Haute Provence	44°	275	220	250	235	210	180	195	180	250	350	405	340	53–58	1
Cactus Peak	36°	255	130	190	160	120	100	70	95	130	200	240	210	48–54	2
Sacramento Peak	33°	20	60	50	50	40	20	10	45	70	75	100	55	58	2
Kitt Peak	32°	35	30	30	50	30	20	15	20	30	55	95	80	64-65	2
Naini Tal	29°	97	66	77	72	51					115	131	120	63–68	3
Mt Abu	25°	44	52	62	64	55	40			72	98	87	73	65–68	4
Tamanrasset	23°	75	105	120	125	112	45	62	82	125	150	200	120	57–58	2
Haleakala	21°	20	25	30	60	60	20	15	20	45	65	102	45	61–65	2
Kerguelen	−49°	65	80	100	105	100	85	95	105	80	105	70	60		5

1. Barbier (1959). 2. Smith and Steiger (1968). 3. Saxena (1970). 4. Rao and Kulkarni (1971). 5. Barbier (1965).

the radiation between 20 and 30° N (summer) and a maximum about twice as large between 40 and 50° S (winter).

2.1.2. The maximum number density occurs typically just above the 90 km level of the atmosphere and is of the order of 10^3 to 10^4 cm^{-3}.

Opinions differ as to how this height of peak concentration varies. Bullock and Hunten (1961) found an apparently significant variation of 3 to 5 km peak to peak, with the maximum in March. Blamont and Donahue (1964) have presented evidence that the height during wintertime is about 5 km higher than in summertime. Observations by Graham *et al.* (1971) support this result, whereas Toroshelidze (1968) states that the wintertime height is lower than the summertime one. The same result was obtained by Gibson and Sandford (1971) from nighttime observations.

Sullivan and Roberts (1968) have found that the difference between the height of maximum concentration measured from evening and from morning twilight varies through the year, being zero at the solstices and about 5 km at the equinoxes.

One obvious comment is that sporadic height variations are of the same magnitude as those given here, and that spurious results are obtained unless a very large number of measurements is included.

A rather striking feature is the very steep topside gradient of the Na distribution. Hunten and Wallace (1967) from their daytime rocket measurements, give a scale height of about 3 km for the Na layer above the height of maximum concentration. The same result is obtained by Donahue and Meier (1967). Gault and Rundle (1969) estimate with refined photometry techniques the twilight topside scale height to be less than 4.5 km, and Sandford and Gibson (1970) give from their laser measurements of the nighttime Na content a scale height between 2 and 3 km for the topside and also for the bottomside of the layer.

2.2. LITHIUM AND POTASSIUM

Considerably less is known about these two alkali metals although they have been extensively studied (see e.g., Hunten, 1967). It has been proven that the atmosphere has been contaminated by artificial injection of Li by atomic bombs and this renders it difficult to determine with certainty the natural column density of this metal. Careful studies by Sullivan and Hunten (1964), and Gault and Rundle (1969) give a probable value of 4 to 5×10^6 cm^{-2} in twilight at wintertime and a seasonal variation roughly of the same relative magnitude as that of Na. Potassium has, according to later corrections (Hunten, 1971), a column density of 3 to 4×10^7 cm^{-2} and shows little, if any, seasonal variation. An attempt of Albano *et al.* (1970) to measure K radiation in dayglow seems to have been negative.

The vertical density distribution profiles of both Li and K are very similar to that of Na, also with respect to the height of maximum density. Gault and Rundle (1969) found the K height to be somewhat lower than the Na height, but Hunten (1971) pointed out that unavoidable uncertainties in the transmission function of the lower atmosphere could be responsible for this difference.

Apart from the great differences in column density, the Na/K/Li ratio being approximately $1000:7:1$, the behavior of the three alkali metals in the atmosphere seems to be fairly similar. Lack of strict conformity in single cases may be attributed to observational difficulties of the fainter radiations or to sporadic local disturbances.

At the present time there seems to be little justification for considering the principal processes governing their distribution in the atmosphere to be radically different. The discussion of such processes therefore will here be restricted to Na, the most studied of the alkali metals.

3. Chemistry

The data deduced from observation of the radiation from alkali metals concern the behavior of neutral atoms and tell nothing about compounds or ionization degree. These metals, on the other hand, are known to be chemically very active and easily ionized, and a great portion of the metal in question may well appear in other forms than neutral atoms. There is one measurement of the concentration of Na^+ with a rocket borne mass spectrometer (Narcisi, 1966), giving a distribution very similar to that of neutral Na and a concentration of about 10% of a typical daytime concentration. Recent mass spectrometer measurements confirm the general shape of the Na^+ distribution. The concentration ratio of Na^+ to Mg^+ is found to be about 1 to 20, giving a maximum Na^+ concentration of 50 to 100 cm^{-3} (Narcisi, 1972, 1973). There seems at the moment to be no possibility of measuring the distribution of the metal compounds, and the discussion of the chemistry of alkali metals in the atmosphere must therefore be based on what is known from laboratory reactions.

The set of reactions needed for the explanation of the daytime and twilight radiation on the one hand and the nighttime radiation on the other, may in principle be different. In the former case it is sufficient that the reactions produce the neutral atom in the ground state (^2S), as the excitation is done by solar radiation, whereas in the latter case the atom must be excited (to the ^2P level) directly by the chemical process or else other excitation processes must be invoked. Approaches on this basis have been done. Another approach could be to try to explain the day and night behavior by the same set of processes, with the addition in daytime of the ionization processes.

A number of authors have studied the photochemistry of Na. Table III gives the reaction processes which have been discussed, their respective rate coefficients k_i (in $cm^3\ s^{-1}$ or $cm^6\ s^{-1}$ for two- or three-body processes), and references to authors. Important processes may still have been overlooked, and the rate coefficients are in most cases rather uncertain. The last three columns of the table contain reaction rates (per unit species itemized in column 6) at the heights 80, 90, and 100 km in the atmosphere. Concentrations of the reactive particles O, O_3, H, and OH have been estimated from the latest studies of Hesstvedt (1971) and Nicolet (1971). The electron concentration [e] is taken from Knecht (1965) and the negative ion concentration $[X^-]$ has arbitrarily been put equal to 0.1 [e].

Chapman (1939) used the reactions 1, 5, and 10 and was able to explain qualitatively the Na layer formation. Hunten (1954) studied the same set of reactions quan-

TABLE III

Reaction processes

Process	No(i)	Ref.	Rate coef. k_i (cm³s⁻¹, cm⁶s⁻¹)	Ref.	Species conc.	Rate per unit species (s⁻¹) 80 km	90 km	100 km
$Na+O+M\rightarrow NaO+M$	1	1	7(−33)	2	[Na]	5(−8)	1(−7)	2(−8)
$Na+O_2+M\rightarrow NaO_2+M$	2	3	2(−33)	2	[Na]	7(−5)	6(−7)	8(−9)
$Na+H+M\rightarrow NaH+M$	3	4	4(−32)	4	[Na]	6(−9)	6(−10)	2(−11)
$Na+O\rightarrow NaO$	4	4	1.2(−14)	4	[Na]	2(−4)	6(−3)	6(−3)
$Na+O_3\rightarrow NaO+O_2$	5	1	6.5(−12)	2	[Na]	5(−4)	3(−4)	3(−5)
$Na+H\rightarrow NaH$	6	4	6(−11)	4	[Na]	2(−2)	2(−2)	6(−3)
$Na+hv\rightarrow Na^+$	7	5	2(−5)	2	[Na]	2(−5)	2(−5)	2(−5)
$Na+Ma\rightarrow Ma'$	8	6	?		[Na]			
$Ma'+Q\rightarrow Na+Ma$	9	6	?		[Ma']			
$NaO+O\rightarrow Na+O_2$	10	1	4(−11)	2	[NaO]	8(−1)	2(1)	2(1)
$NaO+O_3\rightarrow Na+2O_2$	11	8	2(−12)	8	[NaO]	2(−4)	1(−4)	1(−5)
$NaO+hv\rightarrow Na+O$	12	7			[NaO]			
$NaH+OH\rightarrow Na+H_2O$	13	4	2(−15)	4	[NaH]	4(−9)	4(−11)	2(−13)
$NaH+O\rightarrow Na+OH$	14	3	2(−18)	4	[NaH]	4(−8)	1(−6)	1(−6)
$NaH+H\rightarrow Na+H_2$	15	9	4(−17)	4	[NaH]	2(−8)	2(−8)	4(−9)
$NaO_2+hv\rightarrow Na+O_2$	16	7			[NaO_2]			
$NaX^++e\rightarrow Na+X$	17	7			[NaX^+]			
$Na^++X^-\rightarrow Na+X$	18	5	1(−8)	2	[Na^+]	[5(−7)	8(−6)	4(−5)]
$Na^++e\rightarrow Na$	19	5	2(−12)	5	[Na^+]	1(−9)	2(−8)	8(−8)
$Na^++XY\rightarrow NaX^++Y$	20	7			[Na^+]			
$Na^++X+M\rightarrow NaX^++M$	21	10,7			[Na^+]	[As process 1 or 2?]		
$Na^++O_3\rightarrow NaO^++O_2(?)$	22	11	1(−11)	11	[Na^+]	8(−4)	5(−4)	5(−5)
$NaO_2+O\rightarrow NaO+O_2$	23	12	1(−11)	2	[NaO_2]	2(−1)	5(0)	5(0)
$NaO_2+H\rightarrow NaH+O_2$	24	3	3(−12)	7	[NaO_2]	1(−3)	1(−3)	3(−4)
$NaO_2^++e\rightarrow NaO+O$	25	10	3(−7)?	10	[NaO_2^+]	[2(−4)	2(−3)	1(−2)]
$NaO_2^++e\rightarrow Na+O_2$	26	10	3(−7)?	10	[NaO_2^+]	[2(−4)	2(−3)	1(−2)]

1. Chapman (1939). 2. Hunten (1967). 3. Bates and Nicolet (1950). 4. Srivastava and Shukla (1970a, b).
5. Bates (1947). 6. Rundle (1971). 7. Blamont and Donahue (1964). 8. Saxena (1969). 9. Bates (1954)
10. Sullivan and Hunten (1964). 11. Donahue (1966). 12. Bates (1960).

titatively and obtained a formula for the distribution of neutral Na atoms which gave a fair representation of the observations. Later corrections of the rate coefficients (Blamont and Donahue, 1964; Hunten, 1967) have made it fairly obvious that Chapman's reactions cannot alone explain all experimental results. Reaction 1 is even probably of no significance at all as may be seen by comparing corresponding rates of reactions 1 and 5 in the three last columns of Table III.

The problem of ionization was first tackled by Bates (1947), using the reactions 7, 18, and 19. He drew the conclusion that a major part of the Na atoms may well in daytime be in an ionized form. Later authors (Hunten, 1954, 1967; Blamont and Donahue, 1964; Gadsden ,1964b) have arrived at the conclusion that ionization may probably not play a very important role in the Na layer formation. This conclusion is supported in Narcisi's (1973) observation of the Na⁺ concentration.

Bates and Nicolet (1950) introduced reactions 2, 14, and 24 while Bates (1960) initiated reaction 23. Blamont and Donahue (1964) added 12, 16, 17, 20, and 21 to the above mentioned processes in their attempt to explain the variation of Na dayglow, and obtained promising results. Later corrections of rate coefficients render their specific solutions less attractive today, but their method of tackling the problem has general value.

Saxena (1969) studied the processes 1, 2, 5, 10, and 23 supplemented by a new one, process 11, with a view to explaining the emission rate of Na in nightglow. He assumed that NaO is in an excited state and that process 10 is the only one to produce $Na(^2P)$. By a suitable choice of rate coefficients and using probable values for the concentration of O, O_2, M (third body) and Na (from Sullivan and Hunten's (1964) twilight observations), he found emission rates in fair agreement with observed ones. His results depend critically on the choice of rate coefficients, especially the one proposed for reaction 11, which may well be grossly overestimated, and on his method for eliminating O_3 from his expressions.

Srivastava and Shukla (1970a, b) added reactions 3, 4, 6, and 13 to the list of processes. After having studied several of them, they conclude that 1, 2, 5, 10, and 23 are adequate for explaining the volume emission rate of Na in nightglow. They further claim to be able to account for the seasonal variation of nightglow with a set of processes which include the new process 4. There is a dimension error in their expression for the volume emission rate which lends doubt to its correctness.

One of the general results of Chapman (1939), Hunten (1954), and Blamont and Donahue (1964) was that the content of neutral Na at higher levels must be proportional to the total amount of Na (free and bound) in the atmosphere. This implies, when bearing in mind the observed steep topside of the Na layer, that the total amount of Na cannot follow the other constituents of the atmosphere in their vertical distribution. This is rather remarkable in view of the strong turbulent motion at these heights. The only answer to this question seems to be to seek a strong source of Na near its maximum of concentration and a sink higher up. Hunten and Wallace (1967) have suggested that the source and sink are both provided by dust particles distributed with a scale height of 3 km. This thought has been pursued by Donahue and Meier (1967) who found that a very thin layer of aerosols at about the height of maximum density of Na, may be sufficient to explain the observed Na distribution when diffusion and turbulent motion are taken into account. Hansen and Donaldson (1967) have pointed out that the steep topside gradient can be understood by a competition between upward mixing and photoionization. The fate of the ions is unclear, although some suggestions were made.

Gadsden (1968, 1969, 1970, 1971) has studied the consequences of meteor ablation at 92 km with promising qualitative results. Some features of the Na distribution are unexplained, such as the seasonal abundance variation at higher latitudes. This difficulty may, however, be overcome by invoking the injection process, proposed by Hunten and Godson (1967), of Na-containing material into the 90 km level by atmospheric mixing during the polar night. The particle size of the material has to be $\leqslant 0.01\ \mu$ in order to get the correct scale height for its distribution around and above the height of maximum concentration of Na, and the particles may be considered as macromolecules containing $\sim 10^4$ atoms. Rundle (1971) introduced reaction processes of a type similar to processes 8 and 9 in Table III. Here Ma stands for macromolecule and Q for energy.

Castleman (1973) has told us that it is very unlikely that an aerosol layer of Na compounds can give the solution to the Na distribution problem. There is, however, an interesting experiment reported by Allen (1970) of a chemiluminescence of Na by exposing anhydrous sodium sulfate crystals to 'active nitrogen' which may have some relevance in this connection. The Na D lines and the 3914 band of N_2^+ were observed in the strong luminescence from the surface of the crystals.

Keller and Beyer (1971) have investigated the clustering of CO_2 to alkali ions, process 21 with X and M being CO_2. Their drift tube experiments indicate a rate coefficient of about 2×10^{-29} and a rate coefficient for the reverse process of about 10^{-14}. The respective coefficients for O_2 are 5×10^{-32} and 8×10^{-13}. The study is interesting and suggestive, but the reaction processes are still too slow by a factor of 100 to explain the observed small ratio of Na^+ to Na.

Even if the theory of a layer of macromolecules is rather speculative at the moment, it has many attractive features and may, coupled with an adequate transport theory, explain some of the observations of Na, but scarcely all. The last columns of Table III may give an indication of what processes, other than 8 and 9, will probably be operative in a closed system of reactions. Of the 8 loss processes of Na, process 1 and 3 seem to be too slow to compete with the rest. Process 2 is certainly operative at lower heights, but drops out higher up in competition with the well established process 5. Process 4 and 6, if at all possible and if the rate coefficients are about correct, would be effective loss processes. So will process 7 in daytime. The last one has to be coupled with a de-ionization process, 18 or 19, possibly the first, or one or more of 20 to 22 of which we know little. These must in case be followed by 17.

The most probable gain process of Na, apart from the speculative process 9, is process 10 in connection with NaO and process 14 if NaH enters into the system (reaction 6). In this case processes which remove OH must be included and the Na problem will be coupled to the OH and to the O-H balance problem in the atmosphere.

Our knowledge of the chemistry of Na in the atmosphere may not seem to have advanced very much in the last few years. However, some new important experimental results about the behavior of the Na 'layer' have been obtained which may clarify matters. A number of interesting studies have also been done and some new interesting thoughts have been presented. What we mostly want are better determinations of rate coefficients, a closer study of the 'dust' layer possibility, and the solution of a set of equations for a closed system of reaction processes which include the O-H balance and transport phenomena.

Acknowledgments

I am greatly indebted to Dr D. M. Hunten who has read through the manuscript and given valuable comments. I am also indebted to my collaborators Drs G. J. Kvifte,

V. Hansen, and A. Gulbrandsen for help in collecting and processing data which are used here, and to the Norwegian Research Council for Science and the Humanities for financial support.

References

Ahmed, M., Silverman, S. M., and Lloyd, J. W. F.: 1970, *Planetary Space Sci* **18**, 1666.
Albano, J., Blamont, J. E., Chanin, M. L., and Petitdidier, M.: 1970, *Ann. Geophys.* **26**, 151.
Allen, E. R.: 1970, *J. Geophys. Res.* **75**, 2947.
Barbier, D.: 1959, *Ann. Geophys.* **15**, 412.
Barbier, D.: 1965, *Ann. Geophys.* **21**, 299.
Bates, D. R.: 1947, *Terr. Magn.* **52**, 71.
Bates, D. R.: 1954, in G. P. Kuiper (ed.), *The Earth as a Planet*, Univ. of Chicago Press, Chicago,p. 576.
Bates, D. R.: 1960, in J. A. Ratcliffe (ed.), *Physics of the Upper Atmosphere*, Academic Press, New York and London, p. 219.
Bates, D. R. and Nicolet, M.: 1950, *J. Geophys. Res.* **55**, 235.
Blamont, J. E. and Donahue, T. M.: 1964, *J. Geophys. Res.* **69**, 4093.
Bullock, W. R. and Hunten, D. M.: 1961, *Can. J. Phys.* **39**, 976.
Castleman, A. W. Jr.: 1973, this volume, p. 143.
Chamberlain, J. W.: 1961, *Physics of the Aurora and Airglow*, Academic Press, New York and London.
Chapman, S.: 1939, *Astrophys. J.* **90**, 309.
Davis, T. N. and Smith, L. L.: 1965, *J. Geophys. Res.* **70**, 1127.
Delannoy, J. and Weill, G.: 1958, *Compt. Rend. Acad. Sci.* **246**, 2925.
Donahue, T. M.: 1966, *J. Geophys. Res.* **71**, 2237.
Donahue, T. M.: 1969, Paper IAGA General Scientific Assembly, Madrid.
Donahue, T. M. and Meier, R. R.: 1967, *J. Geophys. Res.* **72**, 2803.
Gadsden, M.: 1964a, *Ann. Geophys.* **20**, 261.
Gadsden, M.: 1964b, *Ann. Geophys.* **20**, 383.
Gadsden, M.: 1967, in B. M. McCormac (ed.), *Aurora and Airglow*, Reinhold Publishing Corporation, New York, p. 109.
Gadsden, M.: 1968, *J. Atmospheric Terrest. Phys.* **30**, 151.
Gadsden, M.: 1969, *Ann. Geophys.* **25**, 721.
Gadsden, M.: 1970, *Ann. Geophys.* **26**, 141.
Gadsden, M.: 1971, *Ann. Geophys.* **27**, 401.
Gadsden, M. and Salmon, K.: 1958, *Nature* **182**, 1598.
Gadsden, M. and Purdy, C. M.: 1970, *Ann. Geophys.* **26**, 43.
Gadsden, M., Donahue, T. M., and Blamont, J. E.: 1966, *J. Geophys. Res.* **71**, 5047.
Gault, W. A. and Rundle, H. N.: 1969, *Can. J. Phys.* **47**, 85.
Gibson, A. J. and Sandford, M. C. W.: 1971, *J. Atmospheric Terrest. Phys.* **33**, 1675.
Graham, D. A., Ichikava, T., and Kim, J. S.: 1971, *Ann. Geophys.* **27**, 483.
Hansen, W. B. and Donaldson, J. S.: 1967, *J. Geophys. Res.* **72**, 5513.
Hesstvedt, E.: 1971, in G. Fiocco (ed.), *Mesospheric Models and Related Experiments*, D. Reidel Publishing Company, Dordrecht-Holland, p. 52.
Hunten, D. M.: 1954, *J. Atmospheric Terrest. Phys.* **5**, 44.
Hunten, D. M.: 1967, *Space Sci. Rev.* **6**, 493.
Hunten, D. M.: 1970, *Astrophys. J.* **159**, 1107.
Hunten, D. M.: 1971, in B. M. McCormac (ed.), *The Radiating Atmosphere*, D. Reidel Publishing Company, Dordrecht-Holland, p. 3.
Hunten, D. M. and Godson, W. L.: 1967, *Atmospheric Sci.* **24**, 1.
Hunten, D. M. and Wallace, L.: 1967, *J. Geophys, Res.* **72**, 69.
Hunten, D. M., Vallance Jones, A., Ellyett, C. D., and McLauchlan, E. C.: 1964, *J. Atmospheric Terrest. Phys.* **26**, 67.
Keller, G. E. and Beyer, R. A.: 1971, *J. Geophys. Res.* **76**, 289.
Knecht, R. W.: 1965, in G. M. Brown (ed.), *Progress in Radio Science*, Vol. III, Elsevier Publishing Company Amsterdam, Holland, p. 14.

Kvifte, G.: 1953, Report No 205, Dept. of Physics, Univ. of Oslo.

Kvifte, G.: 1972, in A. Egeland, Ø. Holter, and A. Omholt (eds.), *Cosmical Geophysics*, Universi-tetsforlaget, Oslo, Norway, p. 81.

Kvifte, G. and Wallace, L.: 1970, *Planetary Space Sci.* **18**, 623.

Llewellyn, E. J. and Evans, W. F. J.: 1971, in B. M. McCormac (ed,), *The Radiating Atmosphere*, D. Reidel Publishing Company, Dordrecht-Holland, p. 17.

McNutt, D. P. and Mack, J. E.: 1963, *J. Geophys. Res.* **68**, 3419.

Narcisi, R. S.: 1966, *Ann. Geophys.* **22**, 224.

Narcisi, R. S.: 1972, private communication.

Narcisi, R. S.: 1973, this volume p. 171.

Nicolet, M.: 1971, in G. Fiocco (ed.), *Mesospheric Models and Related Experiments*, D. Reidel Publishing Company, Dordrecht-Holland, p. 1.

Noxon, J. F.: 1967, in B. M. McCormac (ed.), *Aurora and Airglow*, Reinhold Publishing Corporation, New York, p. 123.

Noxon, J. F.: 1968, *Space Sci. Rev.* **8**, 92.

Rao, V. R. and Kulkarni, P. V.: 1971, *Ind. J. Pure Appl. Phys.* **9**, 644.

Rees, M. H. and Deehr, C. S.: 1962, *J. Geophys. Res.* **67**, 2309.

Rundle, H. N.: 1971, in B. M. McCormac (ed.), *The Radiating Atmosphere*, D. Reidel Publishing Company, Dordrecht-Holland, p. 90.

Sandford, M. C. W. and Gibson, A. J.: 1970, *J. Atmospheric Terrest. Phys.* **32**, 1423.

Saxena, P. P.: 1969, *Ann. Geophys.* **25**, 847.

Saxena, P. P.: 1970, *Ann. Geophys.* **26**, 505.

Slipher, V. M.: 1929, *Publ. Astron. Soc. Pacific* **41**, 263.

Smith, L. L. and Steiger, W. R.: 1968, *J. Geophys. Res.* **73**, 2531.

Srivastava, A. N. and Shukla, R. V.: 1970a, *Ann. Geophys.* **26**, 501.

Srivastava, A. N. and Shukla, R. V.: 1970b, *Astrophys. Space Sci.* **8**, 136.

Sullivan, H. M. and Hunten, D. M.: 1964, *Can. J. Phys.* **42**, 937.

Sullivan, H. M. and Roberts, M. G.: 1968, *Nature* **220**, 361.

Thomas, L.: 1972, private communication.

Toroshelidze, T. J.: 1968, *Bull. Acad. Sci. Georgian S.S.R.*, **51**, 579.

PART IV

EXPERIMENTAL RESULTS AND INTERPRETATIONS

MASS SPECTROMETER MEASUREMENTS
IN THE IONOSPHERE

ROCCO S. NARCISI

*Air Force Cambridge Research Laboratories, Laurence G. Hanscom Field,
Bedford, Mass., U.S.A.*

1. Introduction

Notable contributions to our knowledge of the earth's ionosphere have been made from rocket and satellite measurements with ion mass spectrometers in the period 1967–72. Rocket measurements from low to high latitudes have provided information on the diurnal behavior of the D, E and F regions and on the effects of solar eclipses, PCA, aurora, sporadic E, and meteor showers. The long sought D region negative ion composition has been measured. Satellite measurements have yielded detailed data on diurnal, latitudinal, and longitudinal variations in the ion composition of the topside ionosphere giving evidence for solar, geomagnetic, and seasonal control of the ionization, as well as proof of upward plasma flow over the poles. This review is limited to observations with ion mass spectrometers with emphasis on the altitude range 60 to 1300 km.

2. *D* Region

2.1. POSITIVE IONS

Positive ion composition measurements have been made in the D region with rocket-borne mass spectrometers at equatorial, middle, and high latitudes under a variety of conditions (Narcisi, 1967, 1968, 1971; Goldberg and Blumle, 1970; Narcisi and Roth, 1970; Goldberg and Aikin, 1971; Krankowsky *et al.*, 1972a; Johannessen and Krankowsky, 1972; Narcisi *et al.*, 1972a, b). In all cases quadrupole mass spectrometers housed in cryogenically cooled pumping systems or in titanium getter pumps were utilized. Generally, the normal D region is found to be dominated by water-cluster ions, $H^+(H_2O)_n$, (Figure 1) which decrease abruptly above about 82 km in the daytime and above about 86 km at twilight and at night. These altitudes also represent the transition levels above which NO^+ and O_2^+ dominate and metallic ions appear. This situation changes during high latitude ionospheric disturbances. Under conditions of auroral absorption (1.4 to 3.0 dB at 27 MHz) NO^+ and O_2^+ are found to be the major ions down to about 75 km (Krankowsky *et al.*, 1972a). Similarly, day, night, and sunset rocket measurements during a PCA (Narcisi *et al.*, 1972b) with 30 MHz riometer absorptions of 3, 0.4 and 0.7 dB, respectively, showed a suppression of the water-cluster ion cutoff level, and NO^+ and O_2^+ as the major ions above 77 km at night (Figure 2), above 73 km in the daytime, and above 78 km at sunset, the latter two altitudes being lower altitude measurement limits. The PCA results have aided in clarifying the ion chemistry of the disturbed D region (Narcisi, 1972) and have permitted a determination of mesospheric nitric-oxide concentrations (Narcisi *et al.*, 1972c).

B. M. McCormac (ed.), Physics and Chemistry of Upper Atmospheres, 171–183. All Rights Reserved.
Copyright © 1973 by D. Reidel Publishing Company, Dordrecht-Holland.

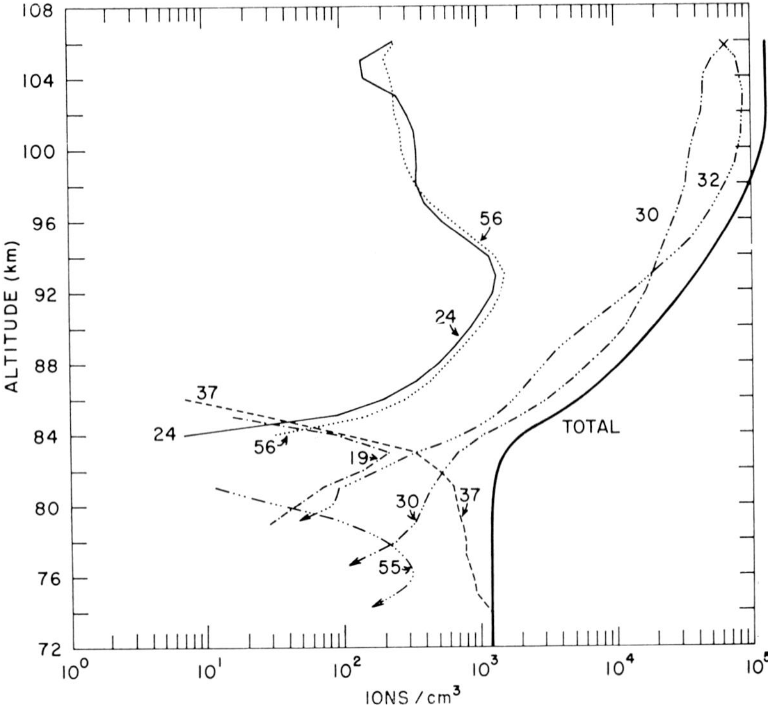

Fig. 1.   Major positive ions in the D and E regions at a 20° solar zenith angle (Narcisi *et al.*, 1972a).

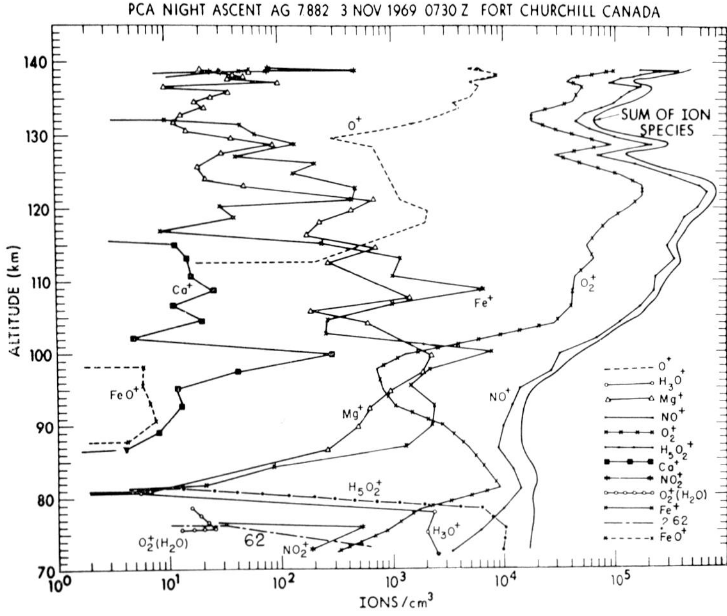

Fig. 2.   Positive ion composition in the nighttime D and E regions during a simultaneous PCA event and auroral event (Narcisi *et al.*, 1972b).

In the quiescent, mid-latitude D region, $H_5O_2^+$ is found to dominate the ion composition down to about 65 km with H_3O^+, $H_7O_3^+$, and $NO^+(H_2O)$ present in smaller amounts. The NO^+ ion is generally much less abundant than $H_5O_2^+$ in the daytime and very little NO^+ is found below 85 km at night and at twilight. Very rough variations with time of day between 75 and 85 km indicate about 10^3 ions cm^{-3} near midday, about 10^2 ions cm^{-3} near midnight, a decay of about 400 to 100 ions cm^{-3} through sunset ($\chi = 88.5$ to $98.6°$) and about a threefold increase at sunrise ($\chi = 102.5$ to $90°$) (Narcisi and Roth, 1970). D region ion concentrations are difficult to determine and subject to large errors; however, if the increase at sunrise ($\chi = 90°$) is correct, then water-cluster ions apparently can be produced by fairly low energy solar radiation. Direct ionization of water conglomerates has been suggested as a possibility by Hunt (quoted in Narcisi and Roth, 1970; Narcisi, 1971).

Another serious problem in all the measurements is that fragmentation of the weakly bound cluster ions can occur through energetic collisions as a result of the electric draw-in fields or by thermodynamic breakup at the increased shock layer temperatures (Narcisi, 1970; Narcisi and Roth, 1970). In such reactions the ion loses one or more water molecules so that the measured composition may not be representative of the ambient. In any case, it seems clear that cluster-ions dominate. Subsonic measurements and experiments with the sampling E field are presently being conducted to resolve these problems.

The present problem in the chemistry of the quiescent D region is to find a fast process which can convert NO^+ to water-cluster ions in the 70 to 86 km region. Rocket measurements during a solar eclipse indicate that the cluster-ion cutoff altitude rises from 82 to 86 km from full sun to totality at the expense of a large decrease in NO^+ concentrations (Narcisi *et al.*, 1972a). This was considered evidence for the conversion process and although the actual reaction could not be definitely determined, possible reactions with water conglomerates were suggested.

The complexity of the D region may be appreciated by the variety of ion species detected. The measured mass to charge ratios and possible identifications are : $19(H_3O^+)$, $21(H_3^{18}O^+)$, $30(NO^+)$, $32(O_2^+, S^+)$, $34(S^+)$, $37(H_5O_2^+)$, $39(H_5^{16}O^{18}O^+)$, $41(Na^+ \cdot H_2O)$, $46(NO_2^+)$, $48(NO^+ \cdot H_2O, SO^+)$, $50(O_2^+ \cdot H_2O, SO^+)$, $55(H_7O_3^+)$, 60 ± 1 (unknown), $63 \pm 1(NO_2^+ \cdot H_2O, SO_2^+, O_4^+)$, $66 \pm 1(NO^+(H_2O)_2)$, $73 \pm 1(H_9O_4^+)$, $80 \pm 2(SO_3^+, H_xSO_3^+)$, $91 \pm 1(H_{11}O_5^+)$, $96 \pm 2(SO_4^+, H_xSO_4^+)$, and $109 \pm 1(H_{13}O_6^+)$. Atomic metal ions have been omitted. The ions 73, 91, and 109 amu were found to be enhanced in the cold high latitude summer mesopause (Johannessen and Krankowsky, 1972), but all the above ion species except the 109 amu ion were also observed at mid-latitudes (Thomas *et al.*, 1972). The S ions and S-containing species appear to be restricted below 86 km, falling off in concentration with the water-cluster ions. If the identification of these ions as S is correct, it is a major problem to explain their presence in the D region (Narcisi, 1971).

2.2. Negative ions

Negative ion measurements from seven rocket flights with cryogenically pumped quadrupole mass spectrometers have been reported. These include nighttime quiescent measurements over Ft. Churchill, Canada (Narcisi *et al.*, 1971), a nighttime measurement from Andoya, Norway during a weak auroral event (Arnold *et al.*, 1971); midday and midnight measurements in a PCA event from Ft. Churchill (Narcisi *et al.*, 1972d); and a measurement near totality of a solar eclipse from Wallops Island, Virginia (Narcisi *et al.*, 1972a). Narcisi and co-workers found heavy ions with masses near 60, 62, 76, 78, 80, 98, 116, 134, and 152 amu in relatively large concentrations between 73 and 92 km and a layer of very heavy negative ion species centered near 88 km (Figure 3). Between 90 and 92 km, the negative ion concentrations decrease by almost two orders of magnitude. These ions have been tentatively identified as $NO_3^- (H_2O)_n$, $n = 0$ to 5, with possible admixtures of $CO_3^- (H_2O)_n$, $n = 0$ to 5 and CO_4^-.

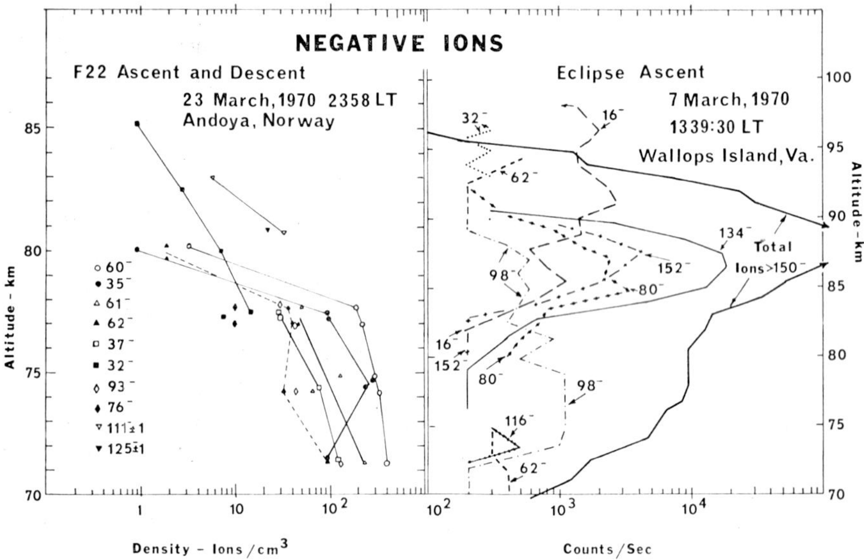

Fig. 3.   Negative ion composition measurements in the *D* region. Left panel from Arnold *et al.* (1971). Right panel from Narcisi *et al.* (1972a).

Cluster ion-fragmentation due to the sampling *E* field was detected and thus both the relative and absolute concentrations measured are questionable. The possible existence of very heavy ions greater than 150 amu was also evident (Figure 3). Lighter ions with masses 16, 32, 35, 37 and 46 amu, probably O^-, O_2^-, Cl^-, and NO_2^- are predominant above 90 km but in very small concentrations. With the exception of O_2^-, these species may be produced from contaminants or by the sampling method.

Drastic differences in the negative ion composition during a PCA were observed in comparison to quiet conditions. Near midday, O_2^- ions were the primary species and were found in large concentrations between 72 and 94 km. At night O^- was generally the major ion between 76 and 94 km; O_2^- was present below 80 km and became larger than O^- below 76 to 74 km. (The mass range of the PCA day spectrometer did not include 16 amu.) In both PCA flights the heavy negative ions measured in other flights were present below 85 km. The negative ion measurements are largely unexplained; details are discussed by Narcisi (1972).

The measurements from the flight by Arnold *et al.* (1971) in Figure 3 are in sharp contrast to the measurements of Narcisi and co-workers. Large amounts of Cl^- ions (35 and 37) were measured below 80 km perhaps as a result of charge transfer with a Cl contaminant. Additionally, CO_3^- (60) was the dominant ion along with HCO_3^- (61) and NO_3^- (62) from 71 to 80 km. Mass 93 ± 1 amu ($CO_4^- \cdot H_2O$ or $NO_2^- \cdot HNO_2$) was found below 80 km instead of mass 98 ± 2 amu. Although O_2^- was similarly enhanced from 78 to 85 km owing to an energetic particle disturbance, the heavy cluster ion mass numbers $111\pm1 (CO_4^- (H_2O)_2)$ and $125\pm1 (NO_3^- (HNO_3))$ are completely different. The drop off in negative ion concentrations above 80 km may have resulted from an increasingly negative vehicle potential with altitude because no negative ions, including contaminants, were measured above 85 km. Arnold *et al.* also reported trace constituents of $68(O_2^- (H_2O)_2)$ and $78(CO_3^- (H_2O))$. The uncertainties in the D region negative ion composition measurements point out the requirement for additional rocket and laboratory work.

3. *E* Region

3.1. MOLECULAR IONS

Rocket measurements of the diurnal ion composition variations in the E region have been reported or reviewed by Keneshea *et al.* (1970), Narcisi (1971), Danilov (1972) and Johnson (1972). NO^+ and O_2^+ generally dominate the E region ion composition (Figures 1, 2, 4, 5 and 6). In the daytime mid-latitude region, measurements of the NO^+/O_2^+ concentration ratio show excursions between 0.5 and 3 from 100 to 140 km. In twilight periods this ratio can become very large (Figure 4) while at night it is about ten (Figure 6). A typical quiet nighttime E region is shown in Figure 6; however, NO^+ concentrations up to 10^4 ions cm^{-3} have also been observed in the vicinity of 110 km and were shown to be caused by atmospheric ion transport processes (Narcisi, 1971). These measurements contributed to the development of a diurnal E region model (Keneshea *et al.*, 1970).

E region measurements with quadrupole mass spectrometers have been made in auroras (Donahue *et al.*, 1970; Swider and Narcisi, 1970; Zipf *et al.*, 1970; Narcisi *et al.*, 1972b). In all cases the NO^+ density was greatly enhanced and the O^+ density was somewhat larger than expected (Figure 2). Zipf *et al.* (1970) measured both the ion and neutral composition in a bright auroral arc and found essentially no O_2^+ ions. This is consistent with their neutral measurements in that the NO density was greater than the O_2 density in the auroral arc. Apparently, NO is enhanced in proportion to the strength of the aurora, but the processes by which this occurs are unknown.

3.2. METEORIC IONS AND SPORADIC *E*

A broad layer composed mainly of Mg^+ and Fe^+ with a peak ion density near 93 km is a permanent global feature of the ionosphere (Figures 1, 2 and 4) and seems to be continuously maintained by ablating meteoroids (Narcisi, 1971). The ion species measured in this layer include $23(Na^+)$, 24-25-$26(Mg^+)$, $27(Al^+)$, $39(K^+)$, 40-42-44 (Ca^+), $52(Cr^+)$, 54-56-$57(Fe^+)$, $59(Co^+)$, 58-$60(Ni^+)$ and $72(FeO^+)$ (Goldberg and Blumle, 1970; Narcisi, 1971; Johannessen and Krankowsky, 1972; Krankowsky *et al.*, 1972b; Narcisi *et al.*, 1972b). The relative abundances of these species are in quite good agreement with the relative elemental concentrations found in chondrite meteorites.

Thin meteoric ion layers are often found up to 140 km and also superimposed on the broad meteoric ion distribution (Figure 2). Above 100 km, these layers may be composed mainly of Si^+, or Mg^+, or Fe^+ (Narcisi, 1971). A Si layer is typically located near 110 ± 5 km in the daytime and has been found to be enhanced on one occasion to produce a sporadic E layer (Figure 4). Intense mid-latitude sporadic E layers composed of meteoric ions have been observed (Narcisi *et al.*, 1967; Young *et al.*, 1967; Narcisi, 1968). In twilight periods and at night, weak sporadic E layers may be composed mainly of either NO^+ or meteoric species. The molecular ion layer densities, however, are limited by electron recombination to a maximum of about 2×10^4 ions cm^{-3} (Narcisi, 1968, 1971; Rosenberg and Zimmerman, 1972). The meteoric species content in the E region was seen to be enhanced during a meteor

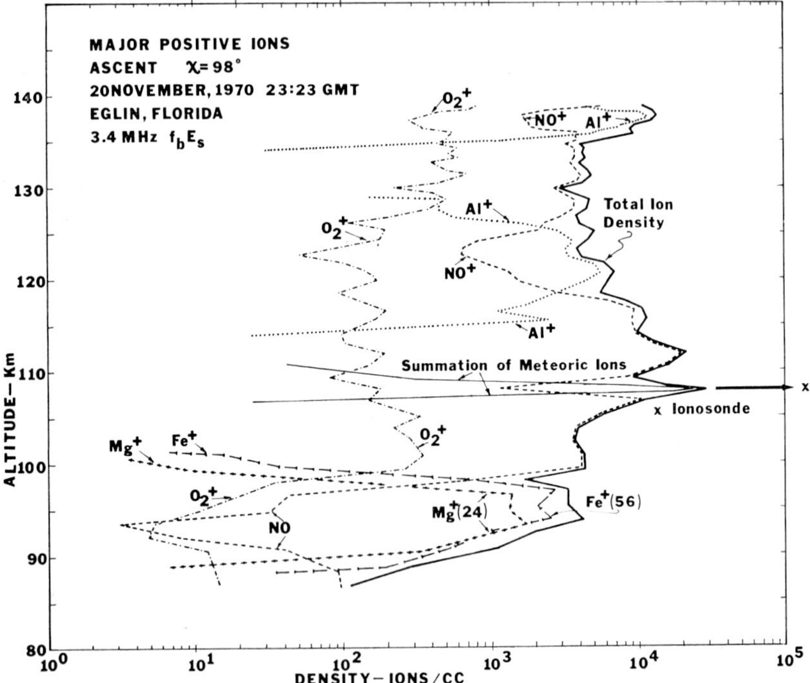

Fig. 4.   Ion composition of the E region near sunset showing a blanketing sporadic E layer. The layer consisted of Si^+ (predominant), Mg^+, Fe^+, and smaller amounts of Na^+, Ni^+, Ca^+. The Al^+ layers resulted from trimethyl aluminum released from an earlier rocket to measure neutral winds. Results are from Narcisi *et al.* (1972e).

shower (Narcisi, 1968), and the reduction of NO^+ and O_2^+ densities in a meteoric ion layer was observed many times in accord with theory (e.g., Figure 4). Both upward and downward movements of thin meteoric layers have also been observed (Narcisi, 1971).

4. F_1, F_2 and Topside Ionosphere

4.1. DAYTIME ROCKET MEASUREMENTS

In the 200 to 1000 km altitude range, the ion species identified for the measured mass to charge ratios include $1\,(H^+)$, $2\,(D^+)$, $4\,(He^+)$, $7\,(N^{++})$, $8\,(O^{++})$, $14\,(N^+)$, $15\,(^{15}N^+)$, $16\,(O^+)$, $18\,(^{18}O^+)$, $20\,(Ne^+)$, $28\,(N_2^+)$, $30\,(NO^+)$ and $32\,(O_2^+)$. Figure 5 shows measured daytime mid-latitude distributions of these species. The ions D^+, N^{++}, $^{15}N^+$, and Ne^+ are not shown since these are generally less than 10 ions cm^{-3} (Hoffman *et al.*, 1969). From their measurements, Ershova and Sivtseva (1971) have suggested that D^+ may possibly be He^{++} above 600 km. In the daytime ionosphere O^+ ions are by far the dominant constituent from 200 to 1000 km, N^+ ions being about 1 to 3% of the O^+ ions over this altitude range. Daytime mid-latitude rocket measurements between 200 and 630 km with a Wien filter by Maier (1969) and with a Bennett mass spectrometer by Brinton *et al.* (1969a) show results for O^+, N^+, and total ion concentrations which

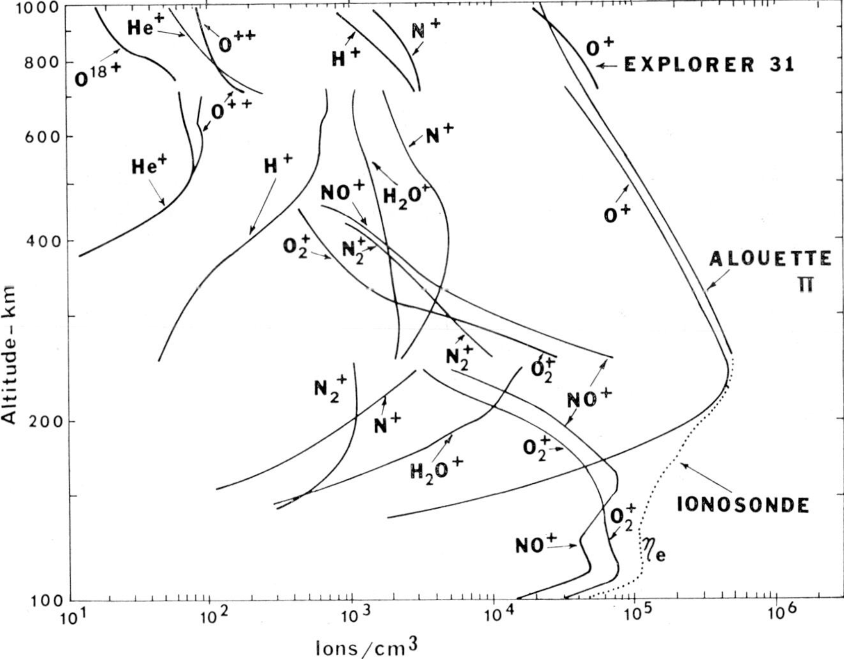

Fig. 5. Ion composition of the mid-latitude ionosphere near solar minimum. Data above 250 km are from Hoffman *et al.* (1969); below 250 to 115 km from Holmes *et al.* (1965); 115 to 100 km from Narcisi (1968). Explorer 31 was near 1000 km at the point of intersection with the rocket trajectory. Electron densities were obtained with a topside sounder and ground based ionosondes. Meteoric ions are not included. H_2O^+ is a contaminant.

are nearly identical to the magnetic sector spectrometer measurements of Hoffman *et al.* (1969) for similar NmF2 concentrations. However, Maier (1969), Brinton *et al.* (1969a), and Ershova and Sivtseva (1971) obtain H^+ and He^+ ion concentrations about a factor of 10 larger than those of Hoffman *et al.* but, interestingly, all show generally similar H^+/He^+ concentration ratios above 300 km. Hoffman (1969) and Brinton *et al.* (1969a) indicate that H^+ and He^+ are apparently in chemical equilibrium below 400 km, but their inferred H concentrations differ by an order of magnitude on account of the factor of 10 difference in their H^+/O^+ concentration ratios. (Support for the lower H^+ and He^+ ion concentrations comes from the satellite measurements by Brinton *et al.* (1969b) in Figure 8.)

Although the NO^+ and O_2^+ scale heights above 180 km are considered to be well matched to those of O_2 and N_2 as required by the ion chemistry, the measured absolute concentrations of the molecular ions also show significant differences as can be seen in Figure 5 by the shift in these species at 250 km. Below about 170 km these molecular ion species become the dominant constituents. Daytime rocket results for NO^+, O_2^+, O^+, N^+, and N_2^+ concentrations by Pharo *et al.* (1971) between 130 and 306 km are within a factor of 2 to 3 of those by Holmes *et al.* (1965) and exhibit a constant NO^+/O_2^+ ratio of about 2. The molecular ion concentrations of Brinton *et al.* (1969a) are also generally within a factor of 2 to 3 of those of Holmes *et al.*, but show instead an NO^+/O_2^+ ratio of 0.8 at 200 km, increasing steadily to about 2 at 375 km. Thus, all these measurements exhibit strikingly similar results with the exception of the relative concentrations of the light ions and molecular ions above 250 km. Whether or not these variations are real or due to the sampling method is difficult to determine.

During periods of enhanced solar activity with $NmF2 = 1$ to 2×10^6 electrons cm^{-3}, daytime rocket measurements were obtained by Goldberg and Blumle (1970) with a pumped quadrupole mass spectrometer between 68 and 303 km, and by Giraud *et al.* (1971) with a Bennett spectrometer between 120 and 400 km. In comparison to solar minimum measurements, the O^+ and N^+ concentrations between 140 and 303 km are generally larger by factors of 2 to 4, but the concentrations of the molecular ions, N_2^+, NO^+, and O_2^+, show very small changes.

Two rockets with Bennett spectrometers measured the response of the ionospheric region between 130 and 290 km during a solar eclipse at 42.5% and 85% solar obscuration (Brace *et al.*, 1972). The total ion concentrations decreased 30% in the *F* region and 50% in the *E* region with little change in relative composition. Arnold *et al.* (1969) and Zhloodko and Klyueva (1970) reported rocket ion composition measurements between about 100 and 230 km in the polar ionosphere utilizing a quadrupole spectrometer and a Bennett spectrometer, respectively. The daytime results of Arnold *et al.* resemble the mid-latitude results of Holmes *et al.* (1965) except that the polar O_2^+ concentrations are much lower in the *E* region. Daytime measurements of Zhloodko and Klyueva were made during a complete radio wave absorption and showed total concentrations of 3 to 4×10^5 ions cm^{-3}, $NO^+/O_2^+ \sim 4$ up to 180 km, and only 1 to 10% O^+ ions. Many more rocket measurements are required to define both the quiescent and disturbed polar ionosphere.

4.2. Nighttime Rocket Measurements

A composite of nighttime results between 90 and 900 km is presented in Figure 6. The maximum decay from the daytime ionosphere, a factor of about 400, occurs near 160 km. At night the NO^+/O_2^+ ratio increases significantly below 200 km but above 200 km there is a reversal to O_2^+ dominance. The molecular ion to O^+ transition takes place at 220 km compared to 170 km during the day, and an O^+ to H^+ transition occurs at 430 km at night. Hoffman's H^+ and He^+ nighttime concentrations (Figure 6) are similar to the daytime measurements of others but are a factor of 10 greater than his daytime results in Figure 5. (Diurnal mid-latitude satellite measurements near

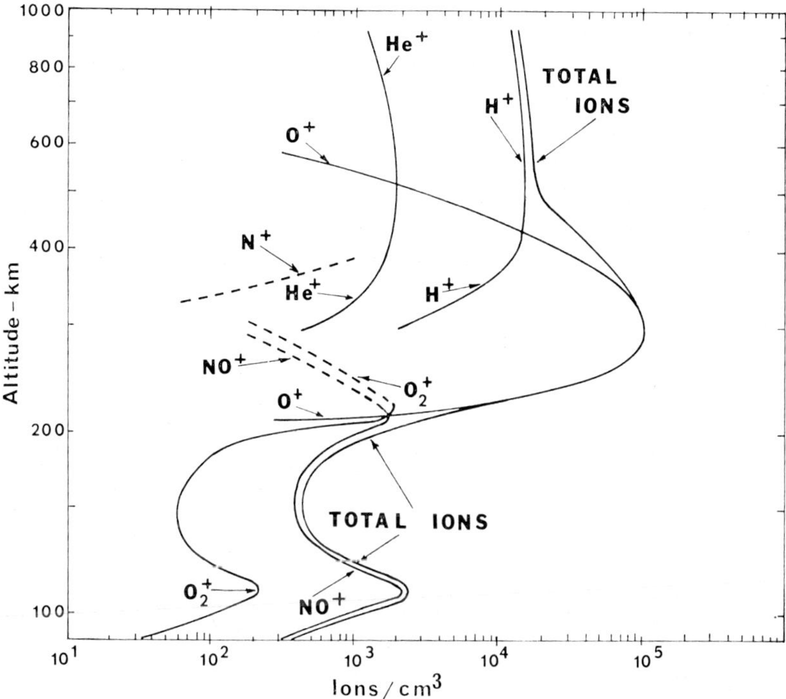

Fig. 6. Ion composition of the nighttime mid-latitude ionosphere. Data above 200 km are from Hoffman (1967); below 220 to 140 km from Holmes *et al.* (1965); and below 140 km from Narcisi) (1971). Dashed curves are from Giraud *et al.* (1971). Meteoric ions are omitted.

1100 to 1300 km (Figure 8) indicate that H^+ ions increase a factor of 5 from day to night (Brinton *et al.*, 1969b).) Giraud *et al.* (1971) obtained nighttime mid-latitude measurements with a Bennett spectrometer between 120 and 400 km (Figure 6.) Nighttime polar ionosphere measurements between 100 and 180 km with a Bennett spectrometer were reported by Zhloodko and Klyueva (1970) but their data suffered from instrumental problems.

4.3. Satellite Measurements

Only a small fraction of the wealth of satellite ion composition data obtained from more than a dozen satellites has been reported. Detailed measurements of the latitudinal variations of ion composition were first made with a Bennett spectrometer on OGO 2 (Taylor *et al.*, 1968). In a nearly polar dawn-dusk orbit, the major ions between 415 and 1525 km are O^+ and H^+; N^+ and He^+ are minor constituents. At 1000 km, O^+ dominates in both northern and southern polar ionospheres yielding to H^+ at lower latitudes. The high-latitude ionosphere has two striking features: (1) a persistent light ion trough in H^+ and He^+, where H^+ concentrations drop one to two orders of magnitude near $\pm 60°$ geomagnetic latitude, and (2) a variable poleward peak in ionization. Poleward of the trough, strong fluctuations occur in ion composition and in both the magnitude and position of the ionization peak.

The light ion trough was subsequently observed with ion spectrometers on Explorer 31 (Hoffman, 1969), Explorer 32 (Brinton *et al.*, 1969b) and OGO 4 (Taylor *et al.*, 1970) which also showed an accompanying trough in O^+ and N^+ concentrations near $-70°$ geomagnetic latitude (Figure 7). The light ion trough has been associated with the plasmapause and the VLF whistler cutoff (Taylor *et al.*, 1969). In the trough regions the H^+ and He^+ concentrations decrease with altitude from 600 to 2500 km where O^+ is the major ion; this is contrary to the plasmasphere where light ion con-

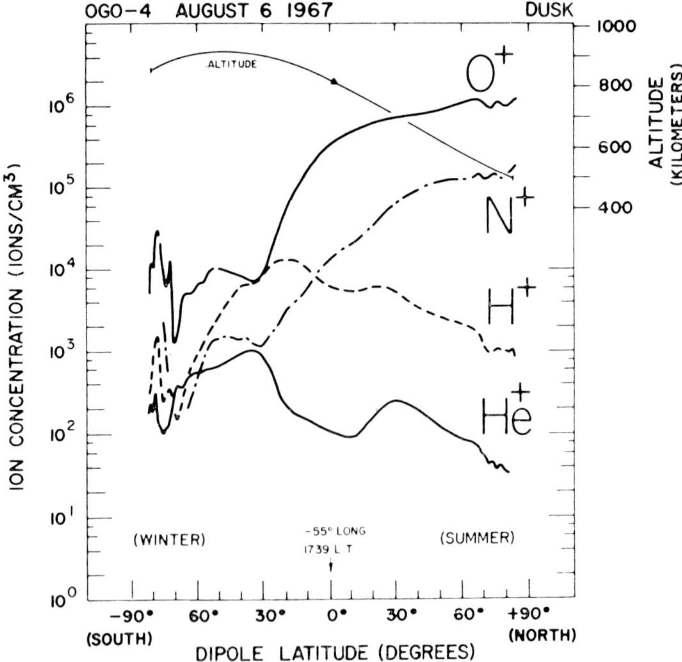

Fig. 7. Latitudinal distributions of ions observed in the interval 20 50 UT to 21 39 UT showing the high latitude and equatorial troughs, after Taylor *et al.* (1970).

centrations increase with altitude (Brinton *et al.*, 1971). This was considered as evidence for upward plasma flow or 'trough wind' by Brinton *et al.* (1971). Hoffman (1969) provided the first experimental observation of the polar wind by interpreting the phase differences between the maxima of the satellite roll modulations of the H^+ and O^+ currents as evidence that H^+ ions are flowing upward with a velocity of 10 to 15 km s^{-1} at 2800 km altitude.

Diurnal and seasonal variations of the ion composition between 280 and 2700 km were observed at mid-to-low-latitudes with the ion mass spectrometer on Explorer 32 and were correlated with solar zenith angle (Brinton *et al.*, 1969b). Figure 8 shows the diurnal and seasonal behavior in O^+, N^+, H^+, and He^+ between 1100 and 1300 km. Detailed results from OGO 4 showed evidence that the geomagnetic field plays an important role in controlling the distribution of upper atmosphere ionization (Taylor, 1971). The altitude of the O^+–H^+ transition level varies markedly with longitude. This behavior was interpreted as being caused by the interaction of atmospheric winds and the geomagnetic field (Brinton *et al.*, 1970) although atmospheric E fields could also explain the behavior (Stubbe and Chandra, 1970). A broad, deep equatorial trough in $n(He^+)$ extending between about $\pm 30°$ geomagnetic latitude (Figure 7)

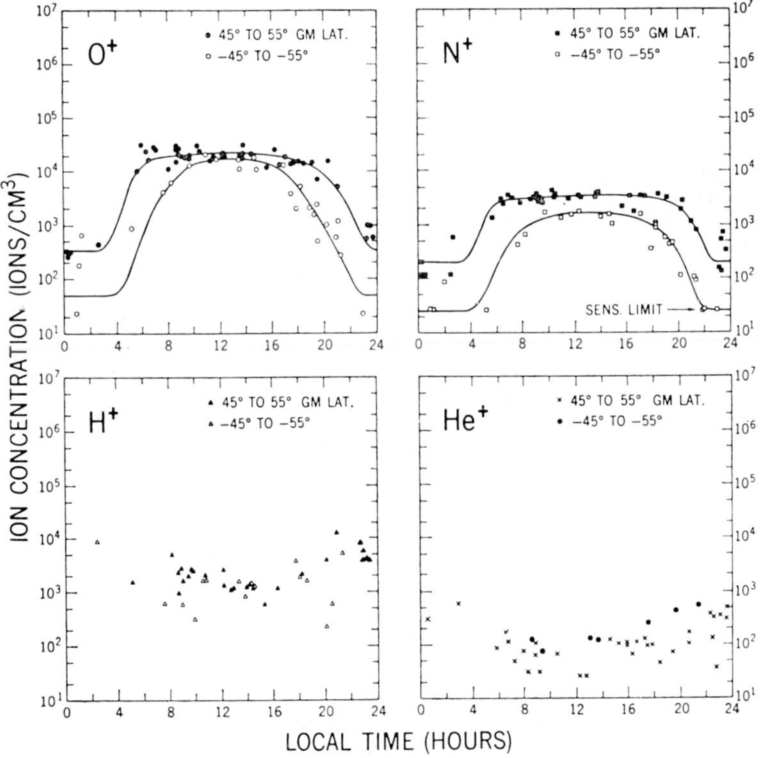

Fig. 8. Variation of the diurnal behavior in the altitude range 1100 to 1300 km between the summer, northern hemisphere and winter, southern hemisphere, from Brinton *et al.* (1969b).

and the winter bulge in $n(\mathrm{He}^+)$ appear as persistent features of the topside ionosphere (Taylor *et al.*, 1970; Taylor, 1971). It seems that past attempts to ascertain solar cycle variations in H^+ and He^+ may not have been conclusive, due to the sampling method and comparison of results at different times and locations (Taylor, 1971). Thermospheric hydrogen concentrations and temporal variations at 350 km were inferred from Explorer 32 ion composition measurements and compared to model predictions (Brinton and Mayr, 1971).

References

Arnold, F., Berthold, W., Betz, B., Lammerzahl, P., and Zahringer, J.: 1969, *Space Res.* **9**, 256.
Arnold, F., Kissel, J., Krankowsky, D., Wieder, H., and Zahringer, J.: 1971, *J. Atmospheric Terrest. Phys.* **33**, 1169.
Brace, L. H., Mayr, H. G., Pharo, M. W., III, Scott, L. R., Spencer, N. W., and Carignan, G. R.: 1972, *J. Atmospheric Terrest. Phys.* **34**, 673.
Brinton, H. C. and Mayr, H. G.: 1971, *J. Geophys. Res.* **76**, 6198.
Brinton, H. C., Pharo, M. W., III, Mayr, H. G., and Taylor, H. A.: 1969a, *J. Geophys. Res.* **74**, 2941.
Brinton, H. C., Pickett, R. A., and Taylor, H. A., Jr.: 1969b, *J. Geophys. Res.* **74**, 4064.
Brinton, H. C., Mayr, H. G., Pickett, R. A., and Taylor, H. A., Jr.: 1970, *Space Res.* **10**, 652.
Brinton, H. C., Grebowsky, J. M., and Mayr, H. G.: 1971, *J. Geophys. Res.* **76**, 3738.
Danilov, A. D.: 1972, in Aeronomy Rep. 48, COSPAR Symposium on D- and E-Region Ion Chemistry, (Ed. by C. F. Sechrist and M. A. Geller), University of Illinois, Urbana, p. 195.
Donahue, T. M., Zipf, E. C., Jr., and Parkinson, T. D.: 1970, *Planetary Space Sci.* **18**, 171.
Ershova, V. A. and Sivtseva, L. D.: 1971, *Space Res.* **10**, 1063.
Giraud, A., Scialom, G., Pokhounkov, A., Poloskov, S., and Tulinov, G.: 1971, *Space Res.* **11**, 1057.
Goldberg, R. A. and Blumle, L. J.: 1970, *J. Geophys. Res.* **75**, 133.
Goldberg, R. A. and Aikin, A. C.: 1971, *J. Geophys. Res.* **76**, 8352.
Hoffman, J. H.: 1967, *J. Geophys. Res.* **72**, 1883.
Hoffman, J. H.: 1969, *Proc. IEEE.* **57**, 1063.
Hoffman, J. H., Johnson, C. Y., Holmes, J. C., and Young, J. M.: 1969, *J. Geophys. Res.* **74**, 6281.
Holmes, J. C., Johnson, C. Y., and Young, J. M.: 1965, *Space Res.* **5**, 765.
Johannessen, A. and Krankowsky, D.: 1972, *J. Geophys. Res.* **77**, 2888.
Johnson, C. Y.: 1972, *Radio Sci.* **7**, 99.
Keneshea, T. J., Narcisi, R. S., and Swider, W., Jr.: 1970, *J. Geophys. Res.* **75**, 845.
Krankowsky, D., Arnold, F., Wieder, H., Kissel, J., and Zahringer, J.: 1972a, *Radio Sci.* **7**, 93.
Krankowsky, D., Arnold, F., Wieder, H., and Kissel, J.: 1972b, *Int. J. Mass Spectrom. Ion Phys.* **8**, 379.
Maier, E. J. R.: 1969, *J. Geophys. Res.* **74**, 815.
Narcisi, R. S.: 1967, *Space Res.* **7**, 186.
Narcisi, R. S.: 1968, *Space Res.* **8**, 360.
Narcisi, R. S.: 1970, *Trans. Amer. Geophys. Union* **51**, 366.
Narcisi, R. S.: 1971, in F. Verniani (ed.), *Physics of the Upper Atmosphere*, Editrice Compositori Publisher, Bologna, p. 12.
Narcisi, R. S.: 1972, in *Proceedings of COSPAR Symposium, AFCRL* Special Reports, in press.
Narcisi, R. S. and Roth, W.: 1970, *Advan. Electron Electron. Phys.* **29**, 79.
Narcisi, R. S., Bailey, A. D., and Della Lucca, L.: 1967, *Space Res.* **7**, 123.
Narcisi, R. S., Bailey, A. D., Della Lucca, L., Sherman, C., and Thomas, D. M.: 1971, *J. Atmospheric Terrest. Phys.* **33**, 1147.
Narcisi, R. S., Bailey, A. D., Wlodyka, L. E., and Philbrick, C. R.: 1972a, *J. Atmospheric Terrest. Phys.* **34**, 647.
Narcisi, R. S., Philbrick, C. R., Thomas, D. M., Bailey, A. D., Wlodyka, L., Baker, D., Federico, G., Wlodyka, R., and Gardner, M. E.: 1972b, in *Proceedings of COSPAR Symposium on November 1969 Solar Particle Event, AFCRL* Special Reports, in press.
Narcisi, R. S., Philbrick, C. R., Ulwick, J. C., and Gardner, M. E.: 1972c, *J. Geophys. Res.* **77**, 1332.
Narcisi, R. S., Sherman, C., Philbrick, C. R., Thomas, D. M., Bailey, A. D., Wlodyka, L. E., Wlody-

ka, R. A., Baker, D., and Federico, G.: 1972d, in *Proceedings of COSPAR Symposium, AFCRL Special Reports*, in press.

Narcisi, R. S., Philbrick, C. R., MacLeod, M. A., and Rosenberg, N. W.: 1972e, *Eos Trans. AGU* **53**, 462.

Pharo, M. W., III, Scott, L. R., Mayr, H. G., Brace, L. H., and Taylor, H. A., Jr.: 1971, *Planetary Space Sci.* **19**, 15.

Rosenberg, N. W. and Zimmerman, S. P.: 1972, *Radio Sci.* **7**, 377.

Stubbe, P. and Chandra, S.: 1970, *J. Atmospheric Terrest. Phys.* **32**, 1909.

Swider, W., Jr. and Narcisi, R. S.: 1970, *Planetary Space Sci.* **18**, 379.

Taylor, H. A., Jr.: 1971, *Planetary Space Sci.* **19**, 77.

Taylor, H. A., Jr., Brinton, H. C., Pharo, M. W., III, and Rahman, N. K.: 1968, *J. Geophys. Res.* **73**, 5521.

Taylor, H. A., Jr., Brinton, H. C., Carpenter, D. L., Bonner, F. M., and Heyborne, R. L.: 1969, *J. Geophys. Res.* **74**, 3517.

Taylor, H. A., Jr., Mayr, H. G., and Brinton, H. C.: 1970, *Space Res.* **10**, 663.

Thomas, D. M., Narcisi, R. S., Bailey, A. D., and Wlodyka, L. E.: 1972, *Trans. Amer. Geophys. Union* **53**, 456.

Young, J. M., Johnson, C. Y., and Holmes, J. C.: 1967, *J. Geophys. Res.* **72**, 1473.

Zhloodko, A. D. and Klyueva, N. M.: 1970, *Space Res.* **10**, 746.

Zipf, E. C., Borst, W. I., and Donahue, T. M.: 1970, *J. Geophys. Res.* **75**, 6371.

OBSERVATION AND INTERPRETATION OF HYDROXYL AIRGLOW EMISSIONS

R. L. GATTINGER and A. VALLANCE JONES

*Astrophysics Branch, National Research Council of
Canada, Ottawa, Canada*

Abstract. The observation and interpretation of a number of interesting features associated with the OH airglow emissions are discussed. Correlations between magnetic activity and OH emission brightness as well as O_2 1.27 μm band brightness are considered. Correlations between the brightness of the O_2 1.27 μm band and the brightness of the OH and OI(5577 Å) airglow emissions are also considered with references made to possible excitation mechanisms. Observational evidence which tends to confirm the possible existence of a nonequilibrium distribution in the population of the rotational levels of excited OH molecules in the atmosphere is presented.

1. Introduction

During the past few years there have been a number of interesting observations and discussions of the complex behavior of the OH airglow emissions. Using data gathered during an extensive synoptical observational program Shefov (1969a) observed a correlation between both the OH emission brightness and temperature, and strong magnetic storms. He also observed a marked seasonal variation in the relative population rates of the various vibrational levels of the emitting OH molecules (Shefov, 1969b). Dick (1972) obtained excellent high resolution spectra of the night airglow OH emissions, and derived from them the intensity and temperature variations throughout the course of a number of nights. A correlation between the OH emissions, the IR atmospheric oxygen band system, and the OI(5577 Å) emission was suggested by Evans *et al.* (1971). Harrison *et al.* (1971) presented results which indicated that some of the newly-formed excited OH molecules do not attain an equilibrium rotational distribution before emission occurs; Nicholls *et al.* (1972) considered quantitatively the mechanism whereby rotational equilibrium could be achieved. However, Shefov (1972) found no evidence for any nonequilibrium rotational distribution in his data.

A number of laboratory measurements dealing with OH formation mechanisms (Charters *et al.*, 1971; Murphy, 1971; Potter *et al.*, 1971) have been made recently. Evans and Llewellyn (1972) extended the results to predict the intensities of the OH bands and the total system in the night airglow.

A number of these experimental observations will be dealt with in more detail in the following sections, and additional experimental observations will be presented. Observations of the O_2 1.27 μm band will also be discussed because of the apparent partial covariation of this band and the OH emissions.

2. Correlation of OH and O_2 1.27 μm Emissions with K_p

Results obtained from the Auroral Observatory at Fort Churchill using a three channel IR photometer (1.23, 1.27, and 1.50 μm) were analyzed to determine if any correlation between the OH emissions and the K_p value existed as observed by Shefov (1969a). Observations were made for a total of about 20 days during two observing periods which were in March 1971 and January 1972. An event similar to the type described by Shefov was observed during the March 1971 period. The results are given in Figure 1. Since this was only a single event the results are far less significant than those given by Shefov. Furthermore, there was another magnetic disturbance about five days earlier which could have affected the results during the period under consideration. In spite of these limitations the large change in OH emission intensity suggests that it was an event of the type discussed by Shefov.

The OH emission intensities correspond to times near local midnight. The observed brightness of the OH 7,4 band near the end of the period (13 kR) agrees well with the value of 11.7 kR derived by Evans and Llewellyn (1972). However, the observed brightness of the OH 3,1 band (40 kR) is considerably lower than their value (63.3 kR). If it is assumed that the average brightness of the OH 3,1 band in the night airglow is about 60 kR then the average OH 7,4 band brightness is about 20 kR.

The intensity variation of the O_2 1.27 μm night airglow emission during the event was similar to that of the OH emissions. Following the onset of the magnetic disturbance the 1.27 μm emission decreased sooner than the OH emissions; Shefov (1969a) noted a similar effect with the O_2 0,1 8645 Å band. On the basis of the OH and $O_2(^1\Delta_g)$ excitation mechanisms referred to in the following section the results suggest that the O_2 1.27 μm emission near the beginning of the disturbance had a significant component arising from a three-body O association mechanism, and that this component became small in about 2 days at which time the O_2 1.27 μm emission arose largely from a formation mechanism associated with O_3 and OH formation. After about 2

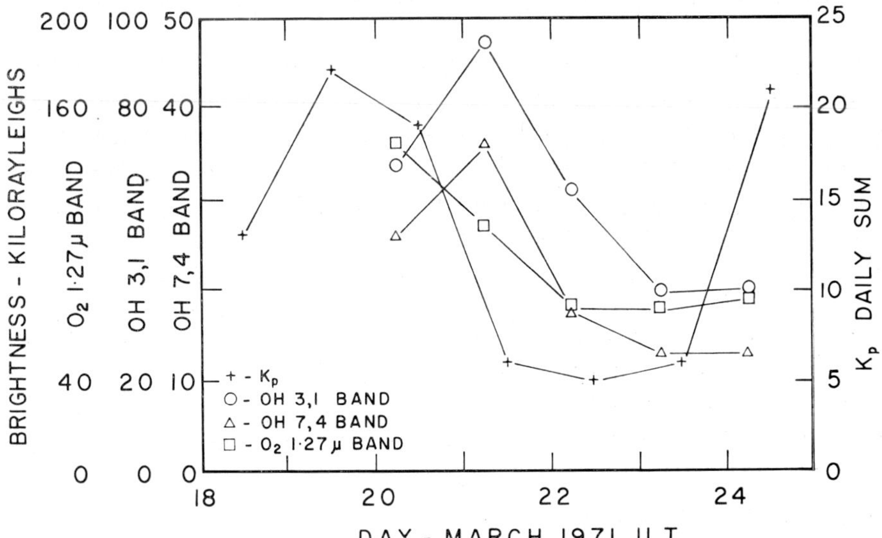

Fig. 1.   Correlation of OH and O₂ 1.27 μm emissions with K_p.

more days the ratio of the O_2 1.27 μm to the OH emissions again increased indicating that the O association mechanism again was the cause of a large portion of the 1.27 μm emission although the total brightness was abnormally low.

It is conceivable that the energy influx associated with the magnetic disturbance could alter the atmospheric mixing processes such that the O concentration in the 95 km region would increase which could cause an increase in the 1.27 μm emission from the O association reaction. This change in the O concentration could propagate downwards to the 85 km level where the subsequent increase in the rate of formation of O_3 and OH could directly increase the intensity of the O_2 1.27 μm band and the OH emissions. When the atmospheric mixing processes returned to normal there would be a decrease in the O association and O_3 and OH formation rates for several days due to the lower O concentrations since the loss rate of O would have been abnormally large for the preceding few days. This could cause a decrease to below normal in the intensity of the OH emissions as observed by Shefov (1969a). The intensity of the O_2 1.27 μm emission would also be abnormally low if it arises from the O_3 and OH formation and O association mechanisms.

3. Correlation of O_2 1.27 μm, OH, and OI(5577 Å) Night Airglow Emissions

Spectrometric and photometric observations of the O_2 1.27 μm, OH, and OI (5577 Å) emissions were made during the NASA 1969 Airborne Auroral Expedition. Evans *et al.* (1971) presented preliminary photometric results which indicated that some degree of correlation existed between the three emissions if only the night airglow component of the OI (5577 Å) emission was considered. Gattinger and Vallance Jones (1972), using IR spectrometric data also obtained a correlation between the O_2 1.27 μm band and the OI (5577 Å) night airglow emission. They compared the average 1.27 μm band intensity for a complete flight with the corresponding OI (5577 Å) night airglow brightness. Spectra obtained during three flights are given in Figure 2; each one was derived by averaging all the scans in a given flight. Intensities for the O_2 1.27 μm band were determined by measuring the area of the spectral feature after subtracting an estimated night airglow OH component and assuming a smooth profile for the remaining unexplained background emissions. The night airglow OI (5577 Å) component was determined using the relationship

$$I_{\text{NIGHT}}^{5577\,\text{Å}} = I_{\text{TOTAL}}^{5577\,\text{Å}} - 5\left(I_{\text{TOTAL}}^{4278\,\text{Å}} - 3I_{\text{TOTAL}}^{4861\,\text{Å}}\right) - 10I_{\text{TOTAL}}^{4861\,\text{Å}}$$

for times when the term in parentheses was below about 10 R. If the term was greater then the previously determined value was retained. The results obtained for nine flights are given in Figure 3. An atmospheric transmission of 22% was assumed for the O_2 1.27 μm band (Evans *et al.*, 1970). Observing periods near sunrise or within several hours after sunset were omitted to exclude any possible contributions from twilight or day airglow excitation mechanisms.

The recent results of Evans *et al.* (1972), and Llewellyn *et al.* (1973) on the vertical profile of the O_2 1.27 μm emission in the night airglow and the excitation mechanisms

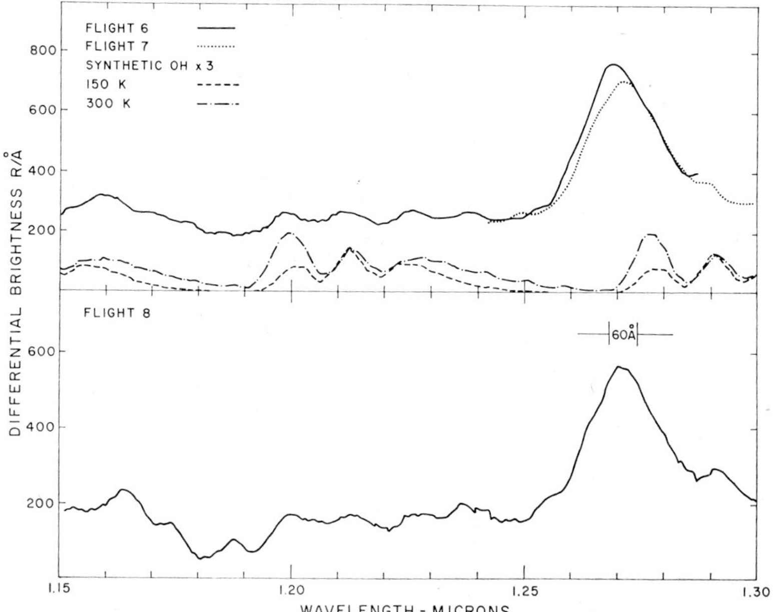

Fig. 2. Averaged spectra obtained during the NASA 1969 Airborne Auroral Expedition showing
the O_2 1.27 μm emission.

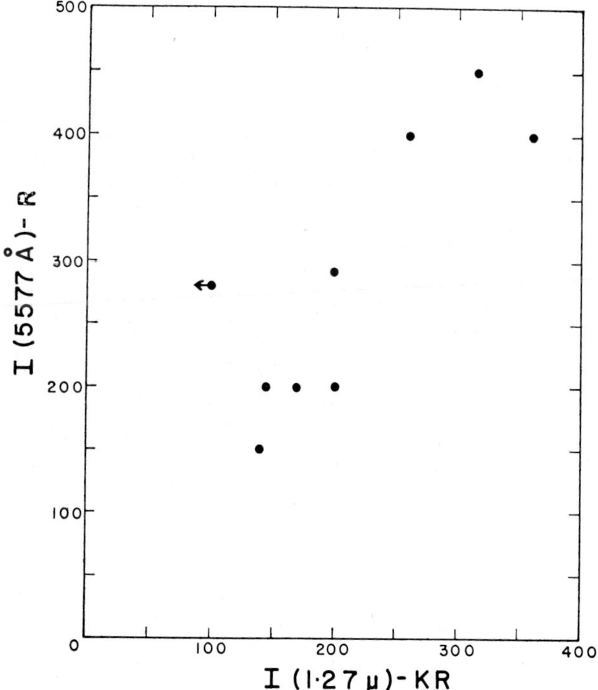

Fig. 3. Correlation between O_2 1.27 μm and OI(5577 Å) night airglow emissions observed during
the NASA 1969 Airborne Auroral Expedition.

they suggested provide important information for the analysis of the data obtained during the NASA 1969 Airborne Auroral Expedition. A more detailed analysis which also included the OH emissions was undertaken. In order to improve the time resolution and still maintain a good signal-to-noise ratio in the O_2 1.27 μm spectra each flight was divided into four approximately equal time intervals; the results obtained for the intensities of the O_2 1.27 μm emission for each of six flights are given in Figure 4. The average intensities of the Q branch of the OH 8,6 band for the same time periods (kindly provided by G. Moreels) are also plotted. A different method from the one described above was used to estimate the corresponding OI(5577 Å) night airglow intensities which are also given in Figure 4. It was assumed that the maxima for the

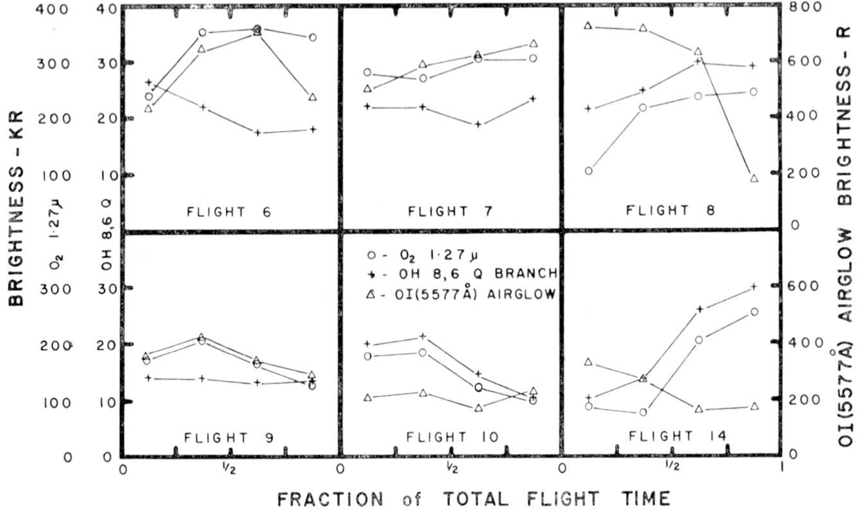

Fig. 4.   Correlation of OH, O_2 1.27 μm, and OI(5577 Å) night airglow emissions observed during the NASA 1969 Airborne Auroral Expedition.

ratios I(4278 Å)/I(4861 Å) and I(5577 Å)/I(4861 Å) in proton excited aurora are 3 and 10, respectively, and that whenever the observed ratio I(4278 Å/I(4861 Å) falls below 3 the ratio I(5577 Å)/I(4861 Å) should be reduced in a linear fashion. This was an attempt to account for the presence of soft proton precipitation in at least a very approximate fashion since the observed ratio I(4278 Å)/I(4861 Å) was frequently considerably less than 3. The ratio I(5577 Å)/I(4278 Å) for electron excited aurora was again taken to be 5; if this component of the 4278 Å emission was less than 50 R a new value for the OI(5577 Å) night airglow intensity was calculated but if it was greater than 50 R the previous value was retained. This procedure was followed for 6 s intervals throughout each flight. It is very difficult to estimate the accuracy of the determination of the OI(5577 Å) night airglow intensity; the errors could be very significant.

Inspection of the results in Figure 4 indicates that a very good correlation between the O_2 1.27 μm and OH emissions existed during flights 10 and 14. Since the OI (5577 Å) emission was relatively weak on these flights, and more importantly, since this emission varied in a totally different manner compared with the other two emissions it is likely that very little of the O_2 1.27 μm emission was due to an O association mechanism.

For flight 6 and flight 9 the O_2 1.27 μm emission correlated much better with the OI(5577 Å) night airglow emission than with the OH emission. Presumably during these two flights the production of $O_2(^1\Delta_g)$ by an O association mechanism was the dominant process. The intensity of the OI(5577 Å) emission was also generally considerably larger than during flight 10 and flight 14. During flight 7 the change in intensity of the three emissions was small so no obvious correlation could be detected, but high the intensity of the O_2 1.27 μm band was again accompanied by a strong OI(5577 Å) night airglow emission. For flight 8 the determination of the OI(5577 Å) night airglow component was very uncertain due to the presence of fairly strong aurora for approximately the first three-quarters of the flight. However, if the results are considered to be sufficiently accurate then it can be seen that the O_2 1.27 μm emission can be correlated with the OH emissions even if a large OI(5577 Å) component is present.

4. Nonequilibrium Rotational Population Distributions in Excited Hydroxyl Molecules

In the $H-O_3$ mechanism for the production of excited OH molecules the energy balance is such that if the OH molecules are produced in the ninth vibration level then there is sufficient energy remaining to populate the rotational levels up to the $X^2\Pi_{3/2}$, $J = 15/2$ level; for initial formation in the eighth vibrational level rotational excitation up to the $X^2\Pi_{1/2}$, $J = 29/2$ level is possible. There is laboratory evidence which indicates that at sufficiently low reaction pressures formation in the highest vibrational levels along with a high degree of rotational excitation are indeed preferred (Charters et al., 1971). At increased pressures an equilibrium rotational distribution is quickly achieved through collisions with other molecules (Nicholls et al., 1972). Harrison et al. (1971) have found that OH night airglow spectra can exhibit a non-uniform rotational temperature across the P branch of a band. Nicholls (1971) found that this effect must be caused by a production of OH molecules with a nonequilibrium population distribution in the rotational levels with subsequent emission before complete thermalization has occurred. Shefov (1972) has not been able to observe this effect, and it is also not apparent in the spectra obtained by Dick (1972). However, Nicholls (1971) observed the effect in the high resolution spectra obtained by Broadfoot and Kendall (1968). In the following discussion an attempt will be made to show that this nonequilibrium effect does occur in the atmosphere but that it is not always present.

Three spectra of the OH emissions in the 1.5 μ region are shown in Figure 5. The upper trace, obtained at Fort Churchill in January 1972, shows a fairly typical spectrum, while the lower pair of traces, obtained near local midnight at Ottawa, Canada on June 9, 1971, both possess an anomalous feature at 1.5145 μm. This feature was

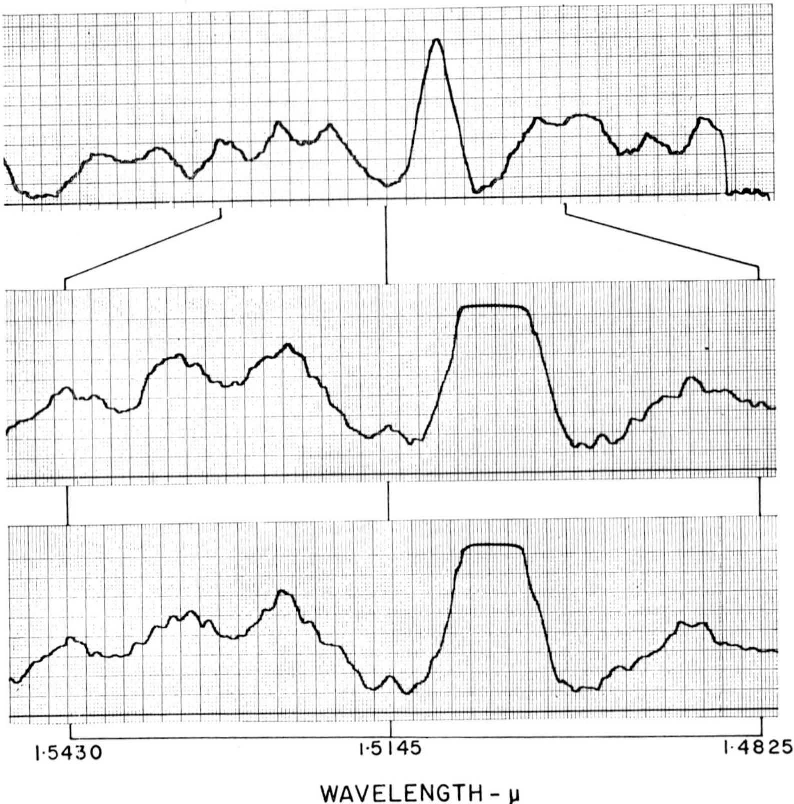

Fig. 5. OH airglow emission spectra in the 1.5 μm region at 60 Å spectral slit width, upper trace showing normal spectrum, lower two traces showing nonequilibrium rotational population distributions.

observable over a period of at least 2 hr but it has not been detected again. The feature was precisely at the wavelength of the OH 2,0 band $P_1(8)$ line. Since the relative intensities of the P branch lines in the OH 3,1 band indicated that the average emission temperature was below 200 K, which is in accordance with what is expected for the 85 km region at 45°N latitude in June (*U.S. Standard Atmosphere Supplements*, 1966), it was assumed that along with the low temperature component there may be a weak high temperature component which could cause the higher rotational levels to be populated to a significant degree. However, by using synthetic spectra it was found that an emission feature at 1.5145 μm could not be produced using two spectra at different temperatures even for a case in which 75% was assumed to come from a 175 K emission and 25% from a 700 K emission; the effect obtained in the 1.5145 μm region was merely a general increase in the emission level in the minimum between the Q branch and the P branch. However, if an atmospheric temperature of 175 K was assumed and a nonequilibrium rotational population was added to the main equilibrium distribution then the emission at 1.5145 μm could be matched by the synthetic spectra. Addition into the two $K' = 7$ levels of about 12% of the population in the $^2\Pi_{3/2}$, $J = 3/2$ level for the 175 K distribution was sufficient to produce the observed emission.

Further evidence for the existence of nonequilibrium population distributions can be seen in a spectrum published by Shemansky and Vallance Jones (1961) as their Figure 2. Emissions in excess of those predicted by their synthetic spectrum are present at numerous wavelengths. Those in the 1.515 and 1.592 μm regions could be due to nonequilibrium emissions arising from the P(8) lines of the 2,0 and 3,1 bands, respectively, as discussed above. However, a much more convincing argument for the existence of nonequilibrium distributions can be presented on the basis of the occurrence of the strong unexplained emission in their spectrum at 1.538 μm. This wavelength corresponds approximately with the 4,2 band R branch head which occurs at about 1.537 μm with $K'' = 10$. All R branch lines with K'' values from 7 to 13 are within about 35 Å of the branch head. By means of synthetic spectra it was found that no reasonable blend of atmospheric temperatures could cause the observed 1.538 μm emission. By assuming an atmospheric temperature of 175 K and an additional nonequilibrium component with about 9, 7.5, 5.5, 4, 3, and 2% of the population of the $^2\Pi_{3/2}$, $J = 3/2$ level for the 175 K distribution to be in each of the $K' = 8, 9, 10, 11, 12,$ and 13 levels, respectively, the synthetic spectrum given in Figure 6 was obtained. The nonequilibrium distribution was chosen to have a maximum in the $K' = 7$ level since initial formation in the ninth vibrational level could produce this effect. Further arbitrary manipulations of the relative populations of the nonequilibrium component could obviously produce a better match between the observed and the synthetic spectra

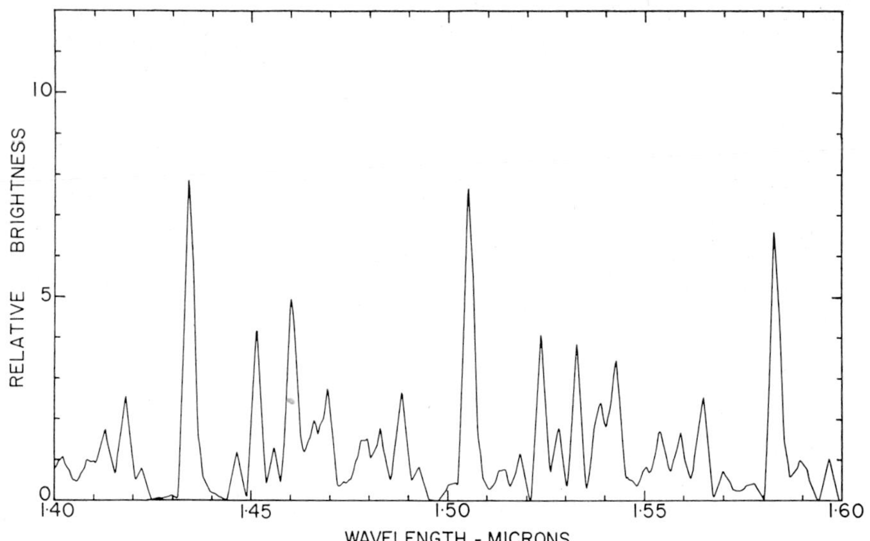

Fig. 6.   Synthetic OH spectrum with 25 Å spectral slit width and T = 175 K showing formation of 4,2 band R branch head at about 1.538 μm caused by nonequilibrium rotational population distribution.

but that would not be justified at this time. It would be better to consider all the vibrational and rotational levels of the OH emission system to see what initial excitation characteristics and subsequent interactions are necessary to reproduce the observed spectra. Formation of excited OH molecules in the eighth vibrational level with rotational exitation up to the fourteenth level would also play an important role in the formation of the R branch head of the 4,2 band.

Acknowledgments

We wish to express our appreciation to Mr G. Moreels for allowing us to use his unpublished data on the OH airglow emissions. We are also deeply indebted to Dr M. Bader and Mr L. Haughney for their assistance during the NASA 1969 Airborne Auroral Expedition. The capable support provided by the staff at the Churchill Research Range is also gratefully acknowledged.

References

Broadfoot, A. L. and Kendall, L.: 1968, *J. Geophys. Res.* **73**, 426.
Charters, P. E., Macdonald, R. G., and Polanyi, J. C.: 1971, *Appl. Opt.* **10**, 1747.
Dick, K. A.: 1972, *Ann. Geophys.* **28**, 149.
Evans, W. F. J. and Llewellyn, E. J.: 1972, *Planetary Space Sci.* **20**, 624.
Evans, W. F. J., Wood, H. C., and Llewellyn, E. J.: 1970, *Can. J. Phys.* **48**, 747.
Evans, W. F. J., Llewellyn, E. J., Moreels, G., and Blamont, J.: 1971, *Trans. Amer. Geophys. Union* **52**, 482.
Evans, W. F. J., Llewellyn, E. J., and Vallance Jones, A.: 1972, *J. Geophys. Res.* **77**, 4899.
Gattinger, R. L. and Vallance Jones, A.: 1972, *Trans. Amer. Geophys. Union* **53**, 470.
Harrison, A. W., Evans, W. F. J., and Llewellyn, E. J.: 1971, *Can. J. Phys.* **49**, 2509.
Llewellyn, E. J., Evans, W. F. J., and Wood, H. C.: 1973, this volume, p. 193.
Murphy, R. E.: 1971, *J. Chem. Phys.* **54**, 4852.
Nicholls, D. C.: 1971, M. Sc. Thesis, Univ. of Saskatchewan, Saskatoon.
Nicholls, D. C., Evans, W. F. J., and Llewellyn, E. J.: 1972, *J. Quant. Spectr. Radiative Transfer* **12**, 549.
Potter, A. E. Jr., Coltharp, R. N., and Worley, S. D.: 1971, *J. Chem. Phys.* **54**, 992.
Shefov, N. N.: 1969a, *Planetary Space Sci.* **17**, 97.
Shefov, N. N.: 1996b, *Planetary Space Sci.* **17**, 1629.
Shefov, N. N.: 1972, *Can. J. Phys.* **50**, 1225.
Shemansky, D. E. and Vallance Jones, A.: 1961, *J. Atmospheric Terrest. Phys.* **22**, 166.
U.S. Standard Atmosphere Supplements: 1966 (U.S. Government Printing Office, Washington, D.C.)

$O_2(^1\Delta)$ IN THE ATMOSPHERE

E. J. LLEWELLYN, W. F. J. EVANS, and H. C. WOOD*

*Institute of Space and Atmospheric Studies,
University of Saskatchewan, Saskatoon*

1. Introduction

The presence of large amounts of O_2 excited to the $a^1\Delta_g$ state was first shown by Vallance Jones and Harrison (1958) who detected the 0-1 band of the IR atmospheric system in the evening twilight. However, since that time the IR atmospheric emissions have been extensively studied using ground based, airborne, and rocket techniques, and have provided significant new information relating to the physics and chemistry of the atmosphere. In the present paper these measurements and the mechanisms responsible for the formation of $O_2(^1\Delta)$ in the atmosphere are considered.

2. Observations

2.1. Integrated intensity measurements

The early observations of the (0-1) band at 1.58 μm (Vallance Jones and Harrison, 1958; Vallance Jones and Gattinger, 1963) and the more recent synoptic observations of the (0-0) band (Evans *et al.*, 1970b; Evans and Llewellyn, 1972a) have been restricted to the twilight airglow so that only limited information may be obtained for the diurnal variation of the emission. However, the use of balloon and airborne instrumentation has permitted the extension of observations throughout the complete diurnal period (Evans *et al.*, 1969) as shown in Figure 1. It is apparent that the emission intensity exhibits a time constant effect due to the 60 min radiative lifetime. This behavior is repeated for all measurements although there is some seasonal variation in the evening twilight decay (Evans and Llewellyn, 1972a). It should be noted that observations at 1.58 μm must be made with high resolution as the spectral blending of the OH Meinel bands, which are known to undergo a twilight variation (Moreels *et al.*, 1970; Pick *et al.*, 1971), makes the interpretation of the observations uncertain. This effect may explain the apparent differences in the twilight variations of the 1.58 μm band at different spectral resolutions (Vallance Jones and Gattinger, 1963; Gattinger, 1968). The high resolution measurements of the 0-1 band are in excellent agreement with the balloon observations and the more recent ground based studies of the 0-0 band (Evans *et al.*, 1969). However, it should be noted that these comparisons have assumed a photon ratio for 0-0/0-1 bands of 46; this value is in good agreement with that from laboratory studies (Becker *et al.*, 1971) although it is significantly smaller than the value from direct atmospheric measurements (Pick *et al.*, 1971).

* Present address: MISU, University of Stockholm, Sweden.

B. M. McCormac (ed.), Physics and Chemistry of Upper Atmospheres, 193–202. All Rights Reserved.
Copyright © 1973 by D. Reidel Publishing Company, Dordrecht-Holland.

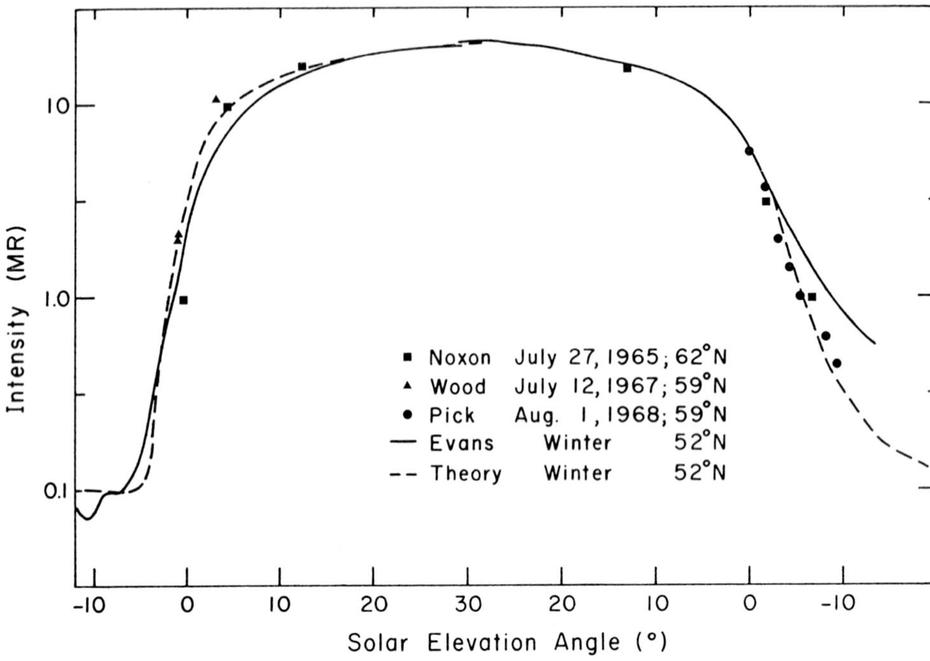

Fig. 1.   Predicted and observed integrated intensities for the 1.27 μm O_2 emission in the airglow.

From the observations of the twilight emission at 1.58 μm Vallance Jones and Gattinger (1963) proposed that the $O_2({}^1\Delta_g)$ molecules were excited through the photolysis of O_3 in the Hartley band (Equation (1)),

$$O_3({}^1A) + h\nu \rightarrow O_2({}^1\Delta_g) + O({}^1D). \tag{1}$$

However, to obtain agreement with the observed twilight variation these authors postulated that quenching of the excited state occurred through both 0 and M. The subsequent 1.27 μm observations (Evans et al., 1969) indicated that quenching of the excited state by O could be neglected. Thus for a static O_3 profile it is a simple matter to derive the emission height profile and the diurnal variation of the total emission intensity. The predicted diurnal variation, which is based on an extrapolation of the Johnson et al. (1954) O_3 height profile, is included in Figure 1. It is apparent that there is a significant difference between the predicted and observed variation during evening twilight; this difference is consistent with either an enhanced O_3 concentration in the region where collisional quenching is negligible or a secondary dayglow mechanism. The nightglow emission has also been studied and these observations have indicated that the nightglow intensity is typically 100 kR, although variations of a factor of 2 are quite common. These observations (Evans et al., 1971) have also indicated that the O_2 emission covaries with the Meinel OH emission.

2.2. HEIGHT PROFILES

2.2.1. *Dayglow*

The dayglow integrated intensity observations have been complemented with measurements of the dayglow height profile (Evans *et al.*, 1968; Megill *et al.*, 1970b; Wood *et al.*, 1970) and the Churchill observations are shown in Figure 2. A comparison of the observed and predicted height profiles (Figure 3) verifies that the primary excitation mechanism in the dayglow is the photolysis of O_3. The presence of a high altitude emission is apparent in the observations and is predicted in the extensive calculations of Wood (1972) for excitation through O_3 photolysis. Conventional O_3 sensing techniques cannot confirm the presence of this second O_3 region, except for solar zenith angles near 90°, although the stellar occultation technique has shown a second O_3 layer, near 80 km, at night (Hays *et al.*, 1972).

The presence of this upper O_3 layer has, however, been confirmed from concurrent measurements of the 1.27 μm emission height profile and the O_3 concentrations in the twilight (Llewellyn and Evans, 1971). The measured O_3 profile (Miller and Ryder, 1973) and that derived from the $O_2(^1\Delta_g)$ measurements are in good agreement (Evans and Llewellyn, 1972a).

2.2.2. *Nightglow*

The possible contribution of excitation mechanisms other than O_3 photolysis to the

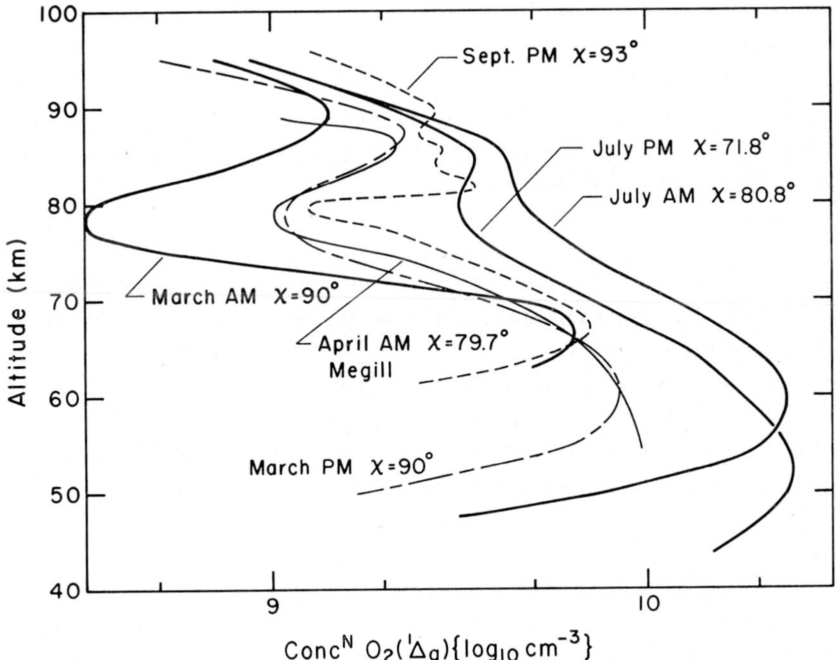

Fig. 2. $O_2(^1\Delta_g)$ concentration height profiles measured over Churchill.

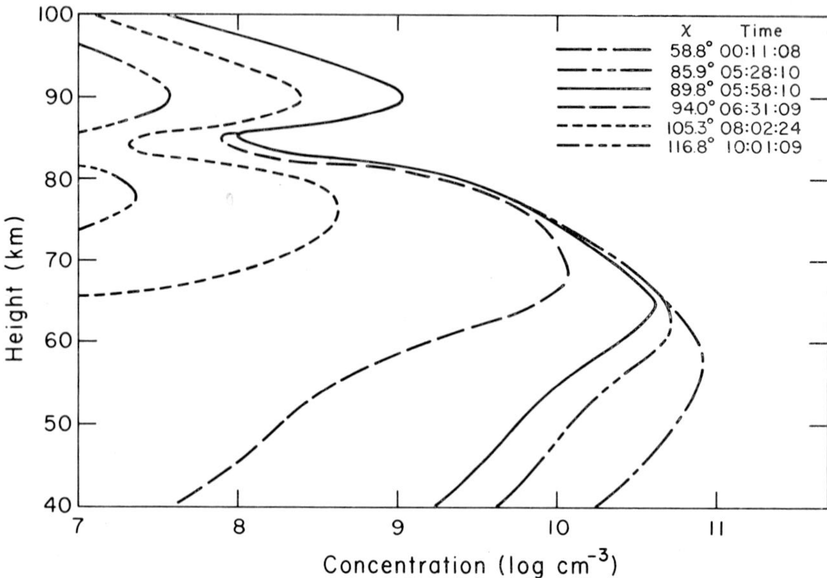

Fig. 3.   The predicted variation of the $O_2(^1\Delta_g)$ emission profile in the evening twilight.

dayglow may be determined from a measurement of the emission height profile at a time when the primary dayglow mechanism is absent, i.e. the nightglow. A recent rocket experiment (Evans *et al.*, 1972) has determined the nightglow profile and the derived volume emission profile is shown in Figure 4. On this occasion the emission intensity was only one half of the typical value, and the increased instrument sensitivity has permitted the identification of structure not seen in the dayglow measurements. This nightglow experiment also measured the Meinel band OH emission in the $\Delta v = 2$ sequence, and the derived volume emission profile is in good agreement with that predicted from model calculations.

It is apparent from this nightglow measurement that the 1.27 μm emission is contained in two layers (Figure 4). The lower layer is in the same altitude region as the day and twilight upper layer, and in very close agreement with the OH profile measured on the same flight. The agreement of the $O_2(^1\Delta_g)$ upper layer and the OH emission profile in the twilight has been previously observed in an experiment using the same instrumentation (Llewellyn and Evans, 1971). The upper nightglow layer has not previously been detected in the dayglow observation, probably because of the low intensity, but is in the same altitude regime as that reported for other O_2 emission features (Reed, 1968; Packer, 1961) and the green line airglow (Greer and Best, 1967).

2.3. SEASONAL VARIATION

The rocket measurements have been made over a range of seasons so that in conjunction with the ground based measurements the seasonal behavior of the emission may be derived. The observations of the main dayglow layer (Figure 2) have indicated that the emission intensity in this layer is independent of season. Hence any seasonal varia-

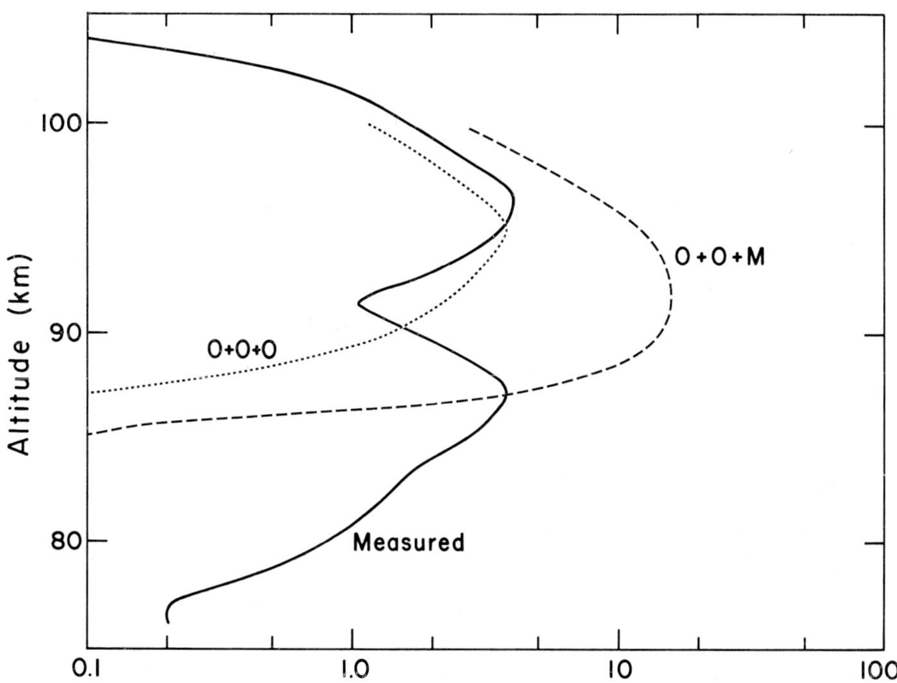

Fig. 4.   Predicted and observed $O_2(^1\Delta_g)$ concentrations in the night airglow.

tions in the O_3 concentrations below 75 km must be small. However the initial studies by Vallance Jones and Gattinger (1963) indicated that the twilight emission intensity exhibited a seasonal variation. This work has been extended (Evans and Llewellyn, 1972a) and it is apparent that the twilight intensities are consistent with an increased decay time constant during the winter months. If it is assumed that the only excitation mechanism for the 85 km layer is the photolysis of O_3 then these twilight variations may be inverted to derive the mesospheric O_3 concentrations and their seasonal variations. A similar seasonal variation in the mesospheric O_3 content has been obtained by Noxon (1972) from aircraft observations of the twilight decay of the emission intensity. However, the inferred O_3 concentration could be affected by seasonal variations in the mesospheric wind patterns and changes in the atmospheric temperature profile which can change the atmosperic transmission. The magnitude of these effects is expected to be much less than the 3:1 winter to summer variation derived from the measurements.

This seasonal variation is of importance as studies of the Meinel OH bands could indicate whether the H atom concentration is also exhibiting a seasonal variation. Hence a combined observational program may provide many new parameters for atmospheric simulations.

2.4. AURORAL STUDIES

Many attempts have been made to study the aurorally associated emission from $O_2(^1\Delta,)$

but the auroral morphology is still uncertain. A series of aircraft observations (Llewellyn *et al.*, 1969) suggested an auroral enhancement but with possible quenching of the excited state; a similar quenching of the excited state is also suggested from observations during the ionospheric modification experiment (Evans *et al.*, 1970a) although the results are much more difficult to understand. Rocket measurements of the $O_2(^1\Delta_g)$ emission in aurora have been reported by Megill *et al.* (1970a) and their interpretation indicated quenching of the excited state. However, a different analysis of these data (Evans and Llewellyn, 1971) has suggested that the measurements were consistent with observing the nightglow layer through a long path length.

More recent observations of the auroral enhancement of the emission (Noxon, 1970; Evans *et al.*, 1971; Gattinger and Vallance Jones, 1972) have indicated that the $O_2(^1\Delta_g)$ molecules appear in a general glow around the auroral precipitation regions. For observations of auroral activity (IBC II) that commences after the time that the atmosphere at 100 km is in the solar UV shadow the auroral enhancement is typically equal to the nightglow intensity (100 kR). For activity commencing during the period of solar illumination the enhancement is typically 500 kR although much larger enhancements have been observed on a few occasions.

3. Discussion

3.1. DAYGLOW EMISSION

The large dayglow intensity observed for the IR atmospheric system of O_2 requires that the $O_2(^1\Delta_g)$ molecules are excited through the photolysis of O_2 by solar UV radiation. An extensive model calculation based on this mechanism has reproduced the observed dayglow profile, as shown in Figure 3. Hence it is apparent that the O_3 concentration profile does exhibit a minimum near 80 km as shown in the simultaneous measurements of the O_2 emission and the O_3 profile. A simple oxygen atmosphere cannot reproduce this minimum but the extensive chemical and transport models (Gattinger, 1968, 1969, 1971; Hesstvedt, 1968; Bowman *et al.*, 1970; Wood, 1972) all indicate a minimum in the O_3 concentration at this altitude. This minimum must be ascribed to the increased association of O through the constituents H and OH (Gattinger, 1968). However, the lack of complete agreement between the observations and the model calculations suggests that many adopted parameters could still be in error.

3.2. NIGHTGLOW EMISSION

Many previous discussions of the role of $O_2(^1\Delta)$ in the atmosphere have commented on the uncertainty of the nightglow emission and the contribution of this mechanism to the dayglow emission. It was shown previously that the nightglow intensity has a typical value of 100 kR although a factor of 2 variation is not uncommon. For the measured nightglow profile (Section 2) the emission appeared in two distinct layers so that at least two mechanisms must be considered. One is related to the OH* mechanism and the other is probably associated with other O_2 emission mechanisms.

As the excitation energy of the $O_2(^1\Delta_g)$ state is 0.98 eV any postulated excitation mechanism must provide at least this energy. Possible reactions which could give $O_2(^1\Delta_g)$ are

$$H + O_3 \rightarrow OH + O_2 + 3.34\ eV, \tag{2}$$

$$O + O_3 \rightarrow O_2 + O_2 + 4.06\ eV, \tag{3}$$

$$O + O_2 + M \rightarrow O_3 + M + 1.04\ eV, \tag{4}$$

$$O + O + M \rightarrow O_2 + M + 5.17\ eV, \tag{5}$$

and

$$HO_2 + O \rightarrow OH + O_2 + 2.38\ eV. \tag{6}$$

Equation (2) was considered by Evans and Llewellyn (1970) from studies of intensity covariations of $O_2(^1\Delta)$ and OH*. These authors suggested that the excitation might occur in either the primary reaction of a subsequent energy transfer. Laboratory investigations have since indicated the energy transfer rate is much too slow and that the primary mechanism excites only the OH (Charters *et al.*, 1971).

Laboratory studies of the reaction between O and O_3 (Equation (3)) have indicated that the rate constant is temperature dependent and an order of magnitude slower than that required at mesospheric temperatures. A direct atmospheric test of this mechanism is not simple as the low mesospheric temperatures occur during summer months when there is only a short duration of UV darkness at 100 km. Thus a significant fraction of the dayglow excited molecules remains even 3 hr after sunset. Equations (3) and (6) may also be rejected on the grounds that reactions of this type are expected to result in vibrational excitation of the newly formed bond. However, recent measurements for the reaction kinetics of HO_2 (Hochnandel *et al.*, 1972) indicate that much of the accepted chemistry should be modified and a comparison with the calculated excitation rates for OH (Evans and Llewellyn, 1972b) suggests that the product from Equation (6) is not excited OH.

Thus the possible remaining reactions for production of $O_2(^1\Delta)$ in the nighttime atmosphere are the three-body recombination reactions of O and the reaction between HO_2 and O. The reaction

$$O + O_2 + M \rightarrow O_3 + M \tag{4}$$

is exothermic to an amount 1.04 eV so that the third-body M can be expected to accept the stabilizing energy. As the excitation energy for $O_2(^1\Delta)$ is 0.98 eV it is apparent that if $M \equiv O_2$ then the third body could be readily excited into the $a(^1\Delta_g)$ state; the formed ozone would then react with H.

It can be shown that if $O_2(^1\Delta)$ is produced in Equation (4) at a rate equal to the measured total rate constant then the predicted intensity ratio for the Meinel OH band and the $O_2(^1\Delta)$ emissions is in good agreement with that observed in the lower nightglow layer (Evans *et al.*, 1972). However, it is probable that the rate for $O_2(^1\Delta)$ production in Equation (4) is much less than the total reaction rate (Schiff, 1973) so that there will be a negligible contribution to the atmospheric $O_2(^1\Delta)$ concentrations.

The similarity of the upper nightglow layer of the $O_2(^1\Delta_g)$ profile with that for other O_2 emissions suggests that the association of O_2 (Equation (5)) is a probable excitation mechanism. For the measured total rate constant, $k = 8 \times 10^{-33}$ cm^6 s^{-1} (Campbell and Thrush, 1967), and an O concentration of 8×10^{11} cm^{-3} (Evans and Llewellyn, 1970) the calculated peak concentration is 5×10^8 cm^{-3}, a factor of 3 larger than observed. However, the rate constant for $O_2(^1\Delta_g)$ production could be smaller than the value used here. Another recombination possibility is the Chapman mechanism in which the third body is O; the measured rate constant is 4.8×10^{-33} cm^6 s^{-1} for O^{1}S production (Felder and Young, 1972) but there is nothing to preclude a total rate constant of 10^{-31} cm^6 s^{-1}, the value required to generate the upper nightglow profile. The calculated height profiles from these mechanisms are included in Figure 4.

It is readily apparent that the nightglow excitation mechanism is still unknown and that the absence of $O_2(^1\Delta)$ molecules in laboratory studies of chemiluminescent reactions is a major difficulty. However, there can be little doubt that the only source capable of supplying the observed energy flux in the nightglow (0.16 erg cm^{-2} s^{-1}) is the association of odd oxygen formed during the daytime.

3.3. Auroral emission

The presence of the upper nightglow layer also suggests that the auroral enhancements may be explained in terms of an increase in the speed of the upper nightglow mechanism rather than electron impact (Cole, 1971; Hasted, 1971). If the Chapman mechanism is the responsible excitation source then enhancements in the O concentration of 25% are necessary to cause a 50 to 100 kR enhancement in the observed emission; such concentration changes could be caused either by a modification in the vertical transport or a variation in the O concentration through the auroral precipitation. For this mechanism the auroral enhancements would tend to be in a general glow around the form in agreement with the observations (Noxon, 1970; Evans *et al.*, 1971). However, there is still difficulty in explaining the observed decay time of the enhancement. For daytime aurora it appears that the auroral precipitation may trigger an increase of the atmospheric O_3 content. A simple calculation shows that the O_3 concentration in the emission region must equal 5×10^7 cm^{-3} to sustain a 500 kR emission over a 10 km height range. Obviously a height profile of a known auroral enhancement would be of great value in elucidating the appropriate excitation mechanisms.

3.4. Atmospheric ionization

The large $O_2(^1\Delta_g)$ concentrations observed in the initial dayglow measurements (Evans *et al.*, 1968) were suggested by Hunten and McElroy (1968) to be the source of O_2^+ ions measured by Narcisi and Bailey (1965) in the D region. These authors noted that the energy of the state, 0.98 eV, would extend the wavelength cutoff so that significant ionization could occur at lower altitudes. It was also suggested that the metastable state would permit the reaction

$$N + O_2(^1\Delta) \rightarrow NO + O \tag{7}$$

to have a rate constant equal to that required for the atmospheric production of NO.

However, subsequent measurement of the appropriate rate constant (Wayne, 1971) showed that, although this reaction is spin allowed, the rate is too small to provide a significant contribution to the measured mesospheric NO densities.

However, there is little doubt that the ionization of $O_2(^1\Delta_g)$ does contribute significantly to the measured ion densities, although recent calculations by Paulsen *et al.* (1972), which include the effects of CO_2 absorption, have reduced the original estimates of the production rates. This has been discussed in detail by Donahue (1972) who has shown that in the region 77 to 85 km the ionization of $O_2(^1\Delta_g)$ is important for both the measured O_2^+ concentrations and the conversion to H_2O cluster ions. Hence a seasonal variation in the ion concentration in this height range might be expected from the observed seasonal variation in the $O_2(^1\Delta_g)$ concentrations. However, the problem is extremely complicated as atmospheric simulations have shown that such a variation is not expected on simple chemical grounds and that a large change in the atmospheric transport parameters must be postulated (Gattinger, 1971).

4. Conclusion

As has been noted the calculation of $O_2(^1\Delta_g)$ concentrations through atmospheric simulations is extremely difficult since the complete O-H chemistry and mass transport must be included in the model. It has been suggested that many of the presently adopted parameters could be in error and this may be particularly significant for the predissociation of O_2 in the S-R bands. Recent calculations (Wood, 1972) have indicated that the vibrational populations in the $O_2(X^3\Sigma_g^-)$ state must be known if the O production rate is to be accurately determined.

Despite these limitations the measurement of the $O_2(^1\Delta)$ volume emission rate does permit the determination of the atmospheric O_3 profile over an extended height range which is significant for atmospheric simulations. It is probable that satellite measurements could provide an improved knowledge of the high altitude O_3 concentrations in the daytime. There is at present no satellite method that can provide measurements for altitudes comparable to those for the nighttime stellar occultation technique.

Acknowledgment

This work has been supported with grants in aid from the National Research Council of Canada.

References

Becker, K. H., Groth, W., and Schurath, U.: 1971 *Planetary Space Sci.* **19**, 1009.
Bowman, M. R., Thomas, L., and Geisler, J. E.: 1970, *J. Atmospheric Terrest. Phys.* **32**, 1661.
Campbell, I. M. and Thrush, B. A.: 1967, *Proc. Roy. Soc.* **A296**, 222.
Charters, P. E., Macdonald, R. G., and Polanyi, J. C.: 1971, *Appl. Opt.* **10**, 747.
Cole, K. D.: 1971, *J. Atmospheric Terrest. Phys.* **33**, 1241.
Donahue, T. M.: 1972, *Radio Sci.* **7**, 73.
Evans, W. F. J. and Llewellyn, E. J.: 1970, *Ann. Geophys.* **26**, 167.
Evans, W. F. J. and Llewellyn, E. J.: 1971, *J. Geophys. Res.* **76**, 7013.

Evans, W. F. J. and Llewellyn, E. J.: 1972a, *Radio Sci.* **7**, 45.

Evans, W. F. J. and Llewellyn, E. J.: 1972b, *Planetary Space Sci.* **20**, 624.

Evans, W. F. J., Hunten, D. M., Llewellyn, E. J. and Vallance Jones, A.: 1968, *J. Geophys. Res.* **73**, 2885.

Evans, W. F. J., Llewellyn, E. J., and Vallance Jones, A.: 1969, *Planetary Space Sci.* **17**, 933.

Evans, W. F. J., Llewellyn, E. J., Haslett, J. C. and Megill, L. R.: 1970a, *J. Geophys. Res.* **75**, 6425.

Evans, W. F. J., Wood, H. C., and Llewellyn, E. J.: 1970b, *Planetary Space Sci.* **18**, 1065.

Evans, W. F. J., Llewellyn, E. J., Moreels, G., and J. Blamont.: 1971, *Trans. Amer. Geophys. Union* **52**, 482.

Evans, W. F. J., Llewellyn, E. J. and Vallance Jones, A.: 1972, *J. Geophys. Res.* **77**, 4899.

Felder, W. and Young, R. A.: 1972, *J. Chem. Phys.* **56**, 6028.

Gattinger, R. L.: 1968, *Can. J. Phys.* **46**, 1613.

Gattinger, R. L.: 1969, *Ann. Geophys.* **25**, 825.

Gattinger, R. L.: 1971, in B. M. McCormac (ed.), *The Radiating Atmosphere*, D. Reidel Publishing Company, Dordrecht-Holland, p. 51.

Gattinger, R. L. and Vallance Jones, A.: 1972, *Trans. Amer. Geophys. Union* **53**, 470.

Greer, R. G. J. and Best, G. T.: 1967, *Planetary Space Sci.* **15**, 1857.

Hasted, J. B.: 1971, in G. Fiocco (ed.), *Mesospheric Models and Related Experiments*, D. Reidel Publishing Company, Dordrecht-Holland, p. 220.

Hays, P. B., Roble, R. G., and Shah, A. N.: 1972, *Science* **176**, 793.

Hesstvedt, E.: 1968, *Geofys. Publik.* **27**, 1.

Hochnandel, C. J., Ghormley, J. A., and Ogren, P. J.: 1972, *J. Chem. Phys.* **56**, 4426.

Hunten, D. M. and McElroy, M. C.: 1968, *J. Geophys. Res.* **73**, 2421.

Johnson, F. S., Purcell, J. D. and Tousey, R.: 1954, in R. Boyd and M. Seaton (eds.), *Rocket Exploration of the Upper Atmosphere*, Pergamon, London p. 189.

Llewellyn, E. J. and Evans, W. F. J.: 1971, in B. M. McCormac (ed.), *The Radiating Atmosphere*, D. Reidel Publishing Company, Dordrecht-Holland, p. 17.

Llewellyn, E. J., Wood, H. C., and Vallance Jones, A.: 1969, *Trans. Amer. Geophys. Union* **50**, 271.

Megill, L. R., Despain, A. M., Baker, D. J., and Baker, K. D.: 1970a, *J. Geophys. Res.* **75**, 4775.

Megill, L. R., Haslett, J. C., Schiff, H. I., and Adams, G. W.: 1970b, *J. Geophys. Res.* **73**, 6398.

Miller, D. E. and Ryder, P.: 1973, *Planetary Space Sci.* **21**, in press.

Moreels, G., Evans, W. F. J., Blamont, J. E., and Vallance Jones, A.: 1970, *Planetary Space Sci.* **18**, 637.

Narcisi, R. S. and Bailey, A. D.: 1965, *J. Geophys. Res.* **70**, 3687.

Noxon, J. F.: 1970, *J. Geophys. Res.* **75**, 1879.

Noxon, J. F.: 1972, *Trans. Amer. Geophys. Union* **53**, 456.

Packer, D. M.: 1961, *Ann. Geophys.* **17**, 67.

Paulsen, D. E., Huffman, R. E., and Larabee, J. C.: 1972, *Radio Sci.* **7**, 51.

Pick, D., Llewellyn, E. J., and Vallance Jones, A.: 1971, *Can. J. Phys.* **49**, 897.

Reed, E. I.: 1968, *J. Geophys. Res.* **73**, 2957.

Schiff, H. I.: 1973, this volume, p. 85.

Vallance Jones, A. and Gattinger, R. L.: 1963, *Planetary Space Sci.* **11**, 961.

Vallance Jones, A. and Harrison, A. W.: 1958, *J. Atmospheric Terrest. Phys.* **13**, 45.

Wayne, R. P.: 1971, in G. Fiocco (ed.), *Mesospheric Models and Related Experiments*, D. Reidel Publishing Company, Dordrecht-Holland, p. 240.

Wood, H. C.: 1972, Ph.D. Thesis, University of Saskatchewan.

Wood, H. C., Evans, W. F. J., Llewellyn, E. J., and Vallance Jones, A.: 1970, *Can. J. Phys.* **48**, 862

OXYGEN AND NITROGEN VIBRATION IN THE THERMOSPHERE

JAMES C. G. WALKER

Dept. of Geology and Geophysics, Yale University, New Haven, Conn., U.S.A.

1. Introduction

There are two factors that contribute to interest in vibrational temperatures in the thermosphere. First, the probability of conversion of vibrational energy into kinetic energy in collisions is generally small, so it is likely that vibrational temperatures differ from kinetic temperatures. Second, the rates of a number of important aeronomic reactions may depend on the vibrational levels of the participating molecules. Unfortunately, there is little information on several potentially important reactions that produce or destroy vibrating molecules, so vibrational temperatures remain largely uncertain.

2. Oxygen

Almost any reaction that includes an O_2 among its products can be a source of vibrational excitation. Useful reviews have been presented by Dalgarno (1963, 1970), and some of the sources have been evaluated by Hudson and Mahle (1972), Lane and Dalgarno (1969), and Vlasov (1971). The most serious problem is uncertainty concerning the process that removes vibrational quanta. It is possible that vibrating oxygen molecules are rapidly quenched by O, either through a process of atom-atom interchange (Bates and Moiseiwitsch, 1956)

$$O + O_2(v) \rightarrow O_2(v' < v) + O \tag{1}$$

or by a direct process resulting from a strong chemical interaction

$$O + O_2(v) \rightarrow O + O_2(v' < v). \tag{2}$$

According to a theoretical study by Breig (1969) quenching occurs on essentially every collision. If this is true, enhanced vibrational temperatures will not occur in the thermosphere.

Contradicting this conclusion, however, is an observation by Reid and Withbroe (1970) based on the attenuation of solar UV radiation. From the decrease with altitude in the effective absorption cross section at 1335 Å these authors conclude that the vibrational temperature exceeds 1100 K at 130 km and 3200 K above 160 km. The problem requires further study.

3. Nitrogen

From the theoretical point of view, the case of N_2 is better, because there is less uncertainty concerning the removal processes. Atomic N is too rare for the atom-

atom interchange process to be important. The problem has been examined by Walker (1968) and Walker *et al.* (1969). It appears that quenching of N_2 vibration occurs at a significant rate only at altitudes of about 120 km and below.* The quenching process is a vibrational exchange collision with CO_2, and because of its large molecular mass the CO_2 density decreases rapidly above the turbopause. Although quenching by O is unusually fast, presumably due to chemical interaction (Breshears and Bird, 1968), this process is not important in the atmosphere. Vibrational quanta produced within the thermosphere are removed by downward diffusion of vibrating N_2 molecules, so that vibrational temperatures may be calculated from a balance between diffusion and the sources of vibrational quanta.

The sources are not well known, however. Among the possibilities are thermal electron impact, for situations of high electron temperature, photoelectron impact, and the reaction

$$N + NO \rightarrow N_2(v) + O. \tag{3}$$

The largest problem, however, is the reaction that quenches metastable oxygen atoms

$$O(^1D) + N_2 \rightarrow O(^3P) + N_2(v). \tag{4}$$

If the energy released by this reaction is transferred largely to N_2 vibration this is the largest source, and vibrational temperatures of the order of 3000 K may be expected at high altitudes (Walker *et al.*, 1969). If, on the other hand, the yield of vibrational quanta is low, vibrational temperatures large enough to produce a significant population of vibrating molecules are not expected (Strobel and McElroy, 1970). Since the density of molecules in vibrational level i decreases as

$$e^{\dfrac{-3400i}{T_V}},$$

vibrational temperatures below about 1500 K are not very interesting.

The vibrational yield of the $O(^1D)$ quenching reaction has been examined theoretically by Fisher and Bauer (1971), who conclude that it is small. It is not clear that this result is correct, so let us assume that N_2 vibration in the thermosphere is still interesting and turn to the question of the vibrational distribution.

Vibrational exchange collisions between N_2 molecules

$$N_2(v_1) + N_2(v_2) \rightarrow N_2(v_1 - 1) + N_2(v_2 + 1) \tag{5}$$

reshuffle vibrational quanta in such a way as to drive the distribution towards the Boltzmann distribution. Disequilibrating processes are the sources of vibrational quanta, for which the distribution over vibrational levels is largely unknown

---

* Recent measurements by McNeal *et al.* (1972) show that quenching by O is fast enough to play a role between 120 and 200 km (Breig *et al.*, 1972). The description given here of the N_2 vibrational energy budget must be revised in the light of this result.

(the sources are not likely to produce a Boltzmann distribution by themselves), and the diffusion of vibrating molecules from one altitude to another. Quenching would be an additional disequilibrating process, but it is not important in the thermosphere.

Because of the anharmonicity of the N_2 molecule, departures from the Boltzmann distribution occur when the vibrational temperature differs from the kinetic temperature even when vibrational exchange collisions are much more rapid than all other processes (Bray, 1968; Fisher and Kummler, 1968; Treanor *et al.*, 1968). The effect is small for vibrational levels below about the fifth, however, so we may neglect it at the low vibrational temperatures that are of interest in the thermosphere. In what follows we shall use a harmonic model of N_2 vibration.

A numerical study of source-induced departures from the Boltzmann distribution has been presented by Schunk and Hays (1971) who find that these departures are small, except at times of very rapid change of the vibrational temperature. The discussion that follows is intended to provide some insight into this result, and to consider also the effects of diffusion. The procedure we shall follow is to estimate the magnitudes of the different terms in the equation of continuity. We use, throughout, the neutral density model of Jacchia (1971) with an exospheric temperature of 1000 K.

4. Departures from the Boltzmann Distribution

The continuity equation for N_2 molecules in vibrational level i may be written

$$\frac{\partial N_i}{\partial t} = Q_i + V_i - L_i - \frac{\partial F_i}{\partial z} \tag{6}$$

where N_i is the density of molecules in level i, $Q_i{}^*$ is the rate of production of molecules in level i by electron impact, quenching of metastable states, or chemical reactions, V_i is the rate of production and L_i is the rate of loss caused by vibrational exchange collisions with other N_2 molecules, F_i is the vertical flux of vibrating molecules resulting from diffusion, t is time, and z is altitude. Let us examine the terms in this equation in order to assess the probability of significant departures from a Boltzmann distribution over vibrational levels. In what follows we shall assume a steady state.

In Equation (6) the terms that may cause departures from the Boltzmann distribution are Q_i and $\partial F_i/\partial z$. The vibrational exchange terms tend to restore the Boltzmann distribution, for which $V_i = L_i$. Thus, if V_i and L_i are very much larger than the other terms we may conclude that the distribution is Boltzmann. Departures can occur only when the exchange terms are minor.

An expression for the exchange terms has been presented by Schunk and Hays (1971) derived from work by Treanor *et al.* (1968), Fisher and Kummler (1968), Rapp and

---

* Because of the rapid quenching of N_2 vibration by O, Q_i should be taken as the difference between the production rate and the quenching rate for molecules in level i. Vibrational exchange is more rapid than quenching at all altitudes of interest.

Englander-Golden (1964), Sharp and Rapp (1965), and Rapp (1960):

$$V_i = k \left\{ (i + 1) N_{i+1} \sum_{s=1}^{\infty} s N_{s-1} + i N_{i-1} \sum_{s=1}^{\infty} s N_s \right\} \tag{7}$$

and

$$L_i = k N_i \left\{ i \sum_{s=1}^{\infty} s N_{s-1} + (i + 1) \sum_{s=1}^{\infty} s N_s \right\}, \tag{8}$$

where $k = 6.5 \times 10^{-17} T^{3/2}$ cm^3 s^{-1} (Schunk, 1972) and T K is the kinetic temperature.

In order to evaluate these terms we shall assume that the distribution is Boltzmann. For the Boltzmann distribution,

$$N_i = \frac{N}{Z} e^{-i\alpha}, \tag{9}$$

where N is the density of N_2, $\alpha = \theta/T_v$ where $\theta = 3400$ K and T_v is the vibrational temperature, and the partition function

$$Z = \sum_{i=0}^{\infty} e^{-i\alpha} = (1 - e^{-\alpha})^{-1}. \tag{10}$$

It is convenient to define $\varepsilon = Z - 1$. Then

$$\sum_{s=1}^{\infty} s N_{s-1} = N (1 + \varepsilon) \quad \text{and} \quad \sum_{s=1}^{\infty} s N_s = N \varepsilon \quad \text{and}$$

$$L_i = k N_i N (i + 2\varepsilon i + \varepsilon). \tag{11}$$

Since $N_{i+1} = N_i e^{-\alpha}$ and $e^{-\alpha} = \varepsilon/(1 + \varepsilon)$ we obtain the same expression for V_i. For a Boltzmann distribution the rate of production of a given level by vibrational exchange collisions is equal to the rate of loss.

In the thermosphere we expect vibrational temperatures of 3000 K or less (Walker *et al.*, 1969), corresponding to $\varepsilon \lesssim 0.5$. Therefore, for vibrational levels above the first we may approximate

$$L_i = A i k N_i N, \tag{12}$$

where A is a numerical factor of the order of 2. Let us compare this, now, with the production term, Q_i. We take the ratio

$$L_i/Q_i = \frac{A i k (N_i/N) N}{Q_i/N} \tag{13}$$

and ask whether it is greater than or less than one. Information on the variation of Q_i with i is lacking for most excitation processes, but we may expect this variation to be much less rapid than that of N_i/N. If this is the case, we can see that for sufficiently large i the ratio will be less than one, and departures from the Boltzmann distribution may occur. At what values of N_i/N is this likely to take place?

We equate L_i/Q_i to one and find

$$iN_i/N = \frac{Q_i/N}{AkN}. \tag{14}$$

Maximum daytime values of Q/N are less than 10^{-5} vibrational quanta s^{-1} (Walker *et al.*, 1969), where $Q = \sum_{i=0}^{\infty} iQ_i$. According to Strobel and McElroy (1970) the values are much less. Let us take $Q_i/N < 10^{-6}$ s^{-1}, $A = 2$, and $k = 2 \times 10^{-12}$ cm^3 s^{-1} for $T = 1000$ K. Then, at 200 km altitude, where $N = 2.6 \times 10^9$ (Jacchia, 1971) departures from the Boltzmann distribution caused by the source term Q_i are possible for

$$i \frac{N_i}{N} < 10^{-4}.$$

At 300 km, where $N = 8.5 \times 10^7$ we find departures for

$$i \frac{N_i}{N} < 3 \times 10^{-3}.$$

From inspection of Equation (13) we see that a large source can cause departures from the Boltzmann distribution at high altitudes, where N is small and vibrational exchange is slow, and at high vibrational levels, where the population is small. The rough calculations we have presented, however, show that the departures can be expected only for vibrational levels containing a negligibly small fraction of the total N_2 density. For levels containing more than 0.1% of the N_2 molecules we do not expect departures at altitudes below 300 km. The number of levels over which the Boltzmann distribution can be expected to extend depends, of course, on the vibrational temperature. At $T_v = 1700$ K, for example, departures may occur at 300 km and above for $i \geqslant 3$. At $T_v = 3400$ K, on the other hand, the distribution will be Boltzmann for $i < 6$. The important conclusion, however, is that levels containing a significant fraction of the population follow the Boltzmann distribution, at least at altitudes below 300 km. At higher altitudes diffusion must be considered.

The diffusive flux of molecules in level i is given by

$$F_i = - D \left(\frac{dN_i}{dz} + \frac{N_i}{H} \right) \tag{15}$$

where

$$D = \frac{5 \times 10^{18}}{N_t} \left(\frac{T}{300} \right)^{0.75} \text{cm}^2 \text{ s}^{-1}$$

(Walker *et al.*, 1969), N_t cm^{-3} is the total ambient number density, and

$$H = \left(\frac{mg}{KT} + \frac{1}{T} \frac{dT}{dz} \right)^{-1} \tag{16}$$

where m is the molecular weight of N_2, g is the acceleration due to gravity, and K is Boltzmann's constant. In order to evalutate the flux and its divergence, let us assume that the vibrational distribution is Boltzmann. Then

$$N_i = \frac{N}{1+\varepsilon}\left(\frac{\varepsilon}{1+\varepsilon}\right)^i \tag{17}$$

and

$$\frac{dN_i}{dz} = \frac{\varepsilon^i}{(1+\varepsilon)^{i+1}}\frac{dN}{dz} + N\frac{\varepsilon^{i-1}}{(1+\varepsilon)^{i+2}}(i-\varepsilon)\frac{d\varepsilon}{dz}. \tag{18}$$

Taking advantage of the fact that, in diffusive equilibrium,

$$\frac{dN}{dz} = -\frac{N}{H} \tag{19}$$

we find

$$F_i = -DN\frac{\varepsilon^{i-1}}{(1+\varepsilon)^{i+2}}(i-\varepsilon)\frac{d\varepsilon}{dz}. \tag{20}$$

In calculating the divergence of the flux we will, for simplicity, ignore the slow variation of $(N/N_t)\,T^{0.75}$ with altitude and will take DN as constant. Then

$$\frac{dF_i}{dz} = -DN\left\{\frac{\varepsilon^{i-1}(i-\varepsilon)}{(1+\varepsilon)^{i+2}}\frac{d^2\varepsilon}{dz^2} + \frac{\varepsilon^{i-2}}{(1+\varepsilon)^{i+3}}(i^2-4i\varepsilon-i+2\varepsilon^2)\left(\frac{d\varepsilon}{dz}\right)^2\right\}. \tag{21}$$

If we express the loss by vibrational exchange collisions as

$$L_i = kN^2\frac{\varepsilon^i}{(1+\varepsilon)^{i+1}}(i+2\varepsilon i+\varepsilon) \tag{22}$$

then

$$\frac{L_i}{dF_i/dz} = -\frac{kN(i+2\varepsilon i+\varepsilon)\varepsilon(1+\varepsilon)}{D}$$
$$\times\left\{(i-\varepsilon)\frac{d^2\varepsilon}{dz^2} + \frac{(i^2-4i\varepsilon-i+2\varepsilon^2)}{\varepsilon(1+\varepsilon)}\left(\frac{d\varepsilon}{dz}\right)^2\right\}^{-1}. \tag{23}$$

In order to evaluate ε and its derivatives we must examine the equation of conservation of vibrational quanta. It is

$$\frac{\partial n}{\partial t} = Q - \frac{\partial F}{\partial z}, \tag{24}$$

where n is the total number density of vibrational quanta

$$n = \sum_{i=0}^{\infty} iN_i = N\varepsilon \tag{25}$$

for a Boltzmann distribution, and F is the total flux of vibrational quanta

$$F = \sum_{i=0}^{\infty} iF_i =$$
$$= -D\left(\frac{\mathrm{d}n}{\mathrm{d}z} + \frac{n}{H}\right) \tag{26}$$

from Equation (15). Using Equations (19) and (25) we obtain

$$F = -DN\frac{\mathrm{d}\varepsilon}{\mathrm{d}z}. \tag{27}$$

In the steady state the conservation equation reduces to

$$\frac{\mathrm{d}}{\mathrm{d}z}\left(DN\frac{\mathrm{d}\varepsilon}{\mathrm{d}z}\right) = -Q \tag{28}$$

which yields

$$\frac{\mathrm{d}\varepsilon}{\mathrm{d}z} = -\frac{1}{DN}\int_z^{\infty} Q(h)\,\mathrm{d}h \tag{29}$$

given zero flux of vibrational quanta at the top of the atmosphere, $F=0$ at $z=\infty$. Let us assume that Q/N is more or less independent of altitude. Then

$$\int_z^{\infty} Q(h)\,\mathrm{d}h = Q(z)\,H \tag{30}$$

where H is the N_2 scale height. Then

$$\frac{\mathrm{d}\varepsilon}{\mathrm{d}z} \sim -\frac{QH}{DN}$$

and

$$\left(\frac{\mathrm{d}\varepsilon}{\mathrm{d}z}\right)^2 \sim \left(\frac{QH}{DN}\right)^2. \tag{31}$$

If, in addition, we neglect the variation of DN with altitude, then

$$\frac{\mathrm{d}^2\varepsilon}{\mathrm{d}z^2} = -\frac{Q}{DN} \tag{32}$$

and

$$\left|\left(\frac{\mathrm{d}^2\varepsilon}{\mathrm{d}z^2}\right)\bigg/\left(\frac{\mathrm{d}\varepsilon}{\mathrm{d}z}\right)^2\right| = \frac{DN}{QH^2}. \tag{33}$$

For $T=1000\,\mathrm{K}$, $D=1.3\times10^{19}/N_t\ \mathrm{cm^2\ s^{-1}}$ and $H=3\times10^6$ cm. To be on the safe side let us take $Q/N < 10^{-4}$ s^{-1}. Then

$$\left| \left(\frac{d^2\varepsilon}{dz^2}\right) \Big/ \left(\frac{d\varepsilon}{dz}\right)^2 \right| > \frac{1.3 \times 10^{10}}{N_t}. \tag{34}$$

Inspection of Equation (23) reveals that the term in $d^2\varepsilon/dz^2$ is larger than the term in $(d\varepsilon/dz)^2$ by approximately the factor

$$\frac{\varepsilon}{i} \left| \left(\frac{d^2\varepsilon}{dz^2}\right) \Big/ \left(\frac{d\varepsilon}{dz}\right)^2 \right| = \frac{\varepsilon}{i} \times \frac{1.3 \times 10^{10}}{N_t} \tag{35}$$

for $i \gg \varepsilon$. If we take $\varepsilon/i > 0.1$ we find that the second term is negligible for $N_t \lesssim 1.3 \times 10^9$, corresponding to altitudes above about 300 km (Jacchia, 1971). It is at these higher altitudes that diffusion is most important, so let us neglect the second term.

Then,

$$\frac{L_i}{dF_i/dz} = \frac{kN^2}{Q} \frac{(i + 2\varepsilon i + \varepsilon)\, \varepsilon\, (1 + \varepsilon)}{(i - \varepsilon)} \sim \frac{AkN\varepsilon}{Q/N}, \tag{36}$$

where A is a numerical factor of the order of 2.

For a daytime situation let us take $Q/N < 10^{-5}$ s^{-1}, $\varepsilon = 0.5$ at high altitudes (Walker *et al.*, 1969), $A = 2$, and $k = 2 \times 10^{-12}$ cm^3 s^{-1}. We find

$$\frac{L_i}{dF_i/dz} > 2 \times 10^{-7}\, N. \tag{37}$$

We conclude that diffusion is less important than vibrational exchange at altitudes below 370 km (Jacchia, 1971). Since ε is approximately linear in Q, this result is not strongly dependent on the value of Q and ε. As Equation (36) shows, the relative importance of diffusion and vibrational exchange is also largely independent of the vibrational level i.

We can summarize these results by saying that daytime production rates are too small to cause departures from the Boltzmann distribution at altitudes below about 300 km for vibrational levels containing a significant fraction of the total population. Diffusion will not significantly perturb the Boltzmann distribution at altitudes below about 370 km. Departures from the Boltzmann distribution at altitudes above 300 km are not likely to have a significant effect on atmospheric processes.

5. Vibrating Ions

The ions of N_2 and O_2 may be vibrationally excited, and vibrational excitation could affect their chemical properties. Photoionization produces a substantial proportion of highly excited oxygen ions (Smith, 1970), and the same may be true of the reaction with atomic oxygen ions

$$O^+ + O_2 \rightarrow O_2^+\,(v) + O \tag{38}$$

(Bates, 1955). There are several processes that might degrade the vibrational energy of the ion (Shin, 1966; Bates and Reid, 1969; Dalgarno, 1970), but the chemical life-time of O_2^+ in the ionosphere is sufficiently short, of the order of 100 s, that these processes may not have time to operate. It is, moreover, possible that the rate of dissociative recombination depends on the vibrational level of the ion (cf. Bardsley and Biondi, 1970), in which case recombination would produce a further distortion of the ion vibrational distribution.

6. Summary

Vibrational temperatures in the thermosphere are not equal to the kinetic temperature. For oxygen, theoretical work suggests little vibrational excitation, but there is ex-perimental evidence for a vibrational temperature in excess of 3000 K. Nitrogen has received more theoretical attention, but it is not yet clear whether high vibrational temperatures can be expected. Departures of the vibrational distribution from the Boltzmann distribution are not significant for N_2 up to altitudes of at least 300 km. The vibrational distribution of oxygen has not yet been examined. If quenching by O occurs at the gas kinetic rate, then diffusion and vibrational-exchange collisions will be negligible, the distribution will be determined by a balance between produc-tion and quenching, and departures from the Boltzmann distribution will be large. Substantial vibrational excitation of the ions of O_2 and N_2 is possible, and their vibrational distributions are likely to be far from Boltzmann.

Acknowledgments

This work has been supported, in part, by the National Aeronautics and Space Ad-ministration under Grant number NGR 07-004-109. It has benefited from discussions with R. W. Schunk, G. P. Newton, and V. B. Wickwar.

References

Bardsley, J. N. and Biondi, M. A.: 1970, in D. R. Bates and I. Esterman (eds.) *Advances in Atomic and Molecular Physics*, Vol. 6 Academic Press, New York, p. 1.
Bates, D. R.: 1955, *Proc. Phys. Soc. London* **A68**, 344.
Bates, D. R. and Moiseiwitsch, B. L.: 1956, *J. Atmospheric Terrest. Phys.* **8**, 305.
Bates, D. R. and Reid, R. H. G.: 1969, *Proc. Roy. Soc. A.* **310**, 1.
Bray, K. N. C.: 1968, *J. Phys. B. Proc. Phys. Soc. Ser. 2*, **1**, 705.
Breig, E. L.: 1969, *J. Chem. Phys.* **51**, 4539.
Breig, E. L., Brennan, M. E., and McNeal, R. J.: 1972, *J. Geophys. Res.*, in press.
Breshears, W. D. and Bird, P. J.: 1968, *J. Chem. Phys.* **48**, 4768.
Dalgarno, A.: 1963, *Planetary Space Sci.* **10**, 19.
Dalgarno, A.: 1970, *Ann. Geophys.* **26**, 601.
Fisher, E. R. and Kummler, R. H.: 1968, *J. Chem. Phys.* **49**, 1075.
Fisher, E. R. and Bauer, E.: 1971, Paper P-789, Inst. for Defense Analyses, Arlington, Virginia.
Hudson, R. D. and Mahle, S. H.: 1972, *J. Geophys. Res.* **77**, 2902.
Jacchia, L. G.: 1971, Smithsonian Astrophysical Observatory Spec. Report 332, Cambridge, Mas-sachusetts.
Lane, N. F. and Dalgarno, A.: 1969, *J. Geophys. Res.* **74**, 3011.

McNeal, R. J., Whitson, M. E., and Cook, G. R.: 1972, *Chem. Phys. Letters* **16**, 507.
Rapp, D.: 1960, *J. Chem. Phys.* **32**, 735.
Rapp, D. and Englander-Golden, P.: 1964, *J. Chem. Phys.* **40**, 573; 3120.
Reid, R. H. G. and Withbroe, G. L.: 1970, *Planetary Space Sci.* **18**, 1255.
Schunk, R. W.: 1972, private communication.
Schunk, R. W. and Hays, P. B.: 1971, *Planetary Space Sci.* **19**, 1457.
Sharp, T. E. and Rapp, D.: 1965, *J. Chem. Phys.* **43**, 1233.
Shin, H.: 1966, *Advances Chem.* **58**, 44.
Smith, A. L.: 1970, *J. Quant. Spectrosc. Rad. Trans.* **10**, 1129.
Strobel, D. F. and McElroy, M. B.: 1970, private communication.
Treanor, C. E., Rich, J. W., and Rehm, R. G.: 1968, *J. Chem. Phys.* **48**, 1798.
Vlasov, M. N.: 1971, *Geomag. Aeron.* **11**, 426.
Walker, J. C. G.: 1968, *Planetary Space Sci.* **16**, 321.
Walker, J. C. G., Stolarski, R. S., and Nagy, A. J.: 1969, *Ann. Geophys.* **25**, 831.

O$\small\text{I}$ EMISSIONS

J. F. NOXON*

Blue Hill Observatory, Harvard University, Cambridge, Mass., U.S.A.

1. Introduction

This paper reviews some recent significant developments in aeronomy which follow from observations of atomic oxygen (O$\small\text{I}$) emissions – in particular the three features at 1304, 5577, and 6300 Å. Other O$\small\text{I}$ emissions as well as those of N$\small\text{I}$ will not receive attention here since they do not figure as prominently in the advances of the last year or two. In general our selection of topics has been dictated by their relevance to atmospheric physics and chemistry, that is to problems of atmospheric composition and energy balance and to the nature and rate of aeronomically significant chemical reactions. Thus the aurora, for example, receives scant attention except where the interpretation of observations bears more upon the target atmosphere than upon the characteristics of the impinging particle flux.

1.1. CHEMILUMINESCENCE IN THE E REGION

The relation of red and green line emission to the F region plasma structure continues to receive attention. Recent emphasis has been upon (a) the quantitative comparison of observed intensity with that calculated using measured electron density profiles, and (b) large scale morphological studies which employ satellite borne photometers.

Thomas and Donahue (1972) have used a horizon scanning narrow beam photometer onboard OGO 6 to obtain height profiles for 5577 Å nightglow. On a number of occasions they compared their results with theoretical profiles computed using electron density data gathered simultaneously at a ground station below the satellite. The satisfactory agreement established that the F region green line nightglow was well explained by the familiar two-step ionic recombination process for which the rate determining step is a reaction of O^+ with O_2. To obtain quantitative agreement they were forced not only to employ the low O_2 model advocated by Jacchia (1971) but also to utilize a rate coefficient for the ion atom interchange reaction only half the value calculated by extrapolation of laboratory measurements to the F region temperature. Thomas and Donahue then show how the observed green line height profile can be inverted to yield the electron density profile below the altitude of maximum electron density; as they remark, their technique provides satellite information on that portion of the F region inaccessible to the radio topside sounder. They present maps of the F region electron density structure as a function of height along the satellite track, although the sensitivity of the photometer only permits this in the equatorial region and not at mid-latitude where the F region green line component is generally very weak.

Schaeffer *et al.* (1972) analyzed a daytime 5577 Å height profile obtained with a

* Aeronomy Laboratory, NOAA, Boulder, Colo., U.S.A.

rocket photometer and isolated the contribution due to F region ionic recombination with the help of theory and simultaneous measurements of electron and O_2^+ density. They found excellent quantitative agreement between predicted and observed intensity using the most recent laboratory determinations of the dissociative recombination rate and efficiency for production of $O(^1S)$.

Isolation of the F region component of the total green line intensity is usually quite difficult from the ground owing to the large contribution from chemiluminescence near 100 km. In the tropics this is made easier by the relatively greater intensity of the high altitude component, the more so, as Hernandez (1971) has shown, if one uses a high resolution interferometer which can distinguish the very different Doppler widths of the two components.

Understanding of the F region 5577 Å nightglow has evidently advanced to the point where it can be utilized as a tool in satellite investigations of the F region plasma structure. Limitations are imposed not only by the weakness of the emission at mid-latitude but also by the necessity to make assumptions concerning the O_2 vertical distribution. To some extent the latter problem can be met by combining green line measurements above the electron density maximum with topside sounder electron density measurements so as to determine the O_2 density in this region; an acceptable extrapolation of O_2 density down through the lower portion of the F region can then doubtless be made.

While the history of F region 6300 Å nightglow is of rather greater antiquity the subject still appears to flourish, thanks to the introduction of satellite and aircraft measurements. Shepherd (1972), Reed and Blamont (1972), and Chandra *et al.* (1972) have all recently reported briefly on satellite observations although full details are not yet available. By combining successive passes these authors have produced maps of 6300 Å intensity over a sizable geographic area. The maps clearly show the enhanced equatorial belts as well as regions where the intensity is augmented by conjugate photoelectrons. Detailed analysis is only just commencing but should yield results of considerable importance in both ionospheric and magnetospheric studies.

In another recent satellite study, Blamont and Luton (1972) report very interesting measurements of the exospheric temperature from Doppler line width profiles of the 6300 Å dayglow using a Fabry-Perot interferometer. Global maps of the day time temperature show major rapid increases in T_∞ at higher latitudes in association with magnetic storms.

Simultaneous measurements of 6300 Å nightglow and F region plasma structure have been performed by Markham *et al.* (1972) using a jet aircraft equipped with an ionospheric sounder. The range of the aircraft is such that the entire equatorial region can be examined in latitude in a few hours. They quantitatively compare observed and predicted 6300 Å intensity and find excellent agreement over the entire equatorial region using the Jacchia (1971) model for O_2 and N_2 with an exospheric temperature in the vicinity of 900 K. No anomalies in either atmospheric composition or reaction rates are associated with the regions of unusually strong emission (400 R); the entire complicated structure of 6300 Å emission over an extended range of latitude seems entirely to reflect corresponding changes in F region plasma structure.

Brown and Steiger (1972) describe further work on a comparison of 6300 Å night-glow with total electron content measured above Hawaii. Despite limitations imposed by lack of detail in the ionospheric data their interpretation appears to imply internal consistency with the ionic recombination source.

Mullaney *et al.* (1972) have investigated the correlation between fluctuations in 6300 Å nightglow and the scintillation in radio signals received from stars or beacon satellites imposed by irregularities in the F region plasma structure. Fluctuations in the nightglow intensity of up to 10% were found in the frequency range 10^{-4} to 10^{-2} Hz, but only when radio source scintillation was high. While it is doubtless not surprising to find a manifestation of plasma irregularities in both quantities the optical measurements served to extend the frequency range accessible to study well below that permitted by the radio measurements alone. It is possible that further studies of this nature may help in understanding the nature and propagation characteristics of the waves which presumably are responsible for generating the plasma irregularities.

1.2. CHEMILUMINESCENCE IN THE LOWER THERMOSPHERE

Both Schaeffer *et al.* (1972) and Thomas and Donahue (1972) report measurements on the lower layer of 5577 Å emission near 95 km, as do Dandekar and Turtle (1971). Schaeffer *et al.* interpret their measurements using the new and much larger rate coefficients reported by Felder and Young (1972) for production of $O(^1S)$ in triple collisions of O atoms and for quenching of $O(^1S)$ by OI. They conclude that the required O atom abundance near 95 km is nearly three times smaller than that given in the Jacchia (1971) model. Although Thomas and Donahue remark on the ability of their technique to isolate the 95 km layer they do not discuss their observations; the technique will clearly yield very interesting results on the world-wide distribution of O near 100 km.

1.3. ARTIFICIAL HEATING OF F REGION PLASMA

Biondi and his collaborators (Biondi *et al.*, 1970; Sipler and Biondi, 1972) and Hernandez (1972) have detected the effect upon 6300 Å nightglow of plasma heating in the F region by a powerful ground based radio transmitter. The first measurements (Biondi *et al.*, 1970) were made under conditions in which the electron heating was insufficient to yield a 6300 Å enhancement from thermal electron impact on O. Instead the heating manifested itself through the dependence of the O_2^+ dissociative recombination rate upon electron temperature. Subsequent changes in the operating characteristics of the transmitter led to a great increase in electron heating and a marked enhancement of 6300 Å by energetic ambient electrons (Hernandez, 1972; Sipler and Biondi, 1972). The principal result of these measurements has been the determination of the quenching rate of $O(^1D)$ by N_2 which follows from an analysis of the rise and fall characteristics of the 6300 Å enhancement. The results agree reasonably well with laboratory determinations of the quenching rate. The absolute magnitude of the enhancement is a convenient monitor of the degree of heating imparted to the ambient electrons despite the uncertainty which exists due to a lack of knowledge of the energy distribution of the electrons above the 2 eV threshold. Accurate measurements of an enhancement in both red and green lines would serve to provide at least a rough estimate of the energy distribution function, owing to the higher excitation threshold (~ 4 eV) for the green line.

1.4. PHOTOELECTRON EFFECTS

In the dayglow, as well as in early twilight, one expects enhancement of the oxygen lines as a result of local photoelectron excitation, as discussed at earlier Institutes. Hays and Sharp (1971) briefly describe a rocket measurement of the 5577 Å dayglow in which the photoelectron component is isolated and Schaeffer *et al.* (1972) infer a photoelectron contribution in their rocket measurements of 5577 Å by subtracting the contribution from dissociative recombination of O_2^+. The local photoelectron contribution to 6300 Å has been inferred during early twilight by Noxon and Johanson (1972). Their paper shows how one can reliably separate the dissociative recombination and photodissociation contributions to $O(^1D)$ production during twilight; having done so one finds a residual for solar depression angles less than $4°$ which may be identified as due to local photoelectron excitation, no other source being then available. Fully half of the total 6300 Å intensity observed in the zenith at sunset appears to be due to local photoelectrons and the method seems to offer some possibilities for routine monitoring of photoelectron production rates. Schaeffer *et al.* (1972) have analyzed a rocket dayglow 6300 Å profile and found on the occasion of their flight a total photoelectron contribution to 6300 Å of about 0.8 kR; the average sunset contribution reported by Noxon and Johanson is comparable. No comprehensive theoretical treatment of the twilight behavior of the 6300 Å photoelectron component has yet been published. While complicated by the need to consider the escape of photoelectrons produced above 300 km such a treatment is needed in order to evaluate the ground based results reported by Noxon and Johanson as a means for monitoring the day to day changes in local photoelectron fluxes.

Several important papers have recently appeared which deal with the enhancement in 1304 and 6300 Å produced by magnetically conjugate photoelectrons as measured by satellites. Meier (1971), and Buckley and Moos (1971) report identification of conjugate effects in 1304 Å while Shepherd (1972) has prepared 6300 Å maps from satellite observations which show a wave of advancing 6300 Å intensity associated with the arrival of conjugate photoelectrons. He notes that the equatorward boundary is not necessarily determined by the geometry of conjugate sunrise but often is rather the equatorward edge of the *F* region ionospheric trough. The explanation appears to be similar to the effect discussed by Noxon and Johanson (1970) in which the conjugate photoelectrons yield a much smaller 6300 Å enhancement when the local ambient electron density is high. In this case the incoming photoelectrons give a far greater portion of their energy to the ambient electrons and proportionately less is available for excitation of $O(^1D)$. In addition there appear to be major effects associated with the total ambient electron content along the magnetic field line connecting conjugate points (Shepherd *et al.*, 1973); a high content reduces the incident low energy electron flux.

Troy (1972) reports simultaneous measurements of 6300 Å, electron temperature, and suprathermal (conjugate) electron flux. The three are well correlated with locations where impinging conjugate photoelectrons are expected. Troy suggests that the major excitation process for 6300 Å involves the heated ambient electrons, not the incoming photoelectrons. While this is doubtless sometimes the case, the older literature offers counter-examples; the relative importance of the two processes probably depends upon the ambient electron density. If the relative concentration of ambient electrons to 0 atoms is high then a major portion of the incident photoelectron energy goes to the ambient electrons and less to direct excitation of $O(^1D)$. If, at the same time, the absolute plasma density is not too high then the ambient electrons may be heated sufficiently to yield significant 6300 Å emission. It is no easy matter do relate the 6300 Å enhancement (or lack of it) to the magnitude of the incoming photoelectron flux; for this purpose measurements of the 1304 or 1356 Å lines offer a less complicated means owing to their higher excitation energy.

1.5. Atmospheric composition and dynamics

Some recent studies of both 5577 and 6300 Å bear upon the atmospheric composition near 100 km and, in some cases, much higher in the thermosphere; the conclusions, in particular, relate to the O/O_2 ratio and its variations. We have already mentioned the ability of the horizon scanning photometer of Thomas and Donahue (1972) to yield height profiles of the 95 km layer of 5577 Å and the inference by Schaeffer *et al.* (1972) that the O density at this altitude is less than in the Jacchia (1971) model. We have also referred to the latitude survey of 6300 Å by Markham *et al.* (1972) which, in effect, provides information concerning the O_2/N_2 ratio. Such measurements all bear on the question of the O and O_2 densities at the base of the diffusive equilibrium regime and their variation in space and time. That such a variation does exist is a notion that becomes less surprising as time passes; it is becoming more apparent that the O/O_2 balance in the lower thermosphere is a quantity which sensitively depends on such factors as the vertical transport rate and other dynamical processes, not just on photochemistry alone.

Large changes in the abundance of thermospheric O_2 have been inferred by Noxon and Johanson (1972) from an investigation of 6300 Å in early twilight. They were able to separate the contribution to $O(^1D)$ production due to photodissociation of O_2 in the Schumann-Runge continuum; the Schumann-Runge portion of the twilight intensity was found to exhibit a three fold seasonal change with a maximum in the summer. Equally large changes were found from day to day and at 27 day intervals after a major magnetic storm. Their analysis suggested that the major portion of such changes had to be due to a variation in the O_2 density at the base of the diffusive regime near 100 km. The seasonal variation which they inferred corresponds well to that measured at 100 km by Roble and Norton (1972) using satellite measurements of solar Ly-α absorption during occultation by the earth.

Other changes in atmospheric composition are implied by observations of low latitude aurora; unpublished analysis of several displays in which simultaneous optical and incoherent radar scatter measurements were performed by J. V. Evans and the author shows that internal consistency is only attainable if one accepts the existence of a major enhancement in the molecular species, O_2 and N_2, in the F region without an accompanying increase in the O density. Similar results have been reported from satellite borne mass spectrometers passing over aurora. These major changes in atmospheric composition are presumably the result of dynamical motions induced by Joule heating near 100 km, although the interpretation is by no means yet clear.

Acknowledgment

This work has been supported by a grant to Harvard University from the National Science Foundation, GA 28371.

References

Biondi, M. A., Sipler, D. P., and Hake, R. D.: 1970, *J. Geophys. Res.* **75**, 6421.
Blamont, J. E. and Luton, J. M.: 1972, *J. Geophys. Res.* **77**, 3534.
Brown, W. E. and Steiger, W. R.: 1972, *Planetary Space Sci.* **20**, 11.
Buckley, J. L. and Moos, H. W.: 1971, *J. Geophys. Res.* **76**, 8378.
Chandra, S., Troy, B. E., Reed, E. I., and Blamont, J. E.: 1972, *Trans. Am. Geophys. Union* **53**, 474.
Dandekar, B. S. and Turtle, J. P.: 1971, *Planetary Space Sci.* **19**, 949.
Felder, W. and Young, R. A.: 1972, *J. Chem. Phys.* **56**, 6028.
Hays, P. B. and Sharp, W. E.: 1971, *Trans. Am. Geophys. Union* **52**, 885.
Hernandez, G. J.: 1971, *Planetary Space Sci.* **19**, 467.
Hernandez, G. J.: 1972, *J. Geophys. Res.* **77** 3625.
Jacchia, L. G.: 1971, Smithsonian Astrophys. Special Report 332.
Markham, T. P., Buchau, J., Anctil, R. E., and Noxon, J. F.: 1972, in preparation.
Meier, R. R.: 1971, *J. Geophys. Res.* **76**, 242.
Mullaney, H., Papagiannis, M. D., and Noxon, J. F.: 1972, *Planetary Space Sci.* **20**, 41, 1982.
Noxon, J. F. and Johanson, A. E.: 1970, *Planetary Space Sci.* **18**, 1367.
Noxon, J. F. and Johanson, A. E.: 1972, *Planetary Space Sci.* **20**, 2125.
Reed, E. I. and Blamont, J. E.: 1972, *Trans. Am. Geophys. Union* **53**, 474.
Roble, R. G. and Norton, R. B.: 1972, *J. Geophys. Res.* **77**, 3524.
Schaeffer, R. C., Feldman, P. D., and Zipf, E. C.: 1972, *J. Geophys. Res.* **77**, 6828.
Shepherd, G. G.: 1972, *Trans. Am. Geophys. Union* **53**, 475.
Shepherd, G. G., Brace, L. H., and Whitteker, J. H.: 1973, *J. Geophys. Res.*, in press.
Sipler, D. P. and Biondi, M. A.: 1972, *J. Geophys. Res.* **77**, 6202.
Thomas, R. J. and Donahue. T. M.: 1972, *J. Geophys. Res.* **77**, 3557.
Troy, B. E.: 1972, *Trans. Am. Geophys. Union* **53**, 473.

VERTICAL RED LINE 6300 Å DISTRIBUTION AND TROPICAL NIGHTGLOW MORPHOLOGY IN QUIET MAGNETIC CONDITIONS

G. THUILLIER and J. E. BLAMONT

Service d'Aéronomie, 91 – Verrières-Le-Buisson, France

Two airglow experiments were carried by the OGO 4 and OGO 6 satellites launched on 27 July, 1967, and 5 June, 1969, respectively. The observations cover a period of maximum solar activity and show latitude and longitude effects in the low latitude nightglow 6300 Å emission. The purpose of this paper is to present an interpretation of the main features of the nightglow 6300 Å morphology in tropical regions in terms of the thermospheric dynamics.

1. The Photometers OGO D 12 and OGO F 11

Both experiments are designed to determine the vertical distribution of the 6300 Å emission in twilight and nightglow conditions by a remote sensing method.

The field of view is $\frac{1}{2}°$ in a vertical plane and $6°$ in a horizontal plane. The emission region is scanned by a stepping mirror, the reflected beam reaches a telescope consisting of a spherical mirror with a slit located at its focus. A two-lens system makes the beam converge on the photomultiplier cathode. Analog output is provided by a linear electrometer. In the airglow spectrum the red line is selected by an interferential filter of 25 Å band width. In the OGO F 11 experiment, a second filter centered at 3914 Å was added. Both photometers have a periodical internal calibration sequence where photomultiplier gain and filter transmission are checked. By grazing the layer, contamination of measurements by extra sources, such as moonlight reflected on clouds, is generally avoided.

2. Data Analysis

2.1. Computation of a vertical profile $I(z)$

For each line of sight, the photometer receives a flux $M\alpha$ corresponding to the integral emission contained in the field of view. This measurement is expressed in Rayleigh using ground and inflight calibration:

$$M\alpha = 10^{-6} \int_{(L)} I(z)\, \mathrm{d}l$$

L being the optical path through the emission layer. This measurement is associated with the direction of view α or with the impact parameter $Z_{\min}$ which is the minimum altitude along the line of sight. By vertical scanning, the photometer records the angular distribution (Figure 1a) and the vertical profile, i.e., emission in photons cm^{-3} s^{-1} is deduced by solving an integral equation (Figure 1b).

2.2. Use of a profile

From a computed profile, we derive four independent parameters: X_1, maximum emission; X_3, altitude of maximum emission; X_2, width at half maximum emission; X_4, scale height over peak emission. Zenithal emission Q is also computed.

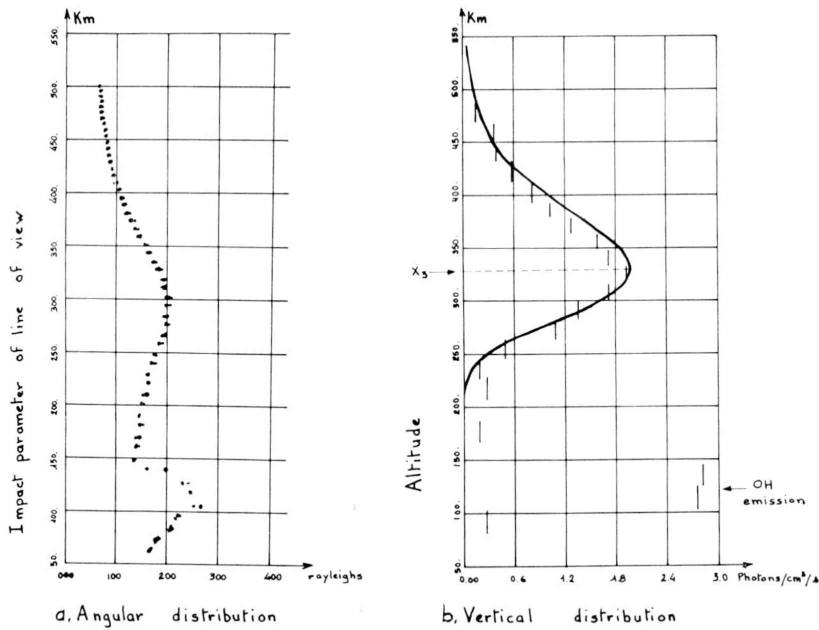

Fig. 1.   6300 Å emission for orbit 121 on June 14, 1969, at 75° E, 20° S and 0506 LT. (a) Angular distribution 3 m. (b) Vertical profile X_3 is the altitude of maximum emissivity.

This reduced data form can be represented by maps of one day of data. Since OGO 4 and OGO 6 satellites are in polar orbit, photometers provide profiles in latitude with great accuracy (a vertical distribution every two degrees latitude). In longitude, the 90 minute period separates two orbits by 22°. Polar orbit also provides an essential property: in one day, all points are at the same local solar time within 10 min maximum difference for the latitude range $-30°$ to $+30°$. Such a map is not a picture of the high atmosphere because all points should be at the same Universal Time. Since maps need 24 hr to be constructed, if solar and magnetic activity are quiet all points are under the same physical conditions. This property allows drawings of iso-lines on maps. Local solar time of observations depends on the direction of view which is related to orbit precession; therefore, season and local time always vary together, and thus for a given season, all local time maps are not available.

The total number of maps is limited for several reasons: twilight period of measurements, spacecraft functioning mode, data quality, and spacecraft altitude over the studied region. In the remaining, we also eliminated maps in disturbed magnetic conditions, and maps at late local time when a contribution by conjugate photoelectrons could be present.

3. Interpretation of Nightglow Tropical Morphology

3.1. THE TROPICAL NIGHTGLOW PROBLEM

In the low latitude region the 6300 Å emission can be ten times more intense than in mid-latitude regions at the same local time. This phenomenon was discovered by Barbier and Glaume (1960) with ground-based measurements in Algeria. This increased emission was named 'arcs' by Barbier. These arcs as observed at Tamanrasset commonly move at night from north to south with an east-west extension. These observations have revealed an apparent correlation between the behavior of the arcs and the Appelton's Equatorial Anomaly in electron density. That is to say both phenomena have their maximum values at about 15 to 20° on either side of the magnetic equator (Barbier *et al.*, 1961; Barbier and Glaume, 1962). In fact, observations were limited to African longitudes, and the conclusion led to the concept that the tropical morphology consists of two symmetrical belts around the earth. From the theoretical point of view, tropical arcs were associated with the basic physics involved in the equatorial anomaly in electronic density (Martyn, 1953).

Figure 2 illustrates some discrepancies between world wide observations and the last conclusion. This figure shows three plots of nightglow and O^+ measurements vs. magnetic latitude. The 6300 Å data are the integrated emissions recorded by the downward looking photometer on OGO 4 (Blamont–Reed, Main body experiment) and the O^+ concentration is recorded by the same satellite (Chandra *et al.*, 1970). Spacecraft perigee is periodically near equatorial regions and allows comparison between 6300 Å data and O^+ measurements. The O^+ variation in magnetic latitude is the well known equatorial anomaly, and except for African longitudes nightglow measurements are not always correlated with electron density. Therefore, this figure sums up the main problem in equatorial morphology, i.e., latitudinal asymmetries and longitudinal effects at the same local time.

Our aim is to explain the main features in equatorial nightglow in the following frame: dissociative recombination process and neutral winds in the F layer. The effect of E fields will also be evaluated.

3.2. DISSOCIATIVE RECOMBINATION PROCESS

The red line 6300 Å has been recognized as coming from the transition between ^1D and $^3P_2$ levels in the OI energy diagram (energy involved 1.96 eV). During the nighttime, F region molecular ions recombine with electrons:

$$O_2^+ + e \rightarrow O^+ + O^+ + 6.96\,eV \qquad (\alpha_1) \qquad\qquad (1)$$

O^+ comes from a charge-exchange reaction

$$O^+ + O_2 \rightarrow O_2^+ + O + 1.53 \, \text{eV} . \qquad (\gamma_2) \qquad (2)$$

Reaction (1) is able to excite O to ^1D and/or ^1S state, but

$$NO^+ + e \rightarrow N + O + 2.76 \, \text{eV} \qquad (\alpha_2) \qquad (3)$$

$$O^+ + N_2 \rightarrow NO^+ + N + 1.09 \, \text{eV} . \qquad (\gamma_2) \qquad (4)$$

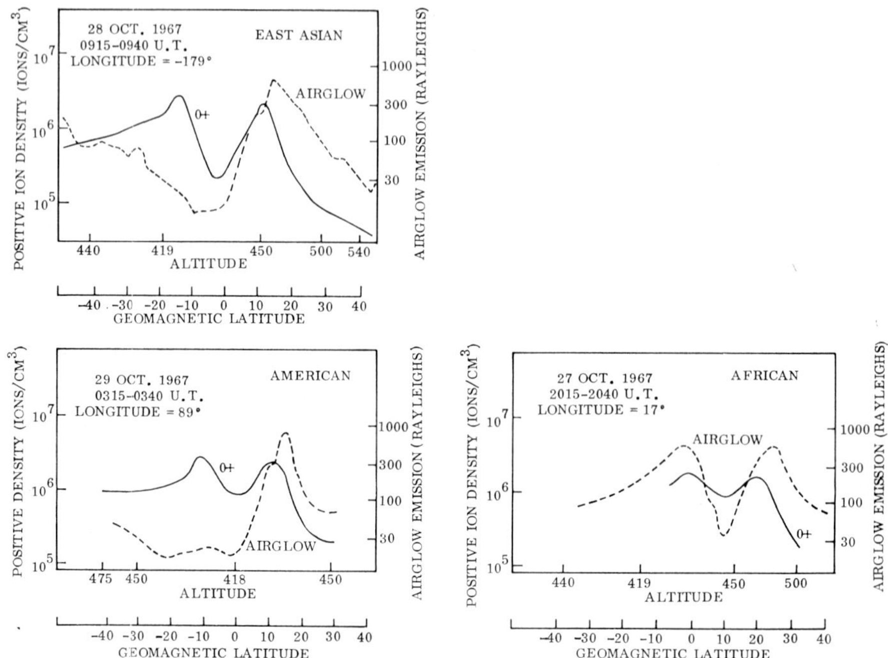

Fig. 2.   6300 Å emission rate and O^+ density vs. latitude for OGO 4 at $\sim 21\,30$ LT.

Reaction (3) is responsible for excitation of N to the ^2D state, but it probably does not contribute significantly to populate the $O(^1D)$ state since spin is not conserved (Dalgarno and Walker, 1964). A relation has been derived taking account of direct ^1D production, cascading from ^1S to ^1D and quenching with N_2 of ^1D state (Peterson *et al.*, 1966).

The volume emission rate is written:

$$\varepsilon(z) = 0.76 K_D \frac{\gamma_1(O_2) \, n_e(z)}{1 + (s_D/A_D)(N_2)} \text{ photons cm}^{-3} \text{ s}^{-1} , \qquad \text{(Noxon, 1971)}$$

$$(5)$$

where K_D is the dissociative recombination process efficiency and s_D is the collisional deactivation coefficient with N_2.

The terms $n_e(z)$ can be described with four independent parameters: N_{max}, maximum electron concentration; h_{max}, altitude of maximum electron concentration; W_e, width parameter; and T_{ex}, exospheric temperature assuming for $n_e(z)$ a diffusive equilibrium over the peak concentration (Rishbeth and Barron, 1960).

As (O_2) and (N_2) need T_{ex} to be described, these four physical quantities are sufficient to describe a vertical profile. By varying each of these parameters we compute the theoretical profile and derive Q, zenithal emission in rayleighs, and X_3, altitude of maximum emission. It can be showed that X_3 is much more controlled by h_{max}

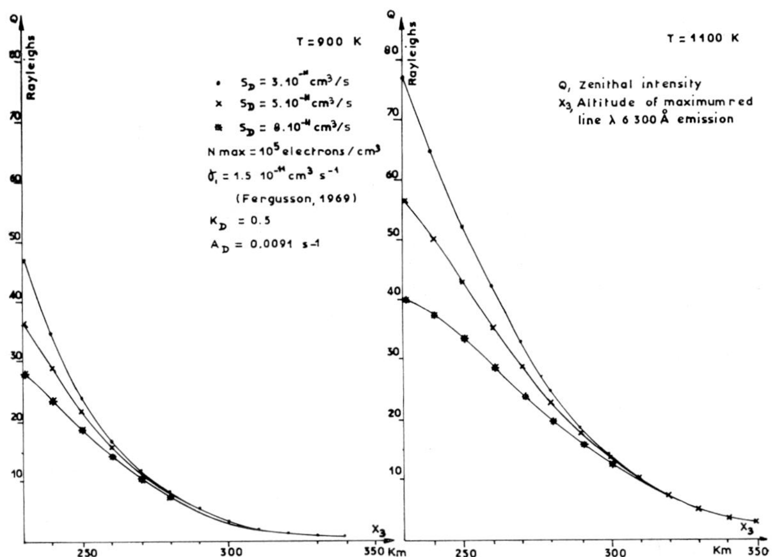

Fig. 3. Zenithal intensity Q vs. the altitude of maximum emission X_3 at 6300 Å for two different temperatures. The volume emission rate is solved using Equation (5) and converted into a zenithal intensity Q using the values shown.

than by T_{ex} or electronic width. Since maps have the property to be at the same local solar time for all longitudes, exospheric temperature variation comes mainly from the latitude effect, so in the studied region no more than 50 K variation should be involved. Therefore X_3 is essentially controlled by h_{max}, electronic width effect being a second order effect. In the same way, it can be showed that zenithal emission is much more dependent on h_{max} than T_{ex} and electronic width. Figure 3 is a plot of Q vs. X_3. Since electronic width has a small effect compared to that of h_{max}, it has not been represented. However, a quenching effect is noticeable. So, h_{max} is the control parameter of the 6300 Å emission.

Therefore, if we want to explain a strong emission with a dissociative recombination process, T_{ex} is certainly unable to produce it, but h_{max} rather than N_{max} can be involved. In particular h_{max} has an exponential effect over 250 km, but the N_{max} effect always remains linear. In this frame, arcs must be explained by low h_{max} values correlated with low X_3 values. Our data show this property.

Figure 4 is map in geographic coordinates obtained for 4 October, 1967 at 2348 LT. The solid line is the magnetic equator. Latitude and longitude effects are apparent.

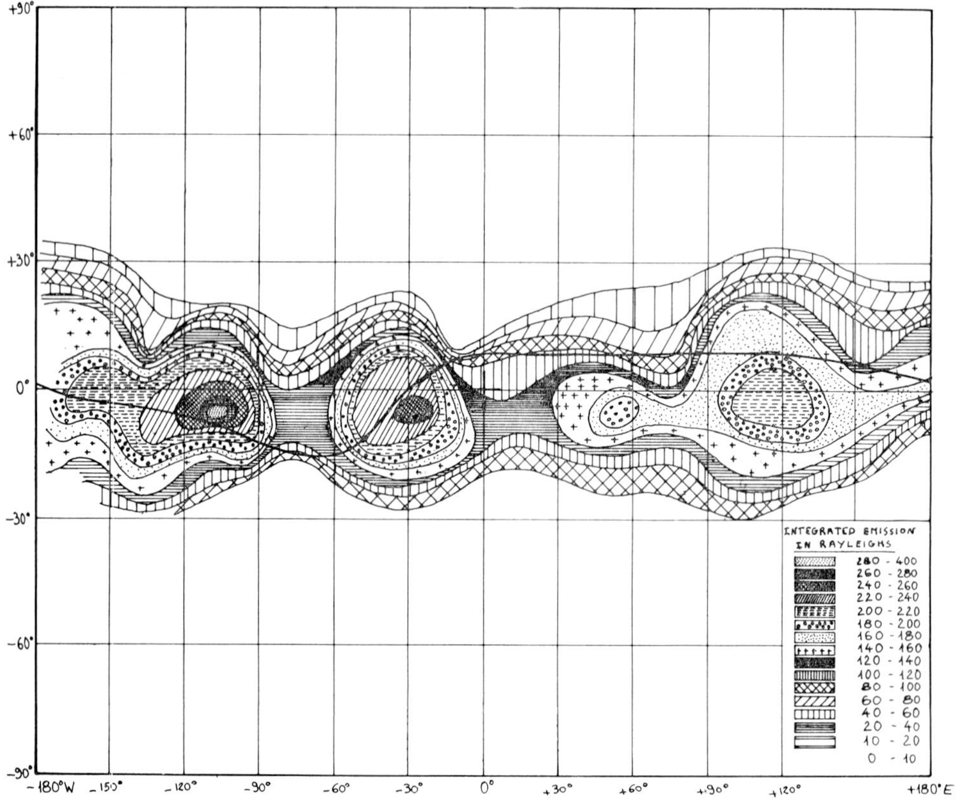

Fig. 4.   Zenithal emission of 6300 Å for 4 October 1967, at 2348 LT.

Figure 5 is a plot in the same coordinates for the same day of the X_3 parameter. The location of the most intense emission is, in general, in good agreement with the location of the lowest X_3 values. This result is consistent with the dissociative recombination process, but a shift can exist for different reasons:

(i) Latitudinal variations of F layer thickness cannot be excluded

(ii) Significant latitudinal changes in electron density related to equatorial anomaly are observed to persist until 200 LT under solar maximum conditions (Lyon and Thomas, 1963)

(iii) Longitudinal effects in electron density have been exhibited (Piggott, 1969).

Therefore, interpretation of nightglow morphology infers the existence of a phenomenon that causes the F layer to be lower in certain locations than in others.

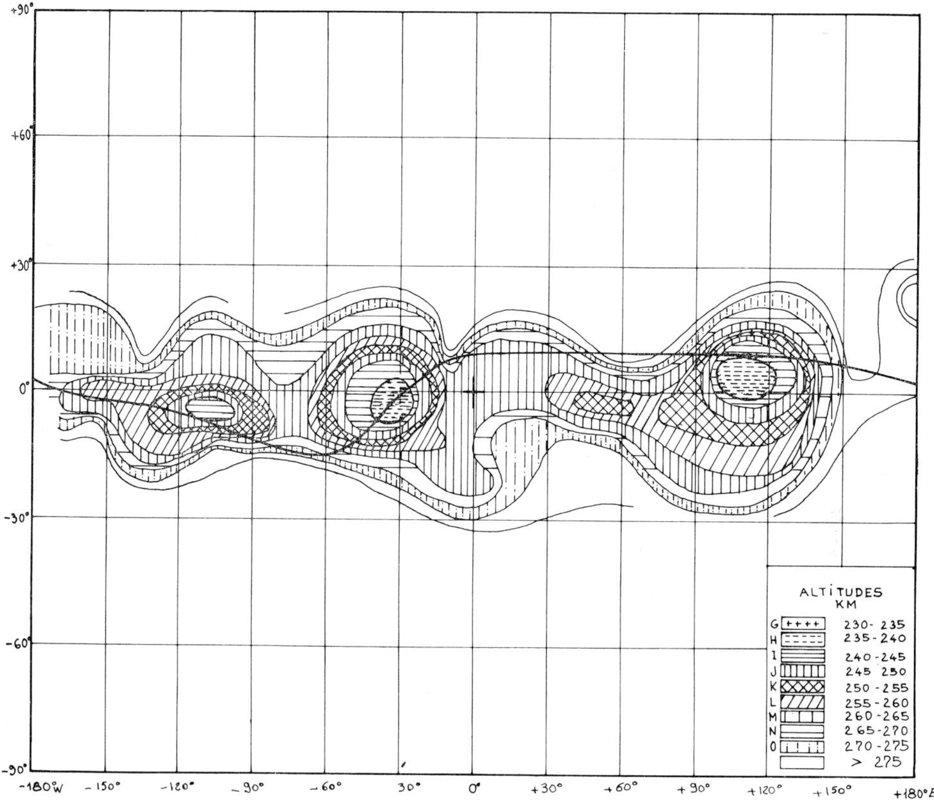

Fig. 5. Altitude of maximum emission of 6300 Å on 4 October 1967, at 23 48 LT.

3.3. PROCESS THAT LOWERS THE F LAYER

Electric fields and thermospheric winds can be involved.

3.3.1. *Electric Field Effects*

Electric fields can provide a movement perpendicular to magnetic field lines. In the nighttime, they are usually westward and produce a convergent and downward movement of ionization toward the magnetic equator. The drift is the most important at the magnetic equator and decreases as latitude increases, but this effect is certainly *symmetrical* with respect to the magnetic equator because E fields at magnetically conjugate points are balanced due to the high conductivity along magnetic field lines.

If we assume uniform E fields in longitude with only a latitudinal dependence, drift can be modified by the earth's magnetic field strength. It will be reduced in the Indian sector with respect to its value in the Pacific sector. Magnetic field strength infers perhaps a light longitudinal effect if other phenomena remain the same in the Indian sector as in the Pacific sector. Therefore, E fields cannot explain the large latitudinal asymmetry and longitudinal effects as observed.

3.3.2. *Neutral Winds*

Thermospheric winds are able to modify the height of maximum electron concentration. Wind is generally decomposed into meridional and zonal winds.

Figure 6a is taken from Bramley and Young (1968). In the day time, without any wind N_{max} and h_{max} latitude distributions are symmetrical with respect to the magnetic equator. As a transequatorial wind blows from north to south, N_{max} and h_{max} distributions become asymmetrical with a reduction at both crests in electron density.

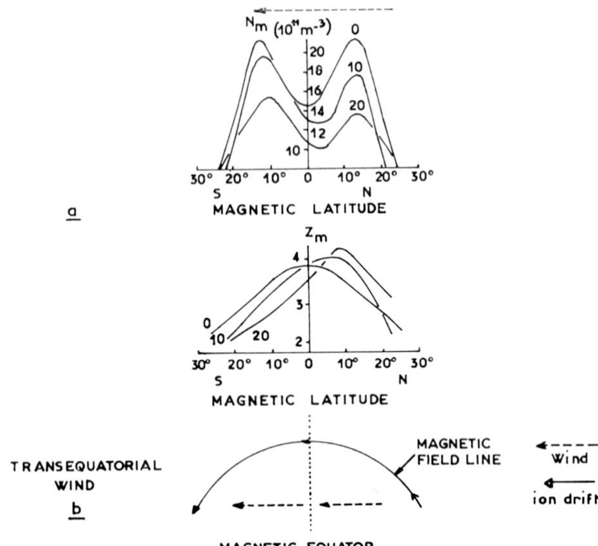

Fig. 6. Equilibrium values of peak electron concentration N_m (top), and reduced height of F2 peak Z_m (middle) plotted against magnetic latitude. The assumed vertical electromagnetic drift velocity at the equator is 4.1 m s⁻¹ (bottom). The curves 0, 10, and 20 respectively assume north-to-south wind speeds of 0, 41, and 82 m s⁻¹ (Bramley and Young, 1968).

Latitudinal h_{max} distribution becomes more asymmetrical as the wind blows than latitudinal N_{max} distribution. Such a wind can be obtained at solstice (June); in the winter hemisphere N_{max} increases and h_{max} decreases with respect to summer distribution. The nighttime computation has been done by Abur-Robb and Windle (1969). Asymmetry in N_{max} is reversed because of the higher recombination rate at lower altitude, but asymmetry in h_{max} remains with the lowest F layer in the winter hemisphere. h_{max} distribution is much more sensitive to winds than N_{max} and has an exponential effect on red line emission. Winds provide asymmetry in latitude and are certainly the main mechanism in nightglow morphology. The conclusion of their studies is that movements are mainly directed along magnetic field lines (Figure 6b) and it can be applied to zonal winds. Their effects depend on magnetic declination δ and we can divide the magnetic equator into three regions:

(i) Indian region from 0 to 150° east; magnetic declination is always near zero, and zonal winds have no effect.

(ii) Atlantic region from -65 west to $0°$; magnetic declination takes its maximum value (20° west) and zonal winds present their maximum effects.

(iii) Pacific region from $+150°$ east to $-65°$ west; magnetic declination is at about 10° east and zonal wind effects are lesser than in the Atlantic region. Uniform meridional winds can produce latitudinal effects, but cannot produce large longitudinal effects because the angle between the wind and magnetic meridian remains small. Zonal winds produce both: being eastward at night, some red line emission is produced in the South Atlantic region, a weaker emission in the North Pacific region, and nothing in the Indian region. Of course, morphology must be explained by composing meridional V_M and zonal winds V_Z. Their combined effect depends on the ratio V_M/V_Z compared to $\tan\delta$. If V_M/V_Z is greater than $\tan\delta$ the zonal wind effect is not apparent on red line emission whose main feature is given by meridional wind. This relation can be verified of phase differences in local time of both winds and because of δ variation along the magnetic equator which introduces a longitudinal and latitudinal effect.

3.4. SAMPLES OF MAPS

Figure 7 shows tropical morphology in summer conditions with a transequatorial

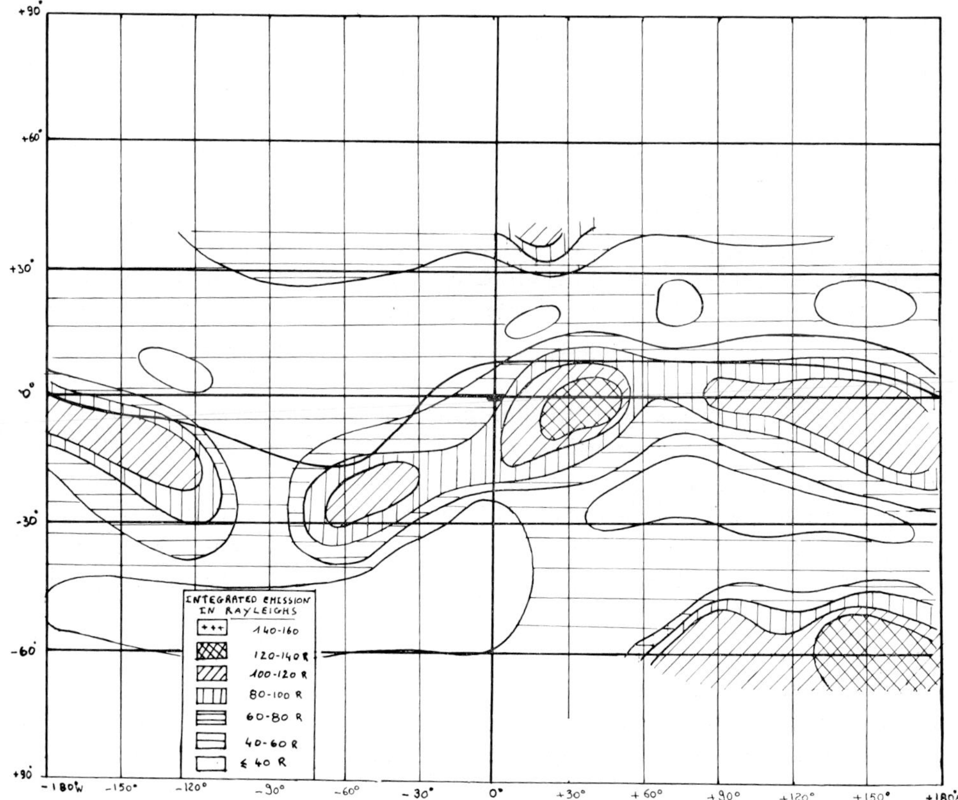

Fig. 7. Zenithal emission at 6300 Å on 23 June 1968, at 2148 LT.

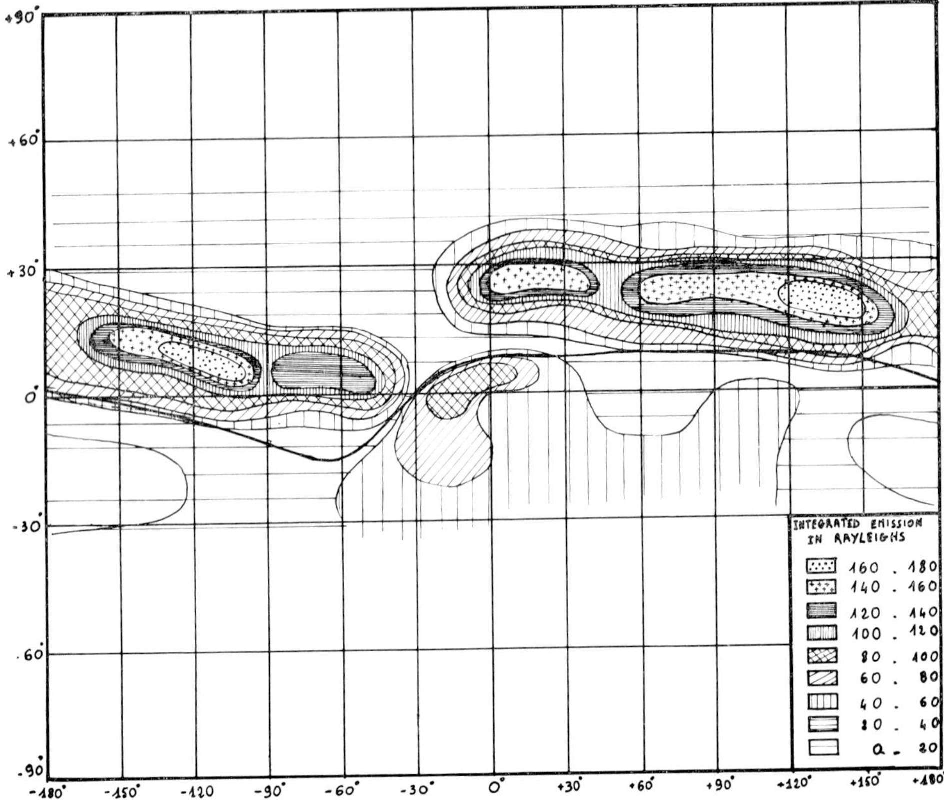

Fig. 8. Zenithal emission at 6300 Å on 7 November 1967, at 2036 LT.

meridional wind from north to south as inferred from the diurnal bulge of temperature (Jacchia's 1971 model). This map was obtained on 23 June 1968, with a 23° solar declination, $A_p = 6$ at 2148 LT. As previously suggested arcs are in the southern hemisphere.

Figure 8 shows a symmetrical situation near winter condition with a meridional wind blowing from south to north and a visible effect of an eastward zonal wind. This map was obtained on 7 November 1967, at 2036 LT with a 16° solar declination and $A_p = 10$. For convergent meridional winds on the geographic equator, the zonal wind effect is centered over the Atlantic region, but in this case the south to North meridional wind mixed with an eastward zonal displaced arc in the northeast direction as shown by this figure.

For Figures 7 and 8 arcs are lying in the winter hemisphere, i.e., in the cold hemisphere except for over the Atlantic region due to zonal wind. Later in the night, the morphology is generally more complex than before midnight. In such conditions, arcs are commonly closer to the magnetic equator than before midnight but they are mainly laying in the cold hemisphere (Figure 9, 2 July 1969 at 218 LT with a 23° solar declination and $A_p = 6$).

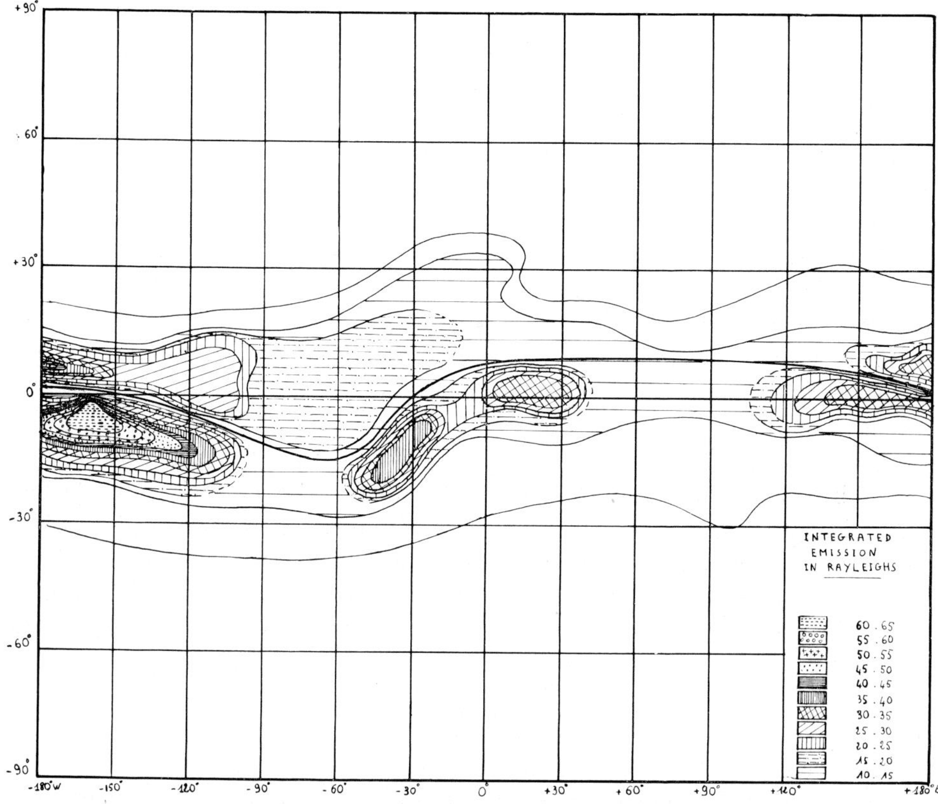

Fig. 9.   Zenithal emission at 6300 Å on 2 July 1969, at 0218 LT.

In equinox condition, by 2200 to 2400 meridional winds are convergent toward the geographic equator. On Figure 4 arcs are lying in the vicinity of the geographic equator i.e., on the side of the magnetic nearest to the geographic equator. Meridional winds are in fact transequatorial with respect to the magnetic equator, and the F layer will go down in the south Indian region and the north Pacific region, therefore in the vicinity of the geographic equator. A small effect of an eastward zonal wind is also visible on Figure 4 and as it is centered over the Atlantic region it is consistent with a symmetrical convergent wind on the geographic equator.

3.5. REQUESTED NEUTRAL WIND SPEEDS

A neutral wind constant with altitude and E fields provides a vertical drift of ionization peak. According to Rishbeth and Barron (1960) and Rishbeth (1966), h_{max} and N_{max} are modified as follows:

$$\left.\begin{array}{l} \dfrac{\Delta h_{max}}{H_i} \sim 0.9 \left(W_{z,h} + W_{z,e}\right) H_i / D'_a \left(h_{max}\right) \\[2ex] \Delta \left(\mathrm{Log}\, N_{max}\right) \sim 0.75 \dfrac{\Delta h_{max}}{H_i} \end{array}\right\} h_{max} < H_i,$$

where H_i is the scale height of ionizable gas (assumed O), $D'_a(h_{max})$, the ambipolar diffusion coefficient, $W_{z,h}$ and $W_{z,e}$ the vertical speed component from neutral wind and E fields, respectively. Since E fields have symmetrical effects on the F layer, we can ignore them in this computation. For two conjugate points the south to north ratio in 6300 Å emission is computed assuming no wind effect on electronic width. For $T = 1000$ K, $h_{max} = 400$ km, and using Jacchia's 1971 model we deduce a horizontal wind speed of 70 m s^{-1} at $\lambda = 15°$ magnetic latitude for a south to north ratio equal to 2.7. This result is certainly acceptable. At lower altitudes for such a wind speed, two opposing phenomena are involved: quenching effect and increased collisional frequency, so the computed wind speed represents a mean value. Larger movements cannot be evaluated with Rishbeth's relations and need time dependent resolution of the ionization continuity equation.

Zonal wind effect is reduced by magnetic declination (a factor 2.7 in Atlantic region), but zonal wind speeds are much greater than meridional wind speed. 200 m s^{-1} are reached in Geisler's 1967 model at night. Rishbeth's (1971) recent theory allows larger zonal wind speed due to the polarization E field at night, in the F region, which makes the ions move more closely with zonal wind. This theory is supported by Woodman's observations at Jicamarca where very large eastward drifts are observed after sunset. Even reduced by the geometrical factor, zonal wind speed is sufficient to explain our observations.

4. Conclusion

We have shown that all phenomena which cause the F layer to go down in the night time can provide arcs in red line emission. Electric fields or neutral winds can produce such motions. Electric fields have latitudinal symmetrical effects on the F layer and cannot explain solstice or equinox asymmetries in latitude as longitudinal effects. Neutral winds, such as transequatorial meridional winds, can explain latitudinal effects. h_{max} is much more dependent on winds than N_{max}, and we have shown that the red line emission is particularly sensitive to h_{max} variations. Zonal winds provide explanation of the main features in longitudinal effects.

More details are needed about wind behavior in low latitude regions to compare with our observations. These observations can also indicate wind variations with local time and season because the red line is affected by h_{max} variation in the frame of the dissociative recombination process.

References

Abur-Robb, M. F. K. and Windle, D. W.: 1969. *Planetary Space Sci.* **17**, 97.
Barbier, D. and Glaume, J.: 1960, *Ann. Geophys.* **16**, 319.
Barbier, D. and Glaume, J.: 1962, *Planetary Space Sci.* **9**, 133.
Barbier, D., Weill, G., and Glaume, J.: 1961, *Ann. Geophys.* **17**, 305.
Bramley, E. N. and Young M.: 1968, *J. Atmospheric Terrest. Phys.* **30**, 99.
Chandra, S., Troy, B. E., Donley, J. L., and Bourdeau, R. E.: 1970, *J. Geophys. Res.* **75**, 3867.
Dalgarno, A. and Walker, J. C. G.: 1964, *J. Atmospheric Sci.* **21**, 463.
Fergusson, E.: 1969, *Ann. Geophys.* **25**, 819.

Geisler, J. E.: 1967, *J. Atmospheric Terrest. Phys.* **29,** 1469.

Lyon, A. J. and Thomas, L.: 1963, *J. Atmospheric Terrest. Phys.* **25,** 373.

Martyn, D. F.: 1953, *Phil. Trans. Roy. Soc.* **A246,** 306.

Noxon, J. F.: 1971, in B. M. McCormac (ed.), *The Radiating Atmosphere,* D. Reidel Publishing Company, Dordrecht-Holland p. 64.

Peterson, V. L., Van Zandt, T. E., and Norton, R. B.: 1966, *J. Geophys. Res.* **71,** 2255.

Piggott, W. R.: 1969, *Use of Satellite Data (Ariel III) for Prediction Purposes* (Ottawa).

Rishbeth, H.: 1966, *J. Atmospheric Terrest. Phys.* **28,** 911.

Rishbeth, H.: 1971, *Planetary Space Sci.* **19,** 357.

Rishbeth, H.: 1972, *J. Atmospheric Terrest. Phys.* **34,** 1.

Rishbeth, H. and Barron, D. W.: 1960, *J. Atmospheric Terrest.* **18,** 234.

INDIRECT EXCITATION PROCESSES IN AURORA

A. VALLANCE JONES and R. L. GATTINGER

Astrophysics Branch, National Research Council of Canada, Ottawa, Canada K1A OR8

Abstract. Indirect excitation processes for auroral emissions are briefly reviewed. New ground based measurements of height effects for the 1, 1 O_2 Atmospheric band in auroral arcs are presented. These measurements are shown to be in quantitative agreement with the hypothesis of Wallace and Chamberlain (1959) that the $v' = 1$ level of $O_2(b^1\Sigma)$ is excited primarily by energy transfer from $O(^1D)$ and deactivated vibrationally in collisions with O_2.

1. Introduction

Most auroral emissions are excited directly in collisions by primary auroral particles or by the secondary (or tertiary) electrons produced in ionizing collisions. The excitation rates for excited particles may be treated, for example, by the method of Rees *et al.* (1969) or the improved technique described by Sharp and Rees (1972).

In a few well known cases indirect excitation processes make a significant contribution. In other cases such processes have been suggested but their contribution is uncertain. In this paper various postulated indirect processes will be discussed briefly, followed by some experimental results relevant to one of the most important energy transfer processes, from $O(^1D)$ to O_2.

Indirect excitation processes may be classified in the following way.

1.1. Dissociative recombination

This undoubtedly plays a role in aurora since the ions O_2^+, NO^+, and N_2^+ are present in the 100 to 200 km region where auroral emission is most intense. O_2^+ is known to give $O(^1D)$ and $O(^1S)$ with yields of 0.9 and 0.1, respectively (Zipf, 1970). Solution of the orthodox ion chemistry steady state equations, as for example in the work of Rees *et al.* (1969) or of Walker and Rees (1965) leads to the conclusion that $[O_2^+]/[NO^+]$ should be about 0.5 near the peak brightness of an auroral arc at 110 km, and consequently that the production of $O(^1S)$ by dissociative recombination should be dominant if the above yields are correct. On the other hand rocket measurements in aurora (Donahue *et al.*, 1970) showed that $[O_2^+]/[NO^+]$ was less than 0.05 at 110 km in similar aurora. Moreover there are difficulties associated with a failure to detect the time constant effects which should be associated with a dominant contribution from dissociative recombination (Vreux and Marette, 1971; Parkinson, 1971)). Parkinson shows however that an 18% contribution would be consistent with his pulsation results. Brekke and Pettersen (1972), from an analysis of pulsating aurora, have also concluded that there could be a significant contribution from a process with a longer time constant, probably dissociative recombination. The observation of abnormally enhanced concentrations of NO within auroral forms by Zipf *et al.* (1970) suggests that there is a serious gap in our understanding of the chemistry of aurora.

Dissociative recombination of NO^+ should lead to $N(^2D)$ but this metastable species is strongly quenched to at least 250 km, so that its contribution to the spectrum is minor although potentially of considerable interest. N_2^+ is produced abundantly but is mostly destroyed in ionic reactions.

An interesting question is the fate of the metastable $O_2^+(a^4 II_u)$ ion which may be produced in significant yields (Dalgarno, 1970; Gérard, 1970).

The reaction

$$O_2^+(a^4 II_u) + e \rightarrow O + O + 10.99 \text{ eV}$$

is sufficient to excite O up to the $3p^3P$ state. It is thus energetically capable of giving a contribution to the 8446 Å, 7774 Å, 1304 Å or 1356 Å permitted O I lines. This contribution would be most significant at heights where dissociative recombination might become dominant over quenching by N_2 (Ryan, 1969).

1.2. THERMAL EXCITATION

Excitation of the $\lambda 6300$ [O I] lines by hot electrons is possibly of some importance in normal aurora as indicated by the calculations of Rees *et al.* (1967). In type-A red aurora there is probably an additional effect from the downward conduction of heat in the electron gas similar to that postulated for SAR arcs by Cole (1965). The excitation of $O_2(^1\Delta)$ in aurora by electrons heated by electric fields has been considered by Noxon (1970b) but does not seem very promising now.

The possibility of excitation in collisions from ions accelerated in a transverse E field has been considered by Cole (1971) and Walker (1970). There is a possibility that O I $\lambda 6300$ and $O_2(^1\Delta)$ could be excited in this manner.

1.3. ENERGY TRANSFER REACTIONS

In cases where significant quenching of an abundantly excited species takes place there is a possibility that the quenching particle may be excited in the process. Numerous processes of this type are energetically possible. Some of the most likely processes are discussed.

1.3.1. $N_2(A^3\Sigma) + NO(X^2II) \rightarrow NO(A^2II) + N_2(X^1\Sigma)$.

Sharp and Rees (1972) have suggested that this process might be the most important one for exciting the NO γ bands in aurora. The presence of these bands has not yet been established with certainty although a peak at 2150 Å corresponding to the 1,0 band has been reported by Duysinx (1970), Sharp and Rees (1972), and Duysinx and Monfils (1972).

1.3.2. $N_2(A^3\Sigma) + O \rightarrow O(^1S) + N_2$.

The possibility that this could be significant in exciting [O I] 5577 Å has been considered by Parkinson and Zipf (1970), Parkinson (1971), and Henriksen (1971). The evidence appears to be against this process except perhaps in a minor role.

1.3.3. $O^+(^2D) + N_2 \rightarrow N_2^+(A^2\Pi)(v' \leq 1) + O(^3P)$.

The possibility of this reaction was originally suggested by Omholt (1957). However, our most recent observations do not indicate any enhancement of the $v' = 0$ or $v' = 1$ levels above what would be expected for electron impact. (Gattinger and Vallance Jones, 1973)

1.3.4. $O(^1D) + O_2 \rightarrow O_2(b^1\Sigma)(v' \leq 2) + O(^3P)$.

This process seems certainly to be important and will be discussed below.

A number of other processes have been suggested including the excitation of $Na(^2P)$ by vibrationally excited N_2 (Hunten, 1965), $O(^1D)$ by $N(^2D)$ (Weill, 1969), and $N(^2D)$ in the reaction between N_2^+ and O (Wallace and McElroy, 1966).

2. Experimental

The observations reported here were made with a 0.5 m Ebert spectrometer having a $4'' \times 5''$, 1200 line/mm grating and using an S-25 extended IR response photomultiplier. The spectrometer was set to scan 7100 to 8800 Å every 12 s at a resolution of 15 to 17 Å. A pulse counting system with data recording on digital magnetic tape was used. The spectrometer was used in conjunction with an 11 channel digital filter wheel photometer which monitored (at 20 samples s^{-1}) 7 auroral emission features and 4 background channels. This system and its calibration have been described in more detail by Gattinger and Vallance Jones (1972), and Vallance Jones and Gattinger (1972a).

The photometer and spectrometer were mechanically coupled to cover almost the same field of view, the field of the photometer being a $4°$ diameter circle while that of the spectrometer is a $4° \times 4°$ square. The elevation angle of the system was recorded automatically by the digital system.

3. Observations

Many sets of spectra were recorded with the spectrometer field centered at various zenith angles relative to that of the lower border of apparently isolated, well defined, discrete homogeneous or rayed arcs. A pair of such spectra, one from the lower border and one from about $20°$ above the lower border of well defined quiet homogeneous arcs is shown in Figure 1. Apart from the change in intensity (which is arbitrary) several striking changes in relative intensity are immediately obvious. OI $\lambda 8446$ and the 1,1 O_2 Atmospheric band are stronger relative to the 2,0 and 3,1 Meinel N_2^+ bands in the upper spectrum while the 0,1 O_2 Atmospheric band at 8644 Å is weaker.

Several sets of the best quality spectra were chosen for analysis and the details of those chosen are set out in Table I. The letters L, M and U in the first column refer to the lowest, middle and upper parts of the auroral forms, respectively; the numbers identify particular arcs studied.

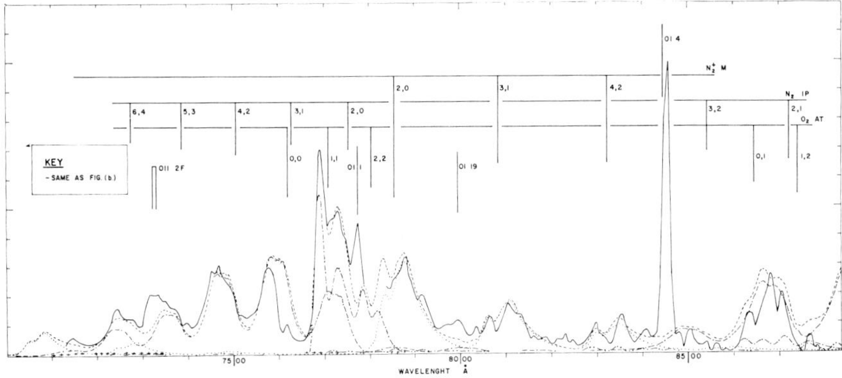

Fig. 1a. Observed spectrum from upper part of auroral arc (4U; see Table I) and matching synthetic spectra.

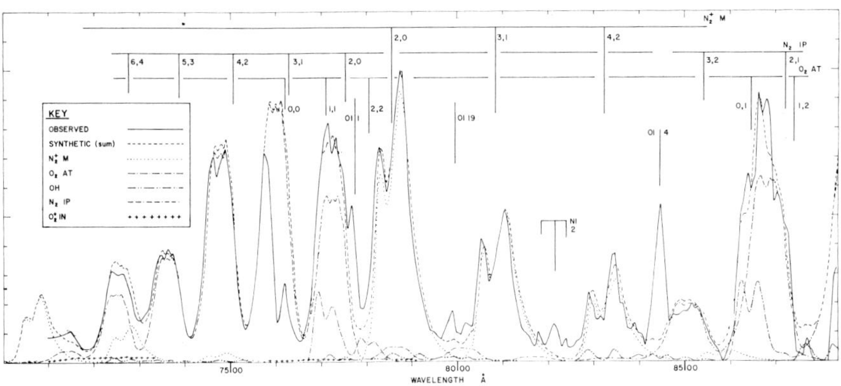

Fig. 1b. Observed spectrum from lower border of arc (Spectrum 1L; see Table I) and matching synthetic spectra.

TABLE I

Auroral forms analysed

Identification number	No. of scans averaged	Description	Zenith angle	Estim. height km
1L	2	Lower border HA	61°	(110)
1M	6	Above 1L	48°	180
2L	5	Lower border Bright active band	67–74°	(90–100)
3M	1	Above RB with LB at 78°	65°	≤ 175
3U	1	Above 3M	55°	≤ 230
4L	1	Lower border RB	58°	(100)
4U	3	Upper part 4L	38°	210

Each spectrum was obtained by averaging one or more individual 12 s scans. The 11 channel photometer gave *inter alia* a continuous record of the $\lambda 5577$ and $\lambda 6300$ intensities of which the former was used to correct the spectra for any small fluctuations in auroral intensity during the scans. In Table I are listed the zenith distance for each spectrum and the estimated geometrical height in the auroral form on the thin sheet approximation. The heights in brackets are the estimated lower border heights

4. Analyses of Spectra

The interpretation of the spectra requires in some cases the determination of the individual band and line intensities from the blends of two or more features. This was done by the synthetic spectrum method, following the procedure described by Vallance Jones and Gattinger (1972b). The synthetic spectra and their sums are shown in Figure 1.

The intensities chosen for the bands of the synthetic spectra were based on theoretical tables of relative band intensities constructed according to the methods discussed by Vallance Jones (1971). Revised versions of the tables in that review were used, based on the population rates and transition probabilities given by Shemansky and Broadfoot (1971) and Shemansky *et al.* (1971).

TABLE II

Absolute emission intensities – kR

Feature	Lower border			Above form		Well above	
	1L	2L	4L	1M	3M	3U	4U
M N_2^+ 2,0	8.34	62.6	6.75	3.52	3.46	2.20	2.14
1P N_2 4,2	6.0	39.4	4.8	2.52	2.48	1.34	1.53
O_2 Atm 1,1	1.48	11.6	1.83	2.0	4.47	1.94	2.07
2,2 [b]	0.49	(3.8)	(0.61)	(0.67)	(1.5)	(0.65)	(0.66)
0,1	2.06	23.7	2.13	0.23	(0.45)	(0.2)	(0.21)
1,2 [b]	0.17	(1.3)	0.21	(0.23)	(0.52)	(0.23)	(0.24)
OI 8446 Å	1.35	8.25	1.98	1.01	2.03	1.97	1.33
OI 7774 Å [a]	0.76	7.6	0.79	0.31	0.52	0.43	0.34
	(1.1)	(7.8)	(0.90)	(0.72)	(1.2)	(0.94)	(0.85)
OI 5577 Å	16	130	24.5	5.6	13.7	6.6	5.9
OI 6300–64 Å	0.89	6.14	0.99	1.13	2.43	4.0	2.86

[a] Values in brackets are upper limits assuming $I(O_2$ Atm; 2,2$) = 0$
[b] Values in brackets are upper limits.

Table II gives the band intensities required to match the synthetic spectra. Generally the relative intensities are very close to the theoretical values and to the values in the work of Vallance Jones and Gattinger (1972b) for bright aurora. The $\lambda 8446$ and $\lambda 7774$ line intensities are also entered in Table II.

The separation of the blend of the 1,1 O_2 Atmospheric band and the 2,0 N_2 1P band is difficult to achieve when consideration is confined to the trace in the region

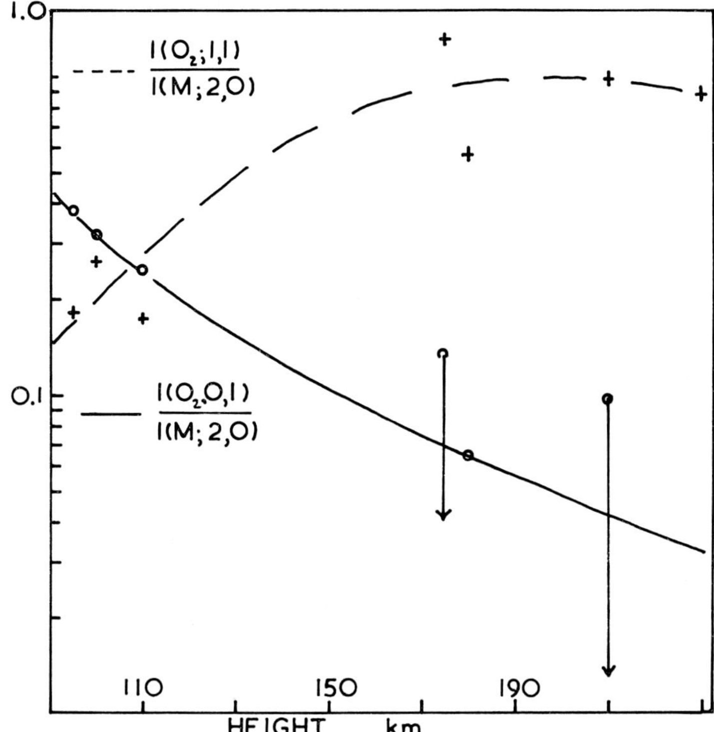

Fig. 2.   Ratios observed for O_2 Atm band intensities to $I(M\ N_2{}^+;\ 2,0)$ vs. estimated height.

of these two bands. If it is assumed that the ratio of $I(1P;4,2)/I(1P;2,0)$ remains constant then the blended feature can be reproduced by varying the intensity and rotational temperature of the 1,1 O_2 band. There is no good reason why this ratio should not remain constant. Moreover the intensity observed for the 2,1 band provides an upper limit to that of the 2,0 band (since the ratio of the intensities of these bands depends only on the transition probabilities).

One important result of this analysis is shown in Figure 2 on which is plotted the ratios of the intensities of 0,1 and 1,1 O_2 Atmospheric bands to those of the 2,0 Meinel N_2^+ bands for each spectrum.

5. Interpretation of O_2 Atmospheric Band Height Variations

As originally suggested by Wallace and Chamberlain (1959) the peculiar height variation of the 1,1 and 0,1 Atmospheric bands (henceforth abbreviated Atm) may be understood in terms of an energy transfer process from the ^1D level of O the energy of which matches closely that of the $v'=2$ level of the $b^1\Sigma$ state of O_2. We will use O* for $O(^1D)$ and $O_2^*(v)$ for $O_2(^1\Sigma, v)$. The energy transfer reaction may thus be written

$$O^* + O_2(0) \rightarrow O_2^*(v \leqslant 2) + O. \tag{1a}$$

The reverse reaction, possible from the $v=2$ level only will be denoted as Equation (1b).

Since $[O^*]$ varies only slowly with height below the quenching height (~ 250 km) the excitation rate of O_2^* would vary as $[O_2]$ so that, in the absence of quenching, the emission rate would fall off as $[O_2]$, that is a little faster perhaps than the N_2 bands. This is possibly true for the 0,1 O_2 band but not for the 1,1 band. Wallace and Chamberlain (1959) suggested that the $v > 0$ levels of O_2 $b^1\Sigma$ could be deactivated by the process

$$O_2^*(v>0) + O_2 \rightarrow O_2^*(v=0) + O_2(v). \tag{2}$$

If this is true, then in the steady state

$$\frac{[O_2^*(1)]}{[O^*]} = \frac{k_{1a}(1)[O_2]}{A_{bx} + k_2[O_2]} = \frac{I(O_2;1,1)}{A_{bx}(1,1)} \cdot \frac{A_{21}}{I(6300)} \tag{3}$$

where $k_{1a}(v)$ denotes the rate constant for Equation (1a) for producing $O_2^*(v)$.

It follows that $I(O_2;1,1)/I(6300)$ should be constant in the region where $k_2[O_2] \gg A_{bx}$ and should vary as $[O_2]$ at greater heights where $A_{bx} \gg k_2[O_2]$. This height variation should provide a means of estimating k_2 and k_{1a}/k_2.

In Figure 3 the ratio $I(O_2;1,1)/I(6300)$ is plotted. It appears that the ratio is fairly constant up to about 180 km and then falls off. If 180 km is indeed the knee height then it follows that $k_2 \cdot [O_2, 180 \text{ km}] = A_{bx}$. Hence $k_2 \simeq 0.8 \times 10^{-10}$ cm^3 s^{-1} since $A_{bx} = \frac{1}{12}$ s^{-1} and $[O_2, 180 \text{ km}] \simeq 10^9$ cm^{-3}. Moreover from the constant ratio region

$$\frac{k_{1a}(1)}{k_2} = \frac{I(O_2;1,1)}{I(6300)} \frac{A_{21}}{A_{bx}(1,1)}$$

$$= 1.7 \times \frac{1}{112} \times \frac{1}{0.67 \times 10^{-1}} = 2.3 \times 10^{-1}$$

so that $k_{1a}(1) \simeq 1.8 \times 10^{-11}$ cm^3 s^{-1}. The overall rate of k_{1a} (Noxon, 1970a; Snelling and Gauthier, 1971) is known to be about 5×10^{-11} cm^3 s^{-1}. It thus appears likely that a significant fraction of O_2^* is produced in the $v=1$ level.

Do these observations prove that energy transfer is a significant process? If we consider the results for the greatest height where quenching of the $v'=1$ level is unimportant then assuming a constant lifetime for the $v'=0$ and 1 levels, the band intensities, the transition probabilities, and the production rates $\eta(v')$ are connected by the relation,

$$\frac{I(1,1)}{I(0,1)} = \frac{\eta(1)}{\eta(0)} \cdot \frac{A(1,1)}{A(0,1)}.$$

Now from Table II, above 170 km $I(1,1)/I(0,1) \gg 4$ so that $\eta(1)/\eta(0) \gg 0.26$. For direct electron impact excitation the ratio would be given by the ratio of the Franck Condon factors $q(1,0)/q(0,0) \simeq 0.08$. It thus appears that some other process capable of populating the $v'=1$ level must be important. The energy transfer process appears to satisfy the requirements.

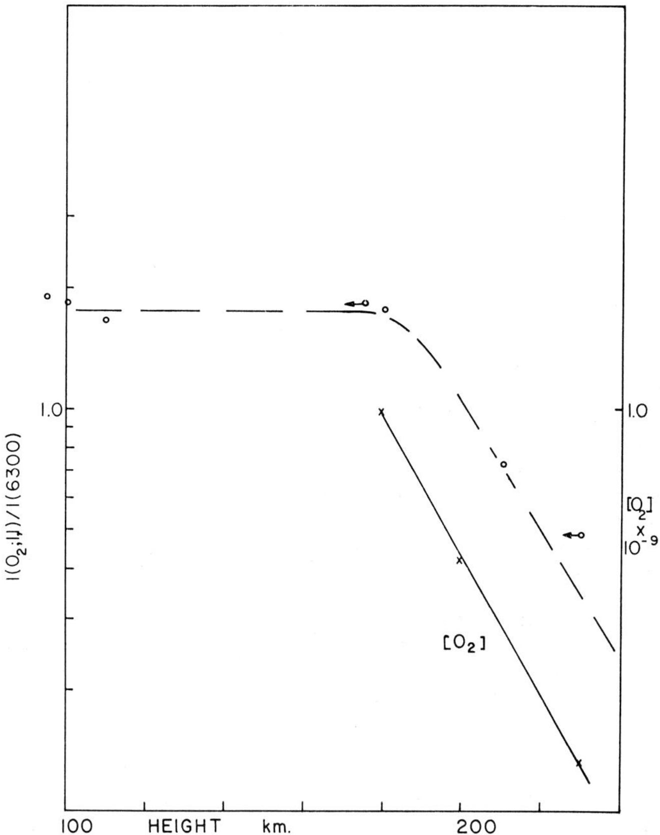

Fig. 3.   Ratio $I(\text{Atm } O_2; 1,1)/I(6300)$ vs. estimated height.

The importance of the indirect process may also be assessed by comparing the total observed emission for the O_2 Atmospheric system with the total population rate from the energy transfer process. An estimate of 540 kR for IBC3 aurora is made by Vallance Jones (1971). This is based on an indirect estimate of the 0,1 band intensity of 23 kR. The total population rate through the energy transfer process may be estimated since the ratio of the production rates $\eta(O^1D)/\eta(O^1S)$ is probably in the range 10 to 20 (since the ratio of the yields from both secondary electron excitation and dissociative recombination are of this order). Therefore since the ratio $[O_2]/[N_2]$ at 100 km is about 1/4.5 and since quenching of $O(^1D)$ by N_2 and O_2 have similar rate coefficients one would expect that for $I(5577) = 100$ kR that $\eta(O_2 \text{ Atm})$ would be about 200 to 400 kR. On the other hand, a calculation by Cartwright *et al.* (1972) gave a ratio of about unity for $I(0,0; \text{Atm})/I(3914)$. Since $I(5577)/I(3914)$ is between 1 and 2 this implies that for IBC3 aurora $I(0,0; \text{Atm})$ should be in the range 100 to 200 kR.

Although subject to some uncertainties these estimates are consistent with the view that a major fraction of the O_2 Atm emission is excited via the energy transfer process.

References

Brekke, A. and Pettersen, H.: 1972, private communication.
Cartwright, D. C., Trajmar, S., and Williams, W.: 1972, *Ann. Geophys.* **28**, 397.
Cole, K. D.: 1965, *J. Geophys. Res.* **70**, 1689.
Cole, K. D.: 1971, *J. Atmospheric Terrest. Phys.* **33**, 1241.
Dalgarno, A.: 1970, *Ann. Geophys.* **26**, 601.
Donahue, T. M., Zipf, E. C., and Parkinson, T. D.: 1970, *Planetary Space Sci.* **18**, 171.
Duysinx, R.: 1970, *Bull. Soc. Roy. Sci. Liège* **39**, 157.
Duysinx, R. and Monfils, A.: 1972, *Ann. Geophys.* **28**, 109.
Gattinger, R. L. and Vallance Jones, A.: 1972, *Ann. Geophys.* **28**, 91.
Gattinger, R. L. and Vallance Jones, A.: 1973, *Can. J. Phys.* **51**, 287.
Gérard, J. C.: 1970, *Ann. Geophys.* **26**, 777.
Henriksen, K.: 1971, private communication.
Hunten, D. M.: 1965, *J. Atmospheric Terrest. Phys.* **27**, 583.
Noxon, J. F.: 1970a, *J. Chem. Phys.* **52**, 1852.
Noxon, J. F.: 1970b, *J. Geophys. Res.* **75**, 1876.
Omholt, A.: 1957, *J. Atmospheric Terrest. Phys.* **10**, 320.
Parkinson, T. D.: 1971, *Planetary Space Sci.* **19**, 251.
Parkinson, T. D. and Zipf, E. C.: 1970, *Planetary Space Sci.* **18**, 895.
Rees, M. H., Walker, J. C. G., and Dalgarno, A.: 1967, *Planetary Space Sci.* **15**, 1097.
Rees, M. H., Stewart, A. I., and Walker, J. C. G.: 1969, *Planetary Space Sci.* **17**, 1997.
Ryan, K. R.: 1969, *J. Chem. Phys.* **51**, 4136.
Sharp, W. E. and Rees, M. H.: 1972, *J. Geophys. Res.* **77**, 1810.
Shemansky, D. E. and Broadfoot, A. L.: 1971, *J. Q Spectr. Radiative Transfer* **11**, 1385.
Shemansky, D. E., Zipf, E. C., and Dohanue, T. M.: 1971, *Planetary Space Sci.* **19**, 1969.
Snelling, D. R. and Gauthier, M.: 1971, *Chem. Phys. Letters* **9**, 254.
Vallance Jones, A.: 1971, *Space Sci. Rev.* **11**, 776.
Vallance Jones, A. and Gattinger, R. L.: 1972a, *Ann. Geophys.* **28**, 85.
Vallance Jones, A. and Gattinger, R. L.: 1972b, *Can. J. Phys.* **50**, 1833.
Vreux, J. M. and Marette, G.: 1971, *Bull. Soc. Roy. Sci. Liège* **40**, 451.
Walker, J. C. G.: 1970, *Planetary Space Sci.* **18**, 1043.
Walker, J. C. G. and Rees, M. H.: 1968, *Planetary Space Sci.* **16**, 459.
Wallace, L. and Chamberlain, J. W.: 1959, *Planetary Space Sci.* **2**, 60.
Wallace, L. and McElroy, M. B.: 1966, *Planetary Space Sci.* **14**, 677.
Weill, G. M.: 1969, in B. M. McCormac and A. Omholt (eds.), *Atmospheric Emissions*, Van Nostrand
  Reinhold Company New York, p. 449.
Zipf, E. C.: 1970, *Bull. Amer. Phys. Res. Soc.* **15**, 418.
Zipf, E. C., Borst, W. L., and Donahue, T. M.: 1970, *J. Geophys. Res.* **75**, 6371.

OBSERVATION OF O(^{1}D) AND N(^{2}D°) EMISSION IN THE POLAR AURORA

J-C. GÉRARD*

Institut d'Astrophysique, University of Liège, 4200 – Cointe Ougrée, Belgium

and

O. E. HARANG

The Auroral Observatory, Tromsö, Norway

1. Introduction

Very little data exist on $N\textsc{i}(^4S°-{}^2D°)$ 5199 Å–5201 Å doublet intensity and variations in high latitude auroras. It is a weak feature of the spectrum and most intensity evaluations were derived from photographic spectra averaging observations over long periods of time. Chamberlain (1961) evaluates its brightness to 1 kR in IBC III aurora, while Vallance Jones (1971) estimates it at 400 R, insisting on its variability. Nightglow and low latitude auroral observations have been extensively made by Weill (1968). Tinsley (1963), and Smith and Webber (1968) recorded the doublet in low latitude auroras. All measurements show a behavior similar to the 6300 Å oxygen forbidden line. Recently, Eather and Mende (1971) using airborne tilting-filter photometers, measured the 5200 Å intensity at various latitudes in the daytime and night-time oval. The 5200/4278 Å ratio varies from 0.05 in 'normal aurora' to 0.4 in the soft zone situated polewards of the hard precipitation oval, thus varying in the same way as the 6300/4278 Å ratio.

The importance of $N(^2D°)$ in the polar ionosphere is twofold:

(a) Because of its long radiative lifetime (26 hr) Wiese *et al.*, 1966) the $^2D°$ level is strongly depopulated by collision reactions with atmospheric species. Knowledge of its absolute intensity and variations with respect to other lines allows an evaluation to be made of the dominant quenching processes.

(b) The problem of excitation and loss of $N(^2D°)$ is of special interest due to its relationship with the presence of large amounts of NO detected in the aurora (Zipf *et al.*, 1970; Duysinx and Monfils, 1971; Sharp and Rees, 1972).

Nitrogen and NO daytime chemistry have been recently discussed (Strobel *et al.*, 1970) and it appears that $N(^2D°)$ metastables play an important role through reactions:

$$NO^+ + e \rightarrow N(^2D°, {}^4S°) + O \tag{1}$$

whose branching ratio is unknown, and

$$N(^2D°) + O_2 \rightarrow NO + O. \tag{2}$$

* Aspirant of the National Foundation for Scientific Research (FNRS).

B. M. McCormac (ed.), Physics and Chemistry of Upper Atmospheres, 241–247. *All Rights Reserved.*
Copyright © 1973 by D. Reidel Publishing Company, Dordrecht-Holland.

It should be noted that due to the long lifetime of $N(^2D^\circ)$, possible deactivation agents are very numerous, including electrons, O_2, NO and possibly O.

Recent developments of low intensity photometry used together with photon counting techniques make observation of weak lines and bands possible, even in the presence of a strong background.

Intensity variation compared to OI 6300 Å line, the equivalent transition in atomic oxygen, can yield information on production mechanisms of excited nitrogen. Determination of $^2D^\circ$ effective lifetime may give a clue to the altitude of the varying parts of the emission.

2. Instrumentation and Data Reduction

The instrument used was a four channel tilting-filter photometer, the principle of which was described by Eather and Reasoner (1969). The filters were periodically tilted between 0 and 10°, it being possible to vary the scan period from 0 to 20 s. The entrance pupil had 2″ diam. and the angle of view reached about 1°45′. The detectors were two EMI 9558-S20 and two EMI 9502-S11 photomultipliers operated in the pulse counting mode. The photon pulses were amplified, shaped, and sent through a scaler of 10 before entering the ratemeters. The ratemeters were eight-bit scalers, read by a buffer at preset intervals, whose lengths depended on expected light level. The buffer was connected to D/A converters monitored by a multichannel recorder and was at the same time read asynchronously by a digital recorder at the rate of 17 samples s^{-1}. The filter temperature was kept constant while the photomultipliers' housing was at ambient temperature (0 to $-20\,^\circ C$). Each scan started with a zero mark allowing accurate determination of the beginning of each spectrum. Absolute sensitivity of each channel was determined by measuring the response when viewing a MgO or $BaSO_4$ screen illuminated by a calibrated tungsten filament ribbon lamp. The observations were made from Skibotn, Norway (IN Lat 67.5N) during a new moon period extending between 9 and 20 December, 1971. More than 60 hr of auroral data were analyzed. Table I lists the observed features together with wavelength and scale factor (the prescaler of 10 being taken into account).

TABLE I

Observed features

Feature	Wavelength (Å)	Scale factor R/Count s^{-1}
N_2^+: (0–1) 1N	4278	0.57
OI: ^3P–^1D	6300	0.90
NI: $^4S^\circ$–$^2D^\circ$	5200	0.40
HI: Hβ	4861	0.37

The following analysis will be primarily devoted to the first three emissions, the Hβ line being essentially different as far as geographical and temporal variations are concerned. The digital data have been processed following two modes, according to the importance of temporal variation of the intensity. Periods with weak intensity variations (less than 20%) were selected and scans were added in phase by computer in order to improve the signal-to-noise ratio. This method made it possible to work with lower intensity data than with individual scans. In particular, the nightglow level was easily reached on [OI] and [NI] doublets. On the other hand, during periods of rapidly varying intensity (generally coinciding with rather strong displays), scans were treated individually to study temporal variations and correlations of the 4 channels. As usual in such measurements, the line or molecular band intensity was obtained by subtracting the contribution of underlying continuum and background from the peak signal. No nightglow contribution has been subtracted from any feature because this procedure carries certain risks and does not necessarily have a physical significance in the polar ionosphere. Discussion of this point will be made in a subsequent section.

3. Intensity Ratios

3.1. $R_1 = I(6300)/I(4278)$

About 50 periods of quiet aurora were selected to study the relative intensity of N_2^+ 4278 Å, [OI] 6300 Å and [NI] 5200 Å features. Figure 1 shows $R_1 = I(6300)/I(4278)$ vs. 4278 Å intensity.* It is clear that R_1 is far from constant and clearly decreases with the total amount of energy of precipitated electrons (as determined by the N_2^+

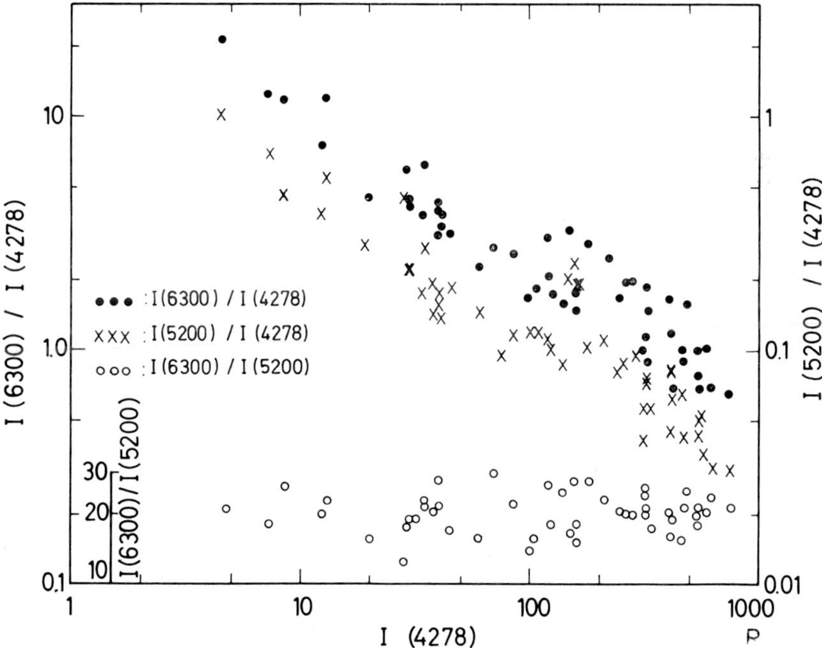

Fig. 1.   Intensity ratios: $R_1 = 6300/4278$ Å; $R_2 = 5200/4278$ Å and $R_1/R_2 = 6300/5200$ Å observed in quiet auroral conditions.

* The ^{3}P–^{1}D transition intensities quoted here do not include the 6363 Å component of the red doublet.

4278 Å intensity), ranging from 20 to 0.7. This variation of R_1 with auroral intensity was indicated by Gattinger and Vallance Jones (1971). A very similar behavior was measured by Eather and Mende (1972) during the 1969 NASA Airborne Expedition. They interpreted this R_1 variation as an indication that an increase in 4278 Å brightness is generally due to a variation in primary electron energy rather than in the flux, in agreement with Hilliard and Shepherd's (1966) observations. Below 50 R, they measured an increase of R_1 with latitude. This fact was interpreted as the indication of a double component in the primary average energy: one is soft (0.5 to 1.5 keV) and increases in energy with latitude, the other is harder (> 1.5 keV) and lies in the position of the oval. It can be speculated that such a break appears in these data between 100 and 150 R on the 4278 Å band.

3.2. $R_2 = I(5200)/I(4278)$

The R_2 ratio has also been plotted in Figure 1 for the same set of data. As could be expected, the general shape is very similar to R_1.

Finally, the $I(6300)/I(5200) = (R_1/R_2)$ ratio is plotted on the lower scale. Although somewhat spread out, on the average the ratio remains close to 20, without showing any systematic dependence on the auroral brightness. It should be noticed that Weill (1968) measured a mean ratio of 27 for 6300/5200 Å in the mid-latitude nightglow. It is obvious that an important part of [NI] and [OI] emission in very weak auroras is due to the nightglow. However, as dissociative recombinations of O_2^+ and NO^+, which probably dominate O^1D and $N^2D°$ production in the nightglow, also contribute to auroral emission, it has been considered preferable not to subtract nightglow contribution.

These data provide a straightforward estimate of the $N(^2D°)$ column density in the aurora: extrapolating for a 4278 Å intensity of 1 kR, the corresponding 5200 Å is 25 R, thus yielding a slant density of $2.5\ 10^{12}$ cm^{-2}.

4. Temporal Variations

Besides the 'static' approach of intensity ratio between [NI] 5200 Å and other auroral features, it is of interest to examine the time variation of the 5200 intensity and its correlation with other emissions. Figure 2 shows an example of such variations. Scan duration was 20 s for all channels; the two points to be seen for each scan correspond to the two measurements made for the backward-forward motion of the filter every 20 s. Because of the lower counting rate, the 5200 Å and Hβ channels clearly show a more important statistical fluctuation than the more intense 6300 Å and 4278 Å channels.

Nevertheless, the following conclusions can be deduced:

(a) Lack of correlation between Hβ and other emissions.

(b) General covariation of N_2^+ and [OI] 6300 Å, through a certain lack of correlation during rapid strong variations.

(c) A decrease of 6300/4278 Å ratio during fast increase of auroral brightness

(Figure 2) in agreement with conclusions of the previous paragraph dealing with R_1.

(d) Better correlation of [NI] with [OI] intensity than with N_2^+ (Figure 2). This last point allows to evaluate the effective lifetime of N_2^+, by using the equation of Omholt (1971):

$$\frac{dn\left(^2D^\circ\right)}{dt} = q\left(^2D^\circ\right) - \frac{n\left(^2D^\circ\right)}{\tau}. \tag{3}$$

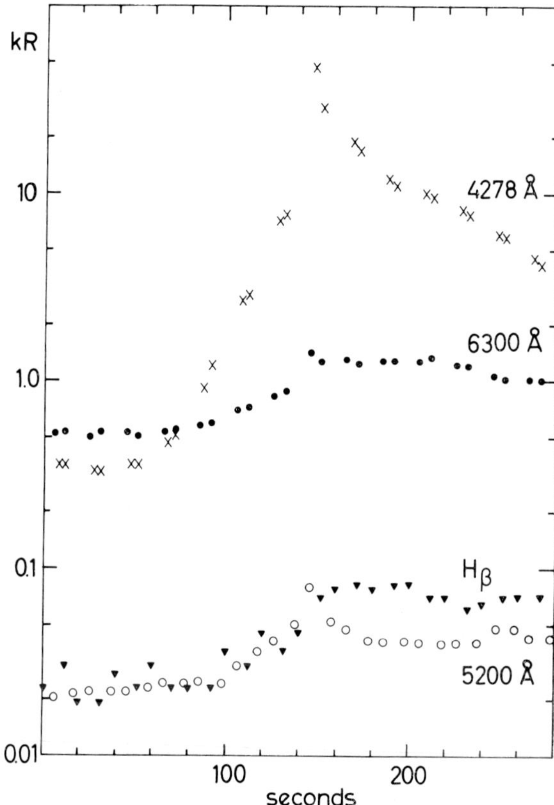

Fig. 2. Example of fast temporal variation.

This is an operational definition of effective lifetime τ, $n\left(^2D^\circ\right)$ being the number density into $^2D^\circ$ level, and $q\left(^2D^\circ\right)$ the production rate. As $I(5200)$ is well correlated with $I(6300)$, one assumes

$$q\left(N^2D^\circ\right) = kq\left(O^1D\right);$$

the following is obtained, assuming short τ for $O\left(^1D\right)$:

$$\frac{dI\left(5200\right)}{dt} = K\,I\left(6300\right) - \frac{I\left(5200\right)}{\tau} \tag{4}$$

with $K = A_{5200}\,k$.

Values of τ most suitable for the observed 5200 Å variations range from 0 to 30 s. Dominant deactivation below 250 km is probable due to chemical reaction with O_2 (rate constant $k = 5 10^{-12}$ cm^3 s^{-1}; Lin and Kaufman, 1970) and to electron (and possibly O) quenching above 250 km.

Limiting the field to the reaction with O_2, the only known process yielding values of τ less than 1 min, the altitude where this lifetime is reached, can be evaluated by using a model atmosphere for O_2 distribution (Jacchia, 1971).

Thus

$$\tau = (5 \times 10^{-12} n(O_2))^{-1}.$$

Values obtained for $n(O_2)$ indicate regions below 150 km. This altitude refers to the origin of the modulation of the intensity, eventually different from the region of the maximum emission. This result seems important, since it questions the validity of Equation (4), $O(^1D)$ emission being essentially due to altitudes above 200 km (Parkinson and Zipf, 1971).

5. Discussion

Both [OI] 6300 Å and [NI] 5200 Å intensities relative to allowed transitions appear strongly dependent on auroral brightness, due to increased quenching by N_2 for $O(^1D)$ and O_2 and electrons for $N(^2D^\circ)$ when electron energy increases.

Strikingly, both forbidden line intensities are well correlated in most cases. The fact that the average 6300/5200 Å ratio in the aurora remains close to that observed in the nightglow suggests similar production mechanisms in normal and disturbed conditions, namely Reaction (1) for $N(^2D^\circ)$ and:

$$O_2^+ + e \rightarrow O(^1D) + O \tag{5}$$

for $O(^1D)$.

The constancy of 6300/5200 Å during variations of auroral brightness can possibly be explained by following considerations. Indeed, if Reactions (1) and (5) govern production and if dominant quenching is attributed to N_2 for $O(^1D)$ and O_2 for $N(^2D^\circ)$ ($\leqslant 250$ km), the ratio of emission rates is:

$$\frac{(6300\ \text{Å})}{(5200\ \text{Å})} \sim \frac{n(O_2^+)}{n(NO^+)} \cdot \frac{n(O_2)}{n(N_2)}.$$

As $n(O_2^+)/n(NO^+)$ is essentially constant (ratio close to 1) above 160 km (Johnson, 1972) the ratio varies little which is in agreement with observations.

Clearly, the knowledge of the efficiency of Reaction (1) and $N(^2D^\circ)$ deactivation coefficient by O is needed before quantitative models can be made.

Acknowledgments

The assistance of Mr A. Tilkin in data processing is gratefully acknowledged.

References

Chamberlain, J. W.: 1961, *Physics of the Aurora and Airglow*, Academic Press, New York, London.

Duysinx, R. and Monfils, A.: 1971, *Ann. Geophys.* **28**, 109.

Eather, R. H. and Mende, S. B.: 1971, *J. Geophys. Res.* **76**, 1746.

Eather, R. H. and Mende, S. B.: 1972, *J. Geophys. Res.*: **77**, 660.

Eather, R. H. and Reasoner, D. L.: 1969, *Appl. Opt.* **8**, 227.

Gattinger, R. L. and Vallance Jones, A.: 1971, *Ann. Geophys.* **28**, 91.

Hilliard, R. L. and Shepherd, G. G.: 1966, *Planetaty Space Sci.* **14**, 383.

Jacchia, L. G.: 1971, 'Revised Static Models of the Thermosphere and Exosphere with Empirical Temperature Profiles', Smithsonian Astrophysical Observatory, Special Report No. 332.

Johnson, C. Y.: 1972, *Radio Sci.* **7**, 99.

Lin, C. L. and Kaufman, F.: 1970, *J. Chem. Phys.* **55**, 3760.

Omholt, A.: 1971, *The Optical Aurora*, Springer-Verlag, New York.

Parkinson, T. D. and Zipf, E. C.: 1971, *Planetary Space Sci.* **19**, 267.

Sharp, W. E. and Rees, M. H.: 1972, *J. Geophys. Res.* **77**, 1810.

Smith, R. W. and Webber, N. J.: 1968, *J. Atmospheric. Terrest. Phys.* **30**, 169.

Strobel, D. F., Hunten, D. M., and McElroy, M. B.: 1970, *J. Geophys. Res.* **75**, 4307.

Vallance Jones, A.: 1971, *Space Sci. Rev.* **11**, 776.

Tinsley, B. A.: 1963, Ph.D. Thesis, University of Canterbury.

Weill, G. M.: 1968, in B. M. McCormac (ed.), *Atmospheric Emission*, D. Reidel Publishing Company, Dordrecht, Holland.

Wiese, W. L., Smith, M. W., and Glennon, B. M.: 1966, *Atomic Transition Probabilities*, Natl. Bur. Standards.

Zipf, E. C., Borst, W. L., and Donahue, T. M.: 1970, *J. Geophys. Res.* **75**, 6371

HYDROGEN AND HELIUM EMISSIONS

P. MANGE

*E. O. Hulburt Center for Space Research, Naval Research Laboratory, Washington,
D.C. U.S.A.*

1. Hydrogen Emissions

There is now a wealth of observational data on the emission lines of H seen in the
atmosphere. The α and β lines of the solar Lyman series at 1216 and 1026 Å excite
the extensive H in the terrestrial atmosphere causing it to re-emit them; they are then
seen from space above the absorbing lower atmosphere. Scattering of the solar Ly-β
line by the atmospheric H also produces the Balmer series Hα line (6563 Å) seen
from the ground. The interpretation of early Ly-α observations has been summarized
(Mange, 1972); discussion is limited here to contributions from 1970 on. Auroral H
emission and its excitation are neglected.

1.1. Atmospheric Ly-α radiation

Distinctive new information (Meier and Mange, 1973) has been obtained from the
OSO-4 satellite in the form of full scans of Ly-α intensity including orientations away
from the upward and downward vertical. Figure 1 shows midnight data from 32
wheel scans over a 1 min interval. The intensity curve was calculated (Meier and
Mange, 1970) by assuming a spherically symmetric model for the H geocorona from
100 km to 12 R_E in which incident solar Ly-α radiation is scattered. The excess of

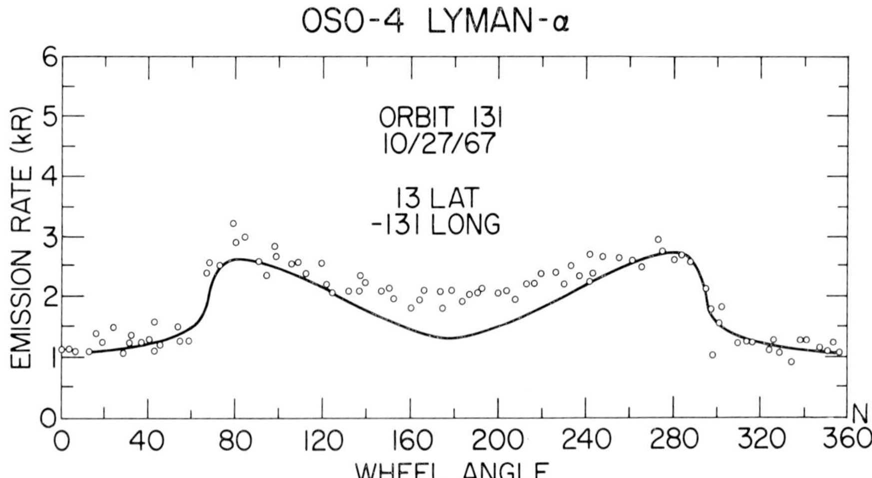

Fig. 1.   Ly-α intensity from OSO 4 in a midnight scan (solar zenith angle 178°) within 20° of zenith
(wheel angle 180°) and of nadir (wheel angle 0, 360°). The curve is the predicted intensity from
theory (Meier and Mange, 1972).

observed intensity over the model near zenith is in part attributed to the presence of extraterrestrial Ly-α background. From an atlas of such scans and their associated curves from model theory, a clear H density maximum (greater emission) is found between 0600 and 0700 in the morning, and a minimum after 1600. The diurnal density variation inferred is 70% in accordance with analysis of OSO-5 results by Vidal-Madjar *et al.* (1972). Latitudinal asymmetry is also suggested by brighter afternoon emission seen in the north in comparison with that from the south as seen from a satellite position over the terminator in the southern hemisphere.

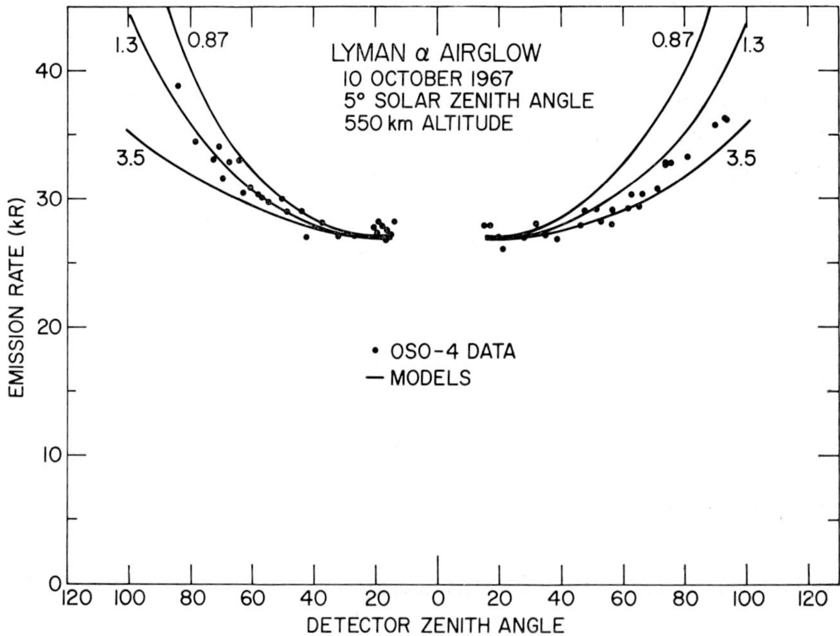

Fig. 2. Daytime Ly-α airglow intensity. Theoretical emission rates are shown for optical depths of 0.87, 1.3, and 3.5 above 650 km (Meier and Mange, 1972).

Figure 2 illustrates the goodness of fit of the model to daytime data. An 1100 K model is used (Kockarts and Nicolet, 1963) corresponding to optical depth 1.3 with some population of satellite exospheric orbits ($R_{sc} = 2.5$ in the notation of Chamberlain, 1963). It has concentrations of 4.1×10^4, 9.0×10^2, and 2.5×10^1 cm^{-3} at 650, 10,000, and 50,000 km, respectively. These concentrations are multiplied by $\frac{2}{3}$ or $\frac{8}{3}$ for the models of 0.87 or 3.5 optical depth.

Blamont and Vidal-Madjar (1971), using data from their H resonance cell detector on OSO 5, showed direct correlation of the Lyman solar line-center intensity with sunspot number. Their more detailed treatment (Vidal-Madjar, 1972) also defines the precise correlation with total solar Ly-α line intensity. Their analysis is found to agree remarkably (Meier and Mange, 1973) with the solar line-center variation deduced from OGO 4 airglow data when adjusted to take account of reduction in geocoronal H with increased solar heating. Moreover, the numerical relationship is supported by the OGO-5 analysis of Thomas and Bohlin (1972). This precision correlation with solar activity permits more exact comparison of observed intensities by different groups under varying observing conditions. Thus, from the observed emission rate, the temperature of the exosphere, and the observing geometry, a computed line-center solar flux is found. The ratio of this to the line-center flux derived from the sunspot number is found to vary greatly: 0.46 and 0.98 for OGO 4 and OSO 4 (Meier and Mange, 1972); 1.8, 0.56, and 0.48 for OGO 4, OGO 5 and OGO 6 (Thomas, 1970, 1972); 0.86 for the resonance filter experiment on OGO 6 (Metzger and Clark, 1970), and 0.63 for the experiment carried on OV1 10 (Clark, 1972). (Nadir observations from the French OGO 5 experiment (Bertaux and Blamont, 1970) were consistent with the zenith OGO-5 data reported by Thomas.) The values cited would be unity if atmospheric models, the solar flux relationship, and the absolute instrument sensitivity were all correct. The spread suggests major differences in absolute sensitivity. Comparison of data from different instruments on the same day supports this. Moreover, it appears that stellar flux observations used to calibrate instruments (Thomas and Krassa, 1971) may be discrepant by as much as a factor of 2 (Carruthers, 1972).

1.2. Balmer-α and Ly-β emissions

Tinsley (1970) has cited Balmer-α observations made since 1958 and, from comparison of his own observations since 1965 with theory, detected a small galactic contribution and an early morning density bulge in H toward the equator. He also argued the desirability of adjusting vertical H models to increase relative content at higher altitudes. Typical intensities at zenith distance $80°$ may range from 30 R looking toward the sun (depressed by 20 to $25°$ below the horizon) to a few R looking away from it. Tinsley and Meier (1971) have conducted a systematic study of Balmer-α emission observed at various sites over a solar cycle, by comparing it with intensities predicted by radiative transfer for the scattering and conversion of solar Ly-β from a H model consistent with Ly-α data. The observed brightness change by factor of 3 is reconcilable with the solar Ly-β cyclic variation, but the reported Ly-β flux at line-center (not well measured) appears too small by a factor of 2 to 5. Weller *et al.* (1971) have described nighttime measurement of Ly-α and Ly-β from a rocket (Young *et al.*, 1971) simultaneous with Balmer-α observation on the ground. The Ly-α data revealed a N–S asymmetry with more H. (greater intensity) toward the south (equator) by comparison with the spherical model calculations. There was agreement of observed Ly-β (27 R zenith intensity inferred) with the model, and also good spatial agreement of Balmer-α intensities. Again, the data require a higher Ly-β solar line-center flux. They have shown that the nighttime intensity ratio of observed Ly-β to Hα is a simple measure of H optical depth without regard to model parameters.

Finally, it may be noted that from the equilibrium relation, $n(H) = 8n(H^+)/9n(O^+)$, and mass spectrometric data, Brinton and Mayr (1971) independently corroborate and precisely determine the time-varying behavior of the H models derived from the emission studies.

1.3. Ly-α FROM THE OUTER GEOCORONA AND THE EXTRATERRESTRIAL BACKGROUND

The time-varying presence of H in satellite orbits above 5 R_E has been detected from OGO 3 (Mange and Meier, 1970; Mange, 1972). Ly-α observations taken as the Mariner-5 spacecraft receded from earth (Wallace *et al.*, 1970) showed greater diminution beyond 7 R_E corresponding to depopulation of satellite orbits. In Vela-4 data taken at altitudes beyond 17 R_E, Chambers *et al.* (1970) found a broad Ly-α intensity maximum of 160 R over the region near the apex of the solar way. Later, Barth (1970) reported a maximum intensity of 370 R seen from beyond Mars with the Mariner-6 scanning spectrometer. Its location was in Ophiuchus near the direction of the sun's motion. Enhancement across the galactic plane, suggested by earlier Mariner-5 results, was not observed.

The OGO-5 spacecraft carried two separate Ly-α photometers which mapped the sky during three spin-up maneuvers designated SU 1, SU 2 and SU 3. The results have been reported by Thomas and Krassa (1971), and Bertaux and Blamont (1971) whose experiments were mutually confirmatory. A maximum of some 530–570 R was seen displaced from the apex of the solar way and paired with a minimum of some 200–250 R in the opposite direction (along the ecliptic).

Nevertheless, an absorption-cell experiment carried on OGO 6 (Metzger and Clark, 1970) shows less than 100 R in the diffuse background. The minimum, whatever its level, establishes an upper limit to any isotropic galactic contribution. Thomas and Krassa made extensive comparison with the observations of other investigators, including earlier Soviet data, as did Tinsley (1971). The theory of Blum and Fahr (1971), now in highly detailed and refined form, is favored. It reasons that scattering from the interstellar H which has penetrated the solar system accounts for the observed celestial intensity pattern and its variation in time. From more detailed review of their OGO-5 data, Thomas and Bohlin (1972) find an antisolar enhancement of geocoronal scattering beyond 11 R_E which they attribute to the presence of a 'geotail' of density 10 to 20 atoms cm^{-3}.

The uncertainties in calibration heavily affect estimates of the relative contributions of airglow and background radiation to the observed intensity at night. Thus, if the OSO-4 solar line-center ratio of near unity cited above is correct, the OGO-5 (Thomas) ratio of 0.56 would imply that the OGO-5 background intensities should be nearly doubled (Meier and Mange, 1972). Consequently, the portion of the midnight Ly-α zenith emission attributed to background would lie in the range 500 to 1200 R. This would be $\frac{1}{4}$ to $\frac{1}{2}$ of the observed intensity (Figure 1).

2. Neutral and Ionized Helium Emissions

2.1. OBSERVING TECHNIQUE

Because the extreme UV neutral and ionized He emissions are much less intense than the far UV H emissions, and because they lie in the shorter wavelength spectral range, observations of them are far less numerous than those of H. Until now no He data have been obtained from satellite instruments. Emissions of neutral He to be discussed have included the 10830 Å 2^3S-2^3P transition seen from the ground and the 584 Å 1^1S-2^1P emission observed above the main absorbing atmosphere. Ionized He is observed in the 304 Å 1^2S-2^2P emission which is the counterpart of the Ly-α line of H. Auroral emissions are not discussed.

Because the photometry of the region is not highly developed, experimenters have not always been able to distinguish fully between contributions from the 304 and 584 Å lines. In large degree, however, the separation can be achieved with thin film filters. Thus, Young et al. (1968) employed an Al film with a scintillating phosphor photomultiplier to cover the range 150 to 750 Å; subsequently, Al/C (150–440 Å) and Al (150–800 Å) films were used with channel multipliers (Young et al., 1971). Ogawa and Tohmatsu (1971) utilized films of aluminum and bismuth in conjunction with channel multipliers to cover the range 300 to 900 Å. The group of Paresce et al. (1970, 1972) and Kumar et al. (1970) have also flown channel multiplier detectors with thin film filters to observe the 304 to 584 Å emissions. Donahue and Kumer (1971) were the first to observe the 584 Å emission; in their rocket-borne instrument photoelectrons emitted from a target exposed to the airglow moved through a retarding potential analyzer. It should be mentioned (Ogawa and Tohmatsu, 1971) that a contribution from dayglow emission of O^+ at 834 Å could contribute to photometer response under appropriate observing conditions.

2.2. THE 584 Å LINE AND ITS ORIGIN

Meier and Weller (1972) have reviewed the extreme UV data so far obtained; their summary of the observational results is reproduced in Table I. In analyzing the 584 Å data cited they evaluated the radiative transfer, just as for the H Ly-α problem, by calculating the full resonant multiscattering of 584 Å solar photons in a spherically symmetric model He atmosphere of appropriate average temperature as deduced by Jacchia (1970). (Calculation of multiple scattering was required because optical depth of order 25 is encountered in penetrating to the 220 km level.) For the geometry of the Donahue and Kumer (1971) flight this yielded a solar line center flux of 3.7×10^{10} photon cm^{-2} s^{-1} Å^{-1}, or a (reasonable) equivalent solar emission width of 0.027 Å based on the solar line emission of 10^9 photons cm^{-2} s^{-1} reported earlier (Hall et al., 1969). Similar calculations for the daytime data of Kumar et al. (1970) yielded a solar line center flux larger by a factor of 2. However, the nighttime data of Ogawa and Tohmatsu (1971) and Paresce et al. (1970, 1972) when evaluated in equivalent fashion produced line-center fluxes larger, respectively, by factors of 43 and two orders of magnitude.

In spite of the large discrepancy in absolute intensity (filtering and calibration in this spectral range is less than ideal), and notwithstanding the fact that the He model atmosphere necessarily did not incorporate the winter He bulge effect, the theoretical reproduction of the data trends was impressive. An example is seen in Figure 3. In it

TABLE I

EUV observations

Investigators	Date	Solar zenith angle, deg	Apogee, km	Local zenith angle, deg	T_{ex} K	Max. 584 Å emission rate, Rayleighs	Max. 304 Å emission rate, Rayleighs
Donahue and Kumer (1971)	July 16, 1966	56–74	1000	22–9	927	700 ± 300	
Paresce *et al.* (1970)	March 10, 1970	150	185	60	1050	15	8
Kumar *et al.* (1970)	June 15, 1970	42	185	0	1260	210 ± 70	9.5 ± 3.1
Ogawa and Tohmatsu (1971)	Sept. 19, 1970	120–145	1000	0–37	1000	100	10 ± 4
Young *et al.* (1971)	Oct. 13, 1969	134	217	0–180	1100	< 4	7.2
Young *et al.* (1968)	Aug. 10, 1967	131	225	40	900	< 12	≤ 4.8

the emission, as calculated, faithfully portrays the observations of Ogawa and Tohmatsu (1971) during ascent and almost all of descent over a range of solar zenith angle. Good matching was similarly obtained for the data of Donahue and Kumer (1971). The upper limits on nighttime 584 Å emission reported by Young *et al.* (1968, 1971) placed no additional constraint on the theory or the observations.

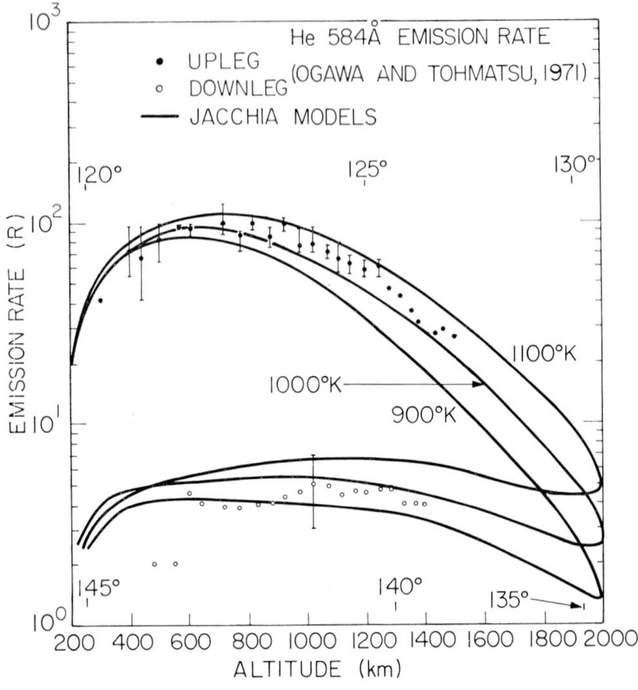

Fig. 3. 584 Å emission rate as a function of altitude. The solar zenith angle change from 120 to 145° is indicated (Meier and Weller, 1972).

Thus, it was concluded by Meier and Weller (1972) that resonant scattering of sunlight is the main source of the radiation, but that secondary sources of 584 Å such as resonant scattering from interplanetary (or interstellar) He which might also be present, could contribute to the discrepancies in absolute magnitude among the experiments. In this connection, Holzer and Axford (1971), following an approach similar to Blum and Fahr (1970) for H, have discussed the formation of a wake of neutral interstellar He behind the sun and the possible enhancement of 584 Å emission by scattering from it.

2.3. Observations of the 10 830 Å line of neutral helium

The emission at 10 830 Å and its related emission at 3888 Å into the metastable 2^3S level 19.7 eV above the ground state, have been observed from the earth's surface. Christensen *et al.* (1971) review the previous observations beginning with the original discovery of the 10 830 Å line in sunlit aurora in 1959, its subsequent observation in twilight and interpretation as resonance scattering of solar photons, the measurement of the 3889 Å line in twilight in 1964, and the various proposals advanced by different authors to explain the excitation of He into the 2^3S level. Their own extensive twilight observations of the 10830 Å line with a grille spectrometer at Socorro, New Mexico showed large dawn/dusk intensity asymmetries in winter, probably as the result of the diurnal He bulge with morning maximum. They relate increases in peak winter intensities, from about 1400 R in 1966–67 to about 4300 R in 1969–70, to long term increase in solar EUV flux observed at wavelengths less than 380 Å. They note that this is in accordance with the original view of McElroy that atmospheric photoelectrons created by such solar photons are responsible for the He metastable production. A pronounced seasonal variation in intensity was attributed to a five-fold variation (at 30° N latitude) in the He present, i.e., the winter He bulge effect. They also utilize the absence of the usual dawn/dusk asymmetry in spring to estimate that 35% of the observed 2^3S excitation is caused by photoelectrons created at the magnetic field line conjugate points. Subsequent observations in Brazil (Christensen *et al.*, 1972) indicate a seasonal variation of He at tropical latitudes in accordance with models of the He bulge determined by geomagnetic rather than geographic latitude.

2.4. THE CELESTIAL GLOW FROM IONIZED HELIUM

Friedman (1960) noted that the 304 Å Ly-α line of ionized He ought to be observable in the sky, since any interplanetary He^+ present would serve as a scattering medium, just as does H in the interplanetary region for its 1216 Å solar line. Although the degree of interplanetary contribution to the 304 Å glow is not well understood at this time, a celestial glow is observed. A listing of the maxima for the five rocket studies conducted to observe 304 Å is given in the last column of Table I. Meier and Weller (1972) have shown that each of these values is a reasonable expectation for single-scattering (small optical depth) of the solar line in an ionized He plasma of typical observed density contained within the plasmasphere. Thus, the 1969 nighttime maximum of Young *et al.* (1971) could arise from scattering of the solar flux of 9.2×10^9 photons cm^{-2} s^{-1} (Timothy and Timothy, 1970) with equivalent width 0.15 Å (Behring *et al.*, 1972) in an He^+ plasmasphere with He^+ concentration of 400 cm^{-3}. Similarly, the nighttime levels of Ogawa and Tohmatsu (1971), and Paresce *et al.* (1972) would require 500 ions cm^{-3} and 1300 cm^{-3}, respectively. The daytime value of Kumar *et al.* (1970) would imply an average of 700 ions cm^{-3}. All these values are within the range of those observed.

Figure 4 (Meier and Weller, 1972) is a contour plot of the nighttime radiation intensity, predominantly of the 304 Å line, observed against the celestial sphere by Young *et al.* (1971). They attributed the shift in the line of symmetry southward from the solar direction as evidence that the He ions responsible for the scattering were within the plasmasphere with its constraining magnetic field. This view contrasted with that of Ogawa and Tohmatsu (1971), who held that since the region of maximum intensity fell close to the solar apex, or the galactic center, near whose location the H Ly-α maximum is also to be found, the source might well be in interplanetary space.

To resolve this question Meier and Weller (1972) calculated the intensity to be expected by direct single scattering from a plasmasphere filled with He$^+$ ions at constant concentration 400 cm^{-3} above 500 km out to shell $L=4.25$. Marked improvement of fit was obtained by the further incorporation of the 'dip' known to exist in the He$^+$ concentrations in the equatorial region. Introduction of the observed diurnal

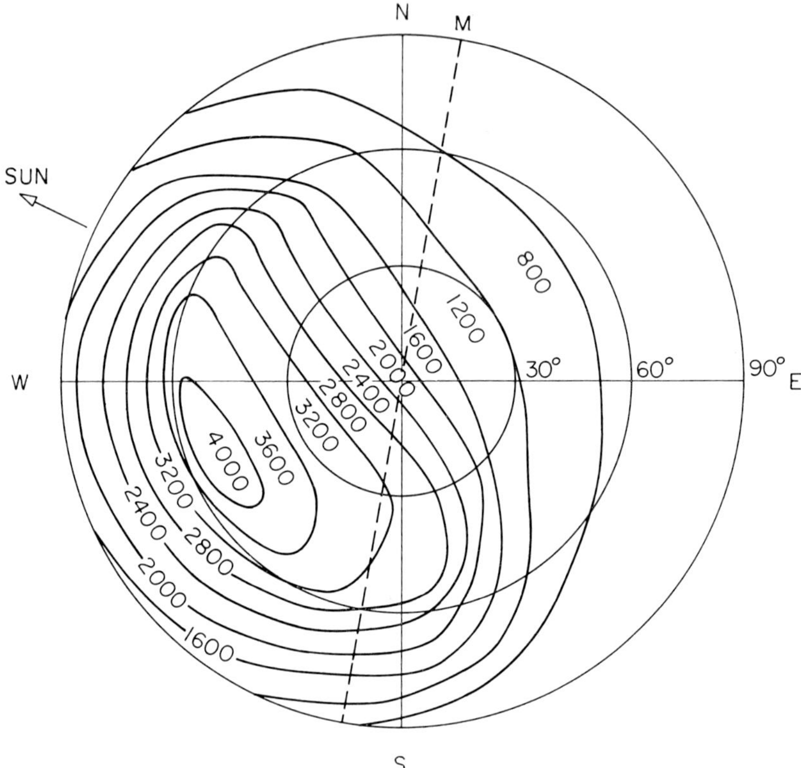

Fig. 4.   Isophotes of A1 photometer data (150–800 Å) from a flight on October 13, 1969. 1R equals 560 counts s^{-1} at 304 Å and 360 counts s^{-1} at 584 Å (Meier and Weller, 1972).

variation of the plasmapause with maximum near 2000 LT provided additional improvement. However, these refinements did not account for the one obvious remaining discrepancy: a substantial minimum signal, in directions away from the sun, where the plasmaspheric model shows no scattering. (First-order scattering into the shadow region could account for only 25% of the emission seen.) The authors point out that the anti-solar signal could arise from column abundances per cm^2 of some 5×10^{10} ions, or atoms (since the detector was broadband), distributed as the scattering medium in the wake of the earth.

The data of Paresce *et al.* (1972) have now also been analyzed more fully, and plotted as an all-sky map. They too find that a constant density plasmasphere model provides a best fit to the observations. However, they still require, as do Meier and

Weller (1972), some form of extended source in the anti-solar region, although it plainly must be 304 Å in character since their detectors were not sensitive to 584 Å.

The present knowledge of the interplanetary medium does not at this time provide a plausible model as a source for the added component He^+ radiation which seems to determine the minimum background level for the observations. The discussion of Holzer and Axford (1971) is a valuable summary of contributions and statements of the possible characteristics of He ionic flow in the interplanetary medium. However, a structured description of an interplanetary source mechanism for the He^+ emission is needed. It may be added that a look at the earth in the light of the He^+ glow from distant vantage point, even as distant as the moon, should also prove most valuable.

3. Observations of Terrestrial Hydrogen and Helium Emissions from the Moon

On April 21, 1972, astronauts on the Apollo 16 mission set up the first astronomical telescope on the surface of the moon; it was a combined camera and spectrograph prepared by George R. Carruthers and Thornton Page of the Naval Research Laboratory. The instrument was an $f/1.0$ Schmidt camera of 7.5 cm aperture which operated in the far UV beyond 1600 Å, either with LiF corrector plate (short wavelength cutoff of 1050 Å), or CaF_2 corrector plate (wavelength limit 1250 Å). It also operated with grating as a spectrograph at wavelengths down to 1050 Å (with LiF corrector plate) or to 500 Å (with no corrector). The camera incorporated the electronographic system first employed by Carruthers to observe the celestial sphere in the UV from rockets. Specifically, the UV is imaged on a KBr photoemitting surface from which the electrons are accelerated through 25000 V and refocused magnetically on nuclear emulsion film to provide a gain factor of some 20 over conventional imaging directly on UV film.

The camera/spectrograph has provided the first pictures of the terrestrial UV airglow, and the first spectral identification of Ly-β, of the 584 Å He line, and of other lines in the terrestrial atmosphere, as reported by Carruthers and Page (1972). Figure 5, taken from their paper, shows the earth in the 1050 to 1600 Å wavelength range with exposure times of 5, 15 and 60 s. The geocoronal glow is well displayed and exhibits reduced intensity on the antisolar side of the earth, just as charted by the lower resolution photometry of earlier experiments. They point out that the dark limb of the earth is seen silhouetted against the far-side geocoronal radiation, an effect verified by radiative transport interpretation (Meier, 1972).

Also evident are the polar auroral zone emissions arising from O and N_2, as well as H (Chubb and Hicks, 1970). The aurora is seen on the day- as well as the nightside of the earth in the 5-s exposure. Almost the entire northern auroral ring is visible, but the southern auroral zone is mostly hidden and appears only as a bright streak on the dark limb.

In the spectrographic mode Carruthers and Page (1972) find a component of the Ly-α radiation from all directions of interplanetary space. In accordance with the earlier photometric observations and theory, the Ly-α line radiation in 10 min ex posure showed a mild concentration toward the sunlit disk. The slit image across the dark limb of the earth showed a step increase outward in the Ly-α spectral line intensity; a similar step increase was seen on the bright limb at shorter exposures. Thus, the earth's disk in the Ly-α line is dark against the background geocorona.

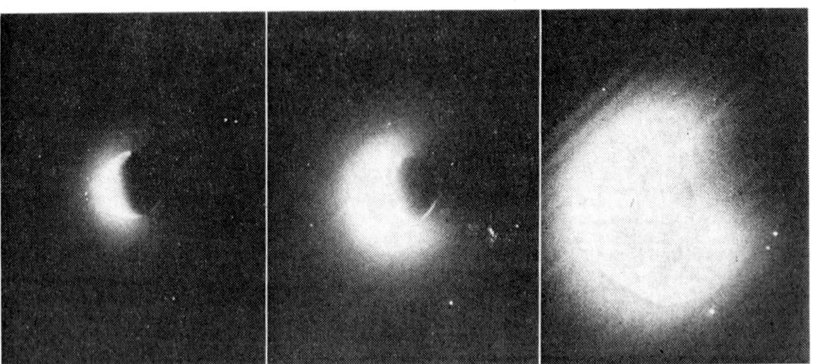

Fig. 5.   The earth photographed in the 1050–155 Å wavelength range, with exposure times of about 5, 15 and 60 s, showing the H geocorona, day airglow, and polar auroras. The diagonal streaks in the longer exposures are instrumental (Carruthers and Page, 1972).

A 30-min spectrographic exposure with no corrector plate revealed (in addition to other emissions) the 584 Å He line, the 834 Å O^+ resonance line, and the 1026 Å Ly-β H line. It is believed that this was the first firm spectrographic identification of these lines in the terrestrial atmosphere.

Acknowledgment

Thanks are offered R. R. Meier and C. S. Weller for valuable discussion.

References

Barth, C. A.: 1970, *Astrophys. J.* **161**, L181.
Behring, W. E., Cohen, L., and Feldman, U.: 1972, *Astrophys. J.* **175**, 493.
Bertaux, J. L. and Blamont, J. E.: 1970, in T. M. Donahue, R. L. Smith, and G. Thomas (eds.), *Space Res.* **10**, North-Holland Publishing Company, Amsterdam, p. 591.
Bertaux, J. L. and Blamont, J. E.: 1971, *Astron. Astrophys.* **11**, 200.
Blamont, J. E. and Vidal-Madjar, A.: 1971, *J. Geophys. Res.* **76**, 4311.
Blum, P. W. and Fahr, H. J.: 1970, *Astron. Astrophys.* **4**, 280.
Blum, P. W. and Fahr, H. J.: 1971, in Bowhill, Jaffe and Rycroft (eds.), *Space Res.* **12**, Akademie-Verlag, Berlin, p. 1569.
Brinton, H. C. and Mayr, H. G.: 1971, *J. Geophys. Res.* **76**, 6198.
Carruthers, G.: 1972, private communication.
Carruthers, G. R. and Page, T.: 1972, *Science*, **177**, 788.
Chamberlain, J. W.: 1963, *Planetary Space Sci.* **11**, 901.
Chambers, W. H., Fehlau, P. E., Fuller, J. C., and Kunz, W. E.: 1970, *Nature* **225**, 713.
Christensen, A. B., Patterson, T. N. L., and Tinsley, B. A.: 1971, *J. Geophys. Res.* **76**, 1764.
Christensen, A. B.; Tinsley, B. A., Teixeira, N. R., and Angreji, P. D.: 1972, *J. Geophys. Res.* **77**, 784.
Chubb, T. A. and Hicks, G. T.: 1970, *J. Geophys. Res.* **75**, 1290.
Clark, M. A.: 1972, private communication.

Donahue, T. M. and Kumer, J. B.: 1971, *J. Geophys. Res.* **76**, 145.

Friedman, H.: 1960, in J. A. Ratcliff (ed.), *Physics of the Upper Atmosphere*, Academic Press, New York and London, p. 133.

Hall, L. A., Higgins, J. E., Chagnon, C. W., and Hinteregger, H. E.: 1969, *J. Geophys. Res.* **74**, 4181.

Holzer, T. E. and Axford, W. I.: 1971, *J. Geophys. Res.* **76**, 6965.

Jacchia, L. E.: 1970, in T. M. Donahue, R. L. Smith, and G. Thomas (eds.), *Space Res.* **10**, North-Holland Publishing Company, Amsterdam, p. 367.

Kockarts, G. and Nicolet, M.: 1963, *Am. Geophys.* **19**, 370.

Kumar, S., Bowyer, C. S., Lampton, M., and Paresce, F.: 1970, *Trans. Amer. Geophys. Union* **51**, 795, (Abstract).

Mange, P.: 1972, in E. Dyer (General ed.), *Solar Terrestrial Physics/1970*: Part IV, D. Reidel Publishing Company, Dordrecht, Holland, p. 68.

Mange, P. and Meier, R. R.: 1970, *J. Geophys. Res.* **75**, 1837.

Meier, R. R.: 1972, private communication.

Meier, R. R. and Mange, P.: 1970, *Planetary Space Sci.* **18**, 803.

Meier, R. R. and Mange, P.: 1973, *Planetary Space Sci.* **21**, 309.

Meier, R. R. and Weller, C. S.: 1972, *J. Geophys. Res.* **77**, 1190.

Metzger, P. H. and Clark, M. A.: 1970, *J. Geophys. Res.* **75**, 5587.

Ogawa, T. and Tohmatsu, T.: 1971, *J. Geophys. Res.* **76**, 6136.

Paresce, F., Bowyer, C. S., Kumar, S., and Lampton, M.: 1970, *Trans. Amer. Geophys. Union* **51**, 795, (Abstract).

Paresce, F., Bowyer, S., and Kumar, S.: 1973, *J. Geophys. Res.*, **78**, 71.

Thomas, G. E.: 1970, in T. M. Donahue, R. L. Smith, and G. Thomas (eds.), *Space Res.* **10**, North-Holland Publishing Company, Amsterdam, p. 602.

Thomas, G. E.: 1972, private communication.

Thomas, G. E. and Bohlin, R. C.: 1972, *J. Geophys. Res.* **77**, 2752.

Thomas, G. E. and Krassa, R. F.: 1971, *Astron. Astrophys.* **11**, 218.

Timothy, A. F. and Timothy, J. G.: 1970, *J. Geophys.* **75**, 6950.

Tinsley, B. A.: 1970, in T. M. Donahue, R. L. Smith, and G. Thomas, (eds.), *Space Res.* **10**, North-Holland Publishing Company, Amsterdam, p. 582.

Tinsley, B. A.: 1971, *Rev. Geophys.* **9**, 89.

Tinsley, B. A. and Meier, R. R.: 1971, *J. Geophys. Res.* **76**, 1006.

Vidal-Madjar, A., Blamont, J. E., and Phissamay, B.: 1972, Paper, XVth. COSPAR Meeting, Madrid.

Wallace, L., Barth, C. A., Pearce, J. B., Kelly, K. K., Anderson, D. E. Jr., and Fastie, W. G.: 1970, *J. Geophys. Res.* **75**, 3769.

Weller, C. S., Meier, R. R., and Tinsley, B. A.: 1971, *J. Geophys. Res.* **76**, 7734.

Young, J. M., Carruthers, G. R., Holmes, J. C., Johnson, C. Y., and Patterson, N. R.: 1968, *Science* **160**, 990.

Young, J. M., Weller, C. S., Johnson, C. Y., and Holmes, J. C.: 1971, *J. Geophys. Res.* **76**, 3710.

GEOCORONAL HYDROGEN

J. L. BERTAUX

Service d'Aéronomie du C.N.R.S., 91 Verrières-Le-Buisson, France

1. Introduction

Atomic hydrogen is a minor constituent of the atmosphere, created mainly by photo-dissociation of H_2O and CH_4 between 60 and 100 km of altitude. It diffuses upwards through the thermosphere (Figure 1) until it reaches the upper boundary of the thermosphere, called the exobase or critical level. Atoms which have at this altitude an upward velocity larger than the escape velocity V_{esc} will leave the earth on hyperbolic trajectories, since they are entering the exosphere which is a collisionless medium. Those with an upward velocity lower than V_{esc} will return to the exobase after a ballistic flight in the exosphere. Eventually a collision in the exosphere will transfer the atom onto a satellite orbit.

At one point of the exobase the outward flux of ballistic particles is determined by the local exospheric temperature T_c and H density n_c, whereas the inward flux is determined by the distribution of T_c and n_c all over the exobase. If these fluxes are not equal, there is a transport of H between different points of the exobase called the lateral flow.

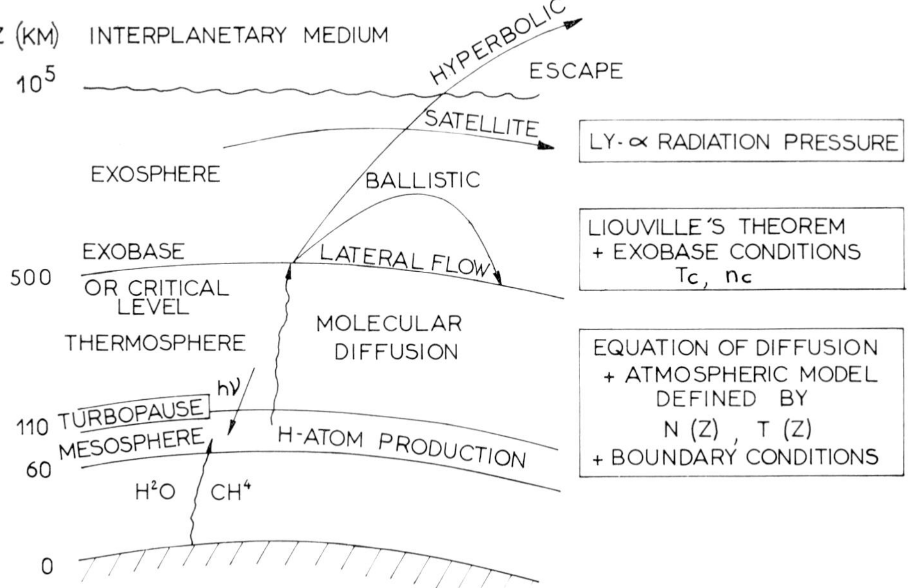

Fig. 1. Structure of H geocorona and physical processes which determines the H concentration distribution.

2. Hydrogen Distribution in the Thermosphere

Here we deal with the distribution of H above the turbopause as a function of altitude z along a vertical with fixed geographical coordinates; the most recent study of this problem was presented by Wallace and Strobel (1972) from which we take the following notations.

The continuity equation is:

$$\partial n/\partial t + \frac{\partial \phi}{\partial z} = 0 \tag{1}$$

where the vertical flux ϕ for transport by molecular diffusion is:

$$\phi = nv = -D\left\{\frac{\partial n}{\partial z} + \left[\frac{(1+\alpha)}{T}\frac{\partial T}{\partial z} + \frac{1}{h}\right]\right\} n \tag{2}$$

in which n, v, and D are the density, diffusion velocity, and diffusion coefficient of H, respectively, $T \equiv T(z)$ is the temperature profile of the background atmosphere and $h \equiv kT/mg$ is the scale height that would have the H distribution if it were alone in the atmosphere. α is the thermal diffusion factor.

At the exobase where the atoms are supposed to have a Maxwellian velocity distribution, the upper boundary condition is given by Jeans' escape. The diffusion velocity must be equal to the effusion velocity v_c, or mean vertical velocity of H. As we consider a constant altitude of the critical level, v_c is only a function of T_c:

$$v_c = (U_c/2\sqrt{\pi})(1 + \lambda_c) e^{-\lambda_c},$$

where $U_c = (2kT_c/m)^{1/2}$ and $\lambda_c = (V_{esc}/U_c)^2$.

The effusion velocity is a strong function of T_c, varying from 5 cm s^{-1} to 2.5×10^4 cm s^{-1} as T_c varies from 600 to 2000 K. In contrast with other atmospheric constituents, the result of this important escape is that the concentration n decreases when T_c increases.

Mange (1961), Bates and Patterson (1971), and Kockarts and Nicolet (1962) have calculated the solution of the diffusion equation in the steady state case ($\partial n/\partial t = 0$ leads in Equation (1) to a constant flux).

Mange (1961) considered the case of a linear increase of the scale height with altitude (Figure 2) and gave an analytical expression of the density variation with altitude. This analytical expression contains the upward vertical flow ϕ as a parameter. If this parameter is higher than a maximum value F_{max}, the density will become negative at some altitude. There is then a physical limit F_{max} to the vertical flow which can be supported by molecular diffusion. The density n_0 at a reference level (here, $z_0 = $ $= 110$ km) is a multiplicative factor, since Equation (2) is homogeneous on n.

Kockarts and Nicolet (1962) computed a set of steady state H distributions for various temperature profiles and exospheric temperature. The H concentration at 500 km is a strong function of exospheric temperature T_c, since a constant flow of 2.5×10^7 atoms cm^{-2} s^{-1} is supposed at 100 km. But short-term variations of T_c, like

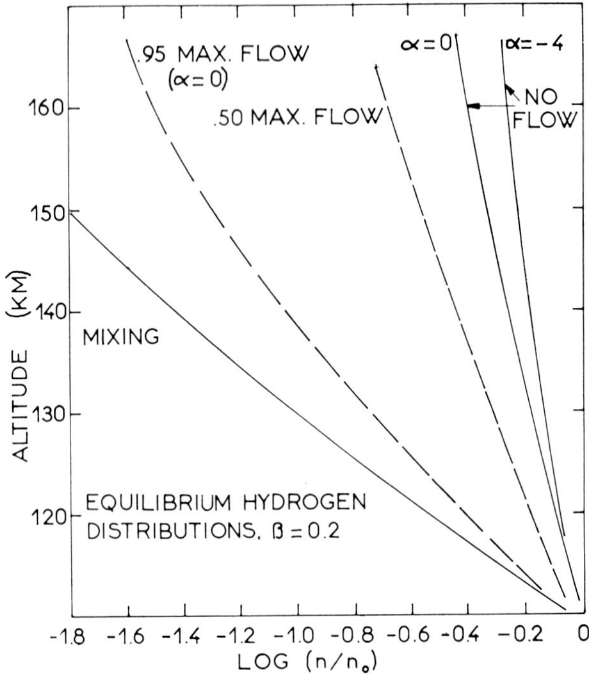

Fig. 2. The 'mixing' curve is the atmospheric model; the two solid curves are computed for the case of no escape, or no upward flow. This is the case of diffusive equilibrium, in which the H distribution would be determined by its own partial pressure and scale height kT/mg. The effect of thermal diffusion is shown ($\alpha = 0$ and $\alpha = -0.4$).

the diurnal variation, will not result in the corresponding change of concentration at 500 km because of the inertia of the system.

In order to study the diurnal variation of H, one has to find the time dependent solution of Equations (1) and (2). This was recently done by Wallace and Strobel (Figure 3), with the model atmosphere of Walker (1965) and the time variation of T_c provided by Jacchia (1965).

At low temperature ($T_0 = 600$ K) the inertia of H is so great that the diurnal variation of the time dependent model is only a few percent. At large temperature ($T_0 = 1800$ K) the inertia is much lower and the time dependent model reaches the steady state model.

3. The Effect of Lateral Flow on the Exobase Density Distribution

So far we have presented solutions of Equation (2) with two boundary conditions: one at the lower boundary, either on the flux or the density, and one at the upper boundary, the exobase, where only the temperature distribution was fixed. The solution leads to a distribution of density at the exobase, which when combined with the temperature distribution produces a transport of H by lateral flow. This lateral flow enhances the

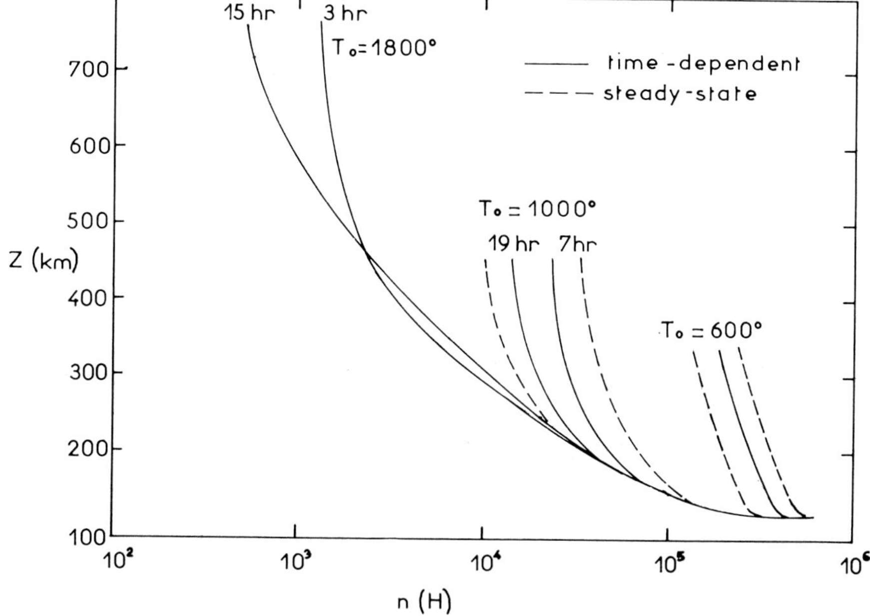

Fig. 3. Comparison of time dependent and steady state model altitude distributions for H at 30°
latitude. The curves, for minimum nighttime temperatures of 600, 1000 and 1800 K, terminate at the
critical level. The time dependent profiles for 1000 and 1800 K are for the time of maximum density
at the critical level and 12 h later; the result for 600 K is shown as a single curve because the diurnal
variation is only $\sim 5\%$. The steady state models are for $T_{min} = 1.01\ T_0$ and $T_{max} = 1.24\ T_0$; they are
not shown for 1800 K since they essentially overlap the time dependent curves. The curves are normal-
ized to a density of 10^7 at 100 km.

concentration in the region of net influx and depletes it in the region of net outflux,
tending to lessen the flows. The atmosphere would tend towards a zero lateral flow
situation. McAfee (1967), Patterson (1970), and Quesette (1972) have conducted
iterative computations to determine the exobase density distribution leading to a zero
lateral flow situation, given an exospheric temperature distribution.

This situation leads to a ratio n_{max}/n_{min} in the same direction and of the same order
of magnitude as the one derived from the time dependent treatment of the equation of
diffusion. Wallace and Strobel indicated that for $T_0 < 1000\,\mathrm{K}$ the lateral flow would
increase the diurnal variation and decrease it for $T_0 > 1000\,\mathrm{K}$. The complete solution
of the time dependent problem including both lateral flow and diffusion is still awaiting
intrepid workers.

4. Distribution of Hydrogen in the Exosphere

Collisions are so infrequent in the exosphere that H atoms follow dynamical trajectories in the earth's gravitational field.

Three kinds of atoms are usually defined, according to the nature of their trajectories. Ballistic particles come from the exobase and return to the exobase; hyperbolic
particles, with a velocity at the exobase higher than the escape velocity V_{esc} leave the

earth and are injected into the interplanetary medium. Satellite particles are created from ballistic or hyperbolic particles suffering an eventual collision, and orbit above the exobase.

Öpik and Singer (1961) and Chamberlain (1963) have established the distribution of H in the exosphere for the case of a uniform exobase, i.e., density n_c and temperature T_c are constant at the constant level R_c of the exobase. In this case density n is only a function of radial distance r. The Liouville's theorem is used, which states that along a dynamical trajectory the density in the phase space is conserved.

Chamberlain (1963) gave analytical expressions for the ballistic and escaping particle number densities with the use of the 'incomplete Γ function.' The number of satellite particles is a result of the balance between creation and loss processes. Chamberlain introduced the concept of a satellite critical level R_{cs} defined as follows: all the perigees of satellite orbits are located below R_{cs}, and at any point between the critical level R_c and R_{cs} the distribution of satellite particles completes the distribution of ballistic particles. The longer the lifetime of a satellite particle against loss proces-

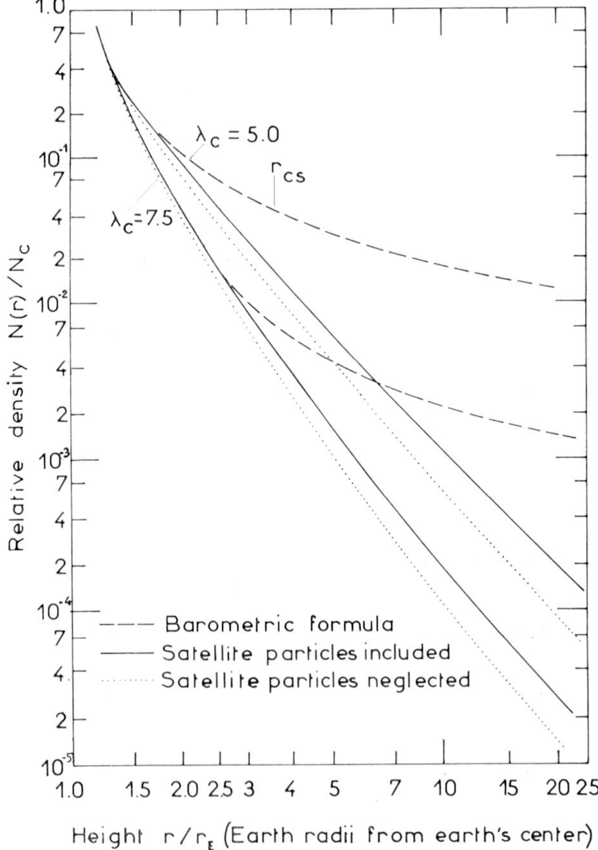

Fig. 4.   Exospheric distributions of H relative density according to Chamberlain for two different temperatures, $T_c = 938$ K and $T_c = 1407$ K. The satellite critical level R_{cs} was 2.5 R_c.

ses, the higher the R_{cs}, but no quantitative relationship was ever given to my knowledge. A usual figure is $R_{cs} = 2.5\ R_c$ when the lifetime is 20 days.

On Figure 4 are shown the distributions of Chamberlain for two different temperatures: $T_c = 938\ K$ and $T_c = 1407\ K$. It is to be remarked that above a few R_E the slope of the curves is quite similar, providing a poor observational criterion to determine the exospheric temperature.

The case of a non-uniform exobase with a given density and temperature distribution was treated by Vidal-Madjar and Bertaux (1972). It was found that up to $\simeq 3000$ km of altitude the density of a non-spherical model is very near the density of the 'local' spherical Chamberlain model, that is to say with exobase conditions equal to the exobase conditions at the point right below the point of interest. A nearly constant density level was found at $\simeq 5000$ km of altitude, above which the higher density is on the higher temperature side.

Concerning the satellite particles, the effect of solar Ly-α radiation pressure on the H distribution was recently investigated by Thomas and Bohlin (1972) and by Bertaux and Blamont (1972). It acts as a repulsive force to be added to the earth's gravitational field. The effect is rather negligible on ballistic and escaping particles, but is very important on satellite particles. The perigee height is decreasing in most configurations of the orbit, and soon the particles return to the exobase. The mean lifetime in orbit was found to be $\sim 4 \times 10^5$ s, much less than the lifetime vs. photoionization and charge exchange with protons ($\simeq 2 \times 10^6$ s). Thus, the number of satellite particles in the exosphere may be less important than previously estimated. Another effect is that the apogee is pushed towards the nightside, sometimes very far away. Ly-α radiation pressure could thus create a 'geotail' as it is suggested by recently published OGO-5 observations (Thomas and Bohlin, 1972).

5. Other Features of the Atomic Hydrogen Distribution

A few features not discussed in the present review would deserve some more theoretical work:

(a) the situation of high latitude regions, and the effect of the so-called 'polar wind'; that is to say that where the magnetic field lines are open, there is an upward escape of protons, and a subsequent depletion of neutral H due to charge exchange reactions, such as:

$$H^+ + O \rightleftarrows H + O^+$$

(b) at chemical equilibrium, this reaction requires the relation:

$$n(H) = \frac{8}{9} \frac{n(H^+)}{n(O^+)} \times n(O).$$

Brinton and Mayr (1971) found that the right side was following a diurnal variation as predicted by the solution of the equation of diffusion for H, for altitudes between 200 and 400 km. Then molecular diffusion of H determines the ratio $n(H^+)/n(O^+)$ in this altitude range.

(c) the breathing velocity of the atmosphere should be taken into account when solving the diffusion equation.

As a conclusion, it should be pointed out that the H problem is not only interesting in itself, but also because it is strongly coupled to the background atmosphere. Learning something about H then is equivalent to learning something about the atmosphere.

References

Bates, D. R. and Patterson, T. N. L.: 1961, *Planetary Space Sci.* **5**, 257.
Bertaux, J. L. and Blamont, J. E.: 1973, *J. Geophys. Res.*, **78**, 80.
Brinton, H. C. and Mayr, H. G.: 1971, *J. Geophys. Res.* **76**, 6198.
Chamberlain, J. W.: 1963, *Planetary Space Sci.* **11**, 901.
Jacchia, J. G.: 1965, *Smithson. Contr. Astrophys.* **8**, 215.
Kockarts, G. and Nicolet, M.: 1962, *Ann. Geophys.* **18**, 269.
Mange, P.: 1961, *Ann. Geophys.* **17**, 283.
McAfee, J. R.: 1967, *Planetary Space Sci.* **15**, 599.
Öpik, E. J. and Singer, S. F.: 1961, *Phys. Fluids* **4**, 221.
Patterson, T. N. L.: 1970, *Rev. Geophys. Space Phys.* **8**, 461.
Quessette, J. A.: 1972, *J. Geophys. Res.* **77**, 2997.
Thomas, G. E. and Bohlin, R. C.: 1972, *J. Geophys. Res.* **77**, 2752.
Vidal-Madjar, A. and Bertaux, J. L.: 1972, *Planetary Space Sci.* **20**, 1147.
Walker, J. C. G.: 1965, *J. Atmospheric Sci.* **22**, 462.
Wallace, L. and Strobel, D. F.: 1972, *Planetary Space Sci.* **20**, 521.

INFRARED OBSERVATIONS OF THE EARTH'S UPPER ATMOSPHERE

J. S. GARING and B. SCHURIN

Air Force Cambridge Research Laboratories (AFSC), Bedford, Mass. 01730, U.S.A.

1. Introduction

In the present paper we shall be describing some representative examples of the more recent IR radiance measurements of the earth's upper atmosphere together with results obtained from these studies. For historical perspective an earlier summary by Howard *et al.* (1965) is available.

From the viewpoint of IR processes the atmosphere can be conveniently divided into two overlapping regions. We consider a lower atmosphere from the surface to roughly 70 km where collisions between molecules are rapid enough to maintain a thermal distribution of vibrational states. For this region of the atmosphere the important radiation mechanisms are simply absorption and thermal emission. In the upper atmosphere collisions between molecules are less effective in maintaining excited vibrational states and radiative equilibrium becomes increasingly important with altitude. For the upper atmosphere there are a large number of competing processes associated with IR airglow and auroral radiation. The problem comes with carrying out measurements of the nonthermal radiation occurring in the upper atmosphere. The radiation levels are extremely low – and, in fact, measurements are presently dependent on development and improvement of existing techniques. Further, these measurements often must be made through the lower atmosphere or with thermal emission as an interfering background. From 5 to 25 μm the thermal emission in the vertical direction lies between 10^{-4} and 10^{-3} W cm^{-2} sr^{-1} μm^{-1} several orders of magnitude greater than the nonthermal radiation. Below 5μm the blackbody curve falls off rapidly enough that we can observe short wavelength radiation such as the OH emission bands.

2. High Resolution Ground-Based Observations

One of the earliest techniques for obtaining data on the upper atmosphere has been the measurement of the IR transmission of the atmosphere, particularly at high resolution, from mountain observatories using the sun as the radiation source, i.e., 'Solar Spectra'. Generally, the purpose of these experiments is to determine the IR spectrum of the sun but almost all of the lines observed are due to the absorption by molecules in the earth's atmosphere.

Ground-based observations are usually restricted to the window regions between the strong absorption bands. The measurements aim at obtaining the highest possible spectral resolution to isolate lines of interest from the intervening atmospheric absorption.

For many years M. Migeotte, L. Delbouille and their co-workers from the University of Liege have been obtaining high resolution solar spectra from the Sphinx Observatory at the 3500 m high Jungfraujoch International Scientific Station in Switzerland using a large grating spectrometer and coelostat to track the sun. The Liege group has published 'Atlases' of these spectra for the 2.8 to 23.7 μm region (Migeotte *et al.*, 1956) and for the 7500 to 12000 Å region (Delbouille and Roland, 1963; Swensson *et al.*, 1970) and are presently gathering data for the visible and near IR portions of the solar spectrum.

More recently Janine and Pierre Connes at the CNRS in France have developed a very high resolution interferometer-spectrometer which they have used at various observatories to make astronomical observations using Fourier transform spectroscopy. They have published an Atlas of these near IR spectra of the planets and the sun (Connes *et al.*, 1969). The resolution of these spectra is 0.08 cm^{-1} and it was from these spectra that HCL and HF were discovered in the Venus atmosphere.

The spectra obtained by Connes *et al.* are one of the most convincing sets of data available which demonstrate the capabilities of Fourier spectroscopy in carrying out measurements of low radiance levels and its advantages over conventional grating instruments. With the Fourier technique an interferogram is recorded by sending light from the source through a two-beam interferometer, usually a Michelson-type instrument, while varying the path difference. The resultant interferogram is the Fourier transform of the source spectrum. In the first instance there is a multiplex advantage realized because there is no sequential scanning of the spectrum; the entire spectral range is collected during each instant of the recording. Secondly, there is a throughput advantage which comes from the cylindrical symmetry of the interferometer. A review of interferometric spectroscopy and some of its applications to atmospheric measurements is available in the report of the Aspen Conference (1970).

Hall (1970) at Kitt Peak National Observatory has been obtaining near IR solar spectra over long atmospheric paths using the sun near the horizon. Measurements are being made in the 1 to 4 μm region with extremely high resolution (roughly 0.02 cm^{-1}) in order to study the minor atmospheric constituents such as the 3 μm O_3 band. They are using considerably updated grating spectroscopy techniques such as large Harrison gratings and working with a completely cooled focal plane in order to reduce detector background noise. In addition, they have developed computer techniques to remove digitally the instrument effects from the observed data – this becomes feasible when the spectral slit width is not much larger than the line width.

3. Observations from Aircraft

Huppie and Stair (1972) have carried out absolute measurements of high altitude atmospheric emission spectra in the 4 to 11 μm region using Michelson interferometers.

For these experiments the radiation was modulated with an optically black liquid

N_2 cooled chopper (external to the aircraft) which was then the cold reference source. Measurements were made viewing vertically from 12 km. The interferometers are operated at the rate of two scans per minute and the resulting spectra were obtained by adding interferograms over a total observation time of 10 min.

All the features of the spectra in Figures 1 and 2 can be identified with thermal emission of known constituents of the atmosphere. The broad feature around 2300 cm^{-1} is the intense v_3 band of CO_2. The features from 2130 to 2050 cm^{-1} are a combination of O_3 and CO_2 emissions, the sharp line at 2080 cm^{-1} is a well defined CO_2 Q-branch. The $3v_2$ band of CO_2 is responsible for part of the structure around 1930 cm^{-1}, the other intense lines being due to the strong 6.3 μm H_2O band. H_2O is responsible for the features to 1340 cm^{-1}. The strong feature centered around 1040 cm^{-1} is the v_3 fundamental of O_3.

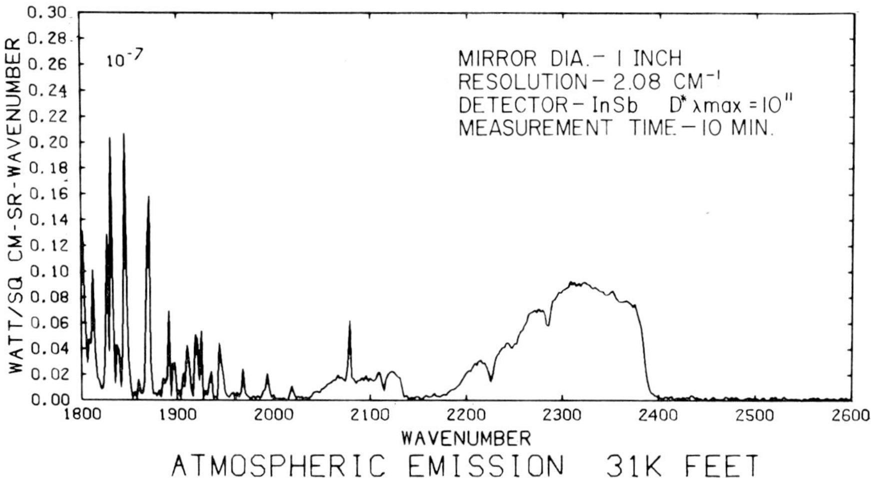

Fig. 1.   Aircraft observations of atmospheric emission at 0° zenith angle (Huppie and Stair, 1972).

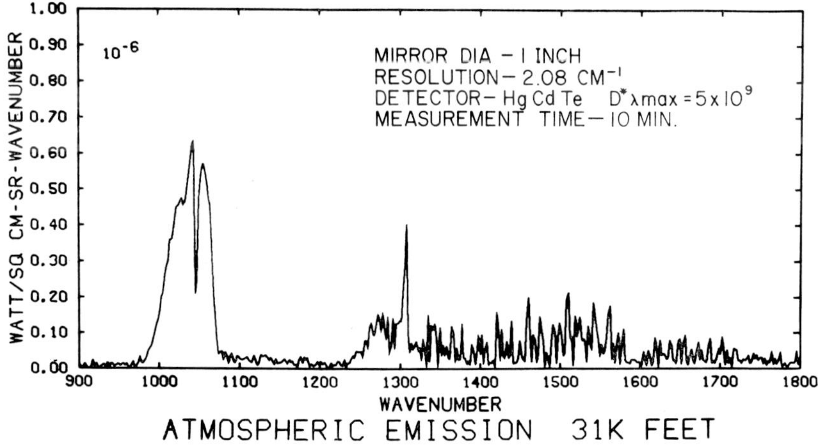

Fig. 2.   Aircraft observations of atmospheric emission at 0° zenith angle (Huppie and Stair, 1972).

J. E. Harries and his co-workers at the National Physical Laboratory in England have observed thermal emission from the stratosphere in the sub-millimeter region, again using Fourier transform spectroscopy (Harries and Burroughs, 1971; Harries *et al.*, 1972). Data were taken from a Comet 2E aircraft at altitudes up to 12.2 km looking at an angle of 10° above the horizontal. This meant that the effective emission path was five to six times the vertical path. The experiments were directed at two problems – measurement of H_2O mixing ratios in the stratosphere and the detection of minor atmospheric constituents.

The stratospheric mixing ratios were obtained by comparing H_2O lines with adjacent magnetic dipole rotation lines of O_2. The intrinsic intensity of the O_2 lines is approximately 10^5 times weaker than that of the H_2O line which makes them particularly suitable for measuring mass mixing ratios which are typically of the order of 10^{-5} to 10^{-6} g/g. For 11 km the authors report a value of $(10 \pm 2) \times 10^{-6}$ g/g of H_2O which is in good agreement with the value expected for a dry lower stratosphere.

The authors were able to obtain a weak indication of the HNO_3 spectrum. For the measurement of such minor atmospheric constituents they recognize the need to lower the viewing angle in order to increase the effective optical path.

4. Balloon-Borne Observations

The most extensive program of balloon-borne observations during the past decade has been carried out by D. Murcray at the University of Denver. Figure 3 is a composite spectrum of balloon-borne data taken by Murcray in the 6 to 20 μm region. The measurements were made with a $\frac{1}{2}$ m Czerny-Turner grating spectrometer used in con-

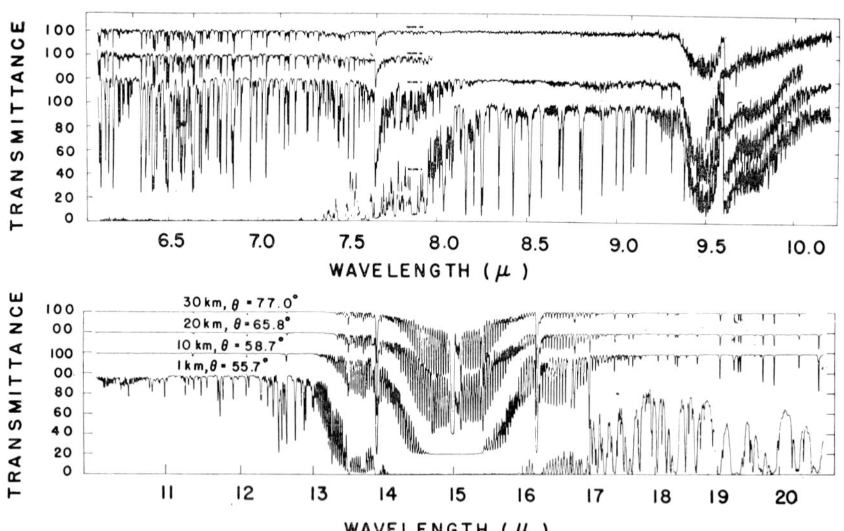

Fig. 3.   Solar absorption spectra of the earth's atmosphere taken from a balloon platform. Balloon altitude and zenith angle indicated (Murcray *et al.*, 1969).

junction with a Ge:Cu detector. The spectral resolution is approximately 0.5 cm^{-1}. The figure is one of the most impressive displays of the vibration-rotation spectra of the atmospheric molecules and indicates the wealth of spectral detail related to temperatures, pressures, and abundances available for analysis. Note the region around 11 μm. At 30 km and a zenith angle less than 90° there is no apparent absorption. In Figure 4 we have the same region, taken with the same instrumentation, but now scanning down to lower altitudes and hence looking through longer path lengths. The spectra are due to the 11 μm HNO_3 band (Murcray *et al.*, 1969) which was observed for the first time from these flights.

A program of balloon measurements has been started by the Liege group (Migeotte, 1972) to supplement their studies from the Jungfraujoch Station. The program is intended for very high resolution solar observations in spectral regions between 1.5 and 10 μm not accessible from the ground – thus primarily in the H_2O and CO_2 bands. Balloon experiments are very successful in rising sufficiently above the H_2O layers to allow solar absorption data to be taken. There is, however, the problem of H_2O contamination from the balloon (or the instrumentation) interfering with the measurements. The Liege report discusses this problem in considerable detail.

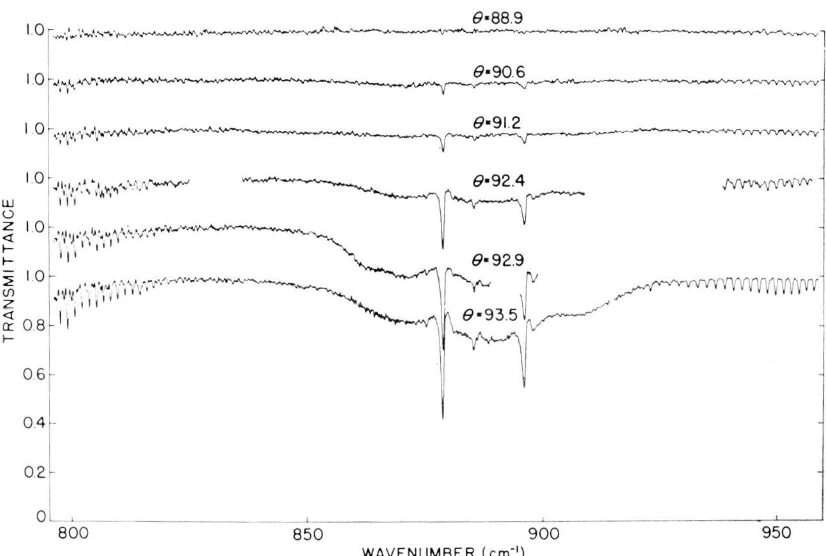

Fig. 4.   Atmospheric transmittance vs. wave number indicating HNO_3 spectra. Spectra were obtained with balloon at 30 km and at the indicated solar zenith angles (Murcray *et al.*, 1969).

5. High Altitude Rocket Measurements

One of the most effective techniques for studying upper atmospheric radiation has been the use of high altitude rockets for carrying out vertical scans through the atmosphere – that is to say, limb radiance measurements.

The MAP program at AFCRL under Walker (1972) uses vertical sounding rockets of the Aerobee type in order to carry measurement equipment to altitudes of 150 to 200 km. The basic instrumentation for limb observations consists of the IR radiometer, stellar aspect sensor, and vehicle attitude control system.

The IR radiometer is a multi-band dual channel instrument. The primary mirror collects IR radiation onto two Ge:Cu detectors mounted in a He cooled dewar. Each detector has an instantaneous field of view of 1.5′ in the vertical by 10′ in the horizontal. Spectral filters are provided for observation in the CO_2 band at 15 μm as well as for the H_2O and O_3 bands. The instrument scans zenith angles from 90° to 110° and back in an approximate triangular wave motion (as the rocket spins) at an average rate of 15° s^{-1}. Calibration from an internal 15 °C blackbody reference source takes place once per scan (every 2.7 s) while the scan is in the extreme down position (110°). Incoming radiation is chopped by a 1 kHz tuning fork chopper, which is temperature stabilized at 50 °C.

Figure 5 shows a typical limb profile observed in the 15 μm CO_2 band. The coordinates are radiance vs. tangent height which is defined as the shortest distance from the surface to the optical line-of-sight. CO_2 limb profiles all show a small but definite limb brightening effect, the peak radiance generally occurring in the 20 to 30 km region.

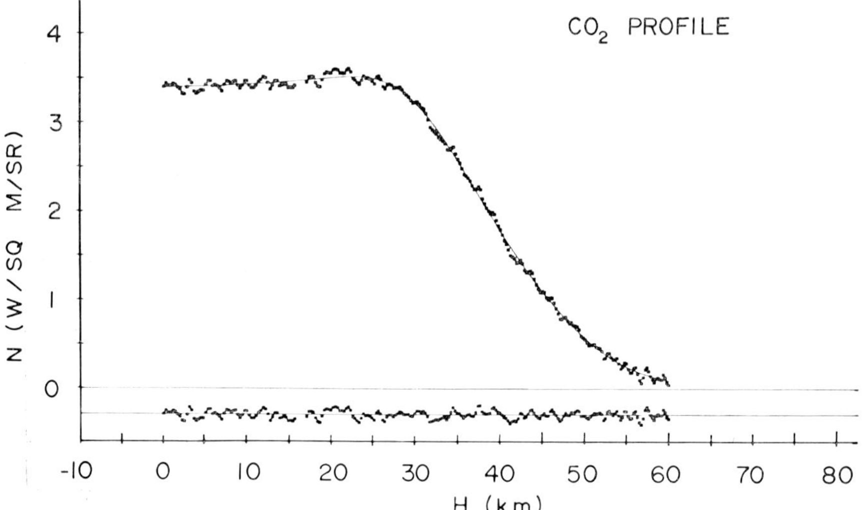

Fig. 5.   Rocket observation of the 15 μm horizon profile. Smooth curve is an analytical fit to the experimental data (Walker, 1972).

Due to the extreme opacity of the CO_2 band, atmospheric layers below 15 km contribute little to the observed radiance, even at small tangent heights. Thus the CO_2 profile is insensitive to lower atmospheric variability. Since the CO_2 mixing ratio is constant with altitude, the shape of the curve above 20 km is determined by the exponential decrease in atmospheric density with increasing altitude, and the altitude-temperature profile in the upper regions. Thus the shape of the CO_2 profile shows only minor variations with season, geographic location, and meteorological conditions.

The strong dependence of the CO_2 limb profile on temperature alone suggests a second application, that is the inference of atmospheric temperature structure by inversion of the radiance integral. Examples of temperature profiles inferred by McKee (1972) from the MAP CO_2 horizon profiles are shown in Figure 6. The solid and dashed lines are the temperatures inferred from the horizon scans. These are compared to ARCUS rocket soundings plotted with the symbols.

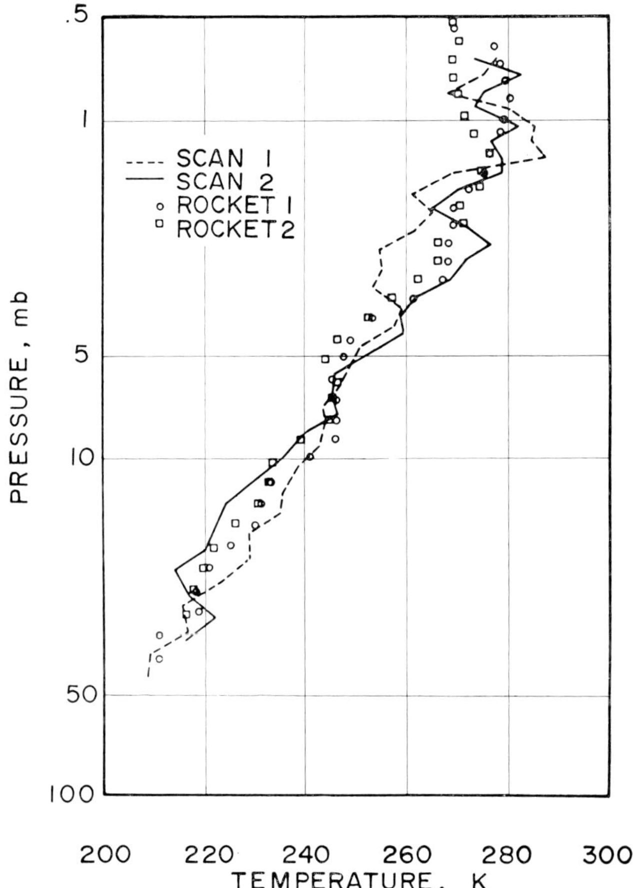

Fig. 6. Comparison of inferred temperature from 15 μm horizon profile data with rocket sonde data (McKee, 1972).

The horizon scans are located on either side of the ARCUS launch site at Pt. Mugu, California. Differences between the two rocket soundings which are six days apart in time are not significantly smaller than the differences between the inferred tempera tures. The inferred profiles certainly appear reasonable for locations on either side o he rocket data as the rocket sounding da ta are seen to lie between the inferred tem peratures at most pressures. Diurnal temperature change is not evident even though the rocket data are near midday and the inferred data at night. Much of the detailed structure evident in the inferred temperatures is due to random noise in the radiometer output. Separation of actual temperature structure from the effects of random radiance noise can be accomplished by correlation of repeated scans.

Horizon profile studies have also been reported by Girard and LeMaitre (1970). In particular, their report discusses the requirements placed by rocket experiments on obtaining high pointing accuracy and some considerations of stellar aspect sensors.

In a different kind of experiment Harwit *et al.* (1971) have carried out measurements of the diffuse celestial background. These measurements essentially set the lower limit to the IR radiance levels one can measure for the earth's atmosphere. They flew an Aerobee 170 rocket from White Sands, New Mexico, which carried a liquid He cooled telescope to a peak altitude of 190 km. Their reported values for the 5 to 23 μm region were in the neighborhood of 1×10^{-11} W cm^{-2} sr^{-1} μm^{-1}.

6. Satellite Measurements

For many years instruments on satellites have been used to observe the upwelling thermal emission from the earth's atmosphere initially with broad band filter radiometers, later with narrow band spectral instruments. The magnitude and spectral variation of the radiation have been used to infer various meteorological parameters of the atmosphere such as the radiation budget, temperature, humidity and O_3 profiles, all as a function of geography and time.

One of the most sophisticated examples of such instruments has been the IRIS interferometers flown by Hanel and Conrath (1970) on the Nimbus satellites. An example of the type of data that can be obtained is shown here in Figure 7. The resolution obtained was equivalent to 1.5 cm^{-1} (unapodized). The field of view of the instrument was 5° or about a 95 km diameter circular area on the ground as seen from 1100 km altitude satellite. In the two upper curves, one can clearly see the absorption due to CO_2, O_3, and H_2O superposed on the nearly gray body emission of the ground. In the Antarctic, the ground is actually colder than the atmosphere and these molecular bands show as emission bands. Note the sharp spike in the middle of the 15 μm CO_2 band; this is the very strongly absorbing R branch of CO_2. Consequently, the observed emission comes primarily from above the tropopause where the atmospheric temperature is again rising and thus this Q-branch always appears as an emission 'line.' The emission in the windows is due primarily to emission from the ground and corresponds to a brightness temperature of about 320 K (47 °C) for the Sahara, 285 K (12 °C) for the Mediterranean Sea, and 190 K (-83 °C) for the Antarctic plateau. From the Nimbus 4 instrument a large data set of calibrated thermal emission spectra, approximately 1.4 million, has been obtained, covering a variety of geophysical and meteorological conditions. From data such as these, profiles of temperature, H_2O, and O_3 are being routinely calculated.

An interesting technique to improve sensitivity has been the selective chopper radio-

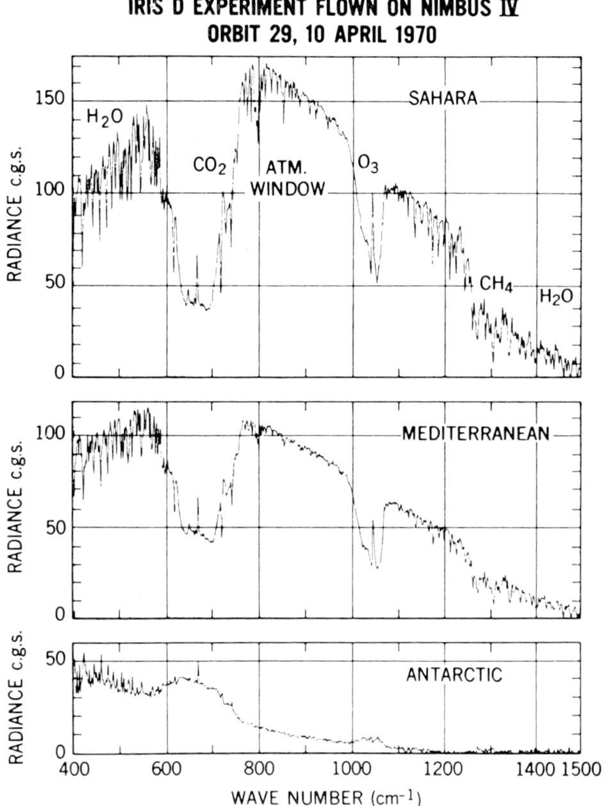

Fig. 7. Selected spectra obtained with the Michelson interferometer flown on the Nimbus-4 satellite (Hanel and Conrath, 1970).

meter developed by Houghton and Smith which has been flown on Nimbus. In this instrument the narrow spectral intervals are selected by narrow band interference filters followed by the 'selective chopper.' The incoming beam is switched between two short cells containing CO_2. The signal is the difference between the fluxes through the two paths. This is equivalent to using a narrow band filter whose pass band is the difference between the absorption of the two CO_2 cells. The height of the emitting layer of the atmosphere being observed is determined by the pass band of the interference filter and the amounts and pressures of CO_2 in the two cells. With this instrument, data on temperatures can be obtained above the 1 mb level (48 km). (Radiation from up to 85 km level can be observed.)

Another use of satellites more pertinent to our discussion of upper atmospheric measurements has been the work of Markov. From 1958 to 1966 Markov and co-workers (1969) from the Lebedev Physical Institute, Moscow, carried out more than thirty launches of radiometric instruments to measure the IR emission of the atmosphere. Included were flights on balloons, rockets, and two satellites, Cosmos 45 and 65.

In general, the radiometers had an effective field of view in the vertical scanning direction of 0.1° which corresponds to approximately 5 km resolution in height at the horizon as seen from an altitude of 200 to 500 km. The radiometers were scanned through 180° from the horizontal through the nadir to the horizontal and back (in a few cases the scan was through 360°). Measurements were obtained at altitudes up to 500 km from rockets and up to 350 km from satellites.

Initially the radiometers had broad regions of spectral sensitivity – 0.8 to 40 μm or 2.5 to 40 μm. Subsequently crystal filters were used to obtain narrower spectral intervals. The response of the instruments is rapid enough so all spectral intervals are covered during the time taken to scan one spatial resolution element. Primarily, these intervals were 0.8 to 4.5 μm, 4.5 to 8.5 μm, and 8.5 to 12.5 μm and were obtained by using differences of signals from overlapping crystal filters.

The most interesting results of these measurements from the rockets and satellites have been the observation of high intensity radiation from relatively thin layers of the upper atmosphere – of the order of 10 to 30 km thick – at altitudes of approximately 150, 280, 430, and 500 km – particularly during geomagnetic perturbations, with emittance values above 400 W m^{-2} having been recorded.

All the rocket launchings were accomplished when the upper atmosphere above 100 km was illuminated by the sun. The satellite measurements were during day and night and at various altitudes. The stratified layers were observed during day and night and at essentially all latitudes.

The maximum radiation is observed in the spectral region from 4.5 to 8.5 μm. The radiation in the region from 0.8 to 4.5 μm is about three times lower (at night, five times lower) while in the long wavelength region ($\lambda > 8.5\ \mu$m) it is very small. On the Cosmos 65 flight, radiation at wavelengths greater than 12.5 μm was observed in two regions at 80 to 150 km. There appears to be a reasonably good correlation between radiance observed and the geomagnetic index, K_ϱ.

Markov argues that the intensity of these layers (comparable to looking down at the earth) can be partially explained if they are considered as thin layers seen edge on. Considering the curvature of the earth, thin layers of 10 or so km in thickness, and the altitudes involved, then it is possible that one is observing the radiation over a path of 500 to 1000 km or more. The equivalent radiation observed in a vertical path is less by a factor of 100 or more. Thus a measurement of 100 W m^{-2} along the layer could correspond to vertical radiation of 10^{-4} W cm^{-2} or 0.07% of the solar constant. This nevertheless still leads a value of roughly 10^{-2} to 10^{-3} erg cm^{-3} s^{-1} for the isotropic density of radiation even assuming this geometry. Where this level of energy input would come from or how it would couple into the observed radiation is still not understood.

References

Aspen International Conference on Fourier Spectroscopy: 1970, (Ed. by G. A. Vanasse, A. T. Stair, and D. J. Baker), AFCRL-71-0019, Special Reports, No. 114.

Connes, J., Connes, P., and Maillard, J. P.: 1969, 'Atlas of the Near Infrared Spectra of Venus, Mars, Jupiter and Saturn', CNRS, Paris, France.

Delbouille, L. and Roland, G.: 1963, 'Photometric Atlas of the Solar Spectrum from λ7498 to λ12016', *Mem. Soc. Roy. Sci. Liege*, Special Volume No. 4, 19.

Girard, A. and LeMaitre, M. P.: 1970, *Appl. Opt.* **9**, 903.

Hall, D. N. B.: 1970, Ph.D. Thesis, Harvard University.

Hanel, R. A. and Conrath, B. J.: 1970, 'Thermal Emission Spectra of the Earth and Atmosphere Obtained from the Nimbus 4 Michelson Interferometer Experiment', Goddard Space Flight Center, X-620-70-244.

Harries, J. E. and Burroughs, W. J.: 1971, *Quart. J. Roy. Meteorol. Soc.* **97**, 519.

Harries, J. E., Swann, N. R., Beckman, J. E. and Ade, P. A. R.: 1972, *Nature* **236**, 159.

Harwit, M. O., Houck, J. R., Jores, B. W., Pipher, J. L., and Soifer, B. T.: 1971, 'Infrared Observations of Diffuse Backgrounds,' AFCRL-71-0327.

Howard, J. N., Garing, J. S., and Walker, R. G.: 1965, in Shea L. Valley (ed.), *Handbook of Geophysics and Space Environments*, Air Force Cambridge Research Laboratories, Chapter 10.

Huppie, R. and Stair, A. T.: 1972, private communication.

Markov, N. M.: 1969, *Appl. Opt.* **8**, 887.

McKee, T. B.: 1972, 'Inference of Stratospheric Temperature and Water Vapor Structure from Limb Radiance Profiles', Colorado State University, Atmospheric Science Paper No. 178.

Migeotte, M. 1972, 'Celestial Spectra and Atmospheric Effects on Radiation Mainly in the Infrared', Final Scientific Report, AFCRL-72-0299.

Migeotte, M., Neven, L., and Swensson, J.: 1956, 'The Solar Spectrum from 2.8 to 23.7 Microns', *Mem. Soc. Roy. Sci. Liege*, Special Volume No. 1.

Murcray, D. G., Kyle, T. G., Murcray, F. H., and Williams, W. J.: 1969, *J. Opt. Soc. Am.* **59**, 1131.

Swensson, J. W., Benedict, W. S., Delbouille, L., and Roland, G.: 1970, 'The Solar Spectrum from λ7498 to λ12016, A Table of Measures and Identifications', *Mem. Soc. Roy. Sci. Liege*, Special Volume No. 5.

Walker, R. G.: 1972, private communication.

BALLOON-BORNE INFRARED MEASUREMENTS

JAMES N. BROOKS, AARON GOLDMAN, JOHN J. KOSTERS,
DAVID G. MURCRAY, FRANK H. MURCRAY, and WALTER J. WILLIAMS
Dept. of Physics, University of Denver, Denver, Colo., U.S.A.

1. Introduction

The previous speakers have discussed the general aspects of atmospheric IR measurements. In this discussion I will emphasize atmospheric IR measurements made using a balloon platform. Currently available balloons will carry 500 kg instrumentation packages to 40 km almost routinely. This weight capability is of considerable advantage for IR measurements, and the balloon is a very good vehicle for making measurements up to these altitudes.

The atmospheric absorption features in the IR are due to the minor constituents since both O_2 and N_2 molecules are almost inactive in the IR. Thus a study of the variation of these absorptions with altitude yields information concerning the altitude distribution of the various minor constituents. Details of the variation of absorption with altitude can be obtained by measuring the variation of the IR solar spectrum or the atmospheric emission spectrum at various altitudes. We have made measurements of both types. The instrumentation required for these measurements is quite different and will be discussed separately.

2. Absorption Measurements

The instrumentation used to obtain the absorption data consists of an IR spectrometer constructed for balloon use, a servo-controlled heliostat and telescope system to image the sun on the spectrometer entrance slit, an onboard digital magnetic tape recording system for recording the data generated by the spectrometer and auxiliary sensors, and the various power supplies, gondola etc. to make the complete balloon borne system. The spectrometer is of Czerny-Turner design and employs a tuning fork chopper to interrupt the radiation so as to give rise to an ac signal from the detector. The radiation is double passed through the system and the radiation is chopped after the first pass. This double-passing increases resolution of the system and chopping after the first pass reduces the stray light contribution to the detector signal. The optical components are mounted to a brazed framework which gives the unit good mechanical rigidity while remaining light in weight. The unit is covered with thin anodized aluminum plates. The whole unit is equipped with heater cards and the temperature is thermostatically controlled at $35\,°C$. All mechanical rotations are accomplished by means of 400 Hz synchronous motors. Ge:Cu has been used as the spectrometer detector on all the recent flights. The tuning fork chopper interrupts the radiation giving rise to 750 Hz ac signal from the detector. This signal is amplified, synchronously rectified, and recorded by means of the digital magnetic tape recording system. This system was also constructed for balloon borne use and operates from 28 vdc, consuming less than 40 W of power. In addition the primary data channels are also telemetered by means of an FM/FM telemetry system. This system is used as a back-up system in case the onboard recorder fails and also as a means of monitoring the system performance during the flight. The heliostat servo control unit employs counter rotating clutch systems for azimuth and elevation orientation. Error signals for the servo systems are derived from a prism beam splitter optical system. The complete system is mounted in a gondola constructed of brazed conduit which provides support and protects the units when they are returned to the ground by parachute.

3. Emission Instrumentation

The sun provides an intense source of IR radiation and as a result it is possible to make high resolution spectral measurements without pushing the detector system to its limiting sensitivity. The atmospheric emission on the other hand is a much less intense and highly variable source and the instrumentation which is used for making observations of the atmospheric emission, particularly in the so called 'window' spectral regions, must be very sensitive. In order to achieve such sensitivity it is necessary that all objects within the detector field of view emit as few photons as possible. This can best be achieved by cooling all optical elements to as low a temperature as possible. For the measurement of the atmospheric emission with altitude provision must also be made to keep various atmospheric gases from freezing on the optical components. For the measurements made under this study these two objectives were accomplished by cooling the whole instrument with liquid N_2 to 80 K. Two instruments have been used; a filter radiometer system and a spectral-radiometer system. The optical system and antifrost system of the spectrometer are shown in Figure 1. The boil off gas from the liquid N_2 used as a coolant is vented into the back of the instrument and flows out through the baffling system in front of the instrument. The incoming radiation is inter-

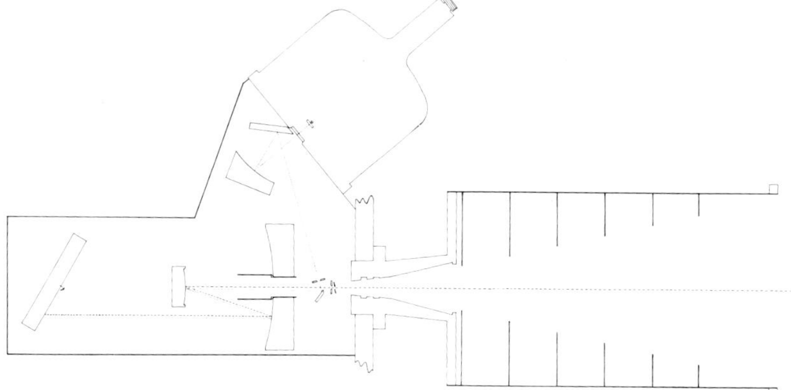

Fig. 1.   Optical system used in the liquid N_2 cooled spectrometer.

rupted by means of a tuning fork chopper at a 156 Hz rate. The ac signal from the detector is amplified and synchronously rectified. Since the spectral radiances to be measured vary by 4 orders of magnitude three different gains are used. The data handling is the same as in the absorption measuring instrument. The filter radiometer is similar in design to the spectrometer except that a filter wheel is placed directly behind the entrance aperture and the entrance aperture is focused onto the detector using an optical system similar to that used to image the exit slit onto the detector. Again the data handling and gondola construction are similar to that employed in the transmission instrument.

4. Results

All of these instruments have been flown several times and data have been obtained concerning the variation in transmittance and emission with altitude for a number of wavelength regions. A complete presentation of the data obtained within the time limits imposed is impossible and I will present only selected results.

4.1. TRANSMITTANCE DATA

Figure 2 shows the variation with altitude of the atmospheric transmittance in the region between 300 cm^{-1} and 600 cm^{-1}. The bottom spectrum was taken from the ground (altitude 1.3 km) and shows the strong absorption due to water vapor which completely absorbs the radiation over much of the spectral region. The next spectrum was obtained at an altitude of 6 km. The rapid decrease in water vapor with altitude

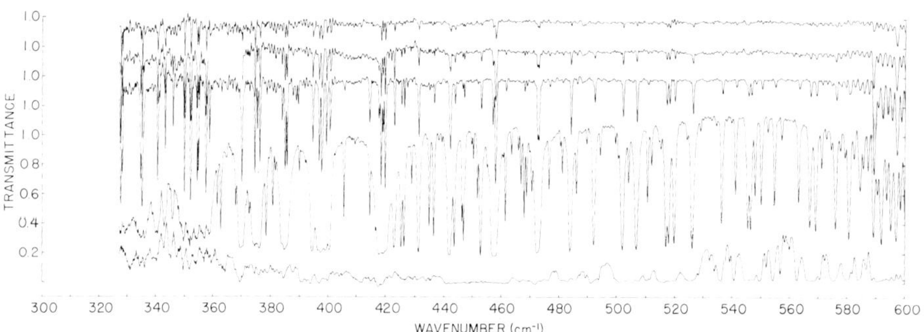

Fig. 2.   IR solar spectra as observed at 1.3, 6, 11, 20, and 29 km. The successive spectra are displaced 0.2 in ordinate for clarity.

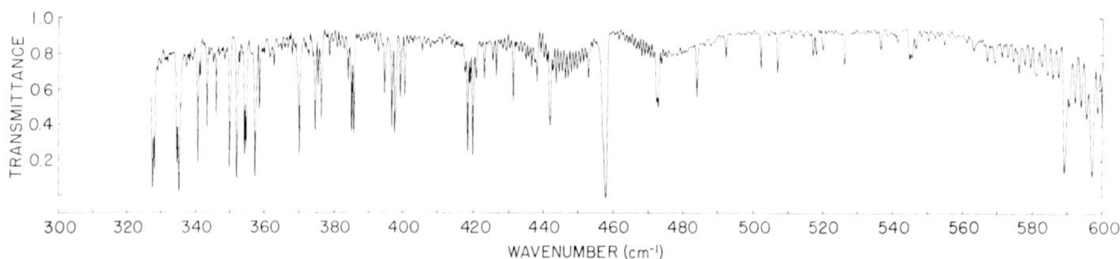

Fig. 3.   IR solar spectrum as observed from 30 km at a solar zenith angle of 93°.

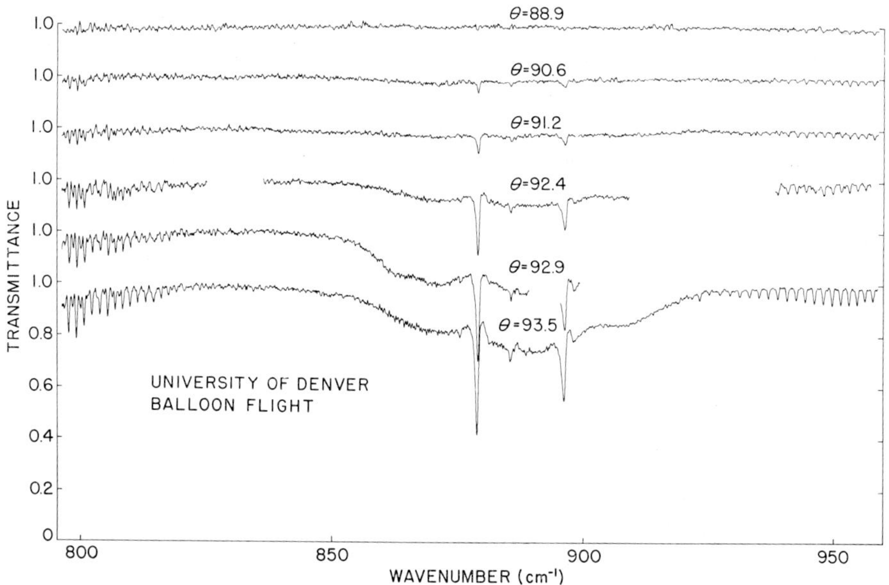

Fig. 4. IR solar spectra as observed from 30 km and at several solar zenith angles. The growth of the HNO₃ absorption is evident.

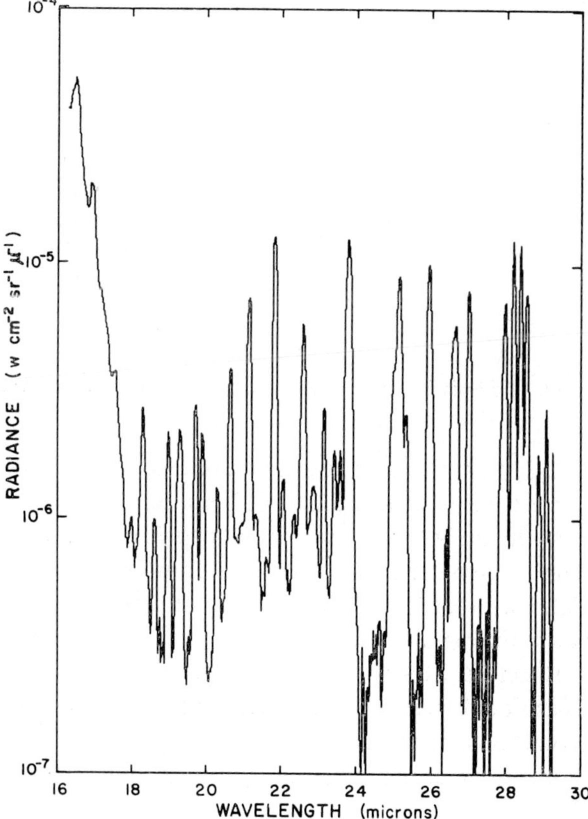

Fig. 5. Atmospheric emission as observed at 12.3 km and an elevation angle of 45°.

is evident in this spectrum which shows a significant transmission over much of the wavelength interval. The spectra obtained at the higher altitudes (11, 20 and 29 km) indicate the dryness of the lower stratosphere since there is very little residual absorption at these altitudes and the change in absorption with altitude is slight. From these data one would infer that water vapor is the only significant atmospheric absorber

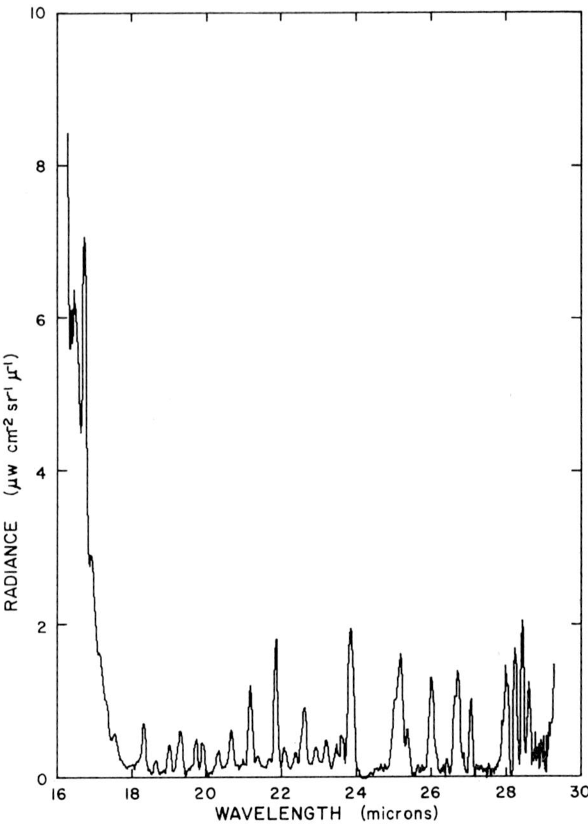

Fig. 6.   Atmospheric emission as observed at 28 km and an elevation angle of 45°.

present at high altitude that produces absorption in this wavelength region. Figure 3 shows that this is an erroneous conclusion. This is a spectrum of the same wavelength region obtained from 30 km but with a solar zenith angle of 93°. The spectrum shows enhanced absorption in the water vapor and CO_2 absorption lines due to the increased path length associated with the increase in optical path length. An additional absorption feature appears in the region between 440 cm^{-1} and 480 cm^{-1} which shows characteristic P, Q, R, structure. This particular absorption feature is due to HNO_3 and is expected on the basis of absorptions observed in other wavelength regions on earlier flights. Figure 4 shows similar sunset results from 800 to 1000 cm^{-1} region. Since this region is relatively free of other absorbers the growth of the HNO_3 absorption as the

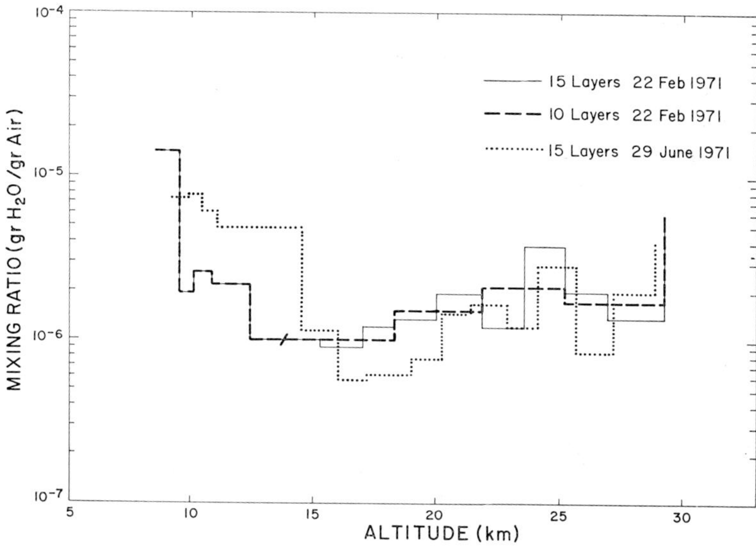

Fig. 7. Water vapor mixing ratio vs. altitude as determined from atmospheric emission data obtained on February 22 and June 29, 1971.

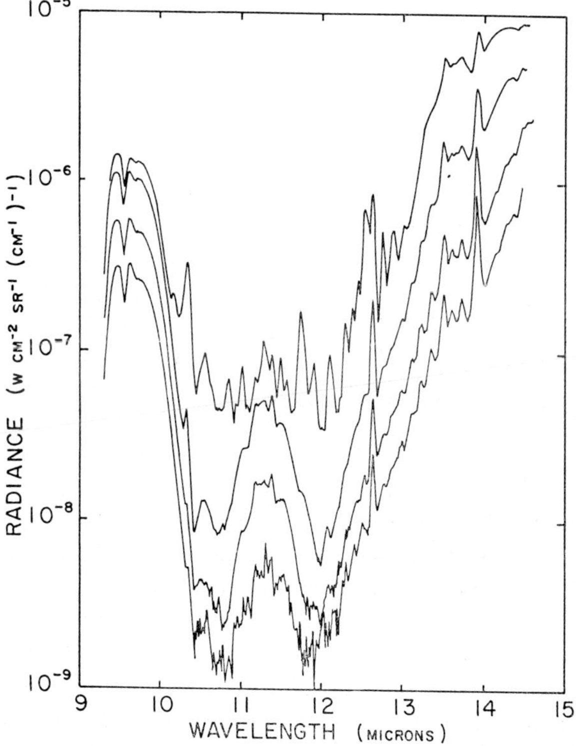

Fig. 8. Spectral atmospheric radiance measured at 3.4, 10.0, 20.1, and 24.7 km on September 15, 1971.

sun sets is very evident in these spectra. These long path spectra are particularly useful in the detection of minor species. The determination of the distribution of the particular species responsible for the absorption features in this type of spectra is a very complex problem because of the rapidly changing path. In addition data can only be obtained at sunrise or sunset. This limits any possible study of diurnal variation of this particular constituent.

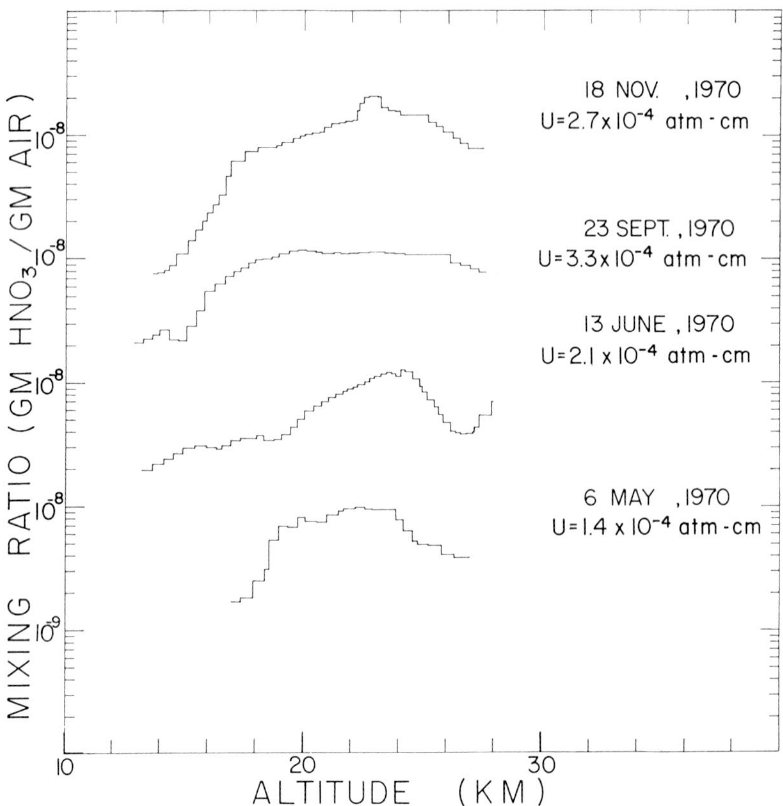

Fig. 9. Distribution of HNO_3 with altitude as determined from atmospheric emission data.

4.2. Emission Data

These shortcomings can be overcome by measuring the radiation emitted by the atmosphere as a function of altitude. Figure 5 shows such emission data for the same spectral region as given in Figure 2. These data were taken at an elevation angle of 45°. Note that while the resolution is considerably less than in the transmittance case the absorption and emission features agree very well. Also note that the presence of emission due to HNO_3 in the region from 21 to 23 μm is evident in this spectrum even though the zenith angle is 45° and not $>90°$ as in the case of the absorption data. Figure 6 shows similar emission data obtained at a higher altitude. The change between the two spectra is evident and is related to the amount of water vapor in the layer between the two altitudes at which the spectra were taken. This change in emission with altitude can be used to determine the distribution with altitude of the constituent responsible for the emission. The results of such an analysis are shown in Figure 7. Figure 8 shows similar emission data for the 10 to 12 μm region obtained at various altitudes. These data were also taken at an elevation angle of 45°. The emission feature between 10.5 and 11.5 μm which shows strongly at the intermediate altitudes, is due to HNO_3.

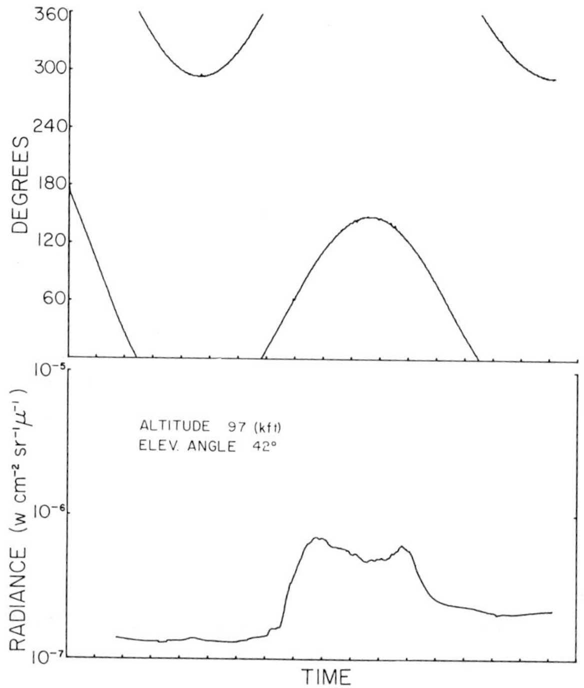

Fig. 10.   Variation in atmospheric radiance with azimuth angle as observed from 30 km. Filter centered at 12.0 μm with 0.2 μm bandpass.

Distributions of HNO_3 with altitude derived from data of this sort are shown in Figure 9. These data are typical of the profiles obtained for HNO_3. The constituent occurs in a definite layer with the maximum mixing ratio occurring in the 20 to 25 km region. All flights indicate a definite decrease in the HNO_3 mixing ratio above 30 km.

   All of the data presented above have been analyzed on the basis that the source of the radiation is isotropic and does not vary significantly with time. These assumptions are not strictly true since on almost all flights there is some variability in the observed atmospheric radiance. The variability generally correlates well with the direction in which the radiometer is looking, i.e., the variations are due to spatial variability. The variations are more pronounced close to the horizon. Figure 10 shows an example of the observed variation in the 12.0 μm region; the cause of this variation has not been determined.

SUNSPOT CYCLE VARIATION IN ATMOSPHERIC DENSITY AT THE LEVEL OF THE SODIUM LAYER

H. DERBLOM

Uppsala Ionospheric Observatory, section of the Research Institute of Swedish National Defence

Abstract. Spectrophotometer studies of the Na layer in twilight, made at Uppsala during 1960–70, have shown that the mean peak height of the layer varies between $\simeq 86$ and $\simeq 92$ km during one 11 yr sunspot period. The lowest values are recorded during sunspot maximum. This height variation is interpreted as a periodic change of the atmospheric density profile. It is concluded that the highest density at the level of Na layer appears during sunspot minimum.

1. Introduction

During recent years considerable information has been obtained about density and temperature in the upper atmosphere. Observations have been made over longer periods at satellite heights, but at lower levels only a limited number of measurements have been made using the falling-sphere and other methods with sounding rockets. References are found in the papers which appeared in CIRA (1965). It is well established from available data that a long term variation exists in the atmosphere above $\simeq 120$ km height, associated with the 11 yr sunspot cycle. It is found that the density at sunspot maximum is higher than at sunspot minimum.

Since the atmosphere around 100 km absorbs a considerable amount of solar radiation, a marked variation of the atmospheric density and temperature is expected, and should be observable by indirect measurements from the ground. Lindblad (1966) studied the hourly rates and heights of meteors and found an apparent variation as a function of the solar cycle. He interpreted this as a periodic variation of the atmospheric density at a height of 90 to 110 km and concluded that the observations indicated a density change by a factor of 2. Such a large change may influence the production of free Na atoms and should be observed as a change in the height of the Na layer during the sunspot cycle.

2. Observations

At the Uppsala Ionospheric Observatory, $59°48'$ N geogr. lat., $17°36'$ E geogr. long., observations of the Na twilight emission have been made since 1960 using a direct-recording spectrophotometer adjusted for $\simeq 1.5$ Å resolution. Records are made in the direction of zenith only, which enables calculations of the height of the peak emission from the shadow height at sunset and sunrise. This method is described by Hunten (1954) and Hunten and Shepherd (1954) and is assumed to give sufficiently comparable values if identical observational techniques are used over a long period.

The obtained heights of the maximum of the Na layer are presented in Figure 1, together with relative sunspot numbers. From about 120 observations made during

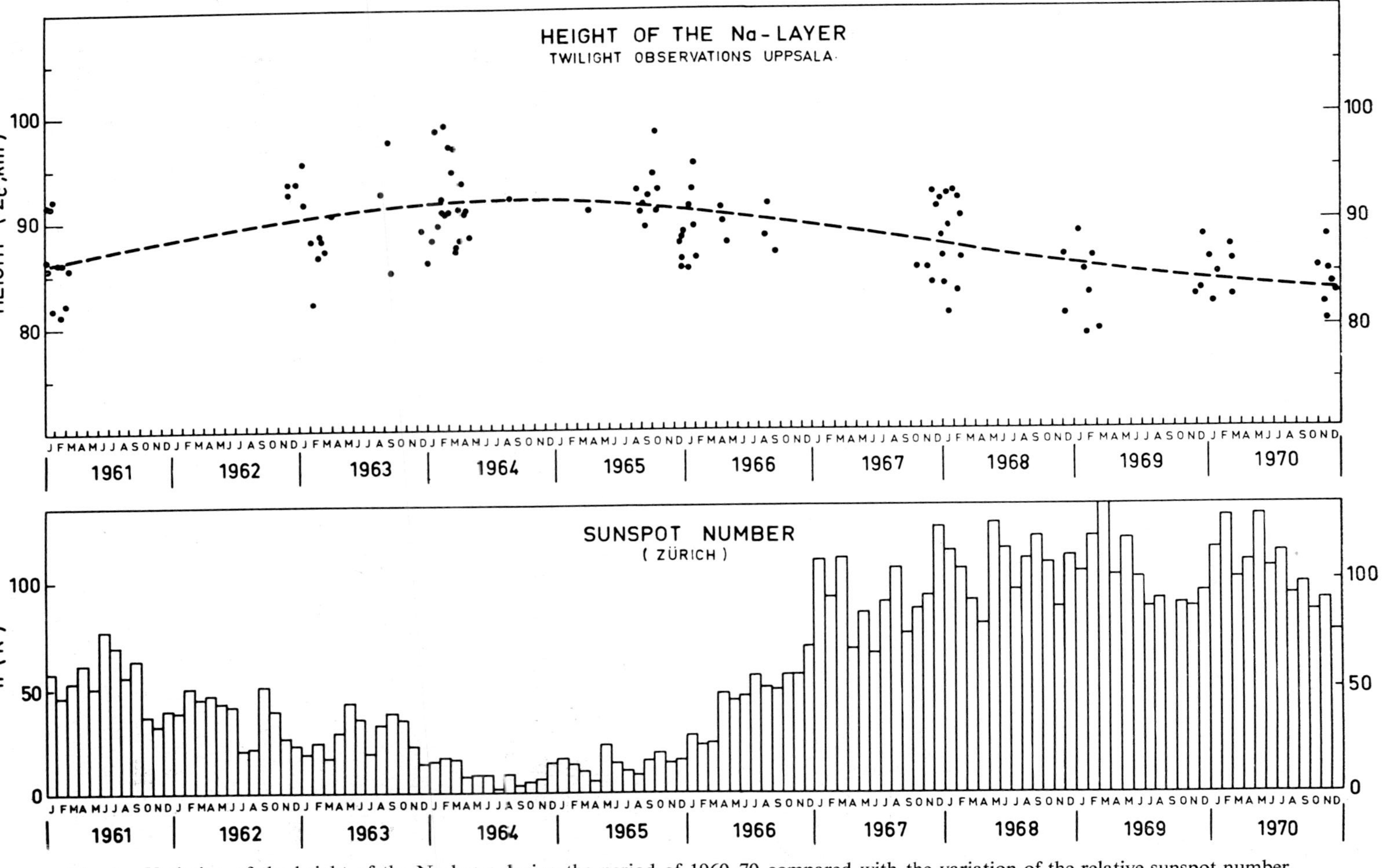

Fig. 1. Variation of the height of the Na layer during the period of 1960–70 compared with the variation of the relative sunspot number.

the period 1960–70 it is found that the mean peak height of the twilight Na layer shows a regular variation in spite of the spreading of the individual values. During 1960 the mean height was about 86 km and increased to a maximum value of about 92 km in 1964. Thereafter the height has shown a successive decrease.

With a resolution of 1.5 Å the doublet 5890/96 Å is sufficiently resolved by the spectrophotometer and the intensity ratio of D_2/D_1 can be evaluated as a measure of the abundance (Blamont and Donahue, 1964; Hunten, 1967). The D_2/D_1 ratio during the whole period is found to be $\simeq 1.47$. Mean ratios for each year are shown in Table I.

TABLE I

Mean D_2/D_1 ratios

Year	1961	62	63	64	65	66	67	68	69	70
D_2/D_1	1.44	1.46	1.45	1.58	1.55	1.54	1.38	1.42	1.51	1.43

These values show a slight increase during years of minimum sunspot activity. However, the corresponding change in abundance is too small to be of importance because of the uncertainty in the observations.

3. Discussion

The most controversial question is whether the maintenance of the Na layer has to be explained by ablation from meteorites or by continuous transport of Na compounds from the ground. In the first case, Na may be liberated directly during the deceleration of meteorites or by action of solar radiation on meteoric dust. In the second case, Na has to be produced by a chemical reaction including atmospheric O.

In the presence of O and O_3, Na atoms become rapidly oxidized. Free atoms are then produced by chemical reactions such as $NaO + O \rightarrow Na + O_2$. Assuming that the total mixing ratio of Na compounds is constant with altitude, any change in the chemistry which releases more Na consequently has to decrease the peak height of the layer. The observations have shown that no pronounced variation is found in the abundance during the period, which implies that the amount of free Na is not determined by chemical processes alone.

From rocket measurements it has been concluded (Donahue and Meier, 1967; Hunten and Wallace, 1967) that the scale height of Na for low latitudes ($<45°$) is about 2 to 3 km. Also at latitudes $>45°$ the scale height of Na seems to be much less than of the ambient atmosphere (Witt, 1972). Such a scale height indicates that the Na supply is distributed in a similar manner having a source which is not in complete mixing with the atmosphere.

Donahue (1966), Hunten and Wallace (1967), Hunten and Godson (1967), and Donahue and Meier (1967) suggested that the source is in the form of dust particles and that Na is liberated from their surface by solar radiation. Gadsden (1968) proposed as an alternative that the ablation from meteors is responsible for the deposition of Na. At these altitudes the deceleration of meteors and the settling velocity of small dust particles are inversely proportional to the atmospheric density. Thus, any change in the apparent height of the Na layer reflects a change in the atmospheric density. In a recent paper by Gadsden (1970) further evidence is given for a deposition of Na from meteors.

According to Lindblad (1966) meteors are decelerated at higher altitudes at sunspot minimum and at lower altitudes at sunspot maximum. It is assumed that during sunspot minimum, when solar radiation is reduced, the upper atmosphere is cooled, accompanied by a contraction and higher density at the 90 km level. During sunspot maximum, when the solar flux is increased, the atmosphere is heated which results in a lower density. At much lower altitudes (in the stratosphere) it is well known that the temperature increases during geomagnetic activity (see Scherhag, 1952; Kriester, 1966). The dependence of the temperature on the sunspot number was established by Schwentek (1971).

The observed density variation at altitudes below 120 km is opposite in phase compared with the variation at higher altitudes. This phenomenon is tentatively interpreted by Lindblad (1966) as a vertical upward transport of air masses due to heating at sunspot maximum and downward motion due to cooling at solar minimum. From this and from different observations it is assumed that an intersection of the density profile obtained during years of minimum activity and the one obtained during maximum activity, has to be found at $\simeq 120$ km. Below the altitude of intersection the increased density during minimum years will result in the observed height increase of the Na layer, i.e., in a greater height at sunspot minimum and a lower height at sunspot maximum.

Consequently it can be concluded that the height variation deduced from Na observations can be explained in the same manner as has been done by Lindblad (1966) from meteor height data. Apparently there exists a periodic variation in the atmospheric density at 90 km level with the highest density occurring at sunspot minimum.

Lindblad (1966) found a density variation by a factor of 2, the present observations have resulted in a factor which is slightly higher.

Acknowledgements

The author is indebted to H. Gunnarsson, who assisted in most of the observations. Further, I want to thank W. Stoffregen for valuable discussions and advice during this work.

References

Blamont, J. E. and Donahue, T. M.: 1964, *J. Geophys. Res.* **69**, 4093.
CIRA: 1965, Cospar International Reference Atmosphere, North-Holland Publishing Company, Amsterdam, Holland.
Donahue, T. M.: 1966, *J. Geophys. Res.* **71**, 2237.
Donahue, T. M. and Meier, R.: 1967, *J. Geophys. Res.* **72**, 2803.
Gadsden, M.: 1968, *J. Atmospheric Terrest. Phys.* **30**, 151.
Gadsden, M.: 1970, *Ann. Geophys.* **26**, 141.

Hunten, D. M.: 1954, *J. Atmospheric Terrest. Phys.* **5**, 44.
Hunten, D. M.: 1967, *Space Sci. Rev.* **6**, 493.
Hunten, D. M. and Shepherd, G. G.: 1954, *J. Atmospheric Terrest. Phys.* **5**, 57.
Hunten, D. M. and Godson, W. L.: 1967, *J. Atmospheric Sci.* **24**, 80.
Hunten, D. M. and Wallace, L.: 1967, *J. Geophys. Res.* **72**, 69.
Kriester, B.: 1966, in K. Rawer (ed.), Proc. NATO Adv. Study Inst., Lindau, North-Holland, Publishing Company, Amsterdam, Holland, p. 81.
Lindblad, B. A.: 1966, in R. Smith-Rose (ed.), *Space Res.* **7**, 1029, North-Holland Publishing Company, Amsterdam-Holland.
Scherhag, R.: 1952, *Ber. d. DWD US-Zone* **6**, 38.
Schwentek, H.: 1971, *J. Atmospheric Terrest. Phys.* **33**, 1839.
Witt, G.: 1972, private communication.

INCOHERENT SCATTER AND VERTICAL
INCIDENCE OBSERVATIONS

P. WALDTEUFEL

Centre National D'Études Des Telecommunications, Issy-les-Moulineaux, 1e, Paris, France

Abstract. This paper summarizes the principles and implementation of the incoherent scatter and ionospheric sounding techniques and discusses some achievements of these techniques with respect to the physics and chemistry of the upper atmosphere.

1. The Ionosphere as a Reflector

Among the most prominent properties of a plasma, such as the ionosphere, one finds the occurrence of collective behavior involving many charged particles which interact between themselves through the space charge E field. A plasma has a natural oscillation frequency, called the plasma frequency f_p:

$$f_p = \frac{\omega_p}{2\pi} = \frac{1}{2\pi}\left(\frac{N_e e^2}{m_e \varepsilon_0}\right)^{1/2}, \tag{1}$$

where N_e is the electron number density, e and m are the electric charge and mass of the electron, ε_0 is the vacuum dielectric constant; M.K.S. units are used.

An electromagnetic wave may only propagate throughout a plasma provided its frequency f is larger than f_p; otherwise, the wave is reflected. This phenomenon has led to the first experimental evidence of an ionized layer in the upper atmosphere; indeed, it has for many years provided most of the available information about it. The *ionosonde* is a pulsed radar which generally transmits and receives vertically. The delay necessary for the pulse to travel up to the reflecting height and back yields a virtual height, which can be used to infer the actual height z for a particular electron density value N_e. Transmitting several different frequencies in turn allows one to determine an electron density profile $N_e(z)$. An interesting although infrequent use of the ionosonde has been described by Hoyle (1957); more information concerning the ionosonde is given by Rawer and Suchy (1967).

Ionospheric soundings have provided a considerable quantity of information about the ionosphere. However they suffer from three severe limitations: There is only one parameter available, N_e; this availability is bounded to the bottom ionosphere; the reduction of virtual heights to true heights is not easy and may lead to significant uncertainties.

The end of the fifties saw the birth of the space era and a considerable development in the probing of the upper atmosphere. *In situ* and space techniques of course played a major part in this development, but the remote radio techniques were not left behind, as is well illustrated by the appearance of the *incoherent scatter* (I.S.) technique.

                    P. WALDTEUFEL

2. Principles and Implementation of I.S.

An electromagnetic wave with f larger than f_{max} (f_p value for the maximum of the
F layer) may carry most of its energy upwards. However, some of it is nevertheless
scattered by the charged particles. Indeed, such particles, when inbedded in an oscillating E field, are accelerated, behave like oscillating dipoles, and scatter away
according to the Thomson differential cross section σ:

$$\sigma = \left(\frac{\mu_0}{4\pi}\frac{e^2}{m}\right)^2 \sin^2\theta, \tag{2}$$

where μ_0 is the vacuum magnetic permittivity, e and m are the electric charge and
mass of the particle, and θ is the angle between the incident field and the scattering
direction.

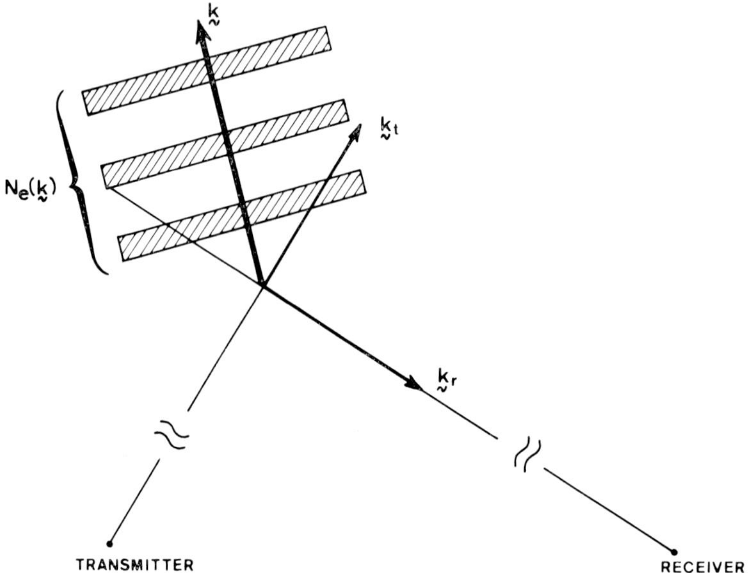

Fig. 1.   Geometry of incoherent scattering.

This formula shows us that scattering by ions will be entirely negligible, due to their
large mass. Even for the electrons, the radiated energy is very weak ($\sigma \approx 8 \times 10^{-30}\mathrm{m}^2$).

Let us now consider a scattering volume V filled by an electron gas (Figure 1).
The scattered field amplitude at the receiving site is the sum of contributions made by
each electron:

$$E_s = \sum E_i, \tag{3}$$

where each elementary field has a phase corresponding to the space location of the
scatterer. If the gas is perfectly homogeneous, for each term it is possible to find

another one with a relative phase shift exactly equal to π, so that these two contributions cancel. The same reasoning can be made again, so that ultimately there is no scattered field at all. This is a simple argument to show why the scattering results but from inhomogeneities in the gas, and not from the bulk of it.

This result can be stated in a more formal way; it is easily shown that one has:

$$E_{\mathrm{s}}(t)\,(:)\,N_{\mathrm{e}}(\mathbf{k}, t) \tag{4}$$

where $\mathbf{k} = \mathbf{k}_i - \mathbf{k}_r$ is called the scattering vector, and $N_{\mathrm{e}}(\mathbf{k}, t)$ is the space Fourier component of the electron density along the $\mathbf{k}$ vector. What this formula says is that the scattered power along $\mathbf{k}_r$ is due to those stratifications of N_{e} in planes perpendicular to $\mathbf{k}$, spaced so that constructive interference results.

Equation (4) is often called the Bragg condition. This condition holds for any kind of scattering and does not depend upon the origin of the inhomogeneities. Several examples of scattering by macroscopic structures obeying this formula can be found in the ionosphere (at equatorial or auroral latitudes in particular).

In the absence of any other factor, inhomogeneities still exist, due to the thermal agitation of the electrons. Since this is a random phenomenon, the scattering field also is random; what matters then is not its amplitude but its frequency power spectrum $S(\omega)$. The above relationship is easily transformed to yield:

$$S(\omega + \omega_0)\,(:)\,\langle |N_{\mathrm{e}}(\mathbf{k}, \omega)|^2 \rangle, \tag{5}$$

where ω_0 is the angular frequency of the incident wave, and the brackets stand for the statistical average.

We now have the principle of an I.S. measurement: derive theoretical spectra as a function of ionospheric parameters, perform a spectral analysis of the scattered signal, and adjust the former to the latter to determine these parameters.

The next task, i.e., the calculation of $\langle |N_{\mathrm{e}}(\mathbf{k}, \omega)|^2 \rangle$ is a complicated plasma physics problem (Evans, 1969, and references of this work). Let us simply investigate here whether or not the collective behavior of the plasma is going to dominate the scattering process. This depends on the ratio between the incident wavelength λ and a characteristic distance λ_{D}, the Debye length of the plasma.

(a) If $\lambda < \lambda_{\mathrm{D}}$, no collective interactions occur and the situation is similar to the scattering by independent electrons. The spectrum corresponds to the electron velocity distribution and therefore exhibits a Gaussian shape with a width depending upon the electron temperature.

(b) If $\lambda > \lambda_{\mathrm{D}}$, collective forces prevail in the structure of $N_{\mathrm{e}}(\mathbf{k}, t)$ and result in a partial control of the electrons by the ionic thermal fluctuations. Since the ions are much slower, the bulk of the resulting spectrum will be much narrower than in the first case. However the very fast (suprathermal) electrons undergo only electrostatic forces between themselves, which tend to build large natural fluctuations at the plasma frequency; hence two narrow spikes (the plasma 'lines') at angular frequencies $\omega_0 \pm \omega_{\mathrm{p}}$. Figure 2 is a sketch of the two possible situations.

The latter case is the most attractive, since

(i) the signal to noise ratio is much higher, as a comparable power is concentrated within a very much narrower bandwidth, and

(ii) the spectrum contains extra information about ions. Therefore the I.S. radars are always designed to have $\lambda \gg \lambda_D$, for the bulk of the ionosphere at least (this implies frequencies smaller than 1300 MHz).

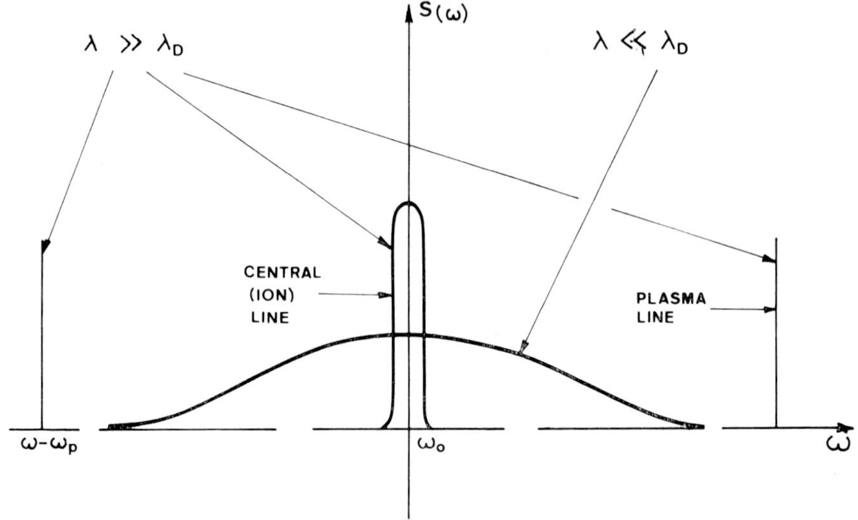

Fig. 2.   General behavior of the scattered spectrum according to the importance of collective interactions.

3. Direct I.S. Measurements

Most of the measurements are obtained through the dependence of the central ('ion') part of the spectrum with respect to ionospheric parameters. The situation is outlined on Figure 3 and qualitatively described hereafter.

(a) The width of the spectrum corresponds chiefly to the electron temperature T_e;

(b) The ratio of spectral power densities in the 'shoulders' and the middle corresponds chiefly to the ratio of T_e to the ion temperature T_i.

(c) Both previous statements hold provided the ion composition is known. Otherwise, changes in this composition are easily misinterpreted as variations in T_e and T_i; one generally has to accept extra hypotheses in order to deal with this case.

(d) The frequency shift of the spectrum as a whole corresponds to a non zero ion velocity along the **k** vector.

(e) The total spectral power (area under the spectrum) is proportional to $N_e(1 + T_e/T_i)^{-1}$ for ionospheric conditions, yielding N_e.

The problem of extracting the parameters values from the measurements (inversion problem) can be solved with various degrees of care. The most elaborate method consists of a non-linear least squares adjustment of the theoretical spectra to the data.

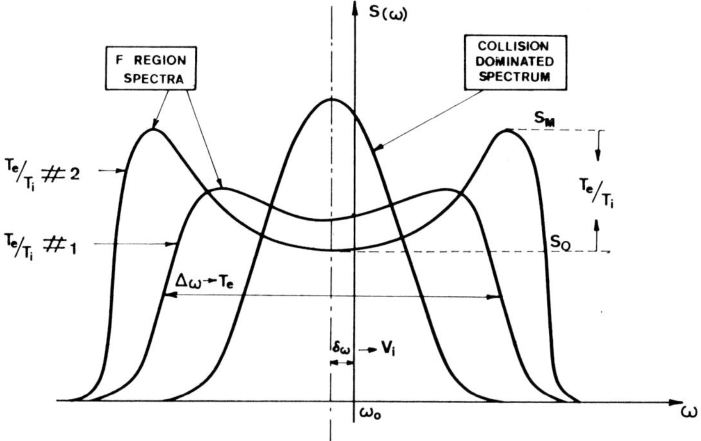

Fig. 3. Shape of the incoherent scatter ion spectrum, with and without collisions; the connection between some geometrical features and ionospheric parameters is indicated.

It must be stressed that, except in the case of N_e (where calibration uncertainties contribute most of the error) errors on the parameters stem only from the random nature of the signal, and can thus be made in principle arbitrarily small, if the averaging time is long enough. Actually, accuracies down to less than 1% in the temperatures, for example, are not uncommon.

The above listed parameters are most frequently obtained in I.S. experiments. They are not the only ones however:

(a) The ion neutral collision frequency v, which can be extracted from the central line when the spectrum becomes collision dominated (see Figure 3), as happens in the lower E region.

(b) The upward and downward photoelectron fluxes, which can be estimated from the plasma lines because they enhance the intensity of these lines well above the thermal level.

The main consideration there lies in the very low value of the Thomson scattering cross section. The energy budget of I.S. is extremely poor, so that any I.S. experiment always needs the use of large antennas, powerful transmitters and low noise receivers. Also, sophisticated equipment is necessary to record, analyze, and process the I.S. signal.

4. Implementation of the I.S. Technique

Every I.S. radar is therefore built along rather similar lines, except in one respect; this is the method chosen to achieve a satisfactory height resolution. The simplest way (but not the cheapest) is to adopt a multistatic geometry: the scattering volume is the volume common to the beams of the receiving and transmitting antennas. Continuous waves are used. This ensures good height resolution and very clean spectra at the same time, at the expense of time resolution.

The other alternative is to use a single antenna, transmit pulses, and define the height resolution as the length traveled by the wave during the pulse. Many altitudes can be explored at the same time; the trouble is that the I.S. spectrum is broadened through convolution with the pulse spectrum up to an extent which sometimes cannot be accepted. However physicists have progressively found the way to increase the actual length of the transmitted signal by using various pulse coding schemes, so that the monostatic configuration can yield quite good spectra.

5. Indirect I.S. Measurements

While the parameters determined directly from I.S. have an intrinsic interest, they can also be combined together so as to obtain further information. As an example, here is an outline of the method devised by Bauer *et al.* (1970) in order to determine parameters of the neutral atmosphere in the F region.

In this region (covering approximately the 250 to 500 km range), the heat budget of the ions can be written quite simply as a local balance between the heat L_{ei} given to ions by the (hotter) electrons and the heat L_{in} given to neutral particles by the ions. The theory of collisions gives:

$$\begin{aligned} L_{ei} \,(:)\, N_e^2 \,(T_e - T_i)^{-3/2}, \\ L_{in} \,(:)\, N_e \,[O]\, (T_i - T_n), \end{aligned} \tag{6}$$

where $[O]$ is the O number density, and T_n is the neutral temperature.

Solving $L_{ei} = L_{in}$ for T_i gives:

$$T_i = (N_e T_e^{-1/2} + C\,[O]\, T_n) \times (N_e T_e^{-3/2} + C\,[O])^{-1}, \tag{7}$$

where C is a known coefficient.

Now T_n does not vary much with altitude and can be expressed in terms of two parameters. The first one is the thermopause temperature T_∞; the other one is a 'slope' parameter S related to the steepness of the $T_n(z)$ profile in the lower thermosphere: $T_n = T_n(z, T_\infty, S)$. Similarly, O obeys diffusive equilibrium and one can write: $[O] = [O]_{z_0} \times f\,(z, T_\infty, S)$ where z_0 is some reference altitude. Finally:

$$T_i\,(z) = T_i\,(z, T_\infty, S, [O]_{z_0}, N_e, T_e) \tag{8}$$

where N_e and T_e are known at each altitude from the measurements. It is now possible to adjust the 'theoretical' T_i curve to the experimental data and to obtain estimates of $[O]_{z_0}$, T_∞, and S. Knowledge of these three parameters is clearly essential for the understanding of fundamental processes in the thermosphere. Note that in the F region T_n is never very different from T_i; in order for the adjustment to be meaningful, one must have very accurate T_i measurements.

There are a number of methods which similarly combine I.S. data into simple models so as to obtain remote parameters such as neutral wind velocities or the neutral composition in the lower thermosphere.

6. Conclusions

The I.S. technique is the most powerful ground based tool available for the study of the upper atmosphere. Its main advantages are that:

(a) It yields several parameters at the same time and location, and

(b) It does so within a very satisfactory accuracy.

Most of the outstanding results achieved so far using I.S. have been in the energetics of the ionosphere (energy budget of the electrons, photoelectrons fluxes), the structure of the thermosphere (systematic variations of the temperature, density, and composition), the dynamics of the upper atmosphere (general circulation, tides, gravity waves). We may expect to see in the future further advances in these areas, and also in the study of other phenomena, particularly in E fields.

The I.S. technique does have limitations; for example it is entirely unsuited to the study of minor constituents. Still, the largest drawback of all is that I.S. radars are very expensive and it alone does not and will not allow global coverage of the thermosphere. As a consequence, we must think of the I.S. facilities in the future neither as isolated observatories, nor as parts of an I.S. network (which is extremely loose) but as imbedded in an heterogenous, global network. Situations have already arisen in which the use of an I.S. radar in connection with a network of ionosondes has proven quite valuable (e.g., in the investigation of gravity waves). It is the opinion of the author that a major step in experimental studies of the upper atmosphere will be taken when the coordination of several I.S. radars with several aeronomical satellites becomes fully operational.

References

Bauer, P., Waldteufel, P., and Alcayde, D.: 1970, *J. Geophys. Res.* **75**, 4825.
Evans, J. V.: 1969, *Proc. I.E.E.E.* **57**, 496.
Hoyle, F.: 1957, *The Black Cloud*, W. Heinemann Ltd, Ed., London.
Rawer, K. and Suchy, K.: 1967, in *Handbuch der Physik* **49**, Springer-Verlag, Berlin.

THE USE OF VLF RADIO WAVES IN
IONOSPHERIC RESEARCH

G. BJÖNTEGAARD and A. EGELAND

The Norwegian Institute of Cosmic Physics, Oslo, Norway

1. Introduction

Due to the long wavelength, radio waves from VLF (3 to 30 kHz) transmitters (which radiate a few tens kW) may propagate halfway around the earth. Since, in addition, the phase is very stable for propagation in the earth-ionosphere waveguide, VLF waves have been extensively used for studying the upper atmospheric parameters as listed in Table I.

The ionospheric parameters that determine the propagation condition are the geomagnetic field, the collision frequency (v), and the electron and ion concentration (N_e, N_i). Geomagnetic irregularities are not directly detectable on VLF records. VLF propagation, however, is sensitive to changes in electron distribution and collision frequency in the lower ionosphere, i.e., the altitude interval from 50 to 100 km. In the earth-ionosphere waveguide it is customary to consider v to be known (Thrane and Piggott, 1966), whereas $N_e(h)$ is the free variable. The mass of the ions is too large to influence VLF propagation in the lower ionosphere during normal conditions (Thomas, 1969).

2. Propagation in the Earth-Ionosphere Waveguide

2.1. Propagation over paths less than 1000 km

The propagation is usually described by geometrical optics. The geometrical optics approximation requires; (a) that the geometrical dimension of the waveguide is large compared to the wavelength, and (b) that the waveguide is plane. Neither of these conditions are well satisfied for VLF waves, but the approximation is fairly good up to about 600 km (Björntegaard, 1972a).

Four standard reflection and conversion coefficients ($_\parallel R_\parallel$, $_\perp R_\perp$, $_\parallel R_\perp$, and $_\perp R_\parallel$) (Budden, 1961) have been used to describe ionospheric reflection. Most recent calculations of reflection coefficients have used the 'full wave solution' technique described by Pittaway (1965). For given values of angle of incidence, magnetic field, wave frequency, collision frequency, and electron density profiles, the complex reflection and conversion coefficients are calculated for a plane wave incident on a plane, stratified ionosphere. By Pittaway's technique it is also possible to calculate the wave's electromagnetic fields as a function of altitude. These calculations clearly show that VLF waves are reflected from a broad (10 to 20 km) region in the lower ionosphere.

B. M. McCormac (ed.), Physics and Chemistry of Upper Atmospheres, 298–305. All Rights Reserved.
Copyright © 1973 by D. Reidel Publishing Company, Dordrecht-Holland.

TABLE 1

Schematic picture of the parameters measured by different VLF techniques

	Electron density in regions:					Resolution		Coverage	
	C–D	E	F	Above		Time	Space	Time	Space
Earth-Ionosphere waveguide:									
Short paths (200 < d > 1000 km)	—	—				XXXX	XX	XXXX	X
Long paths (d > 1500 km)	—	—				XXXX	X	XXXX	XXX
Inversion (see text)	—	—				XXXX	XXX	XX	XXX
Whistler mode:									
VLF Doppler	—	—	—			XXXX	XXXX	X	X
Electron whistlers				—	$\}$ H^+ and He^+ Density, Temp. and	XXX	XX	X	XX
Ion whistlers				—	$\}$ Magnetic field	XXX	XX	X	XX

The length of the line indicates the coverage in various regions. The X density indicates the quality of the data (4X's are highest).

Deeks (1966) used this technique to calculate electron density profiles from observations by Bracewell *et al.* (1951) and Belrose (1957). Electron density profiles which could explain the observed field strengths for different seasons and solar activity were obtained by trial and error. Some of the profiles deduced are shown in Figure 1.

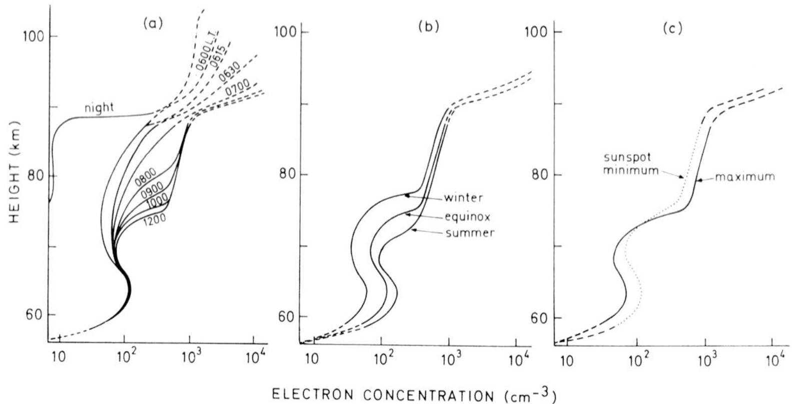

Fig. 1.   The height distribution of electrons parametric in:
(a) LT, (b) season, and (c) solar activity (Deeks, 1966).

Deeks concluded that at most heights the electron concentrations could not be altered by more than $\pm 20\%$ from the distributions in Figure 1 and still account for the observations. Bain and May (1967) have done some similar VLF modeling. Considering that it is difficult to obtain electron densities in the lower ionosphere, these kinds of calculations are valuable and must be regarded as some of the best for the D region. With refined techniques, summing up the contributions from larger areas of the reflecting ionosphere including horizontal gradients, even more accurate models can be set up (Bjöntegaard, 1972b).

VLF techniques also have the advantages of being inexpensive and producing continuous data. It is therefore possible to study, for example, the 'build-up' process (cf. Figure 1) during sunrise. This is important in order to understand electron production and loss mechanisms. For moderate density variations, the recorded VLF phase can, through minimum calculations, be interpreted as changes in reflection heights. But if the density profile is rather 'curly' or the gradient changes noticeably it might be misinterpreted.

2.2. Propagation over longer paths

For distances longer than 1000 km the waveguide mode theory is used (Budden, 1961; Wait, 1962). Each mode can be characterized by a standing wave pattern in the waveguide which fulfills the boundary conditions in the waveguide. For distances longer than 3000 km the propagation can be well approximated by the first order mode.

At night the second order mode can give a significant contribution (Belrose, 1968). By including the first 6 to 8 modes, propagation can be described down to 300 to 400 km.

The measured relative phase and amplitude can then be interpreted as changes in electron density profiles. It should be kept in mind that proposed changes are not necessarily the only solution.

Again, simplifications can be done if exponential electron density profiles are anticipated. It is usual to assume that the gradient stays approximately constant during night and day. Phase and attenuation changes can then be interpreted directly as changes in reflection height when horizontal density gradients are disregarded (cf. e.g., Chilton *et al.*, 1963). Long paths can thus be used to study phenomena with large horizontal extent (comparable to path length).

3. Normal VLF Propagation Conditions

The periodic variations of the ionosphere such as diurnal, seasonal, and variations with solar cycle will be briefly mentioned. VLF waves are well suited for those kinds of long term studies due to continuous recording.

The diurnal variations of phase and amplitude for VLF waves usually indicate reflection heights of about 70 km during the day and 90 km at night (Straker, 1955; Egeland and Riedler, 1964). At the reflection height the electron concentration is from 50 to 500 cm^{-3}. VLF waves are thus reflected from the D layer during the day and from the bottom of the E layer at night.

For N–S paths, where the sun rises and sets at approximately the same time all along the path, relative phase as a function of solar zenith angle is obtained. One can therefore study the time variation of phase velocity and thereby the reflection height as a function of zenith angle. By an inversion procedure (Pierce, 1968; Bjöntegaard, 1972b) phase velocity as a function of solar zenith angle can be derived even if the path is not N–S. By trial and error, electron density profiles which fit the phase velocity variations can be obtained. The density profiles obtained in this way fit well with similar curves obtained by cross modulation (cf. Bjöntegaard, 1972b).

Long term variations of VLF amplitudes and reflection coefficients have been studied by Belrose (1968).

4. Variations During Disturbed Ionospheric Conditions

VLF waves are well suited for detecting disturbances in the lower ionosphere. The strength of VLF is to study statistically parameters like:
(a) Start time relative to the ionizing source onset. (b) Distribution of events with season and solar cycle. (c) Distribution of events over geographic locations. (d) Strength and development of events with the same parameters as (a)–(c).

The weakness of the VLF approach is again the difficulty of obtaining quantitative information of ionospheric parameters for the single events.

Disturbances like PCA, SID, and SPA have been investigated in detail (cf. Wester-lund *et al.*, 1969).

4.1. Effects of X-rays and X-ray stars

Figure 2 shows the phase shift observed on August 2, 1967, on the path of NAA-TOKYO together with the observed flux of X-rays in different frequency ranges.

The correlation between the shape of the X-ray curves and the phase curve is striking. The onset of the SPA and X-ray bursts corresponds to within a minute whereas the peak of the SPA is delayed by 5 to 10 min. The main reason for this time

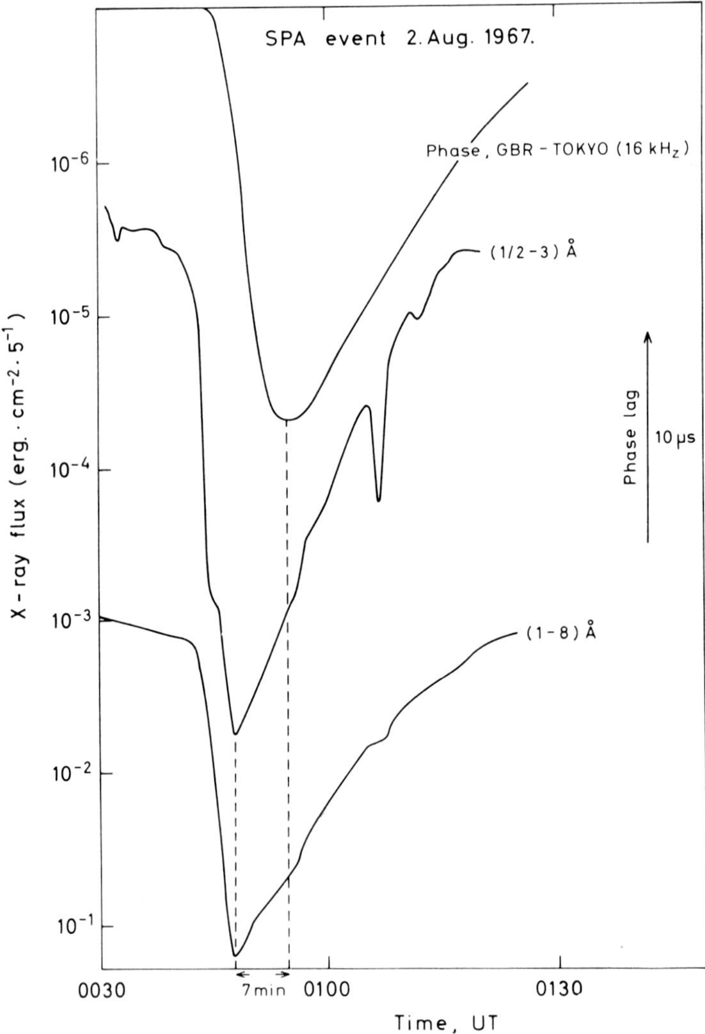

Fig. 2.   VLF phase advance together with measured X-ray fluxes outside the ionosphere in two frequency bands (data from Explorer 37).

lag is introduced by recombination of ions and electrons in the lower ionosphere. This relaxation time has been used extensively to estimate the effective recombination coefficient, α_{eff} (cf. e.g., Mitra, 1966).

Recently, Svenneson *et al.* (1972) investigated the effects of X-ray stars on VLF signals. They found that X-rays in the frequency range 2 to 16 Å may contribute significantly to the ionization at 80 to 90 km altitude during night when the star is around midnight culmination.

This is the kind of problem where VLF measurements show their strength. Those effects can only be studied on a long term basis.

5. Propagation in the Whistler Mode

5.1. VLF Doppler measurements

This is a method for evaluating the electron density in the D, E, and F regions by measuring the Doppler shift in a rocket of VLF waves transmitted from the earth. When the rocket parameters are known, the Doppler shift is a unique function of the phase velocity and thereby the electron density. This method, therefore, gives the local electron density vs. altitude. Figure 3A shows the electron density obtained by Egeland *et al.* (1970) using this VLF Doppler technique during quiet conditions.

The same technique has been used to obtain small scale variations in the electron density during a PCA event. An example of the VLF Doppler shift and the corresponding electron density is shown in Figure 3B.

5.2. Whistlers as diagnostic tool for the magnetosphere

By considering 'cold' electrons (magneto-ionic theory) and assuming further that quasi-longitudinal approximation is valid and that collisions can be neglected, it can be shown that the whistler travel time is given by

$$T = \frac{1}{2c} \int \frac{\omega_p \cdot \omega_c}{\sqrt{\omega(\omega_c - \omega)^3}} \, ds,$$

where ω is the wave frequency, ω_c is the gyrofrequency, and ω_p is the plasmafrequency. T has a minimum for $\omega = \omega_c/4$ (the nose frequency). ω_{nose} is a measure of the minimum magnetic field along the path. If we assume that the whistlers observed have traveled along one field line, the nose frequency gives approximately the location of the path. The dispersion curve will then give ω_p and the electron density in the regions around the magnetic equator can be deduced.

This is mainly the idea behind the work of Carpenter and Smith (1964) on the electron density in the plasmapause and the 'knee' outside (cf. also Carpenter, 1970). The work continues with studies of the variation in time of the plasmapause which is interpreted in terms of sunward surges of the bulge plasma during substorms.

Ion whistlers are observed in the ELF (30 to 3000 Hz) band and are therefore strongly influenced by resonances due to light ions. Stix (1962) describes the propagation of whistlers in a plasma with more than one ion species. According to him there

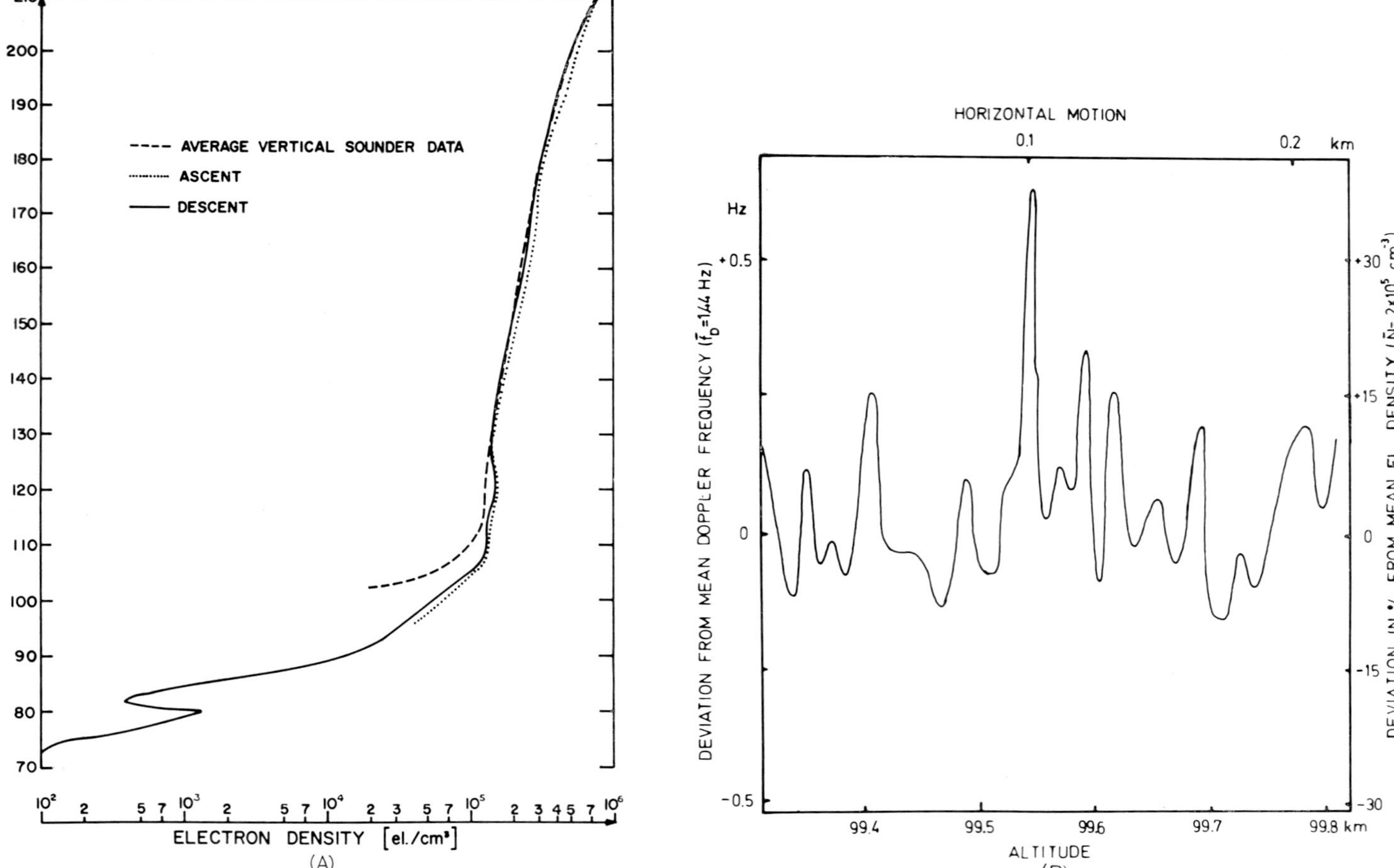

Fig. 3.   Electron density profiles deduced from VLF Doppler technique. (A) is deduced from a rocket launched from Wallops Island, and (B) shows the fine structure from a rocket launched at Andöya (Egeland *et al.*, 1970, 1972).

exists a frequency band below each ion gyrofrequency where propagation in the left hand mode (ion cyclotron mode) is possible. Moreover, there exists a frequency (crossover frequency) between every two ion gyrofrequencies where both the right hand polarized whistler mode and the ion cyclotron mode are linearly polarized so that coupling can take place. The magnitude of the crossover frequency depends on the ratio between the concentrations of the ions and electrons.

The absorption of the proton whistlers – that is the gap between the cutoff frequency and the gyrofrequency – is a function of the proton temperature. Gurnett and Brice (1966) found in this way proton temperatures ranging from 600 to 1050 K with inaccuracies of ± 100 K. Quemada *et al.* (1970) pointed out that those values are somewhat small due to poor approximation methods in calculating the dispersion curve.

References

Bain, W. C. and May, B. R.: 1967, *Proc. IEE* **114**, 1593.
Belrose, J. S.: 1957, Ph.D. Thesis, Cambridge University, England.
Belrose, J. S.: 1968, Paper, AGARD Lecture Series XXIX.
Björntegaard, G.: 1972a, Paper 6-1, in *Proceedings from the VLF-Symposium*, Sandefjord, Norway, NICP Report 7201.
Björntegaard, G.: 1972b, Paper 10-1, in *Proceedings from the VLF-Symposium*, Sandefjord, Norway, NICP Report 7201.
Bracewell, R. N., Budden, K. G., Ratcliffe, J. A., Staker, T. W., and Weekes, K.: 1951, *Proc. IEE* **98**, 221.
Budden, K. G.: 1961, *The Wave-Guide Mode Theory of Wave Propagation*, Logos Press, New York.
Carpenter, D. L.: 1970, *J. Geophys. Res.* **75**, 3837.
Carpenter, D. L. and Smith, R. L.: 1964, *Rev. Geophys.* **2**, 414.
Chilton, C. J., Steele, F. K., and Norton, R. J.: 1963, *J. Geophys. Res.* **68**, 5421.
Deeks, D. G.: 1966, *Proc. Roy. Soc.* **A29**, 413.
Egeland, A. and Riedler, W.: 1964, *J. Atmospheric Terrest. Phys.* **26**, 351.
Egeland, A., Björntegaard, G., and Aggson, T. L.: 1970, *J. Atmospheric Terrest. Phys.* **32**, 1191.
Egeland, A., Paulson, K. V., Naustvik, E., Karlsen, N., and Russel, B. H.: 1972, Paper 14-1, in *Proceedings from the VLF-Symposium*, Sandefjord, Norway, NICP, Report 7201.
Gurnett, D. A. and Brice, N. M.: 1966, *J. Geophys. Res.* **71**, 3639.
Mitra, A. P.: 1966, *Space Res.* **6**, 558, MacMillan, London.
Pierce, J. A.: 1965, *IEEE Trans. Aerospace Electronic Systems*, **AES-1**, 206.
Pierce, J. A.: 1968, 'Measurement and Prediction of Group Velocity at Very Low Radio Frequencies', Harvard Tech Rep. 535.
Pittaway, M. L. V.: 1965, *Phil. Trans. Roy. Soc.* **257**, 219.
Quemada, D., Velut, P. M., and Vigneron, J.: 1970, *Plasma Waves in Space and Laboratory*, Part 2 p. 427.
Stix, T. H.: 1962, *The Theory of Plasma Waves*, McGraw Hill, New York.
Straker, T. W.: 1955, *Proc. IEE*, Monograph No. 114.
Svennesson, J., Reder, F., and Cronchley, J.: 1972, *J. Atmospheric Terrest. Phys.* **34**, 49.
Thomas, L.: 1969, *J. Atmospheric Terrest Phys.* **31**, 991.
Thrane, E. V. and Piggott, W. R.: 1966, *J. Atmospheric Terrest. Phys.* **28**, 721.
Wait, J. R.: 1962, *Electromagnetic Waves in Stratified Media*, Pergamon Press, London.
Westerlund, S., Reder, F. H., and Åbom, C.: 1969, *Planetary Space Sci.* **17**, 1329.

PART V

OTHER PLANETS

THE ATMOSPHERE OF MARS

JOHN C. McCONNELL*

Harvard University, Cambridge, Mass., U.S.A.

1. Introduction

The main emphasis of this review will be on Martian aeronomy, here defined to be the interaction of solar UV radiation with the atmosphere of Mars. A general outline of contemporary knowledge of the chemical composition and of the thermal structure of the Martian atmosphere, necessary for aeronomical calculations, is given in Sections 2 and 3 respectively. The aeronomy of the upper neutral atmosphere and ionosphere is discussed in Section 4. The problems of the stability of the Martian atmosphere and H_2O escape are reviewed in the section on the lower atmosphere (Section 5).

More general reviews of the Martian atmosphere are included in the recent articles by Ingersoll and Leovy (1971) and Hunten (1971). A pre-Mariner survey of the Martian atmosphere is included in the review by Rasool (1963). Reviews of the dynamics of planetary atmospheres, which include discussions specific to Mars, have been given recently by Goody (1969), Gierasch (1970), and Gierasch *et al.* (1970).

In this article rate coefficients for a given reaction will be denoted by k with the reaction number as a subscript.

2. Atmospheric Composition

2.1. CO_2

Prior to 1963 it was considered that Mars had a surface pressure of 85 ± 10 mb, and that the main constituent of the atmosphere was probably N_2. CO_2 was then believed to be a minor constituent. The present consensus of earth-based spectroscopy indicates that the atmosphere of Mars is mainly CO_2 with a partial pressure of 5.5 ± 0.8 mb (Belton *et al.*, 1968; Giver *et al.*, 1968; Carleton *et al.*, 1969). This value is based on an analysis of the vibrational-rotational bands of CO_2 at 1.05 and 1.038 μm. The total pressure of atmospheric gases has also been determined by earth-based spectroscopy and radio occultation data to be such that the fractional abundance of CO_2 is between 0.8 and 1.0 (Kliore *et al.*, 1965, 1972; Kaplan *et al.*, 1969; Rasool *et al.*, 1970).

The above pressure is an average value. Measurements of CO_2 pressures as a function of planetary location indicate that the CO_2 pressures are variable. This is due to large scale topographical variations as first indicated by the radar measurements (Pettingill *et al.*, 1969; Rogers *et al.*, 1970). Measured variations in pressure have been inverted to yield topographical information. Successful experiments include pressure measurements using the 1 μm CO_2 band (Belton and Hunten, 1969, 1971; Wells,

* Now at York University, Downsview, Ontario, Canada.

1969), 2 μm CO_2 band (Herr *et al.*, 1970) and the UV reflection spectrum (Barth and Hord, 1971; Pang and Hord, 1971; Hord, 1972).

On consideration of the heat balance of Mars, Leighton and Murray (1966) suggested that the polar caps were comprised mainly of CO_2. This was confirmed by the measurement of Mariner 7, at the polar cap, of a surface temperature of 148 K, the temperature at which the partial pressure of CO_2 is 6.5 mb (Neugebauer *et al.*, 1969, 1971). Leighton and Murray (1966) also suggested that the polar cap may contain more CO_2 than is in the atmosphere. Due to the observed seasonal variation of the polar caps this could imply a variation in the partial pressure of CO_2 of several millibars. This result has not been confirmed.

2.2. CO

Kaplan *et al.* (1969) detected CO in the Martian atmosphere by the appearance of the (2, 0) and (3, 0) bands of the principal isotope in a high resolution Martian spectrum. They obtained a column abundance of 5.6 cm atm using the (3, 0) band which implies a CO/CO_2 volume mixing ratio (volume/volume) of 8×10^{-4}. Young (1971), using the (2, 0) band obtained $12(+8, -6)$ cm atm. Carleton and Traub (1972) remeasured the (3, 0) band from the Connes atlas and obtained 7.3 ± 1.0 cm atm. They also pointed out that the recent measurements of line strength and broadening parameters by Tubb and Williams (1972) indicate that the value of Young should be lowered.

2.3. O_2

Absorption due to Martian O_2 has recently been observed in the reflected solar spectrum at 7635 Å by several investigators (Carleton and Traub, 1972; Barker, 1972). The observations were made when the relative velocities of Mars and the earth were large enough to produce a Doppler shift of ± 0.34 Å or more. Carleton and Traub (1972) obtained 10.4 ± 1.0 cm atm O_2 in agreement with the results of Barker (1972) of 9.5 ± 0.6 cm atm. These measurements are consistent with the earlier lower limit of O_2 abundance given by Belton and Hunten (1968). The measured O_2/CO ratio is thus about 1.4.

2.4. H_2O

Lines attributed to H_2O absorption have been observed in the solar reflection spectrum near 8200 Å. As for O_2 the measurement must be made when the planetary Doppler shift is at a maximum, due to the strong absorption by terrestrial H_2O. The amount observed is variable, usually between 10 and 40 μm precipitable water ($1 \mu m = 3.3 \times 10^{18}$ molecules cm^{-2}) and frequently falls below the detection threshold of about 5 μm (Schorn *et al.*, 1969; Schorn, 1971). If it is assumed that the H_2O is confined to the first few kilometers, then the mixing ratio of H_2O/CO_2 is about 0.1%.

2.5. O_3

When passing over the south polar cap in 1969 the Mariner 7 UV spectrometer observed an absorption feature at 2550 Å which is most simply explained by the presence of about 10^{-3} cm atm of O_3 (Barth and Hord, 1971). No O_3 was detected over the rest of the planet. The laboratory experiments of Broida *et al.* (1970) suggest that this amount of O_3 may be trapped on the polar cap and need not be present in the atmosphere. Present upper limits on the abundance of O_3 in the Martian atmosphere are set from the reflection spectrum in the region 2550 to 3500 Å obtained by a rocket-borne telescope spectrometer (Broadfoot and Wallace, 1970). The upper limit for the O_3 abundance they obtained is 2×10^{-4} cm atm.

The UV experiment on the Mariner 9 orbiter (Lane *et al.*, 1972) has observed O_3 due to its absorption in the Hartley continuum. They did not observe O_3 over the south polar cap initially. However they did detect O_3 absorption northward of 45° latitude yet south of the north polar cap. These results are suggestive that the observed absorption may be due to O_3 in the atmosphere, rather than trapped on the surface.

2.6. N_2

To date N_2 has not been observed, although the Mariner 6 and 7 UV experiment allows an upper limit of about 5% to be placed on the abundance of N_2 (Dalgarno and McElroy, 1970) since the Vergard-Kaplan bands of N_2 were not detected by the Mariner spectrometer. A similar result is obtained assuming that less than one half of the 1356 Å O emission is due to the 1354 3-0 band of the LBH N_2 system (see later).

2.7. H_2

Molecular hydrogen has not been detected as yet. However an investigation of the H_2O escape problem (Hunten and McElroy, 1970; McElroy and Donahue, 1972) allows fairly stringent limits to be placed on H_2. These results indicate that the mixing ratio of H_2 is of the order of 5×10^{-5}.

2.8. ARGON

On the earth, Ar is a product of radioactive decay. Assuming that the rate of production of Ar on Mars is similar to that of the earth, one might expect a partial pressure of Ar to be about 25% of that of CO_2.

2.9. OTHER CONSTITUENTS

Various limits to other constituents are given in the review by Kuiper (1952). The most definitive upper limit on NO_2 is that of Marshall (1964). From the uncertainty in the total pressure it is possible that 20% Ar could be present in the atmosphere.

At first sight, Mars contains much more CO_2 relative to H_2O and N_2 than the earth. However, this applies only when the atmospheric balance sheet of the earth is considered. On the earth, CO_2 is precipitated as carbonate. Estimates (Rubey, 1951) of the total amount of CO_2 precipitated on the earth yield 70 ± 30 kg cm^{-2} while Mars has 17 ± 0.2 gm cm^{-2} in the atmosphere. On the earth the ratios (by volume) of H_2O/CO_2 and H_2/CO_2 are 44 and 0.07, respectively. If the atmospheres of Mars and the earth have a similar origin, then one would expect both of these ratios on Mars to be similar to those on the earth. The above upper limit on N_2 is just compatable with such a view (but see also Brinkman, 1971; McElroy, 1972a). When allowance is made for the escape of H_2O over 4.5×10^9 yr, the H_2O/CO_2 ratio for Mars suggests that the planets earth and Mars may have had similar relative out-gassing rates (McElroy, 1972a).

3. Temperature Structure

3.1. Introduction

A knowledge of the temperature structure of an atmosphere, and the mean molecular mass, enables the number density of the atmosphere to be specified at each altitude. There have been many theoretical calculations of the temperature structure of Mars (cf. Hunten, 1971; Ingersoll and Leovy, 1971).

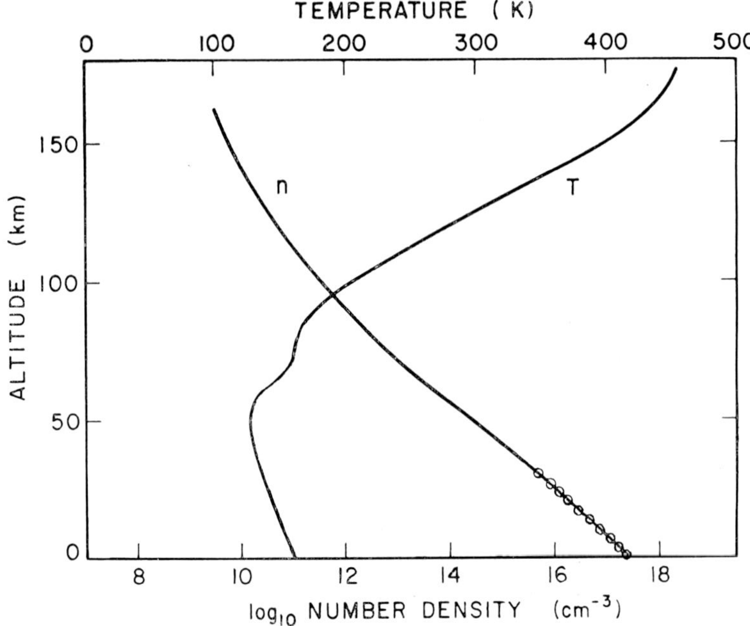

Fig. 1.   A pure CO₂ model of the Martian atmosphere is compared with Mariner 4 measurements (McElroy, 1969).

The main features of these calculations are heating by absorption of solar UV radiation by CO_2 in the upper atmosphere, and by absorption of the solar near IR radiation by the 2.7 and 4.3 μm bands of CO_2. Redistribution of energy is affected by allowance for molecular conduction and radiation transfer in the 15 μm band of CO_2. The steady state and radiative equilibrium assumptions are normally made. A typical temperature profile taken from McElroy (1969), which is in fair agreement with the available experimental data, is shown in Figure 1.

3.2. UPPER ATMOSPHERE

The temperature of the upper atmosphere may be inferred from measurements of ionospheric and airglow profiles (see later). Present estimates indicate that the exospheric temperature may vary from 300 to 400 K. (cf. Barth *et al.*, 1972). The exospheric temperature is sensitive to the heating efficiency of solar photons capable of ionizing CO_2 ($\lambda < 900$ Å). On the basis of the then available experimental data, Henry and McElroy (1968) calculated that the neutral heating efficiency of this radiation was quite high $\sim 55\%$. At that time it was not known that CO_2 was an important source of planetary airglow. An analysis of the Mariner 6 and 7 airglow data by Stewart (1972) showed that much of the incoming photon energy is carried away as airglow and that the heating efficiency is probably between 0.2 and 0.3 over the wavelength range of interest. A similar heating efficiency for a CO_2 atmosphere has been adduced by Dickinson (1971) in a study of the circulation and thermal structure of the Venusian thermosphere.

In spite of the apparent agreement of these studies, it is still not clear how complete our knowledge of the heating in the thermosphere is, since both these studies used solar fluxes which, on the basis of ionospheric data, may be too low by a factor of 3 (R. W. Stewart, 1971; A. I. Stewart, 1972). If the solar fluxes are higher, this would imply that the heating efficiency is of the order of 0.1 which seems to be an impossibly low value. It is possible that this dilemma may be due to the neglect of cooling by eddy conduction. It has been suggested that this process is important on the earth (Johnson and Wilkins, 1965; Johnson and Gottlieb, 1970), and also on Mars (Johnson, 1968). On the basis of a comparison of thermal relaxation time constants and eddy diffusion time constants, McElroy (1967) concluded that eddy conduction would not be important unless eddy diffusion coefficients K, exceeded 10^8 cm^2 s^{-1}. Since it now appears that such K's are required in the thermosphere (McElroy and McConnell, 1971a) perhaps eddy cooling is the answer to the thermospheric heat budget problem.

3.3. LOWER ATMOSPHERE

The ratio S-band occultation experiments of the Mariner 4, 6, 7, fly-bys and the Mariner 9 orbiter provides information on the temperature and pressure of the lower atmosphere of Mars (Kliore *et al.*, 1965, 1972; Fjeldbo *et al.*, 1970; Rasool and Stewart, 1971). Figure 2 shows the atmospheric data from Mariners 6 and 7. A comparison of the atmospheric surface temperature with the ground temperature measured by the Mariner IR radiometer experiment (Neugebauer *et al.*, 1969, 1971) indicates an air-ground temperature discontinuity. The study by Gierasch and Goody (1968) shows that the discontinuity arises because the Martian atmosphere is optically thin and radiatively decoupled from the ground. The dominant process for heat exchange is by turbulent transfer during daylight hours.

A feature of the temperature profiles shown in Figure 2, not satisfactorily reproduced by any theory to date, is the very cold temperatures above about 20 km (Rasool and Stewart, 1971). However, this is also the region where the occultation data are the most inaccurate due perhaps to the persistence of a Martian D region

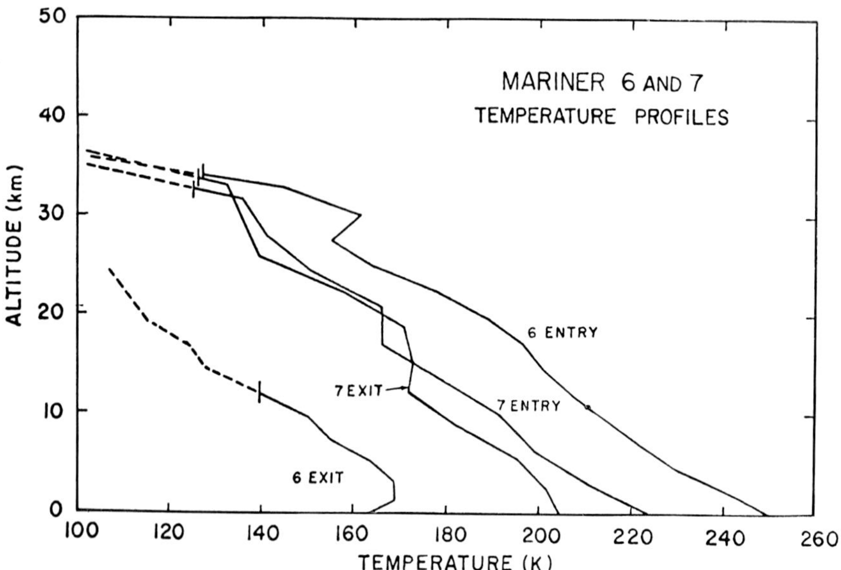

Fig. 2. Vertical temperature distributions in the atmosphere of Mars at the four occultation points of Mariners 6 and 7. The temperature near the ground should be accurate to within 5 K while near 30 km the estimated error is ± 20 K (Rasool *et al.*, 1970).

(Hunten, 1968; Rasool *et al.*, 1970). Nevertheless there are other data which tend to support the existence of a region cold enough so that condensation of CO_2 occurs. The Mariner 6 and 7 IR radiometer experiments recorded a reflection spike at 4.3 μm on crossing the bright limb. For the Mariner 7 crossing they observed the spike, which they attributed to solid CO_2, at about 25 ± 7 km above the limb (Herr and Pimentel, 1970). In addition, the Mariner 6 and 7 television pictures show thin detached hazes above the limb from 5 to 50 km. The lowest hazes were probably not frozen CO_2, but it is quite possible that the higher hazes were frozen CO_2 (Leovy *et al.*, 1971). Preliminary results from Mariner 9 show a thin detached haze at about 60 km above the surface (Masursky *et al.*, 1972).

The initial results of atmospheric temperature obtained from the Mariner 9 S-band occultation experiment (Kliore *et al.*, 1972) and the IR spectroscopy experiment (Hanel *et al.*, 1972) showed that during the dust storm the atmospheric temperature was almost isothermal. This result appears to be most simply explained by assuming that about 10% of the incoming solar radiation is absorbed in an atmospheric dust layer rather than on the ground (Gierasch and Goody, 1972).

4. Upper Atmosphere

4.1. INTRODUCTION

All our knowledge of the upper atmosphere of Mars (> 90 km) is derived from interpretations of the UV airglow experiment of Barth and his associates (Barth *et al.*, 1969,

1971, 1972) and the S-band radio occultation experiments (Kliore *et al.*, 1965, 1972; Fjeldbo *et al.*, 1970; Rasool and Stewart, 1971). The interpretation of the airglow emissions observed by the UV spectrometer is discussed in Section 4.2. In Section 4.3 the ionospheric results are reviewed.

4.2. NEUTRAL ATMOSPHERE – AIRGLOW

Figure 3 shows a typical Mariner 6 and 7 spectrum (Barth *et al.*, 1971). The main features observed were the CO_2^+ A and B bands, the 1304, 1356, and 2972 lines of O, the CO fourth positive and cameron bands systems, the CO^+ first negative bands, the 1657 and 1561 lines of C and the H 1216 Å Ly-α line. No emissions were observed due to N_2 or NO.

In the region of the atmosphere in which the airglow was observed it was expected that emissions associated with constituents such as O and CO would exhibit a scale

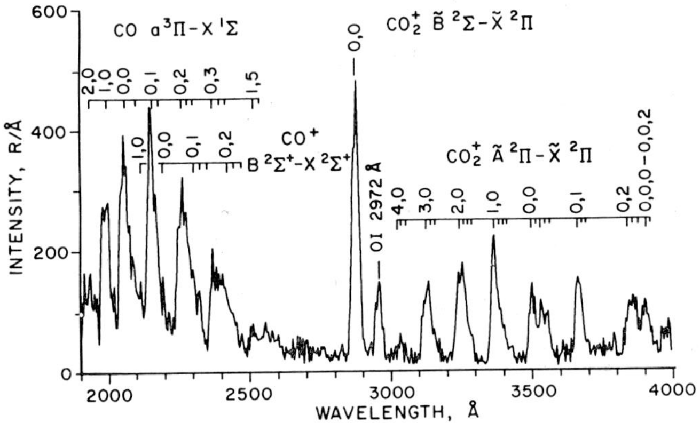

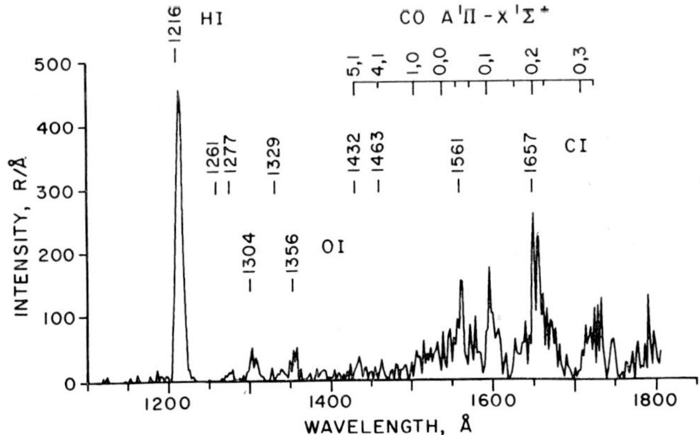

Fig. 3. Ultraviolet spectra of the upper atmosphere of Mars, 1100–1800 at 10 Å resolution and 1900–4000 at 20 Å resolution. The spectra shown are the result of the summation of 4 individual observations (Barth *et al.*, 1971).

height dependence characteristic of the mass of the emitting molecule. However, from the observed height variation of some of the O and CO emissions it was apparent that they were due mainly to electron and photon impact on CO_2. This result enables limits to be placed on the abundance of O and CO in the upper atmosphere as discussed below. The conclusion is that the upper atmosphere of Mars is essentially undissociated. This confirms earlier hypotheses of an undissociated CO_2 upper atmosphere (McElroy, 1967; Cloutier *et al.*, 1969) which, however, were based on results that were by no means universally accepted.

The main excitation mechanisms in the upper atmosphere of Mars are resonance scattering and fluorescence,

$$X + hv \rightarrow X^* \rightarrow X + hv'$$

where $X = CO$, CO_2^+, O, and H; dissociative excitation by electron and photon impact,

$$CO_2 + hv \rightarrow CO^* + O' \tag{1}$$
$$CO_2 + e \; \rightarrow CO^* + O' + e \tag{2}$$
$$CO_2 + hv \rightarrow C^* \; + O' + O'' \tag{3}$$
$$CO_2 + e \; \rightarrow C^* \; + O' + O'' + e \tag{4}$$

where the asterisks and primes indicate that the products may be in excited states; excitation and ionization

$$CO_2 + hv \rightarrow CO_2^{+*} + e \tag{5}$$
$$CO_2 + e \; \rightarrow CO_2^{+*} + 2e; \tag{6}$$

recombination

$$CO_2^+ + e \rightarrow CO^* + O' \tag{7}$$
$$O_2^+ + e \rightarrow \; O^* + O'. \tag{8}$$

The electrons in Equation (1) to (6) are energetic photoelectrons produced by photo-ionization of CO_2 (Henry and McElroy, 1968). Figure 4 shows the variation of the CO_2 cross section and solar flux with wavelength. Also shown are some of the energy thresholds for processes described by Equations (1) to (6). In the following few paragraphs we discuss some of the more important emissions, and compare calculated profiles with the experimental data. A typical model atmosphere used, which is generally consistent with all of the data, is shown in Figure 5. Reference to any data not quoted in the text may be obtained from the articles by Dalgarno *et al.* (1970), McConnell and McElroy (1970), McElroy and McConnell (1971a, b), Thomas (1971), Stewart (1972), and Strickland *et al.* (1972).

The Mariner UV spectrometer viewed the bright limb of Mars tangentially (Barth *et al.*, 1971). The calculation of airglow intensities has taken this geometry into account. The intensities so calculated will be referred to, in the figures, as slant intensities.

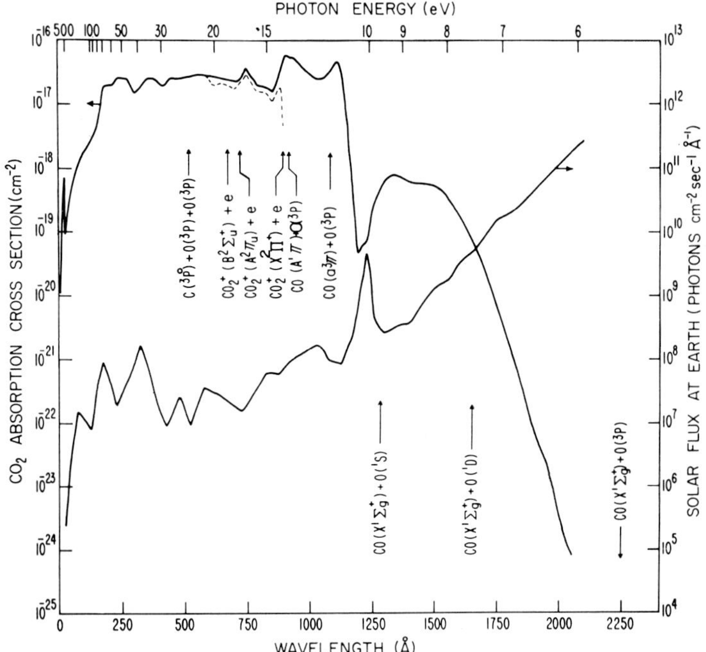

Fig. 4. Variation of the CO_2 absorption cross section and solar flux with wavelength. The solar flux has been averaged over 50 Å intervals. Also shown are the thresholds for various dissociative processes. The dotted line is the photoionization cross section for CO_2.

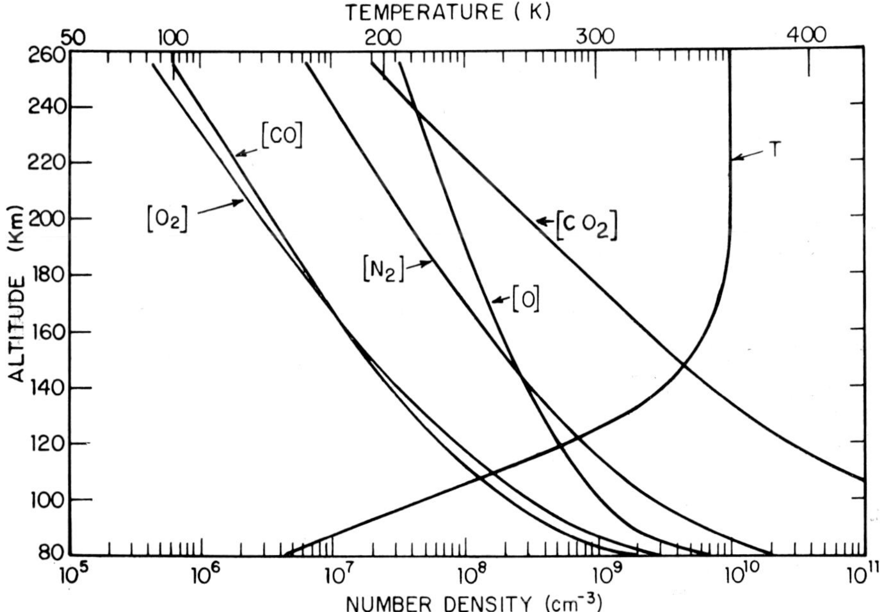

Fig. 5. Upper atmospheric model used in interpreting the airglow data similar to that used by McElroy and McConnell (1971a).

4.2.1. $CO(a^3\Pi - X^1\Sigma)$

An analysis of the cameron band emission provides information on the temperature
structure of the upper atmosphere, and potentially provides information on the
abundance of CO_2^+ in the ionosphere. A reasonable fit to the Mariner 9 data is shown
in Figure 6 obtained using the solar flux data of Hall and Hinteregger (1970) and
recent data obtained on the photo- and electron-excitation cross sections for $CO(a^3\Pi)$
(Freund, 1971; Lawrence, 1972; Wells *et al.*, 1972). It is seen that the processes given
by Equations (1), (2), and (7) are equally important. However, it is well to bear in mind

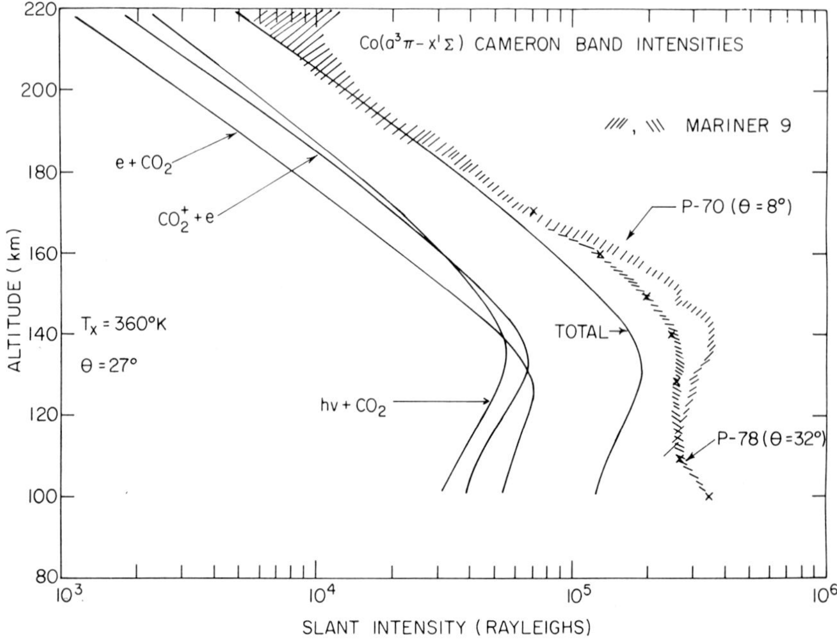

Fig. 6. A comparison of calculated CO cameron band slant intensities with the Mariner 9 results
(cf. Barth *et al.*, 1972). The solar zenith angle used in the calculations was 27.

that several uncertainties exist in interpreting the data. The major uncertainties are in
(a) the amount of solar EUV flux producing photoelectrons, (b) the probability that
process 7 yields a $CO(a^3\Pi)$ molecule (the calculation assumes unity), and (c) the
CO_2^+ densities. The Mariner 9 measurements of the $CO(a^3\Pi)$ intensities (Barth *et al.*,
1972) indicate that the exospheric temperature may vary from between 300 and 400 K.

4.2.2. $CO(A^1\Pi - X^1\Sigma)$

The relative contributions to the fourth positive system airglow are shown in Figure 7.
The photo and electron excitation data were estimated using the recent measurements
of Gentieu and Mentall (1972) and Mumma *et al.* (1971), respectively. In view of the
uncertainty in the solar fluxes, reasonable agreement is obtained.

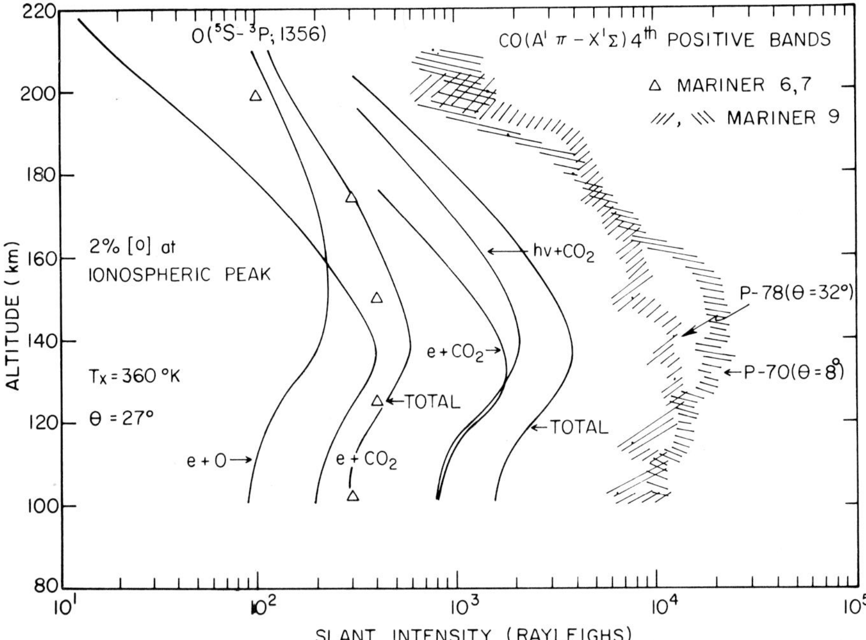

Fig. 7. CO fourth positive slant intensities calculated as described in the text. Also shown are measurements of Mariner 9. The 1356 Å O$_I$ emission calculated for 2 % O is compared with the Mariner 6 and 7 results (Barth *et al.*, 1971).

The information obtained thus far on the $CO(A^1\Pi)$ emission pertains only to the basic physical properties of CO_2. However, an analysis of the vibrational distribution of the airglow provides limits on the abundance of CO (Thomas, 1971). The upper limit Thomas obtained was 1% at 135 km. An absolute lower limit may be obtained by assuming that all the emission observed in the (1, 0) band is due to resonance fluorescence. This provides a lower limit to the CO mixing ratio of 0.3% at 135 km. These limits depend on the value of the solar flux in the region 1500 Å which is at present uncertain to within a factor of 3 (Widing *et al.*, 1970; Parkinson and Reeves, 1969).

4.2.3. CO_2^+ $(A^2\Pi-X^2\Pi)$

These emission processes were first investigated by Dalgarno and Degges (1971), and Dalgarno *et al.* (1970). Theoretical calculations of the CO_2^+ (A–X) emission are shown in Figure 8. It was assumed that the branching for the process given by Equation (5) in regions where no experimental data are available could be estimated by comparison with the known branching for the process in Equation (6) using the measurements of McConkey *et al.* (1968) and Rapp and Englander-Golden (1965). The importance of resonant fluorescence enables estimates of CO_2^+ densities to be made which in turn provides information on the mixing ratio of O in the upper atmosphere (Section 4.4). The deduced abundance is about 2% at 135 km (Stewart, 1972).

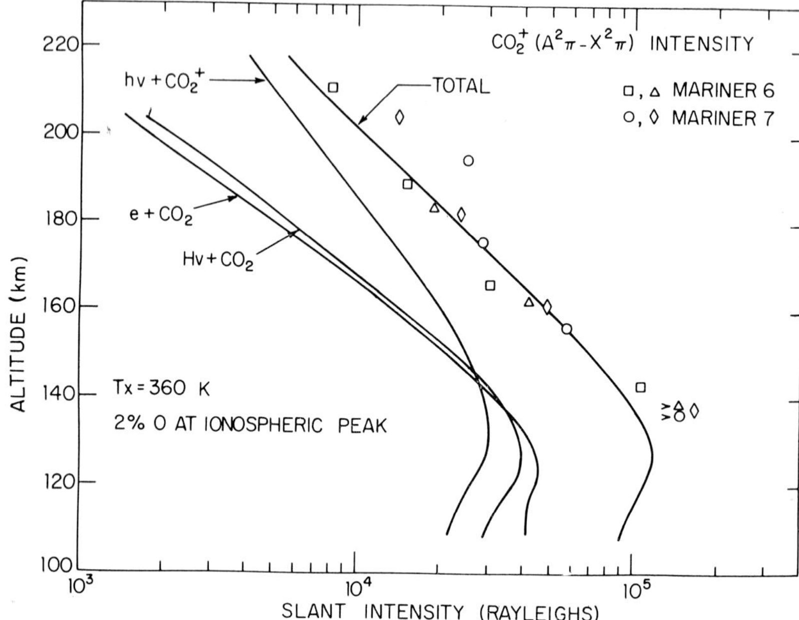

Fig. 8.   Calculated contributions to the CO$_2^+$ (A–X) slant intensity are compared with the Mariner 6 and 7 measurements.

4.2.4. OI(1304)

Although the 1304 Å OI triplet is produced by dissociative excitation, Equations (3) and (4), the main excitation process is resonance scattering of solar photons by O. There is also some contribution due to electron excitation of O but this is less important. The surprising result of the analysis of the Mariner 6 and 7 data is that the upper atmosphere of Mars is almost completely undissociated. The mixing ratio of O/CO$_2$ calculated at 135 km by McElroy and McConnell (1971a) was 0.5 to 1% for an atmosphere characterized by an exospheric temperature of 364 K. A more careful statistical analysis of the data by Strickland *et al.* (1972) yielded similar results. Bearing in mind the uncertainty in the measurements and theoretical estimates this result is in reasonable agreement with that deduced from the CO$_2^+$ emissions.

4.2.5. OI(1356)

The O(^{5}S) state is excited mainly by electron dissociative excitation of CO$_2$ and also by electron excitation of O as shown in Figure 7. The e–O cross section was estimated by assuming a shape, and then obtaining a magnitude by comparing theoretical and experimental terrestrial 1356 Å airglow profiles. The theoretical profiles were calculated using the analysis of Dalgarno *et al.* (1969). The shape of the 1356 Å CO$_2$ cross section adopted was that of Ajello (1971). On the basis of the Mariner airglow data, a cross section with a maximum 2×10^{-18} cm^2 at 100 eV was used. This value is somewhat smaller than that suggested by Wells *et al.* (1972).

The 1356 O$_I$ emission overlaps that of the (3, 0) 1354 Å band of the N$_2$ LBH system. Dalgarno and McElroy (1970) have calculated the intensity of the N$_2$ electron excited airglow expected in the Martian atmosphere. Using their estimates and assuming that, at most, half of the emission comes from N$_2$ (the rest from O and CO$_2$) we obtain an upper limit to the N$_2$ mixing ratio of 4%.

4.2.6. H(1216)

The UV spectrometer on Mariner 6 and 7 observed the 1216 Å H line from 24000 to 200 km. The radiative transfer analysis of Anderson and Hord (1971) implies an exospheric temperature of 350 ± 100 K and a density of H of $3 \pm 1 \times 10^4$ cm^{-3} at 250 km. The H escape rate associated with this temperature is 1.8×10^8 cm^{-2} s^{-1}. It is believed that the ultimate source of the H atoms in the upper atmosphere is photolysis of H$_2$O in the lower atmosphere (Hunten and McElroy, 1970).

4.3. STABILITY OF THE UPPER ATMOSPHERE

McElroy and McConnell (1971a) investigated the implications of the result of low CO and O abundances in the upper atmosphere. They calculated production rates of O and CO shown in Figure 9 and allowed for transport using molecular and eddy diffusion (cf. Colegrove *et al.*, 1966). The values required for eddy diffusion coefficients

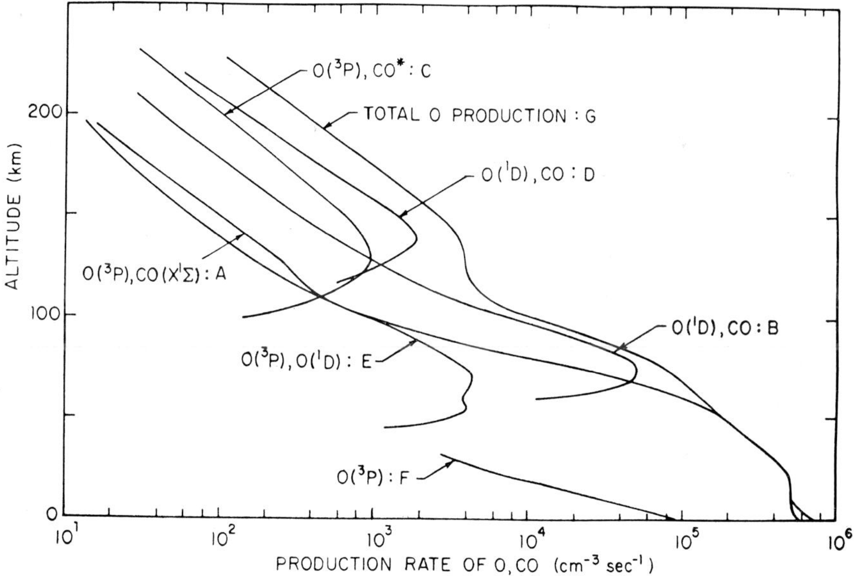

Fig. 9. Productions rates of O and CO on Mars for a solar zenith angle of 60° and with solar fluxes reduced by a factor of 2 for day-night averaging (McElroy and McConnell, 1971a). A: O(^{3}P) and CO($^1\Sigma$) from CO$_2$ by electron and photodissociation with $2270 > \lambda > 1670$ Å. B: O(^{1}D) and CO($^1\Sigma$) from CO$_2$ by photodissociation for $1670 > \lambda > 1080$ Å assuming that the O(^{1}S) produced for $\lambda < 1270$ Å radiates to O(^{1}D). C: O(^{3}P) and excited triplet states of CO from CO$_2$ by electron dissociation and by photolysis in the region $\lambda < 1080$ Å. D: O(^{1}D) and CO from ionospheric reactions. E: O(^{3}P) and O(^{1}D) from photolysis of O$_2$ for $\lambda < 1750$ Å. F: O(^{3}P) from photolysis of O$_2$ in the Herzberg continuum. G: Total O production rate.

were between 10^9 and 10^8 cm^2 s^{-1}, i.e., one or two orders of magnitude larger than those in vogue for comparable regions of the earth's atmosphere. At present such fast mixing processes are not understood, although the model of lateral transport of dissociation products of CO_2, applied to the Venusian thermosphere by Dickinson (1971), may be appropriate for Mars.

Such high mixing coefficients would not be required if there was an efficient *in situ* chemical scheme for recombining CO and O. Three body processes are much too inefficient in the upper atmosphere. At present no scheme is available to recombine CO and O *in situ*. The scheme proposed earlier using O^1D and a CO_3 intermediate (McElroy and Hunten, 1970) is no longer tenable due to more recent experimental data becoming available (Clark, 1971).

4.4. IONOSPHERE

The results from the UV airglow measurements imply that the upper atmosphere is mainly CO_2. This considerably simplifies the analysis of the electron density profiles obtained by the S-band occultation experiment (Kliore *et al.*, 1965, 1972; Fjeldbo *et al.*, 1970; Rasool and Stewart, 1971).

The main processes that occur in the upper atmosphere are photoionization and electron ionization of CO_2

$$h\nu\,(e) + CO_2 \rightarrow CO_2^+ + e\,(+\,e) \qquad \lambda \leqslant 900 \text{ Å } (13.7 \text{ eV}) \qquad (9)$$

$$\rightarrow CO^+ + O\ \ + e\,(+\,e) \qquad \lambda \leqslant 637 \text{ Å } (19.5 \text{ eV}) \qquad (10)$$

$$\rightarrow CO\ \ + O^+ + e\,(+\,e) \qquad \lambda \leqslant 650 \text{ Å } (19.1 \text{ eV}). \qquad (11)$$

As suggested by McElroy and Stewart (cf. Fehsenfeld *et al.*, 1970) the presence of O in the upper atmosphere changes CO_2^+ to O_2^+ via the series of ion-molecule reactions,

$$O\ \ + CO_2^+ \rightarrow O_2^+ + CO \qquad (12)$$

$$\rightarrow O^+ + CO_2 \qquad (13)$$

$$O^+ + CO_2\ \ \rightarrow O_2^+ + CO: \qquad (14)$$

CO^+ produced from dissociative ionization of CO_2 is rapidly converted to CO_2^+

$$CO^+ + CO_2 \rightarrow CO + CO_2^+ \qquad (15)$$

(McElroy, 1969, McElroy and McConnell, 1971b; Stewart, 1972). It should be noted that even if there is no O present, O^+ produced by dissociative ionization implies substantial amounts of O_2^+ as shown in Figure 10. CO_2^+ and O_2^+ are removed finally by recombination

$$CO_2^+ + e \rightarrow CO + O \qquad (16)$$

$$O_2^+ + e \rightarrow\ \ O + O. \qquad (17)$$

Figure 10 shows the change in the density of the major ions O_2^+ and CO_2^+ at a fixed height with variation in the O mixing ratio at 135 km. Thus, knowledge of CO_2^+ densities coupled with an electron density profile enables an estimate of the amount of O present in the upper atmosphere to be made. As mentioned earlier, the mixing ratio

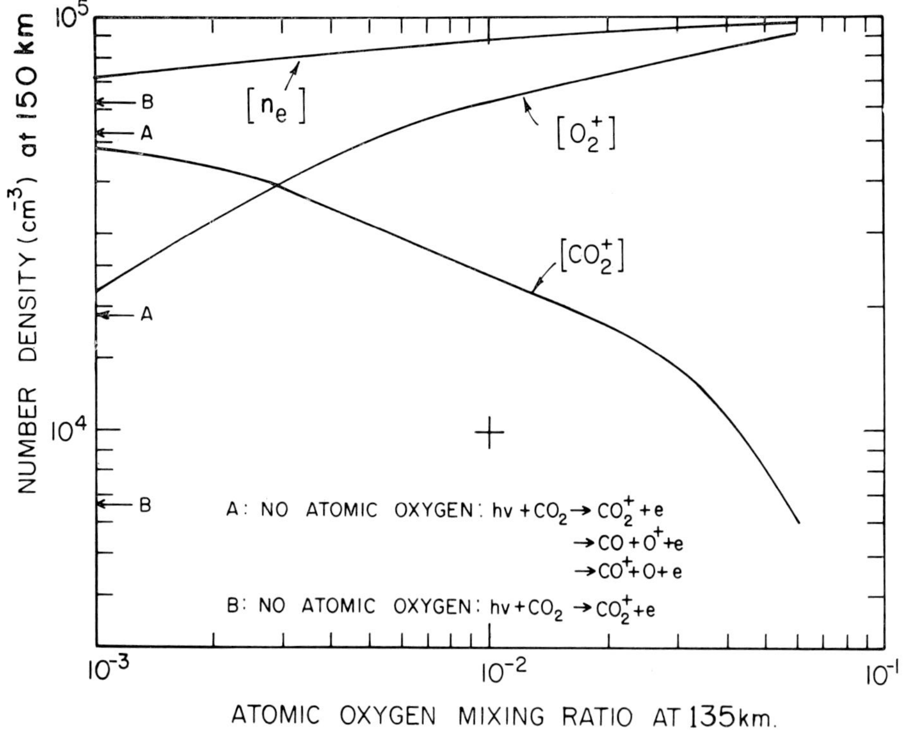

Fig. 10. Variation of O_2^+, CO_2^+, and electron density n_e at 150 km with O mixing ratio at 135 km. Also shown are the O_2^+ and CO_2^+ densities for no O present; A includes dissociative processes, while B does not. The upper arrow with each letter indicates CO_2^+ while the lower indicates O_2^+.

of O/CO_2 at 135 km required to yield the CO_2^+ densities estimated from Figure 8, is about 2%.

In terrestrial ionospheric terminology, the ionosphere of Mars is an *F*1 region, i.e., local production of ions balanced by local electronic recombination. This assumption provides qualitative agreement with the Mariner electron density profiles. However, the problem remains of reproducing the magnitude of observed electron densities, using the solar flux measurements of Hall and Hinteregger (1970) and laboratory measured rate coefficients. The calculated electron densities are too low. This may be due to (a) larger solar fluxes being more appropriate, or (b) the O_2^+ recombination coefficient in the upper atmosphere being smaller than the laboratory value that is used in the ionospheric calculations (R. W. Stewart, 1971; A. I. Stewart, 1972). Figure 11 shows computed ionspheric profiles with 2% and 6% of O at 135 km, for the model atmosphere shown in Figure 5. The solar fluxes of Hall and Hinteregger have arbitrarily been increased by a factor of 2.7 to produce agreement in the absolute electron density.

It has been suggested by O'Malley (1969) that for certain conditions the cross section for recombination in vibrational states $v \neq 0$, may be much smaller than for

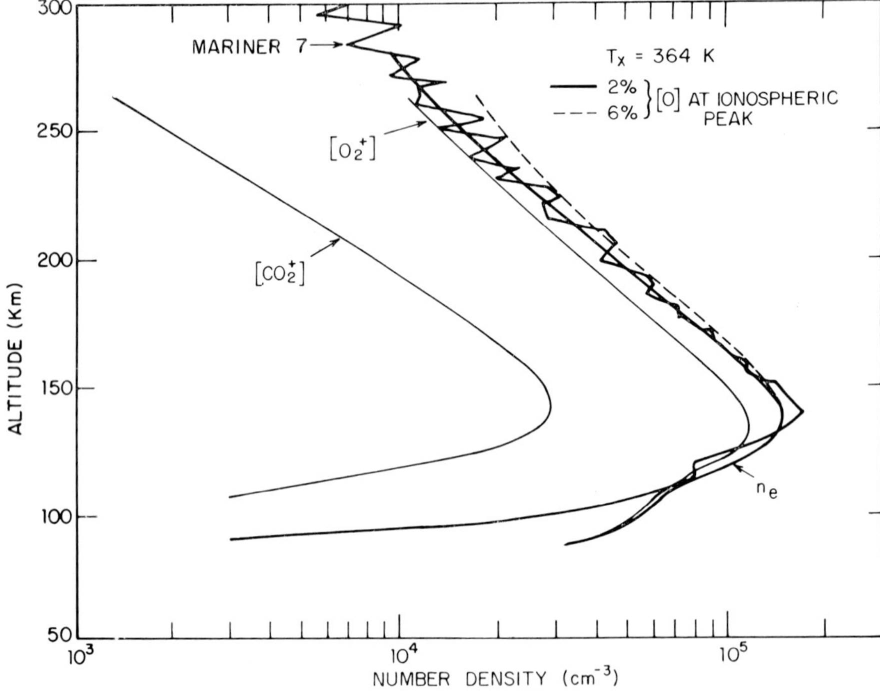

Fig. 11.   Calculated electron density profile for the atmosphere shown in Figure 5. The solar fluxes have been arbitrarily increased by a factor of 2.7. The Mariner 7 results are from Rasool and Stewart (1971). Also shown are the O_2^+ and CO_2^+ concentrations for 2 % O at 135 km, and the electron density calculated for 6 % O at 135 km.

the ground state $v=0$. In addition, it is possible that some of the O_2^+ produced by the ion molecule reactions given by Equations (12) and (14) may be in vibrationally excited states since the reactions are exothermic by more than 1 eV (McElroy, 1972b). Thus if the relaxation time for the $v \neq 0$ states is long compared to the loss time constant, the effective O_2^+ recombination rate may be slower than the laboratory value normally used in atmospheric calculations.

It has been suggested by Cloutier *et al.* (1969) that the solar wind penetrates the lower atmosphere of Mars resulting in a depression of the ionosphere, such that the ionosphere and neutral atmosphere have the same scale height. Such an interaction does not appear to be called for by the recent Mariner results in which the scale height of the ionosphere is approximately twice that of the neutral atmosphere as indicated by the airglow data. This is the case for an *F*1 interpretation. However this does call for the unusually cold exospheric temperature of about 270 K (Stewart, 1971) at the time of Mariner 4. As mentioned earlier this low temperature is perhaps attained due to the influence of eddy cooling.

4.5. Atmospheric escape

A neutral species may escape from a planetary atmosphere if it has sufficient energy, at densities low enough such that it makes no collisions. The escape energies on Mars for CO, C, N, O, and H are 3.49, 1.49, 1.74, 1.99, and 0.12 eV. The exosphere, or the region when the mean free path of a molecule is larger than a scale height, is at about 220 km for the model atmosphere presented in Figure 5. Hydrogen escapes by thermal evaporation of high energy molecules in the tail of the Maxwell distribution. As noted earlier, on the basis of the observed 1216 Å airglow, the H escape flux was estimated to be about 1.8×10^8 molecules cm^{-2} s^{-1}. Hunten and McElroy (1970) pointed out the D/H ratio in the exosphere will be enhanced. For an exospheric temperature of 360 K the enhancement is about a factor of 45 over that in the lower atmosphere.

The physical parameters of Mars, and the low exospheric temperature imply that thermal escape of other atomic species is trivial. However, Brinkman (1971) and McElroy (1972a) have pointed out that atoms may escape by nonthermal processes, such as photodissociation

$$N_2 + h\nu \rightarrow N + N + \text{kinetic energy (K.E.)} \tag{18}$$

where the N atoms have sufficient kinetic energy to escape. Recombination of molecular ions is also important:

$$N_2^+ + e \rightarrow \ N + N + K.E. \tag{19}$$
$$CO_2^+ + e \rightarrow CO + O + K.E. \tag{20}$$
$$O_2^+ + e \rightarrow \ O + O + K.E. \tag{21}$$

McElroy (1972a) calculated the rate of O escape to be of the order of 6 to 8×10^7 molecules cm^{-2} s^{-1} or about half that of H. He also argued that this should be the case since the source of H is water vapor photolysis near the Martian surface (Hunten and McElroy, 1970) and that in a steady state net water destruction should be balanced by H and O escape.

In investigating the evolution of Martian N_2, McElroy concludes that the upper limit on N_2 is 1%, but that it could be as low as $3 \times 10^{-5}\%$ the exact value depending on (a) whether the out-gassing of Martian N_2 was rapid or at a slow uniform rate, and (b) the strength of mixing in the upper atmosphere.

In addition to the above escape mechanisms, it has been noted by Cloutier *et al.* (1969) and McElroy (1972a) that sweeping up of planetary ions by the solar wind is a potentially large source of escaping atoms.

5. Lower Atmosphere

5.1. Stability

The column abundance of CO_2 in the atmosphere is about 2×10^{23} molecules cm^{-2}. From Figure 4 it is clear that CO_2 is dissociated* by sunlight right to the surface, as

---

* It has been suggested that CO_2 does not dissociate. However recent work (cf. Clark, 1971) indicates that the quantum yield is unity. The work of Krezenski *et al.* (1971) on the quantum yield of CO_2 at 2139 Å does not apply directly to the lower atmosphere of Mars since they worked at pressures of between 300 and 600 Torr.

distinct from the case of O_2 on the earth, where dissociation stops at 25 km. The time required to dissociate all of the CO_2 present in the Martian atmosphere is only $\sim 3.5 \times 10^3$ yr (McElroy and McConnell, 1971a; McElroy and Donahue, 1972). In addition, the observed amounts of CO and O_2 could be supplied in only 3 and 8 yr, respectively. In a pure C-O atmosphere, recombination of the dissociation products of CO_2, CO and O, by

$$CO + O + M \rightarrow CO_2 + M \tag{22}$$

is very slow (Stuhl and Niki, 1971; Simonaitis and Heicklen, 1972; Slanger *et al.*, 1972), compared with the competing reaction

$$O + O + M \rightarrow O_2 + M. \tag{23}$$

Thus a CO_2 atmosphere would be expected to have a much larger abundance of CO and O_2 than is observed and to have an O_2/CO mixing ratio of 0.5 rather than the observed value for Mars of 1.4 (McElroy and McConnell, 1971a; cf. Section 2). Alternatively, if surface reactions were important in maintaining the low CO and O_2 levels (however, cf. Clark, 1971), the column abundance of O_3 expected would be $\sim 10^{17}$ molecules cm^{-2}, depending on the surface temperature. This is much larger than observed amounts (cf. Section 2.5). Thus CO and O must be catalytically recombined.

5.2. Water-catalyzed CO_2 recombination

The most likely catalytic scheme for recombining CO_2 on Mars involves catalysis by radicals produced from H_2O and H_2 dissociation – H, OH, HO_2 – which may be formed as follows:

$$h\nu + H_2O \rightarrow OH + H \tag{24}$$
$$O^1D + H_2O \rightarrow 2OH \tag{25}$$
$$O^1D + H_2 \rightarrow OH + H \tag{26}$$
$$H + O_2 + M \rightarrow HO_2 + M. \tag{27}$$

The required efficiency of any catalytic scheme involving H_2O and H_2 may be judged by noting that CO_2 is being dissociated at a rate of 1.7×10^{12} molecules cm^{-2} s^{-1} while the production rate of hydrogeneous radicals is only $\sim 2 \times 10^9$ cm^{-2} s^{-1} (Hunten and McElroy, 1970; McElroy and Donahue, 1972). Thus, each H radical must go through the cycle 1000 times before reforming H_2O or H_2.

One scheme for recombining CO_2 is due to Reeves *et al.* (1966). They suggested catalysis by the series of reactions,

$$CO + HO_2 \rightarrow CO_2 + OH \tag{28}$$
$$OH + O_3 \rightarrow HO_2 + O_2 \tag{29}$$
$$\underline{O + O_2 + M \rightarrow O_3 + M} \tag{30}$$
$$CO + O \rightarrow CO_2 \quad .$$

A modification of this scheme

$$CO + HO_2 \rightarrow CO_2 + OH \qquad (28)$$
$$CO + OH \rightarrow CO_2 + H \qquad (31)$$
$$\underline{H + O_2 + M \rightarrow HO_2 + M} \qquad (27)$$
$$2CO + O_2 \rightarrow 2CO_2$$

was applied quantitatively to Mars by Hunten and McElroy (1970). They noted that although the scheme could work, the catalysis depended critically upon the rate constant for k_{28} for which there were very divergent estimates. It now appears that k_{28} is very slow and this scheme is no longer tenable (cf. Clark, 1971; McElroy and Donahue, 1972).

The stability of the Martian atmosphere has been discussed recently by McElroy and Donahue (1972) and by Parkinson and Hunten (1972). Since the papers of McElroy and Donahue, and Parkinson and Hunten employed different values for the rate constants and the photodissociation rate of CO_2, I have repeated the calculations of Parkinson and Hunten using the data given by McElroy and Donahue. The latter set of data is probably more appropriate since it takes into account the greater efficiency of CO_2 as a third body and also uses more reliable solar flux information. Thus, the comments I shall address to the Parkinson and Hunten scheme will be to my interpretation of it.

McElroy and Donahue (1972) solved the continuity and flow equations for O, describing diffusion in the lower atmosphere by means of an eddy diffusion coefficient K. They approximated the full set of H_2O and H_2 continuity and flow equations by assuming that the odd hydrogen, viz. H, OH, HO_2 was mixed in the lower atmosphere. Treating the odd hydrogen in this manner is reasonable since the time constants for interconversion of the various forms of odd hydrogen and the flow time constants are short compared to the time constants for removal. The odd hydrogen mixing ratio and eddy coefficient were adjusted to obtain a solution subject to the following constraints:

(1) zero flux of O at infinity and at the ground,
(2) a balance in production and loss of O_2,
(3) calculated O_3 abundances less than the observational upper limits, and
(4) 'reasonable values' for the production of H_2O and H_2.

They found that eddy coefficients $K \sim 10^8$ cm^2 s^{-1} were required, which are in accord with the predictions of Gierasch and Goody (1968). The densities of the various minor constituents calculated are shown in Figure 12. In the scheme of McElroy and Donahue, most of the O was transported to the lower atmosphere and recombined via the series of reactions

$$\left.\begin{array}{l} O + HO_2 \rightarrow OH + O_2 \\ CO + OH \rightarrow CO_2 + H \\ \underline{H + O_2 + M \rightarrow HO_2 + M} \end{array}\right\} I \qquad \begin{array}{l} (32) \\ (31) \\ (27) \end{array}$$

$$CO + O \rightarrow CO_2 \quad .$$

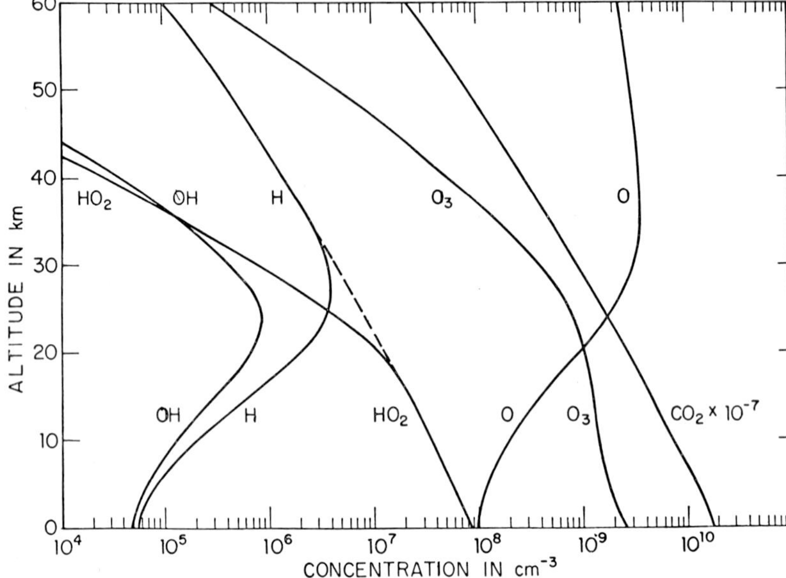

Fig. 12.   Concentration of principal constituents in the Martian atmosphere. Surface temperature is 220 K. Eddy diffusion coefficient is 1.5×10^8 cm^{-1} s^{-1}. The curve for odd H shown as a broken line between the portions labeled HO_2 and H, is plotted as 5×10^{-10} the CO_2 density. The O_2 and CO densities are, respectively, 1.3×10^{-3} and 8×10^{-4} times the CO_2 densities (McElroy and Donahue, 1972).

This scheme maintains O densities at a low level, which in turn keeps the O_3 formed via Equation (30) below the observed limit. The catalytic cycle

$$
\left.
\begin{array}{l}
CO + OH \rightarrow CO_2 + H \\
H + O_3 \rightarrow OH + O_2 \\
O + O_2 + M \rightarrow O_3 + M
\end{array}
\right\} \text{II}
$$

$$\overline{\qquad\qquad\qquad\qquad}$$

$$CO + O \rightarrow CO_2 \quad .$$

(31)
(33)
(30)

was found to be less important. The O_2 balance was due to formation via

$$O + OH \rightarrow O_2 + H \tag{34}$$

while loss of O_2 was by photolysis in the Herzberg continuum

$$O_2 + h\nu \rightarrow 2O \tag{35}$$

and by photolysis of peroxide

$$
\left.
\begin{array}{ll}
 & 2HO_2 \rightarrow H_2O_2 + O_2 \\
 & H_2O_2 + h\nu \rightarrow 2OH \\
2x & CO + OH \rightarrow CO_2 + H \\
2x & H + O_2 + M \rightarrow HO_2 + M
\end{array}
\right\} \text{III}
$$

$$\overline{\qquad\qquad\qquad\qquad}$$

$$2CO + O_2 \rightarrow 2CO_2 \quad .$$

(36)
(37)
(31)
(27)

(This latter series of reactions has the net effect of breaking an O_2 bond.) Photolysis of H_2O_2 contributed about 1/3 to the destruction of O_2.

The rate of formation of H_2O was found to be $\sim 2 \times 10^9$ molecules cm^{-2} s^{-1}, which is in reasonable agreement with previous estimates of the rate of destruction of H_2O (Hunten and McElroy, 1970). The H_2 formation rate was constrained to lie between limits set by a consideration of the H flow problem (next section).

Parkinson and Hunten (1972) considered a similar chemical scheme, but they assumed that photochemical equilibrium was adequate. This method results in the O produced from CO_2 forming O_2 via k_{23} and k_{34}. Thus reformation of CO_2 must proceed with CO and O_2, i.e. scheme III. In this case, O_2 is destroyed mainly by the peroxide chain, photolysis of O_2 being much less important. Such large values of peroxide obtained by Parkinson and Hunten require larger odd hydrogen mixing ratios than those required by McElroy and Donahue (1972), which results in production rates of H_2O and H_2 which come close to violating the upper limits mentioned above. It is important to note that the rate of production of H_2 is critically dependent on the rate of k_{40}. Lower limits for k_{40} range from 2×10^{-13} to 3×10^{-12} while it may be as fast as 3×10^{-11}. With the latter value the photochemical model of Parkinson and Hunten (1972) produces 10 times more H_2 than may be destroyed by O_3, while for $k_{40} = 3 \times 10^{-12}$ the production and destruction of H_2 balance in their model. With such small values of k_{40} I estimate that the model of McElroy and Donahue will switch to the H_2O_2 chain.

The main differences between the models as they stand are: (1) The flow model estimates an O_3 abundance of 2×10^{-4} cm atm while the photochemical scheme estimates 5×10^{-5} cm atm of O_3; (2) The altitude distribution of O_3 is different for both models, and (3) The photochemical model requires a larger photolysis rate of H_2O. Better limits on O_3 abundances and distribution, H_2O photolysis rates, and also the measurement of the critical rate coefficients should help resolve the question as to which model is more appropriate, which will in turn provide information or mixing in the lower atmosphere. Also required is a more complete model of the CO_2, H_2O, and H_2 chemistry and flow.

5.3. Water Chemistry and Escape

The chemistry of H_2O in the Martian atmosphere has been discussed by Hunten and McElroy (1970). The abundance of H_2O appears to be controlled by the vapor pressure curve. The main destruction mechanisms for H_2O are photolysis (Equation (24)) and reaction with O^1D (Equation (25)) which is produced by photolysis of O_3

$$O_3 + h\nu \rightarrow O_2\,(^1\!\varDelta) + O^1D \tag{38}$$

in the Hartley continuum. H_2O is reformed mainly by

$$OH + HO_2 \rightarrow H_2O + O_2. \tag{39}$$

Not all the hydrogenous radicals reform water; H_2 is also formed by

$$H + HO_2 \rightarrow H_2 + O_2. \tag{40}$$

In the lower atmosphere H_2 is destroyed by O^1D (26). Reaction with OH

$$OH + H_2 \rightarrow H_2O + O \tag{41}$$

is much less important due to the low temperatures which prevail over most of the Martian lower atmosphere. The water chemical cycle in the lower atmosphere is summarized in Figure 13 which shows the sources and sinks for odd hydrogen as well as the main reactions for interchanging odd hydrogen.

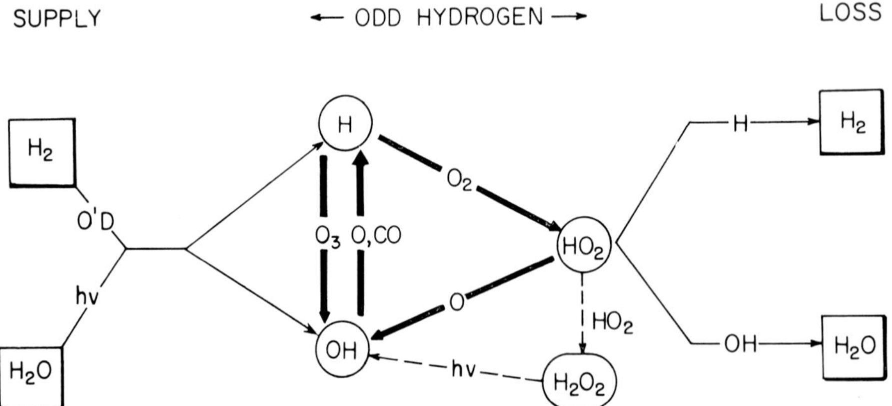

Fig. 13.   A summary of the water chemical scheme in the lower atmosphere. The boxes represent long-lived species while the circles indicate short-lived radicals.

The cycle is continued in the upper atmosphere where H_2 is converted into 2 H atoms by reaction with CO_2^+

$$H_2 + CO_2^+ \rightarrow CO_2H^+ + H \tag{42}$$

followed by

$$CO_2H^+ + e \rightarrow CO_2 + H \tag{43a}$$
$$\rightarrow CO + OH \tag{43b}$$

and Equation (34). The H atoms produced in the ionosphere flow mainly upwards and escape from the exosphere. Depending on the amount of mixing, some will flow down to the chemical sink in the lower atmosphere.

The above chemical scheme allows limits to be set on (a) the rate of formation of H_2 in the atmosphere and (b) the mixing ratio of H_2 in the atmosphere (McElroy and Donahue, 1972).

If it is assumed that no odd hydrogen flows up from the lower atmosphere to supply the escaping H atoms, then they must be supplied *in situ* by ionospheric destruction of H_2 (Equations (42) (43a, b)). Using the CO_2^+ densities discussed earlier this implies an H_2 mixing ratio $\sim 5 \times 10^{-5}$ and at the same time implies that H_2 must be formed at least as fast as $\frac{1}{2}$ the H escape rate i.e.

$$P = \int k_{40}(\text{H})(\text{HO}_2)\,dz \geq 8 \times 10^7 .$$

Since the main mode of H_2 destruction in the lower atmosphere is by O^1D from O_3 we may obtain the maximum H_2 destruction by evaluating

$$D = J_{38}\text{N}(\text{O}_3)\,k_{26}f/k_q , \tag{44}$$

where J_{17} is the average photolysis rate of O_3, f is the H_2 mixing ratio, $N(O_3)$ is the upper limit to the O_3 abundance (column) and k_q is the O^1D–CO_2 quenching rate. The maximum H_2 destruction $D=P$ is $\simeq 10^9$ molecules $\text{cm}^{-2}\,\text{s}^{-1}$. As noted earlier the limits on P

$$8 \times 10^7 \leq P \leq 10^9$$

provide a firm constraint on calculations of H_2O catalysis of CO_2 recombination.

The above outline of H_2O chemistry implies a net destruction of H_2O, resulting in production of H_2 and $\frac{1}{2}O_2$ in the lower atmosphere. H_2 produced in the lower atmosphere flows to the ionosphere and eventually escapes as H atoms. Recombination escape of O atoms prevents a buildup of O_2 (Section 4.5.) In 5×10^9 yr an escape flux of 10^8H atoms $\text{cm}^{-2}\,\text{s}^{-1}$ is equivalent to a layer of ice 2.5 m thick. This amount of H_2O could have been retained in a permafrost layer (Leighton and Murray, 1966; Smoluchowski, 1968). Figure 14 is a schematic diagram of the various flows involved in the escape of H_2O from Mars.

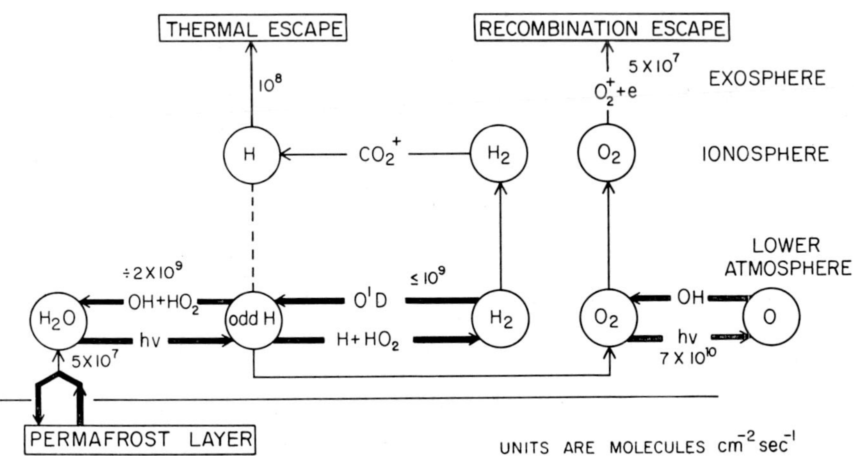

Fig. 14. Schematic diagram of the various chemical source and sink terms and flows associated with escape of H_2O from Mars.

                JOHN C.MCCONNELL

Acknowledgments

Some of the results presented here are from a study of the upper atmosphere of Mars undertaken in collaboration with M. B. McElroy to whom I am indebted for permission to publish in advance of publication.

This work was supported in part by the Atmospheric Sciences Section of the National Science Foundation under Grant GP 13982 to Harvard University and in part by York University NRC Grant # A6844.

References

Ajello, J. M.: 1971, *J. Chem. Phys.* **55**, 3169.
Anderson, D. E. Jr. and Hord, C. W.: 1971, *J. Geophys. Res.* **28**, 6666.
Barker, E. S.: 1972, *Nature* **238**, 447.
Barth, C. A. and Hord, C. W.: 1971, *Science* **173**, 197.
Barth, C. A., Fastie, W. G., Hord, C. W., Pearce, J. B., Kelly, K. K., Stewart, A. I., Thomas, G. E.. Anderson, G. P., and Raper, O. E.: 1969, *Science* **165**, 1004.
Barth, C. A., Hord, C. W., Pearce, J. B., Kelly, K. K., Anderson, G. P., and Stewart, A. I.: 1971, *J. Geophys. Res.* **76**, 2213.
Barth, C. A., Hord, C. W., Stewart, I. A., and Lane, A. L.: 1972, *Science* **175**, 309.
Belton, M. J. S., and Hunten, D. M.: 1968, *Astrophys. J.* **153**, 963; Errata, 1969, *Astrophys. J.* **156**, 717.
Belton, M. J. S. and Hunten, D. M.: 1969, *Science* **166**, 225.
Belton, M. J. S. and Hunten, D. M.: 1971, *Icarus* **15**, 204.
Belton, M. J. S., Broadfoot, A. L., and Hunten, D. M.: 1968, *J. Geophys. Res.* **73**, 4795.
Brinkmann, R. T.: 1971, *Science* **174**, 944.
Broadfoot, L. and Wallace, L.: 1970, *Astrophys. J.* **161**, 303.
Broida, H. P., Lundell, O. R., Schiff, H. I., and Ketcheson, R. D.: 1970, *Science* **170**, 1402.
Carleton, N. P. and Traub, W. A.: 1972, *Science* **177**, 988.
Carleton, N. P., Sharma, A., Goody, R. M., Liller, W. L., and Roesler, F. L.: 1969, *Astrophys. J.* **155**, 323.
Clark, I. D.: 1971, *J. Atmospheric Sci.* **28**, 847.
Cloutier, P. A., McElroy, M. B., and Michel, F. C.: 1969, *J. Geophys. Res.* **74**, 6215.
Colegrove, F. D., Johnson, F. S., and Hanson, W. B.: 1966, *J. Geophys. Res.* **71**, 2227.
Dalgarno A. and Degges, T. C.: 1971, in C. Sagan, T. C. Owen, and H. J. Smith (eds.), *Planetary Atmospheres IAU, Symp.* **40**, 337.
Dalgarno A., Degges, T. C., and Stewart, A. I.: 1970, *Science* **167**, 1490.
Dalgarno, A. and McElroy, M. B.: 1970, *Science* **170**, 167.
Dalgarno, A., McElroy, M. B., and Stewart, I. A.: 1969, *J. Atmopheric Sci.* **26**, 753.
Dickinson, R. E.: 1971, *J. Atmospheric Sci.* **28**, 885.
Fehsenfeld, F. C., Dunkin, D. B., and Ferguson, E. E.: 1970, *Planetary Space Sci.* **18**, 1267.
Fjeldbo, G., Kliore, A., and Seidel, B.: 1970, *Radio Sci.* **5**, 381.
Freund, R. S.: 1971, *J. Chem. Phys.* **55**, 3569.
Gentieu, E. C. and Mentall, J. E.: 1972, *Trans. Am. Geophys. Union* **53**, 459 (Abstract).
Gierasch, P. J.: 1970, *Earth Extraterr. Phys.* **1**, 171.
Gierasch, P. J. and Goody, R. M.: 1968, *Planetary Space Sci.* **16**, 615.
Gierasch, P. J. and Goody, R. M.: 1972, *J. Atmospheric Sci.* **29**, 400.
Gierasch, P. J., Goody, R. M., and Stone, P.: 1970, *Geophys. Fluid Dyn.* **1**, 1.
Giver, L. P., Inn, E. C. Y., Miller, J. H., and Boese, R. W.: 1968, *Astrophys. J.* **153**, 285.
Goody, R. M.: 1969, *Ann. Rev. Astron. Astrophys.* **7**, 303.
Hall, L. A. and Hinteregger, H. E.: 1970, *J, Geophys. Res.* **75**, 6959.
Hanel, R. A., Conrath, B. J., Hovis, W. A., Kunde, V. G., Lowman, P. D., Pearl, J. C., Prabhakara, C., Schlachman, B., and Lenn, G. W.: 1972, *Science* **175**, 305.

Henry, R. and McElroy, M. B.: 1968, in J. C. Brandt and M. B. McElroy (eds.), *The Atmospheres of Venus and Mars*, Gordon and Breach, New York, p. 251.

Herr, K. C. and Pimentel, G. C.: 1970, *Science* **167**, 47.

Herr, K. C., Horn, D., McAfee, J. M., and Pimentel, G. C.: 1970, *Astrophys. J.* **75**, 883.

Hord, C. W.: 1972, *Icarus* **16**, 253.

Hunten, D. M.: 1968, in J. C. Brandt and M. B. McElroy (eds.), *The Atmospheres of Venus and Mars*, Gordon and Breach, New York, p. 147.

Hunten, D. M.: 1971, *Space Sci. Rev.* **12**, 539.

Hunten, D. M. and McElroy, M. B.: 1970, *J. Geophys. Res.* **75**, 5989.

Ingersoll, A. P. and Leovy, C. B.: 1971, *Ann. Rev. Astron. Astrophys.* **9**, 147.

Johnson, F. S.: 1968, in J. C. Brandt and M. B. McElroy (eds.), *The Atmosphere of Venus and Mars*, Gordon and Breach, New York, p. 181.

Johnson, F. S. and Gottlieb, B.: 1970, *Planetary Space Sci.* **18**, 1707.

Johnson, F. S. and Wilkins, E. M.: 1965, *J. Geophys. Res.* **70**, 1281.

Kaplan, L. D., Connes, J., and Connes, P.: 1969, *Astrophys. J.* **157**, 187.

Kliore, A., Cain, D. L., Levy, G. S., Eshleman, R., and Drake, F. D.: 1965, *Science* **149**, 1243.

Kliore, A. J., Cain, D. L., Fjeldbo, G., Seidel, B. L., and Rasool, S. I.: 1972, *Science* **175**, 313.

Krezenski, D. C., Simonaitis, R., and Heicklen, J.: 1971, *Planetary Space Sci.* **19**, 1701.

Kuiper, G. P.: 1952, in G. P. Kuiper (ed.), *Atmospheres of the Earth and Planets*, University Chicago Press, Chicago, Ill, p. 306.

Lane, A. L., Barth, C. A., Hord, C. W., and Stewart, A. I.: 1972, *Icarus*, submitted.

Lawrence, G. M.: 1972, *J. Chem. Phys.* **56**, 3435.

Leighton, R. B. and Murray, B. C.: 1966, *Science* **153**, 136.

Leovy, C. B., Smith, B. A., Young, A. T., and Leighton, R. B.: 1971, *J. Geophys. Res.* **76**, 297.

McConkey, J. W., Burns, D. J., and Woolsey, J. M.: 1968, *J. Phys. B* **1**, 71.

McConnell, J. C. and McElroy, M. B.: 1970, *J. Geophys. Res.* **75**, 7290.

McElroy, M. B.: 1967, *Astrophys. J.* **150**, 1125.

McElroy, M. B.: 1969, *J. Geophys. Res.* **74**, 29.

McElroy, M. B.: 1972a, *Science* **175**, 443.

McElroy, M. B.: 1972b, private communication.

McElroy, M. B. and Donahue, T. M.: 1972, *Science* **177**, 986.

McElroy, M. B. and Hunten, D. M.: 1970, *J. Geophys. Res.* **75**, 1188.

McElroy, M. B. and McConnell, J. C.: 1971a, *J. Atmospheric Sci.* **28**, 879.

McElroy, M. B. and McConnell, J. C.: 1971b, *J. Geophys. Res.* **76**, 6674.

Marshall, J. V.: 1964, *Comm. Lunar Planet. Lab.* (*Ariz.*) **2**, 167.

Masursky, H., Batson, R. M., McCauley, J. F., Soderblom, L. A., Weldey, R. L., Carr, M. H., Milton, D. J., Wilhelms, D. E., Smith, B. A., Kirby, T. B., Robinson, J. C., Leovy, C. B., Briggs, G. A., Duxbury, T. C., Acton, C. H., Murray, B. C., Cutts, J. A., Sharp, R. P., Smith, S., Leighton, R. B., Sagan, C., Veverka, J., Noland, M., Lederberg, J., Levinthal, E., Pollack, J. B., Moore, J. T., Hartmann, W. K., Shipley, E. N., de Vaucouleurs, G., and Davies, M. E.: 1972, *Science* **175**, 294.

Mumma, M. J., Stone, E. J., and Zipf, E. C.: 1971, *J. Chem. Phys.* **54**, 2627.

Neugebauer, G., Münch, G., Chase, S. C. Jr., Miner, E., and Schofield, D.: 1969, *Science* **166**, 98.

Neugebauer, G., Münch, G., Kieffer, H., Chase, S. C., Jr., and Miner, E.: 1971, *Astrophys. J.* **76**, 719.

O'Malley, T. F.: 1969, *Phys. Rev.* **185**, 101.

Pang, K. and Hord, C. W.: 1971, *Icarus* **15**, 443.

Parkinson, T. M. and Hunten, D. M.: 1972, *J. Atmospheric Sci.*, **29**, 1380.

Parkinson, W. H. and Reeves, E. M.: 1969, *Solar Phys.* **10**, 342.

Pettingill, G. H., Counselman, C. C., Rainville, L. P., and Shapiro, I. I.: 1969, *Astron. J.* **74**, 461.

Rapp, D. and Englander-Golden, P.: 1965, *J. Chem. Phys.* **43**, 1464.

Rasool, S. I.: 1963, *A.I.A.A.* **1**, 6.

Rasool, S..I. and Stewart, R. W.: 1971, *J. Atmospheric Sci.*, **28**, 869.

Rasool, S. I., Hogan, J. S., Stewart, R. W., and Russell, L. H.: 1970, *J. Atmospheric Sci.* **27**, 841.

Reeves, R. R., Harteck, P., Thompson, B. A., and Waldron, R. W.: 1966, *J. Phys. Chem.* **70**, 1637.

Rogers, A. E. E., Ash, M. E., Counselman, C. C., Shapiro, I. I., and Pettingill, G. H.: 1970, *Radio Sci.* **5**, 465.

Rubey, W. W.: 1951, *Bull. Geol. Soc. Am.* **62**, 111.

Schorn, R. A.: 1971 in C. Sagan, T. C. Owen, and H. J. Smith (eds.), *Planetary Atmospheres, IAU Symp.* **40**, 223.
Schorn, R. A., Farmer, C. B., and Little, S. H.: 1969, *Icarus* **11**, 283.
Simonaitis, R. and Heicklen, J.: 1972, *J. Chem. Phys.* **56**, 2004.
Slanger, T. G., Wood, B. J., and Black, G.: 1972, *J. Chem. Phys.* **57**, 233.
Smoluchowski, R.: 1968, *Science* **159**, 1348.
Stewart, A. I.: 1972, *J. Geophys. Res.* **77**, 54.
Stewart, R. W.: 1971, *J. Atmospheric Sci.* **28**, 1069.
Strickland, D. J., Thomas, G. E., and Sparks, P. R.: 1972, *J. Geophys. Res.* **77**, 4052.
Stuhl, F. and Niki, M.: 1971, *J. Chem. Phys.* **55**, 3943.
Thomas, G. E.: 1971, *J. Atmospheric Sci.* **28**, 859.
Tubb, L. D. and Williams, D.: 1972, *J. Opt. Soc. Amer.* **62**, 423.
Wells, R. A.: 1969, *Science* **166**, 862.
Wells, W. C., Borst, W. L., and Zipf, E. C.: 1972, *J. Geophys. Res.* **77**, 69.
Widing, K. G., Purcell, J. D., and Sandlin, G. D.: 1970, *Solar Phys.* **12**, 52.
Young, L. D. G.: 1971, *J. Quant. Spectr. Radiative Transfer* **11**, 385.

THE UPPER ATMOSPHERE OF VENUS: A REVIEW

RONALD G. PRINN

Dept. of Meteorology, Massachusetts Institute of Technology, Cambridge, Mass., U.S.A.

1. Introduction

The upper atmospheres of Venus and Mars have, for a number of years now, been the subject of a detailed study by aeronomers. These atmospheres have presented a number of intriguing photochemical and dynamical problems. A number of these problems still remain unsolved, particularly on Venus, for which much less space-probe data are available than on Mars.

Useful data on the atmosphere of Venus has come from a number of sources. Ground-based observations have been carried out over many years and these have been augmented by spectroscopic observations from aircraft), balloons and more recently from rockets. The series of Russian atmospheric entry probes Veneras 4, 5, 6, and 7 provided basic information on the lower atmosphere but gave little information on upper atmospheric structure. A considerable quantity of the available information on the upper atmosphere of Venus was obtained from the American fly-by probe Mariner 5.

No attempt will be made to fully discuss the results obtained by these various methods. We will begin by simply reviewing the structural model which has arisen from all these studies. The bulk of this paper will be devoted to detailed discussions of ion and neutral photochemistry, and upper atmospheric circulation.

2. A Review of the Atmospheric Structure

2.1. TROPOSPHERE

The Venus atmosphere is predominantly CO_2 with up to a few percent N_2 possible. The minor species CO, H_2O, HCl, and HF have been observed spectroscopically with mixing ratios (Belton, 1968) of 2×10^{-4}, 10^{-4} to 10^{-6}, and 2×10^{-8} respectively. Interestingly, O_2 has not yet been detected. Mariner 5 measured vertical temperature profiles between 35 and 90 km (Fjeldbo *et al.*, 1971). At an altitude of ~ 66 km, which we will consider as the tropopause, the pressure is ~ 100 mb and the temperature ~ 245 K. Below this level the atmosphere is approximately in dry adiabatic equilibrium right down to the surface where Venera 7 measured pressures ~ 90 b and temperatures ~ 750 K. A temperature profile is shown in Figure 1. The top of the uppermost thick cloud layer which completely covers the planet is probably around 66 km in agreement with the ~ 200 mb pressures measured spectroscopically within these clouds (Belton, 1968). There is also a small decrease in lapse rate at ~ 47 km which might imply a further thick cloud layer below the visible one. We presently have no definite evidence concerning the composition of these visible clouds or any layers below them, although a number of theoretical predictions have been made.

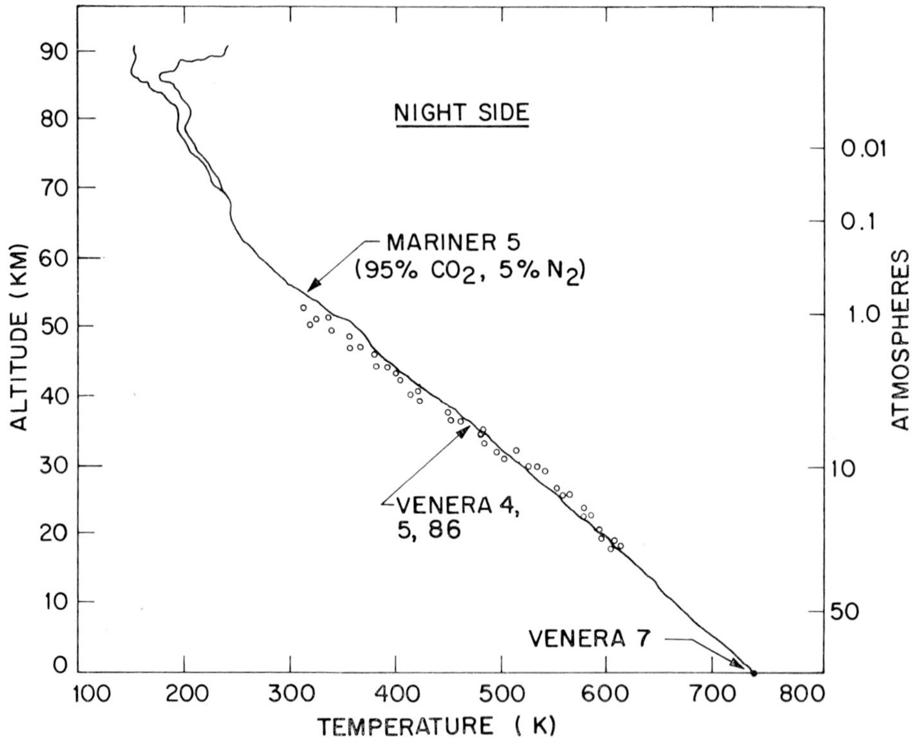

Fig. 1.   Temperature profile of Venus (Rasool and Stewart, 1971).

2.2. STRATOSPHERE-MESOSPHERE

Above ~ 66 km the vertical temperature lapse rate becomes sub-adiabatic with the temperature decreasing to 190 ± 50 K at 90 km. The mesopause is probably not much higher than 90 km, if at all, and a reasonable estimate of its position is ~ 100 km. It is important to note that there is no division of this part of the atmosphere into a definite stratosphere and mesosphere. Consequently, much more rapid transport from the tropopause to the mesopause is possible on Venus than on Earth. There is both photographic (Boyer and Newell, 1967) and spectroscopic (Hansen and Arking, 1971) evidence for a thin transient haze in this stratosphere-mesosphere region at pressures of a few tens of millibars corresponding to altitudes up to ~ 77 km. There is now a considerable body of evidence (Lewis, 1972) to suggest that this high altitude haze is composed of droplets of aqueous HCl solution. Finally, it is most interesting to note that where a comparison is possible (between 35 and 90 km), there appears to be little discernable difference between the dayside and nightside temperature profiles throughout both the troposphere and stratosphere-mesosphere.

2.3. THERMOSPHERE AND EXOSPHERE

As the temperature profile above 90 km has not been measured directly the expected structure of this region of the atmosphere is somewhat model-dependent. We will discuss this region in more detail later and simply state here that temperatures should rise with altitude above the mesopause to values ~ 700 K on the dayside and ~ 300 K on the nightside. These will be appropriate temperatures to describe the exosphere. The turbopause, if present at a comparable background number density to that on earth ($\sim 10^{12}$ cm^{-3}), will lie at the ~ 130 km level. Above the turbopause H$_2$ molecules become the dominant species and photodissociation of these produces H atoms.

2.4. IONOSPHERE

Mariner 5 measurements (Eshleman, 1970) show that both the day and nightside ionospheres possess a prominent main peak at ~ 140 km. The peak electron number densities are $\sim 5.5 \times 10^5$ and 2×10^4 cm^{-3}, respectively. The dayside has an additional secondary peak at ~ 125 km with a peak electron number density of $\sim 2 \times 10^5$ cm^{-3}. The topside dayside ionospheric scale height is approximately twice that of CO$_2$ at ~ 700 K. Above the dayside main layer there is also an extensive ionized region with electron number densities decreasing very slowly from $\sim 3 \times 10^4$ cm^{-3} at ~ 275 km with a sharp cutoff at ~ 550 km. Above the nightside main layer very low electron number densities ($\sim 10^2$ to 10^3 cm^{-3}) exist to very high altitudes (~ 1000 km).

3. Ion Photochemistry

3.1. DAYSIDE

The dayside electron density profile was adequately explained by McElroy (1968a, 1969) and Stewart (1968) using a pure CO$_2$ photochemical steady-state model with the two reactions

$$CO_2 \rightarrow CO_2^+ + e^- \qquad (\text{wavelength} < 902\,\text{Å}),$$
$$e^- + CO_2^+ \rightarrow CO\,(^1\Pi_g) + 0.$$

The main electron peak observed at 140 km was explained by the absorption of EUV radiation and the smaller peak at 125 km by the absorption of soft X-rays. These two layers can usefully be compared to the $F1$ and E layers in the terrestrial ionosphere. The observed and predicted topside electron scale heights agree if the Venusian thermospheric temperature is ~ 700 K.

Although the simple scheme above adequately describes the electron densities the predominant ion is probably not CO$_2^+$. Other primary photoionization reactions are possible such as

$$CO_2 \rightarrow O^+ + CO + e^-$$
$$\rightarrow CO^+ + O + e^-$$

with the O$^+$ producing O$_2^+$ by the reaction

$$O^+ + CO_2 \rightarrow O_2^+ + CO.$$

In addition, Stewart (1971) and Donahue (1971) have pointed out that if the O mixing ratio at ionospheric levels on Venus is similar to the value of ~ 0.03 observed on Mars then most of the CO_2^+ produced from CO_2 will be immediately converted to O_2^+ by the reaction

$$O + CO_2^+ \rightarrow O_2^+ + CO \, .$$

Strickland (1972) has analyzed some recent rocket measurements of the $^3P \leftrightarrow {}^3S$ and $^3P \leftrightarrow {}^5S$ resonant emissions from Venus and concluded that a significantly greater O concentration is required on Venus than on Mars. It is most probable then that the dominant positive ion in the Venusian ionosphere is O_2^+. As McElroy pointed out, the exact nature of the positive ion has little effect on computed electron densities since O_2^+, CO_2^+, and CO^+ all have comparable recombination rates with electrons.

The simple model discussed above qualitatively describes the Venusian ionosphere but a more detailed inspection indicates we require 2 to 4 times the presently accepted solar EUV fluxes in order to obtain an exact fit. Stewart (1971) suggests this could imply a contribution by the solar wind. We may also question our knowledge of the true solar EUV fluxes and their variation with the solar cycle.

The ionospheric profiles may also be used to place upper limits on certain minor constituents in the Venus upper atmosphere. If more than a few percent of O was present, photoionization of O atoms below 900 Å would supply an additional source of electrons above the $F1$ peak. Electron recombination at these high altitudes is controlled by downward diffusion of O^+ and e^-. An additional ionized layer comparable to the terrestrial $F2$ layer is produced and the topside electron scale height becomes twice that of O. If more than a few percent of N_2 was present there would be a shift of the $F1$ peak to higher altitudes and the topside ionospheric scale height becomes twice that of N_2 due to photoionization of N_2 at wavelengths < 790 Å. Using such arguments McElroy (1969) and Stewart (1968) showed that O was less than a few percent and N_2 less than 10% of the total number density in the Venus upper atmosphere.

Electron densities in the ionized region extending from ~ 200 km to an apparent plasmapause at ~ 550 km are too large to be produced by photoionization of H_2 and He alone. Venus has no measurable magnetic field so that solar wind protons may possibly penetrate through the plasmapause into this region. This may provide an additional source of ionization but may also serve to sweep ions around the planet or back into the atmosphere. Eshleman (1970) suggested that this upper region may effectively absorb the solar wind protons and allow the formation of the well-behaved $F1$ peak below. In any case electron densities in the $F1$ peak are probably sufficiently high to be unaffected by the solar wind (Cloutier *et al.*, 1969) even if it penetrated to these low levels.

3.2. NIGHTSIDE

The nightside Venusian ionosphere exhibits a pronounced peak at a comparable altitude to the $F1$ dayside peak. The peak electron concentration is $\sim 3\%$ of that on the dayside. This requires a source of ionization on the nightside and McElroy and Strobel

(1969) have suggested that this source may arise from the rapid transport of He^+ or H_2^+, produced above the turbopause on the dayside, around to the nightside. Horizontal winds of ~ 200 m s^{-1} at the 350 km level are required. Such intense winds may not be implausible at these altitudes as we will discuss later.

4. Neutral Photochemistry

We will discuss the neutral atmosphere under four convenient headings: thermal structure, HLy-α emission, H sources, and finally stability of CO_2.

4.1. THERMAL STRUCTURE

McElroy (1968b) has calculated the expected temperature profile in the upper atmosphere for a pure CO_2 model. He took into account the absorption by CO_2 averaged over the planet of solar EUV and IR photons, the fraction of this absorbed energy converted into local heating, the cooling by CO_2 emission at 15μ m, and the vertical conduction of heat by molecular and eddy diffusion. His calculations yielded an exospheric temperature of ~ 700 K, agreeing very well with Mariner 5 measurements of the electron scale height in the topside dayside ionosphere and the scale height for H atoms above the dayside turbopause. His calculated temperatures at lower altitudes (below 90 km) are, however, considerably greater than those measured by Mariner 5. Evidently, cooling rates at these altitudes exceed those in a pure CO_2 atmosphere.

4.2. HLy-α EMISSION

A most puzzling result from Mariner 5 was the observed emission of HLy-α radiation from the planet. About 25 kR of radiation could be attributed to resonant scattering of solar HLy-α by H atoms with a scale height corresponding to a temperature of ~ 650 K. However, there was also an additional ~ 65 kR of radiation from a source with apparently one half the scale height of H at 650 K. The origin of this unexpected additional radiation has been extensively debated.

Barth (1968) proposed that dissociation of H_2 (at 650 K) below 845 Å yields H(2p), and hence an additional source of HLy-α with the required scale height. This explanation was strongly opposed by Donahue (1968, 1969), and McElroy and Hunten (1969). They pointed out that in a realistic photochemical model including vertical diffusion the H_2 number densities required by Barth ($\sim 10^{10}$ cm^{-3} at the turbopause) produce H atom densities much greater than those observed. These workers suggested instead that the second component was D at 650 K. They argued that for a sufficiently small eddy diffusion coefficient ($\lesssim 10^5$ cm^2 s^{-1}) the different molecular diffusion and escape rates of H and D above the turbopause could increase the D:H ratio at the escape level to 100 times that in the lower atmosphere. This is sufficient to yield the large D densities required to emit ~ 65 kR of HLy-α and still enable a D:H ratio <0.1 in the lower atmosphere demanded by spectroscopic observations. In an attempt to verify this model Wallace *et al.* (1971) obtained rocket spectra with sufficient resolution to detect the 0.33 Å difference between scattering by H and D of the 1 Å wide solar HLy-α line.

No scattering by D was even observed. A third alternative was proposed by Barth (1968, 1970). This model has an exospheric temperature of $\sim 325\,K$ with 'cold' H atoms at this temperature providing the 65 kR component and 'hot' H atoms at $\sim 650\,K$ providing the 25 kR component. The 'hot' atoms were produced by H_2 absorption below 1108 Å. However, an exospheric temperature of 325 K would provide a very poor fit to the observed ionospheric structure.

It is further perplexing to note that rocket measurements from Earth show only the 25 kR component. Moos *et al.* (1969) obtained rocket spectra during the same epoch as the Mariner 5 flight and measured only 18 ± 9 kR of HLy-α. More recently, Wallace *et al.* (1971) reported values of 26 and 24 kR, the latter value communicated to them by Moos. Unless remarkable conditions were prevailing on Venus at the time of Mariner 5 we are forced at present to conclude that the 65 kR component was simply some instrumental artifact.

4.3. HYDROGEN SOURCES

Two sources for the H corona observed on Venus have been suggested. McElroy and Hunten (1969) showed that, providing no cold traps existed in the atmosphere at sufficient pressures for H_2O condensation to occur, H_2O photodissociation is a plausible source. More recently, I have suggested (Prinn, 1971) that HCl photodissociation is a more potent source than H_2O and is not restricted by possible cold traps. McElroy and Hunten used McElroy's calculated temperature profile to argue against the possibility of low H_2O mixing ratios in the Venus upper atmosphere. As we mentioned earlier these calculated temperatures are too high below 90 km and we must therefore re-examine their arguments.

The Mariner 5 measurements give a temperature of $\sim 193\,K$ at the 10 mb (77 km) level. We mentioned earlier that condensation of droplets of aqueous HCl at this level and below is entirely probable. At 193 K the saturated vapor pressures of HCl and H_2O over such droplets (weight % HCl $\sim 25\%$) imply maximum mixing ratios of $\sim 5 \times 10^{-7}$ and $\sim 5 \times 10^{-5}$, respectively, for these species at the 10 mb level. At temperatures less than 188 K, precipitation of pure ice crystals and solid HCl-hydrate crystals occurs. At the 1 mb (86 km) level where the temperature is perhaps as low as 160 K, the saturation mixing ratios for HCl and H_2O over these solid phases would be $\sim 10^{-8}$ and $\sim 10^{-6}$, respectively. With such small maximum mixing ratios at and above the 10 mb level neither H_2O nor HCl can absorb significantly at wavelengths less than the CO_2 cutoff at ~ 1950 Å. Below the 10 mb level, where their mixing ratios can attain the values observed spectroscopically, I have shown that the HCl absorption rate will be 12 to 1200 times that of H_2O for H_2O mixing ratios of 10^{-4} to 10^{-6}. Even if we allow H_2O to maintain a mixing ratio of 10^{-4} up to very high altitudes, the total column absorption rate of HCl is at least an order of magnitude greater than that of H_2O.

Mixing ratios for H_2 around 8×10^{-6} at 50 mb are predicted from my calculations on HCl steady-state photochemistry. This H_2 will be transported rapidly to the turbopause. A study of H_2 transport and H escape will be required to determine the actual

H_2 mixing ratio at the turbopause. If Earth and Venus were formed under similar conditions in the primitive solar nebula, we must apparently deplete over geologic time an amount of H_2 from Venus equal to the amount of H_2 contained in water on Earth. Photodissociation of HCl, which has been produced by reactions between H_2O and NaCl and other minerals at the very hot surface, could enable sufficient rates of H escape even after the predominantly CO_2 atmosphere had evolved.

4.4. Carbon Dioxide Stability

In the atmospheres of both Venus and Mars we are confronted with the problem that the spin-allowed reaction between two $O(^3P)$ atoms is much faster than the spin-forbidden reaction between $O(^3P)$ and $CO(^1\Sigma_g^+)$. Consequently, when exposed to dissociating radiation, these CO_2 atmospheres should finally be converted to predominantly CO, O_2 atmospheres. For more details on the history of this problem the reader is referred to McConnell's (1973) review of the Mars atmosphere. To briefly summarize this history: a two-body scheme in which CO_2 itself catalyzed the recombination of CO and O failed when CO_3 was found to have too short a lifetime; a later three-body scheme in which H atoms catalyzed the oxidation of CO by O_2 failed when the reaction between HO_2 and CO was found to be extremely slow.

In order to provide an answer to this problem on Venus, I explored the possibility (Prinn, 1971, 1972) that certain products of HCl photochemistry might be important catalysts for the oxidation of CO to CO_2 in a 'recombination region' at pressures of several tens of millibars. Modest eddy diffusion coefficients are required to retain a predominantly CO_2 atmosphere up to the turbopause providing such a recombination region exists. I proposed a number of catalytic cycles which may exist in this recombination region. The first cycle involved catalysis by Cl atoms ($M \equiv CO_2$):

$$Cl + O_2 + M \rightarrow ClOO + M,$$
$$ClOO + CO \rightarrow ClO + CO_2,$$
$$ClO + CO \rightarrow Cl + CO_2.$$

A second cycle involved catalysis by H atoms:

$$H + O_2 + M \rightarrow HO_2 + M,$$
$$HO_2 + H \rightarrow OH + OH,$$
$$HO_2 \rightarrow OH + O \quad (\text{wavelength} < 3825\,\text{Å}),$$
$$OH + CO \rightarrow H + CO_2.$$

A third cycle involved catalysis by both Cl and H atoms:

$$H + O_2 + M \rightarrow HO_2 + M,$$
$$Cl + HO_2 \rightarrow ClO + OH,$$
$$ClO + CO \rightarrow Cl + CO_2,$$
$$OH + CO \rightarrow H + CO_2.$$

Using my assumptions for rate constants the first cycle was about ten times more efficient than the second cycle. The third cycle involves the reaction between the two most predominant radicals, Cl and HO_2, in the recombination region. This third cycle will be very important if this reaction is reasonably fast. The first cycle, operating even at 1% efficiency, could remove CO and O_2 from the upper atmosphere at rates equal to their production rates from CO_2 photodissociation. Even if this first cycle is not operative the second or third cycles should be sufficient.

It was also pointed out that the HO_2 radicals in my scheme could react yielding H_2O_2 which would photodissociate to give two OH radicals. Very recently, Parkinson and Hunten (1972), and McElroy and Donahue (1972) proposed the following catalytic cycle on Mars with H_2O as the source of H and OH, and CO_2 the main source of O:

$$H + O_2 + M \rightarrow HO_2 + M\,,$$
$$O + HO_2 \rightarrow OH + O_2\,,$$
$$H + HO_2 \rightarrow OH + OH\,,$$
$$HO_2 + HO_2 \rightarrow H_2O_2 + O_2\,,$$
$$H_2O_2 \rightarrow OH + OH \quad (\text{wavelength} < 3825\,\text{Å})\,,$$
$$OH + CO \rightarrow CO_2 + H\,.$$

This scheme appears feasible for Mars where O_2 mixing ratios appear to equal or exceed those of CO. However, on Venus O_2 has never been detected and the upper limit for its mixing ratio is nearly two orders of magnitude less than that observed on Mars. On Mars there is an important source of O_2 from reactions between odd oxygen and odd hydrogen such as

$$O_3 + H \rightarrow O_2 + OH\,,$$
$$O + HO_2 \rightarrow O_2 + OH\,,$$
$$O + OH \rightarrow O_2 + H\,.$$

This additional source will not be present on Venus if odd hydrogen production occurs in a region where odd oxygen concentrations are very low. I believe the present inability to even detect O_2 on Venus reinforces my conclusion, stated earlier, that there is very little H_2O (or HCl) photodissociation in the region where CO_2 is photodissociating on Venus. In my recombination region O atoms were assumed to be of minor importance. These matters will be discussed in more detail elsewhere.

5. Upper Atmosphere Dynamics

5.1. THERMOSPHERE

The circulation of the Venusian atmosphere above the mesopause has been discussed by Dickinson (1971). He calculated the local solar radiational heating, thermal radiational cooling, and adiabatic expansion cooling. Apparently some of the solar heating occurs at a sufficiently high level so that it cannot be balanced *in situ* by radiational and adiabatic cooling. The energy must be conducted down by molecular diffusion (he ignores eddy diffusion which probably invalidates his model at the lowest levels).

Day-to-night temperature perturbations of a few hundred degrees Kelvin about a global thermospheric mean temperature of $\sim 450\,K$ are predicted. The perturbation temperature varies roughly as the cosine of the zenith angle. Temperature perturbations of less than this planetary scale are damped by more efficient *in situ* cooling. Day- to-nightside winds still play some part in cooling the dayside thermosphere. A convection cell is set up with horizontal winds at the terminator of a few hundred meters per second and vertical winds, upward at the subsolar and downward at the anti-solar point, of around $1\,m\,s^{-1}$. These winds are of the magnitude required by McElroy and Strobel (1969) to sweep H_2^+ and He^+ around to the nightside and provide the nightside ionosphere.

It is important to note that in order to maintain the day and nightside thermospheres at the same temperature (i.e., make horizontal advection of heat the main cooling mechanism for the dayside thermosphere), horizontal winds at the terminator of around $10\,km\,s^{-1}$ are required. Dickinson's calculations indicate that horizontal winds are not the dominant cooling mechanism. This supports Hogan and Stewart's (1969) contention that solar heating should be averaged over both hemispheres on Mars but not on Venus. McElroy averaged over both hemispheres on Venus but his results are unaffected since he assumed a solar heating efficiency about twice Hogan and Stewart's empirically derived value.

5.2. STRATOSPHERE-MESOSPHERE

The circulation of the visible atmosphere of Venus has presented a very interesting dynamical problem. Photographs taken in the near UV of the high altitude haze on Venus indicate that these haze clouds are rotating retrograde with a 4-day period (Boyer and Guerin, 1969). Smith (1967) noted that this is rather surprising in view of radar observations which indicate that the solid surface of the planet is rotating retrograde with a 244-day period. Schubert and Whitehead (1969) suggested that the fast rotation of these high altitude clouds was caused by a rapid zonal flow induced by the very slow motion of the sun relative to the bulk of the atmosphere and in a direction opposite to this motion. Evidence for the possibility of strong zonal flows on a very slowly rotating planet came from 'moving flame' experiments and from theoretical considerations. The basic mechanism by which such a flow arises is presumed to be that a simple Hadley cell becomes tilted on Venus producing a vertical transport of horizontal momentum or Reynolds stress. The dynamics of such a flow regime for Venus have been discussed by Schubert and Young (1970), Malkus (1970), and Gierasch (1970). Tilting could result if the top of the cell (above the cloud tops) is in phase with the sun while the lower levels, due to weak radiative or convective heat conduction, lag behind. Thompson (1970) suggested that the tilting could also be a result of a special nonlinear instability in day to nightside Hadley circulations. Whichever of these theories is correct, it is apparent that the visible atmosphere of Venus is rotating with a 4-day period corresponding to zonal winds of $\sim 100\,m\,s^{-1}$. Such strong zonal motions may help to explain the apparent similarity between dayside and nightside temperature profiles on Venus.

Finally, it should be noted that the strong horizontal winds predicted throughout the Venus upper atmosphere may enable rapid sweeping of CO_2 photolysis products around to the darkside and consequent mixing to lower levels. Shimizu (1969) pointed out that horizontal winds of ~ 100 m s^{-1} are equivalent to vertical eddy diffusion coefficients of $\sim 10^7$ cm^2 s^{-1} for removing photolysis products from the dayside upper atmosphere.

Acknowledgement

Contribution #63 from the MIT planetary Laboratory.

References

Barth, C.: 1968, *J. Atmospheric Sci.* **25**, 564.
Barth, C.: 1970, in C. Sagan, T. Owen, and H. Smith (eds.), *Planetary Atmospheres*, D. Reidel Publishing Company, Dordrecht, Holland, p. 17.
Belton, M.: 1968, *J. Atmospheric Sci.* **25**, 596.
Boyer, C. and Newell, R.: 1967, *Astron. J.* **72**, 679.
Boyer, C. and Guerin, P.: 1969, *Icarus* **11**, 338.
Cloutier, P., McElroy, M., and Michel, F.: 1969, *J. Geophys. Res.* **74**, 6215.
Dickinson, R.: 1971, *J. Atmospheric Sci.* **28**, 885.
Donahue, T.: 1968, *J. Atmospheric Sci.* **25**, 568.
Donahue, T.: 1969, *J. Geophys. Res.* **74**, 1128.
Donahue, T.: 1971, *J. Atmospheric Sci.* **28**, 895.
Eshleman, V.: 1970, *Radio Sci.* **5**, 325.
Fjeldbo, G., Kliore, A., and Eshleman, V.: 1971, *Astron. J.* **76**, 123.
Gierasch, P.: 1970, *Icarus* **13**, 25.
Hansen, J. and Arking, A.: 1971, *Science* **171**, 669.
Hogan, J. and Stewart, R.: 1969, *J. Atmospheric Sci.* **26**, 332.
Lewis, J.: 1972, *Astrophys. J.* **171**, L75.
Malkus, W.: 1970, *J. Atmospheric Sci.* **27**, 529.
McConnell, J. C.: 1973, this volume, p. 309.
McElroy, M.: 1968a, *J. Atmospheric Sci.* **25**, 574.
McElroy, M.: 1968b, *J. Geophys. Res.* **73**, 1513.
McElroy, M.: 1969, *J. Geophys. Res.* **74**, 29.
McElroy, M. and Donahue, T.: 1972, *Science* **177**, 986.
McElroy, M. and Hunten, D.: 1969, *J. Geophys. Res.* **74**, 1720.
McElroy, M. and Strobel, D.: 1969, *J. Geophys. Res.* **74**, 1118.
Moos, H., Fastie, W., and Bottema, M.: 1969, *Astrophys. J.* **155**, 887.
Parkinson, T. and Hunten, D.: 1972, *J. Atmospheric Sci.* **29**, 1380.
Prinn, R.: 1971, *J. Atmospheric Sci.* **28**, 1058.
Prinn, R.: 1972, *J. Atmospheric Sci.* **29**, 1004.
Rasool, S. and Stewart, R.: 1971, *J. Atmospheric Sci.* **28**, 869.
Schubert, G. and Whitehead, J.: 1969, *Science* **163**, 71.
Shubert, G. and Young, R.: 1970, *J. Atmospheric Sci.* **27**, 523.
Shimizu, M.: 1969, *Icarus* **10**, 11.
Smith, B.: 1967, *Science* **158**, 114.
Stewart, R.: 1968, *J. Atmospheric Sci.* **25**, 578.
Stewart, R.: 1971, *J. Atmospheric Sci.* **28**, 1069.
Strickland, D.: 1972, *Bull. Am. Astron. Soc.* **4**, 363.
Thomson, R.: 1970, *J. Atmospheric Sci.* **27**, 1107.
Wallace, L., Stuart, F., Nagel, R., and Larson, M.: 1971, *Astrophys. J.* **168**, L29.

THE JOVIAN UPPER ATMOSPHERE

DARRELL F. STROBEL

Kitt Peak National Observatory, Tucson, Ariz. 85717, U.S.A.

1. Introduction

The upper atmosphere of Jupiter represents a refreshing change from the aeronomy of the terrestrial planets. The solar fluxes incident on it are smaller by a factor of 27 and the temperature and temperature gradients are correspondingly less than on the terrestrial planets. In such a cold atmosphere chemical reactions with high activation energies are of negligible importance. Unlike the terrestrial planets the major constituent H_2 is extremely light and all constituents except H and He fall off very abruptly above the turbopause (with the assumption that the concepts of eddy diffusion and turbopause are applicable to Jupiter).

The standard references on the upper atmosphere of Jupiter are the articles by Gross and Rasool (1964) and Hunten (1969). In addition to Hunten's article, the Proceedings of the Third Arizona Conference on Planetary Atmospheres (1969) is an excellent source of current knowledge on the Jovian planets. Of major importance for upper atmospheric studies are the measurements of the planetary albedo at Ly-α by Moos *et al.* (1969) and Moos and Rottmann (1972). Hunten (1969) has pointed out that useful information on the photochemistry of H_2 and the magnitude of the eddy diffusion coefficient in the vicinity of the turbopause can be obtained from this datum.

For the purposes of definition, the upper atmosphere of Jupiter is that region of the atmosphere where the total particle density is less than 10^{18} cm^{-3}. The NH_3 clouds on Jupiter typically occur at a level where the particle density is $\sim 10^{20}$ cm^{-3}.

2. Composition

The most important information required to investigate any planetary atmosphere is its composition. The composition of the Jovian atmosphere is still somewhat uncertain since the interpretation of spectroscopic observations of Jupiter is not straightforward and many previously interpreted abundances are questionable (McElroy, 1969). Two excellent reviews on the composition of the Jovian atmosphere, as determined by planetary spectroscopy, are available (McElroy, 1969; Hunten, 1971).

Molecular hydrogen is undoubtedly the major constituent. In a recent article Hunten and Münch (1973) argue that the He/H_2 ratio on Jupiter is consistent with and is given by the solar abundance of He, He/$H_2 = 0.11$ by number. Similarly the spectroscopically derived abundances for methane are consistent with the solar abundance of C, C/H $\sim 3.5 \times 10^{-4}$ by number (McElroy, 1969). On Jupiter the CH_4/H_2 mixing ratio is thus $\sim 7 \times 10^{-4}$. For NH_3 the situation is complicated by

B. M. McCormac (ed.), Physics and Chemistry of Upper Atmospheres, 345–353. All Rights Reserved.
Copyright © 1973 by D. Reidel Publishing Company, Dordrecht-Holland.

the fact that it is probably saturated over an extended altitude range and its concentration in the upper atmosphere will be extremely sensitive to the thermal structure near the tropopause. Based on a thermal model of Hogan *et al.* (1969) (model 3, cf. Figure 2), the NH_3 mixing ratio is approximately 3×10^{-7}. The photodissociation and photoionization of H_2, photoionization of He, and photodissociation of CH_4 result in the production of H atoms which emerge as the major constituent in the upper thermosphere. The photolysis of CH_4 can produce small concentrations of CH_3, C_2H_2, C_2H_4, and C_2H_6 in the upper atmosphere. Strobel (1969) has estimated from a study of CH_4 photochemistry that the C_2H_6 mixing ratio is $\sim 10^{-8}$–10^{-7}. The composition of the Jovian upper atmosphere is summarized in Figure 1.

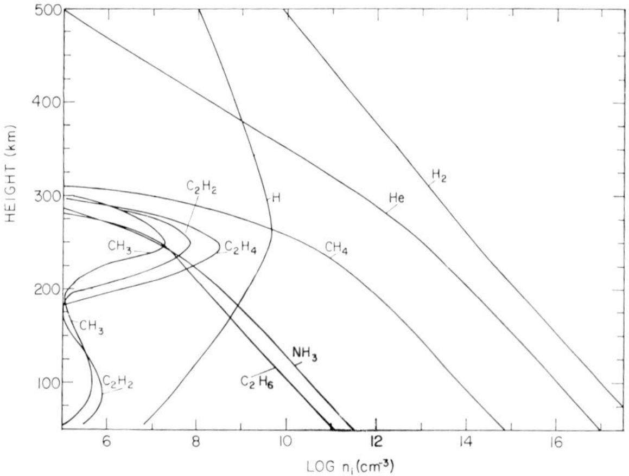

Fig. 1. A model of the density distribution of various constituents in the Jovian atmosphere with $T = 150\,\mathrm{K}$, $K = 3 \times 10^5\ \mathrm{cm^2\ s^{-1}}$ and $\phi_H = -7 \times 10^8\ \mathrm{cm^{-2}\ s^{-1}}$ at 350 km. The concentrations of the hydrocarbons are based on a model with CH formed in 25 % of the CH_4 dissociation events at Ly-α (Strobel, 1972). The NH_3 mixing ratio is based on curve 3 of Figure 2. (The height scale is arbitrary.)

With the reduced solar fluxes incident on the Jovian atmosphere the density profiles of constituents like CH_4 and NH_3 are not appreciably perturbed by photodissociation. The major source of atomic H is photodissociation and photoionization of H_2 high in the atmosphere and its sink is three body recombination in the mesosphere. In between these regions no appreciable production or loss of H atoms occurs and the flux of H atoms is approximately conserved. The approximate steady-state continuity equation for CH_4, NH_3, and H which describes their density distribution is $d\phi_i/dz \simeq 0$, or $\phi_i = $ constant, where for an isothermal atmosphere

$$\phi_i = -(D_i + K)\frac{dn_i}{dz} - \left(\frac{D_i}{H_i} + \frac{K}{H_{av}}\right) n_i \tag{1}$$

is the flux, D_i is the average diffusion coefficient, n_i is the number density, and H_i is the scale height of the ith constituent. In addition K is the eddy diffusion coefficient (Colegrove *et al.*, 1966), T is the temperature, z is the height, and H_{av} is the scale height of the mixed atmosphere. If it is assumed that H_{av} is constant, i.e. the gravitational constant g and the mean mass m_{av} remain constant with height in an isothermal atmosphere, and that $D_i = b_i/N$, where b_i is a constant binary collision coefficient for constituent i with H_2 and N is the total particle number density, then an analytic solution to Equation (1) is

$$n_i = \beta(z)\, n_0 + \{\beta(z) - \exp(-z/H_{av})\}\, \frac{H_{av} H_i}{K(H_{av} - H_i)}\, \phi_0 \tag{2}$$

where

$$\beta(z) = \left[\frac{2}{1 + \exp(z/H_{av})}\right]^{H_{av}/H_i - 1} \exp(-z/H_{av}) \tag{3}$$

and n_0 is the density at the turbopause (defined as where $b_i/N = K$), ϕ_0 is the constant value of ϕ_i, and $z = 0$ is the location of the turbopause. With $\phi_0 = 0$, the density distributions of CH_4 and NH_3 in Figure 1 are adequately given by Equation (2) with $n_0 = 2.1 \times 10^8$ and 1.3×10^5 cm^{-3}, respectively. Similarly for atomic H with $\phi_0 = -7 \times 10^8$ cm^{-2} s^{-1} and $n_0 = 4.4 \times 10^9$ cm^{-3}, the H density profile is adequately described above its turbopause.

With $\phi_0 = 0$, the asymptotic expressions for n_i are

$$n_i(z \to \infty) \to 2^{H_{av}/H_i - 1}\, n_0 \exp(-z/H_i) \tag{4}$$

$$n_i(z \to -\infty) \to 2^{H_{av}/H_i - 1}\, n_0 \exp(-z/H_{av}). \tag{5}$$

Thus at high altitudes the constituent is in diffusive equilibrium and it is mixed deep in the atmosphere. These asymptotes meet at $z = 0$ where $n_i = 2^{H_{av}\backslash H_i - 1} n_0$. For CH_4, $n \simeq 2^6 n_0$ or 64 times greater than the actual CH_4 density at the turbopause. Thus the use of asymptotes to describe the density distribution of minor constituents is not valid for the Jovian atmosphere.

3. Thermal Structure

Gillett *et al.* (1969) observed Jupiter in the spectral range 2.8 to 14 μ and deduced a temperature inversion in the middle atmosphere with a temperature minimum $\leqslant 115$ K and a temperature increase above this level to about 145 K. They suggested that the temperature inversion could be due to near IR solar heating in the 3.3 μm band of CH_4. Detailed thermal calculations by Hogan *et al.* (1969) showed that this mechanism can indeed produce a temperature inversion in the Jovian atmosphere (cf. model 3 of Figure 2).

From the occultation of Beta Scorpii by Jupiter, Hubbard *et al.* (1972) were able to obtain mean scale heights for the mesosphere and lower thermosphere. Based on a solar abundance for the He/H_2 ratio on Jupiter, the temperature range is 130 to 300 K, with the higher values representative of the thermosphere. The mean mesospheric temperature is approximately 160 K, although some data suggest values as high as 205 K.

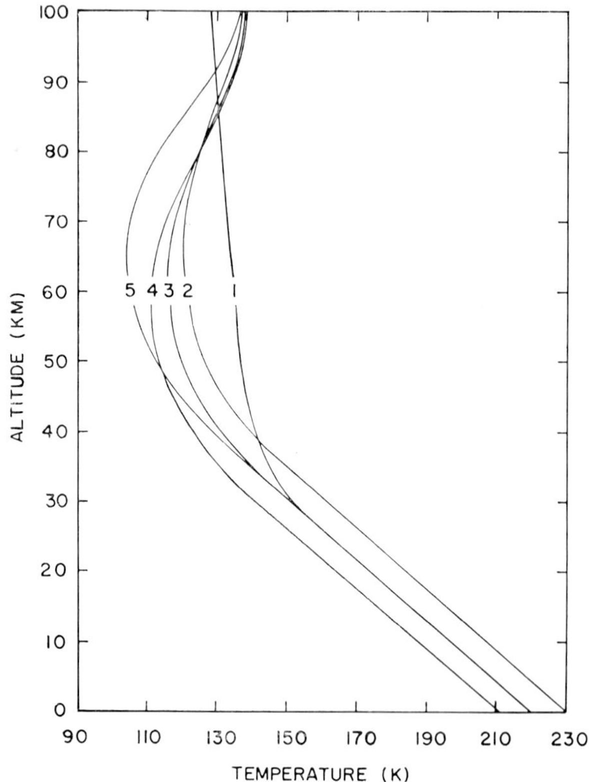

Fig. 2. Thermal models of the Jovian atmosphere from the calculations of Hogan *et al.* (1969) for various H_2 pressures and temperatures at the cloud tops. Model 3 is in agreement with the observations of Gillett *et al.* (1969).

Henry and McElroy (1969) have studied the deposition of EUV energy in the thermosphere of Jupiter and conclude that approximately 86% of this energy is locally converted to heat energy. Shimizu (1971) has calculated exospheric temperatures of 160 to 270 K at various latitudes for a mesopause temperature of 140 K and different periods of the solar cycle. In his calculation IR cooling was neglected. McGovern and Burk (1972) included IR cooling by CH_4 in their calculation but underestimated the cooling term by a factor of the local particle density divided by Loschmidt's number. For a typical mesopause density $\sim 10^{14}\ cm^{-3}$ this would result in an underestimate of the cooling term by 10^{-5}. Their range of exospheric temperatures was 150 to 200 K. For moderate solar activity, the average exospheric temperature is thus approximately 160 to 175 K, although neither calculation has properly included the IR cooling by polyatomic molecules. Preliminary thermal calculations with IR cooling by the author indicate a cooler thermosphere than calculated by either Shimizu (1971) or McGovern and Burk (1972), i.e. only 15° temperature rise above the mesopause temperature for moderate solar activity. Thus to a good approximation the upper atmosphere is isothermal and 150 K. If the high thermospheric temperatures reported by Hubbard *et al.* (1972) are correct, then euv solar heating cannot explain these measurements unless the sun was extremely active around the time of their measurements. However, the larger thermospheric scale heights were obtained on only two of their three best occultation events and could be due, in part, to systematic errors.

4. Photochemistry of CH_4 and NH_3

The photochemistry of CH_4 is important because hydrocarbons are formed which could play an important role in thermal energy budget of the mesosphere. Also it produces H atoms which can influence the interpretation of Ly-α albedo measurements. Approximately 70% of the photons absorbed by CH_4 are at Ly-α. The primary processes for CH_4 photolysis at Ly-α are

$$CH_4 + h\nu\,(\text{Ly-}\alpha) \rightarrow {}^1CH_2 + H_2 \qquad (6)$$
$$\rightarrow CH + H + H_2$$

where ${}^1CH_2$ denotes the singlet state of CH_2. ${}^1CH_2$ reacts with H_2 to form CH_3 while CH reacts with CH_4 to produce C_2H_4, unless the turbopause is very low in the atmosphere; then the reaction $CH + H_2 + M \rightarrow CH_3 + M$ is more probable. Unfortunately, no definitive laboratory data are available for the relative probabilities of these two dissociation paths (Equation (6)). Strobel (1969) has suggested that CH may be produced 25% of the time. CH_3 and H react to recycle CH_4 while CH_3 reacts with itself to produce C_2H_6. Since $n(H) \gg n(CH_3)$, (cf. Figure 1) and the rate coefficients are comparable, only a small fraction of CH_3 produces C_2H_6. The major source of heavier hydrocarbons is the formation of C_2H_4 by $CH + CH_4$. Some C_2H_4 will react with H to give C_2H_5 which in turn reacts with H to produce $2CH_3$. Also C_2H_4 will dissociate to give C_2H_2. The analysis can be extended to include more constituents and more reactions but the conclusion still remains that some heavier hydrocarbons will be produced and CH_4 cannot be completely recycled. To maintain the CH_4 abundance on Jupiter the heavier hydrocarbons diffuse down to the hotter, dense regions of the planet where they undergo thermal decomposition and react with H_2 to produce fresh CH_4 which flows upward to replenish the dissociated CH_4. An alternate explanation for the stability of CH_4 was proposed by McNesby (1969) and involves hot-molecule chemistry. He proposed that absorption of long wavelength UV radiation by CH_3 may excite it sufficiently to overcome the high activation energy of the reaction $CH_3 + H_2 \rightarrow CH_4 + H$ and thus recycle CH_4. This does not solve the problem of breaking the carbon-carbon bond formed in the production of C_2H_4. It can be seen that the stability of CH_4 is heavily dependent on its dissociation path at Ly-α.

Below 945 Å, CH_4 is ionized by the solar EUV flux. However most of this radiation is absorbed by H_2, He, and H before it reaches the CH_4. The small amount of CH_4 ionized will, through subsequent reactions, form polymers (Saslaw and Wildey, 1967).

The photochemistry of NH_3 in the Jovian atmosphere has been discussed by McNesby (1969). The primary processes of NH_3 photolysis are

$$NH_3 + h\nu \rightarrow NH_2 + H \tag{7}$$
$$\rightarrow NH + H_2.$$

Since NH will probably react with H_2 to form NH_2, the relative probabilities of the dissociation paths of Equation (7) are of little importance. McNesby suggests that the NH_2 will intensely absorb sunlight between 4300 to 9000 Å and be excited to the A state, from which it may overcome the high activation energy of $NH_2 + H_2$ and recycle NH_3. If these processes occur, then it is unlikely that much NH_3 will be irreversibly destroyed in the upper atmosphere. In addition most of the NH_3 is photolyzed low in the atmosphere where association reactions, e.g., $H + NH_2 + M \rightarrow \rightarrow NH_3 + M$, are relatively fast.

5. Photochemistry of H_2

Molecular hydrogen has a dissociation continuum below 845 Å and an ionization continuum below 804 Å (Cook and Metzger, 1964). In addition, discrete absorption in the Lyman and Werner bands followed by fluorescent dissociation leads to the production of H atoms (cf. Field *et al.*, 1966; Stecher and Williams, 1967). These processes may be summarized by

$$H_2\left(X^1\Sigma_g^+, v = 0\right) + h\nu\left(\lambda < 1109\,\text{Å}\right) \rightarrow H_2\left(B^1\Sigma_u^+, v = v'\right) \rightarrow$$
$$\rightarrow H_2\left(X, v = v''\right) + h\nu. \tag{8}$$

If $v'' > 14$, then the Lyman band emission leaves the molecule in the vibrational continuum of the ground state and hence dissociated. A similar equation for the Werner bands (the $C^1\Pi_u$ state) holds for $\lambda < 1009$ Å. For each H_2 molecule and He atom ionized two H atoms will be produced (Gross and Rasool, 1964). With the more recent Hinteregger (1970) EUV solar flux measurements an estimate of the H atom production rate similar to Hunten's (1969) estimate can be made. Dissociation of H_2 and ionization of H_2 and He by photons with $\lambda < 845$ Å results in the production of 2.3×10^9 atoms $\text{cm}^{-2}\,\text{s}^{-1}$. Absorption in the Lyman and Werner bands followed by fluorescent dissociation gives a production rate of $\sim 3 \times 10^8\ \text{cm}^{-2}\,\text{s}^{-1}$ and $\sim 2 \times 10^6\ \text{cm}^{-2}\,\text{s}^{-1}$, respectively. For a rapidly rotating planet the global average is one-fourth of the total production rate or $\sim 7 \times 10^8\ \text{cm}^{-2}\,\text{s}^{-1}$, in good agreement with Hunten's estimate of $10^9\ \text{cm}^{-2}\,\text{s}^{-1}$ based on the larger fluxes of Hinteregger *et al.* (1965). The H atoms are transported down by molecular diffusion and eddy diffusion to the denser regions of the atmosphere where they recombine to form H_2. A representative H density profile is shown in Figure 1.

The planetary albedo at Ly-α measured by Moos *et al.* (1969) and by Moos and Rottmann (1972) is in simple terms a measurement of the column density of H atoms above the absorbing CH_4 layer. This column density is a sensitive function of the eddy diffusion coefficient. Hunten (1969) estimated from the Moos *et al.* (1969) measurement that the eddy diffusion coefficient is $5 \times 10^6\ \text{cm}^2\,\text{s}^{-1}$ near the turbopause. In a more recent and detailed analysis of their measurements Wallace and Hunten (1973) find a much lower eddy diffusion coefficient, $\sim 3 \times 10^5\ \text{cm}^2\,\text{s}^{-1}$.

6. Ionosphere

The formation of the Jovian ionosphere has been discussed by Gross and Rasool (1964) and Hunten (1969), who concluded that H^+ is the major ion throughout the ionosphere. Dalgarno (cf. Hunten, 1969) suggested that, in view of the slow removal of ionization by radiative recombination of H^+ and e, radiative association of H^+

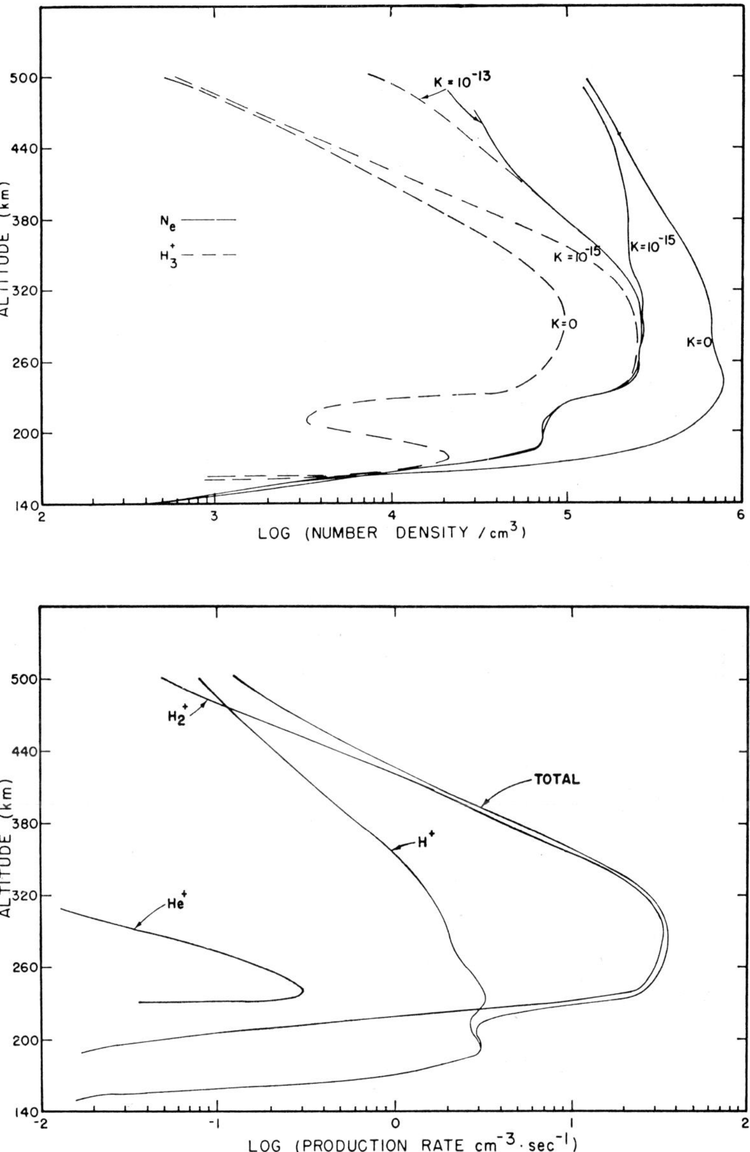

Fig. 3.   In the upper figure the height profiles of H_3^+ ion and electron densities are given for selected values of the radiative association rate of H^+ to H_2. The lower figure gives the primary production rates of H^+, H_2^+, and He^+ (Prasad and Capone, 1971.)

with H_2 to form H_3^+, which is rapidly removed by dissociative recombination, could be important. In a recent study of the Jovian ionosphere, Prasad and Capone (1971) investigated the possible importance of radiative association and found that H_3^+, rather than H^+, would be the major ion and that a factor of 4 reduction in the electron density is possible. In Figure 3 their production rates of ionization for a model atmosphere with 50% H_2 and 50% He are given. Essentially all the photoions are H_2^+; they react with H_2 and H to produce H_3^+ and H^+, respectively. Their results for the Jovian ionosphere are illustrated in Figure 3 as a function of the radiative association rate of H^+ with H_2. However, McElroy (1973) has questioned on quantum-mechanical grounds whether radiative association of H^+ with H_2 is possible and instead suggests that H^+ will react with vibrationally excited $H_2(v)$ to form H_2^+. This reaction is exothermic for $v \geqslant 3$. Calculations indicate that the H_2 vibrational temperature is $\sim 1000\,K$ in the Jovian ionosphere. In addition, McElroy has noted that dissociative ionization of H_2 by photoelectrons is the largest primary source of H^+; a source which has been omitted in all previous models. McElroy also points out that the Jovian ionosphere would terminate abruptly at the turbopause if H^+ could charge transfer with CH_4 at a rate faster than 10^{-18} cm^3 s^{-1}. Thus the present Jovian ionospheric models are quite unsatisfactory. Henry and McElroy (1969) and Prasad and Capone (1971) have studied the energy balance of the electrons and ion gases. They conclude that the electrons, ions, and neutrals have equal temperatures.

7. Summary

In the past the greatest uncertainty about the Jovian atmosphere was the He/H_2 ratio There is now increasing evidence which points to a solar abundance for He/H_2. The Jovian atmosphere is thus composed of 90% H_2 and 10% He with small amounts of CH_4 and NH_3 in the upper atmosphere, volume mixing ratios of $\sim 7 \times 10^{-4}$ and 3×10^{-7}, respectively. To a good approximation the upper atmosphere is isothermal at 150 K. Of great importance would be thermal structure calculations for the mesosphere. The planetary albedo at Ly-α implies an eddy diffusion coefficient $\sim 3 \times 10^5$ cm^2 s^{-1} with the atomic H turbopause at a particle density of 1×10^{14} cm^{-3}. Laboratory work is needed to determine the probabilities of CH_4 dissociation paths at Ly-α. A study of fundamental physical and chemical processes probable in the Jovian ionosphere is required. The photochemistry of NH_3 should be put on a quantitative basis.

Note added in proof. For more recent discussions of the thermal structure of the Jovian thermosphere and the photochemistry of hydrocarbons and ammonia consult the author's papers in *J. Atmos. Sci.* **30**, 1973.

References

Colegrove, F. D., Johnson, F. S., and Hanson, W. B.: 1966, *J. Geophys. Res.* **71**, 2227.
Cook, G. R. and Metzger, P. H.: 1964, *J. Opt. Soc. Am.* **54**, 968.
Field, G. B., Somerville, W. B., and Dressler, K.: 1966, *Ann. Rev. Astron. Astrophys.* **4**, 207.

Gillett, F. C., Low, F. J., and Stein, W. A.: 1969, *Astrophys. J*, **157**, 925.

Gross, S. H. and Rasool, W. I.: 1964, *Icarus* **3**, 311.

Henry, R. J. W. and McElroy, M. B.: 1969, *J. Atmospheric Sci.* **26**, 912.

Hinteregger, H. E.: 1970, *Ann. Geophys.* **26**, 547.

Hinteregger, H. E., Hall, L. A., and Schmidtke, G.: 1965, *Space Res.* **5**, 1175, North-Holland, Amsterdam.

Hogan, J. S., Rasool, S. I., and Encrenaz, T.: 1969, *J. Atmospheric Sci.* **26**, 898.

Hubbard, W. B., Nather, R. E., Evans, D. S., Tull, R. G., Wells, D. C., VanCitters, G. W., Warner, B., and Vanden Bout, P.: 1972, *Astron. J.* **77**, 41.

Hunten, D. M.: 1969, *J. Atmospheric Sci.* **26**, 826.

Hunten, D. M.: 1971, *Space Sci. Rev.* **12**, 539.

Hunten, D. M. and Münch, G.: 1973, *Space Sci. Rev.,* in press.

McElroy, M. B.: 1969, *J. Atmospheric Sci.* **26**, 798.

McElroy, M. B.: 1973, *Space Sci. Rev.,* to be published.

McGovern, W. E. and Burk, S. D.: 1972, *J. Atmospheric Sci.* **29**, 179.

McNesby, J. R.: 1969, *J. Atmospheric Sci.* **26**, 594.

Moos, H. W. and Rottman, G. J.: 1972, Paper, American Astronomical Society, Division of Planetary Sciences, 2nd Annual Meeting, Hawaii.

Moos, H. W., Fastie, W. G., and Bottema, M.: 1969, *Astrophys. J.* **155**, 887.

Prasad, S. S. and Capone, L. A.: 1971, *Icarus* **15**, 45.

Proceedings of the Third Arizona Conference on Planetary Atmospheres: 1969, *J. Atmospheric Sci.* **26**, 795.

Saslaw, W. C. and Wildey, R. L.: 1967, *Icarus* **7**, 85.

Shimizu, M.: 1971, *Icarus* **14**, 273.

Stecher, T. P. and Williams, D. A.: 1967, *Astrophys. J.* **149**, L29.

Strobel, D. F.: 1969, *J. Atmospheric Sci.* **26**, 906.

Strobel, D. F.: 1972, paper presented at NATO Advanced Study Institute on Planetary Atmospheres, Istanbul, Turkey. See paper in *J. Atmospheric Sci.* **30**, 1973.

Wallace, L. and Hunten, D. M.: 1973, *Astrophys. J.,* to be published.

PART VI

SUMMARY AND CONCLUSIONS

SUMMARY

GILBERT WEILL

Institut d'Astrophysique, Paris, France

1. Introduction

The program of the Summer Advanced Study Institute in 1972 reflected the evolution of interests in the broad field of physics and chemistry of the atmospheres. The pending problems of aurora and airglow received less attention but the complex and important aspects of chemistry and dynamics in the mesosphere and stratosphere received well deserved treatment and so did the young science, born with the space age, of the atmospheres of other planets in the solar system.

The concluding session was planned to produce an overall summary of the results reported during the two weeks, through the work of panelists.

Some session chairmen (Bowhill, Ortner, and Thomas) could not be present but asked other panelists to present their summaries. The present chapter was edited with the panelists from a transcript of the presentations and discussions at the meeting.

The neutral atmosphere is clearly organized into problem areas and reviewed by Vallance Jones; Thomas reviews the ionosphere, including the important developments in the chemistry of water cluster ions in the D region.

Berning further deals with reaction chemistry and atmospheric models. In his session, the role of NO chemistry and aerosols in the lower stratosphere is assessed.

Rishbeth and Egeland reviewing physical processes give in a nutshell a summary of the complex theories of minor constituents diffusion and of electric field and currents in the ionosphere. The following two sessions deal with experimental results and their interpretation. These are summarized by Llewellyn and Schiff from quite different viewpoints. From both reviews, a clear list of open questions can be derived, the solution of which will require considerable work in the field, in the laboratory and at the theoretical level.

In a review of reviews, Hunten outlines the aeronomic problems of Mars, Venus and Jupiter, touching Titan on the way.

A very few papers have escaped coverage in the reviews. This is the case, in particular, for lectures of tutorial types (introductions by Bowhill and Schiff).

I think it has been an achievement of the Institute to outline the path of future research and we may look forward to hear, in 1974, which of the questions now standing have been solved.

2. The Neutral Atmosphere*

For a number of years it has been recognized that a major task to be completed is to be able to describe, understand and specify the properties of the neutral atmosphere

---

* Presented by A. Vallance-Jones.

B. M. McCormac (ed.), Physics and Chemistry of Upper Atmospheres, 357–377. All Rights Reserved.
Copyright © 1973 by D. Reidel Publishing Company, Dordrecht-Holland.

and their variation with geophysical conditions. Composition, temperature and dynamics are closely related while interactions with the ionized components of the atmosphere are being recognized as increasingly important.

2.1. DENSITY AND TEMPERATURE

The atmosphere is heated by solar UV, interactions with the magnetosphere, and gravity waves from below. This energy input determines, in conjunction with cooling, photochemical and transport processes, the distribution of temperature, density and composition.

Measurements of density (and of temperatures derived from them) are very extensive. Such inferred exospheric temperatures depend on the assumption of diffusive equilibrium above a chosen reference level. On the other hand direct measurements of temperatures by such methods as incoherent scatter, 6300 Å line profiles, diffuse resonance and other techniques are becoming available. A 2 to 3 hr difference has been found between the diurnal density maximum at 1400 hr and the temperature maximum measured by incoherent backscatter. Diffusive equilibrium may need to be replaced by a time variable solution or models including wind effects.

Seasonal density variations at 100 to 120 km are important and individual measurements sometimes show evidence of wave effects. There is a semi-annual variation having an amplitude of at least 7% at 90 km. Geomagnetic storm effects on density have been shown to be largely localized and to occur with a time lag as short as an hour at the high latitudes where they are centered.

Direct temperature measurements are not in complete agreement with the atmospheric models. For example at the equinox the lowest temperatures are found at high latitudes rather than at the equator after midnight. At the solstice the winter pole is coldest. This trend is modified under disturbed geomagnetic conditions as shown by Rees; a $300°$ temperature rise is found at a latitude of $70°$ at 170 km for $Cq=8$. This is deduced from chemical release temperature measurements at heights between 140 and 160 km which show several hundred degree increases as Cq varies from 0 to 8.

2.2. COMPOSITION

There are still uncertainties regarding atmospheric composition. At 150 km the measurements described by von Zahn have led to the acceptance of a value of $n(O)/n(N_2)$ of unity near 150 km rather than the CIRA 65 value of about 0.4. The new value is reflected in CIRA 1970, shortly to appear. However, at 400 km, OGO-6 mass spectrometer experiments give results which do not seem to be consistent with the new 150 km values under the assumption of diffusion equilibrium. Of special importance is the observation mentioned by von Zahn and Brace of significant changes in $[N_2]$ at high latitudes in response to large magnetic storms.

In the troposphere and stratosphere major components are fully mixed but minor components such as H_2O, CO, and CH_4 which may be oxidized or condensed show variations with height. Some of these components have been measured by the elegant sampling technique described by Martell.

2.3. Atmospheric models

There was considerable discussion of recent atmospheric models i.e. Jacchia 1971 on which CIRA 1970 is based. The time independent base has been lowered to 90 km. It appears, however, that the models are still not completely satisfactory and especially will not be correct in the auroral oval region.

Photochemical models of the mesosphere and lower thermosphere were briefly reviewed by von Zahn.

2.4. Winds

Rees' review of winds in the neutral atmosphere presented a very fully documented account of this topic. Incoherent scatter measurements and chemical release techniques from rockets are providing observations of wind systems in the 100 to 300 km regions. These measurements make it possible to check the predictions of models in which the diurnal wind patterns are calculated from a thermospheric pressure distribution (from an atmospheric model), electron density distributions and a model of the horizontal electric field in the thermosphere. The results of several models were described. At mid-latitudes they predict winds rotating clockwise, directed SE at midnight and NW at midday. The predicted velocities for models neglecting electric fields are typically 200 to 300 m s^{-1} by night and 100 m s^{-1} or less by day. The effects of including the dynamo electric field from the S_q magnetic field modifies the results, with reductions in velocity and changes in direction. Observed winds often behave qualitatively in the manner predicted with considerable differences in detail. Rees suggests that detailed comparative measurements of E and F region winds may make it possible to solve the wind problem and that of the sources of the S_q ionospheric current system.

At higher latitudes geomagnetic disturbances produce large effects on the wind patterns. Westward winds are observed of the order of 500 m s^{-1} between 130 and 150 km in association with 200 to 300 γ positive bays. For negative bays the eastward wind is about five times less in proportion to the field. Winds in association with substorms are presumably generated by the combined effects of heating and of ion motions produced by magnetospheric electric fields. Such winds are also observed above 200 km in association with substorms.

As pointed out by von Zahn thermospheric winds may be of considerable importance in determining atmospheric composition. The variations in O/N$_2$ ratio suggest a circulation upward at poles and downward at the equator. Von Zahn suggested that the global distribution of minor constituents may be sensitive to winds. Moreover the distinctive winter poleward He bulge (a 10 $\times$ excess over the value of Jacchia 1965 at 53° geomagnetic) appears to be due to a global circulating system transporting air from the summer to the winter pole region with vertical flow at high latitudes.

2.5. Aerosols and particles

Link presented evidence from optical scattering measurements indicating the presence of an aerosol layer near or just above 80 km. This layer is apparently due to meteor debris and balloon measurements show it is enhanced after meteor showers. Link has reached conclusions as to particle sizes from the characteristics of the scattered light. Further information on aerosols appears in Section 4.

2.6. GEOCORONAL HYDROGEN

Bertaux presents an excellent review of the theory governing the distribution of H in the upper thermosphere and exosphere. Unlike other constituents H escapes at the top of the atmosphere so that its vertical concentration adjusts itself to the escape rate. Above the exobase H may move laterally in ballistic orbits reentering the exobase elsewhere. Eventually an overall steady state may be established. Recent work described has extended the vertical diffusion process to the time dependent case and Bertaux suggests that an overall time dependent solution including lateral flow is a challenge for the future. Recent developments on the theory of the exosphere are the inclusion of the effect of solar Lyα radiation pressure on the satellite orbit particles. This has been shown to be considerable, shortening the lifetime of particles in such orbits significantly compared to photo-ionization and proton charge exchange. It could also increase apogees on the nightside and produce the 'geotail' suggested by recent OGO-5 observations.

2.7. ELECTRON TEMPERATURES

The relationship between neutral, electron, and ion temperatures was discussed by Brace. There is still some disagreement between incoherent scatter and probe measurements of T_e from 100 to 120 km. The latter are often as high as 2000 K while the former are consistent with expected neutral temperatures. There is also an unresolved discrepancy above 350 km but in between the methods agree.

Brace discusses the theoretical reasons for higher F region T_e's at equator by day in comparison with the minimum at night. Models for diurnal T_e variations with latitude and season are described.

2.8. CONCLUSIONS

The variability of the atmosphere and its dynamic effects are becoming increasingly of concern in the interpretation of many phenomena.

Developing experimental techniques, including numerous incoherent scatter facilities, 120 km satellites, dye laser methods of following metallic vapor trails by night, optical temperature measurements and satellite photometry are providing the basis for testing realistic theories and descriptions of the neutral atmosphere. It is to be expected that much better models of the atmosphere will become available. At the same time, more experiments can be planned where the necessary parameters are measured rather than assumed from a model. This is particularly true of the atmosphere at high latitudes which may be highly variable. Such information is becoming increasingly necessary for a detailed understanding of the details of auroral and airglow emissions.

3. Ionized Regions of the Atmosphere (< 1000 km)*

In this session the greatest attention was devoted to the positive-ion and negative-ion composition of the D region, below about 90 km, and the chemical processes which lead to this composition. Information on the role of high-energy particles as a source of ionization and on the height distribution of electron concentration in this part of the ionosphere, as deduced from observations of VLF transmissions, was also described. Very little interest was shown in the E region, except for the anomalous ion and neutral composition observed under auroral conditions, and the meteoric ions recorded by mass-spectrometer observations at mid-latitudes. For F region heights the most interesting results described were those from satellite-borne mass-spectrometers and ion-traps. It will be convenient to describe the points of interest relating to each of the regions in turn.

3.1. D REGION

3.1.1. *Positive Ions*

Narcisi gave a comprehensive review of rocket-borne observations with quadrupole mass-spectrometers. It appears that the transition from water cluster ions, $H^+(H_2O)_n$, to NO^+ and O_2^+ occurs under normal conditions at about 82 km in daytime and 86 km at night. However, under auroral conditions the conventional ions NO^+ and O_2^+ predominate down to about 75 km, and observations during a PCA also reveal a significant lowering of the transition level. Narcisi also mentioned the approximate variation of cluster-ion densities with time of day, in which a three-fold increase in concentration occurred during the pre-sunrise period, when the solar zenith angle decreased from 102.5 to 90°. This latter result implies that cluster ions seem to be produced by fairly low-energy solar radiations, since shorter wavelengths are absorbed at this time. Narcisi mentioned a previous suggestion by Hunt that these cluster ions might arise from the direct ionization of water conglomerates. He also presented evidence for a fast NO^+ to water cluster ion conversion process from measurements obtained during a solar eclipse.

Reid described previous attempts to synthesize numerically the water cluster ion distribution in the D region. He pointed out that the reaction scheme beginning with O_2^+, as devised by Ferguson and Fehsenfeld, was capable of reproducing the observations qualitatively but the production rate of this primary ion under normal conditions was insufficient to yield the observed cluster-ion concentrations. Narcisi mentioned that under disturbed conditions, the larger O_2^+ production rates (from the ionization process or from charge transfer from N_2^+ ions produced simultaneously) were sufficient to yield the cluster-ion concentrations observed. Reid emphasized the need to devise a scheme for producing the water cluster ions from NO^+ since this represents the major ion produced during normal conditions, and this question was also consider-

* Presented by L. Thomas.

ed by Heimerl and Isaksen. The following indicate the suggestions made by the different speakers:

Narcisi: Direct two-body reaction of NO^+ with water conglomerates;

Reid: Charge transfer from NO^+ to vibrationally excited O_2 molecules:
$NO^+ + (O_2)^* \rightarrow NO + O_2^+$; and

Heimerl: Estimated that the time constant for conversion of NO^+ to $H^+.(H_2O)_2$ needs to be about 70 s for 80 km and suggested that only reactions having smaller time constants need to be considered. The scheme he proposed was as follows:

$$NO^+ \xrightarrow{(N_2, M)} NO^+.N_2 \xrightarrow{(CO_2)} NO.CO_2 \xrightarrow{(H_2O)} NO^+.H_2O \xrightarrow{(OH \text{ or } HO_2)}$$
$$H^+.(H_2O) \xrightarrow{(N_2, M)} H^+.(H_2O)N_2 \xrightarrow{H_2O} H^+.(H_2O)_2.$$

The critical step in this scheme is either of the conversions of $NO^+.H_2O$ to $H^+.(H_2O)$:

$$NO^+.H_2O + OH \rightarrow H^+.(H_2O) + NO_2$$

or

$$NO^+.H_2O + HO_2 \rightarrow H^+.(H_2O) + NO + O_2$$
$$\rightarrow H^+.(H_2O) + NO_3.$$

In order to satisfy the 70 s criterion the rate coefficients of either of these two reactions need to be about 10^{-9} cm^3 s^{-1}. That involving OH is exothermic but the bond energies of HO_2 and NO_3 are uncertain and it is not known whether the reactions involving HO_2 are exothermic or endothermic.

Isaksen: Similar to that of Heimerl except that $NO^+.H_2O$ is formed directly from $NO^+.N_2$ rather than via $NO^+.CO_2$.

Burke pointed out in the discussion of these processes that the direct clustering of water molecules to $NO^+.(H_2O)_m$ or $H^+.(H_2O)_n$, which proceed by three-body mechanisms at temperatures prevailing in laboratory measurements, may occur by two-body reactions at the low-D region temperatures:

$$NO^+.(H_2O)_m + H_2O \rightarrow NO^+.(H_2O)_{m+1} \quad (m \geq 0)$$
$$H^+.(H_2O)_n + H_2O \rightarrow H^+.(H_2O)_{n+1} \quad (n \geq 1).$$

3.1.2. *Negative Ions*

In a review of negative-ion composition measurements reported in the literature, Narcisi mentioned that the AFCRL and Heidelberg groups had carried out a total of 7 rocket-borne mass-spectrometer experiments. However, the results of the two groups had been very different. At Churchill under quiescent conditions, and near totality of a solar eclipse at Wallops Island, the AFCRL measurements had indicated heavy ions of masses 60 to 152 between 73 and 92 km and a layer of these and heavier ions near 88 km. These had been tentatively identified as $NO_3^-(H_2O)_n$, with $n = 0 - 5$, with possible admixtures of $CO_3^- (H_2O)_n$, $n = 0 - 5$, and CO_4^- By contrast, the Heidelberg measurements had indicated large concentrations of Cl^- below 80 km with CO_3^-, HCO_3^- and NO_3^- also being dominant from 71 to 80 km.

Narcisi also drew attention to a pronounced change in negative-ion composition during a PCA event. During daytime O_2^- was the predominant species, with large concentrations between 72 and 94 km. At night O^- was the major ion between 76 and 94 km. Such a change in negative-ion composition was predicted in Reid's numerical studies, according to which O_2^- should be the major ion down to 70 km during a PCA event, with significant concentrations of CO_4^-, whereas NO_3^- was expected to predominate below about 77 km under normal conditions. However, Narcisi pointed out that the relatively large concentrations of O^- and O_2^- measured in the PCA event above 80 km cannot be explained since these ions are apparently destroyed rapidly by associative detachment with atomic oxygen. Reid also mentioned that in connection with the photodetachment of electrons from negative ions, a value for the electron affinity of NO_3^- (3.9 eV) was now available.

3.1.3. *Electron Concentrations and Sources of Ionization*

Egeland described how observations of VLF transmissions could be used to deduce D region electron concentrations and to provide a monitor of disturbed conditions. Potemra presented examples of VLF recordings during different degrees of magnetic disturbance, and also argued from an analysis of high energy particle data and an examination of various sources of ionization that such particles represent an important source for the D and E regions during disturbed and undisturbed conditions at night. They could also be important during the daytime for conditions of high magnetic activity or large solar zenith angles. At this point it seems worth mentioning that there still exists a long-standing and important problem in aeronomy: namely, whether ionization by energetic particles is important at mid-latitudes even above 100 km.

3.2. E REGION

Narcisi also gave a review of O_2^+ and NO^+ concentration measurements at E region heights and Reid outlined the chemistry involved in understanding these measurements. Narcisi mentioned that the relative concentrations of metal ions in the E region were generally in agreement with those of the corresponding elements in chondrite meteorites. Reid described the current ideas on the source and sink of these metal ions but emphasized the need for a more detailed examination of the chemistry involved.

3.3. F REGION

In reviewing mass spectrometer measurements at high altitudes, Narcisi mentioned that the satellite-borne experiment of Taylor *et al.*, Hoffman, and Brinton *et al.* had shown a trough in the He^+ and H^+ concentrations near $\pm 60°$ geomagnetic latitude with a variable peak of ionization on the high latitude side of the trough. The observations of the topside ionosphere by Taylor *et al.* had also shown an equatorial trough in the concentration of He^+, located between about $\pm 30°$ geomagnetic latitude, and also enhanced concentrations of this ion in the winter hemisphere.

Hanson outlined the processes believed to control the equatorial F region and also showed relevant plasma temperature and concentration data derived from the ion-trap experiment on OGO-6 satellite. A particularily interesting feature was the latitude variation of temperature which showed lower values in summer than in winter and very low values, below the neutral gas temperature, near the magnetic equator. These results Hanson attributed to inter-hemispheric plasma flow, quasi-adiabatic expansion occurring as the plasma rises on the summer side, and compression as it descends on the winter side.

The ion-trap data indicates that the equatorial F region generally shows a regular smooth variation but a strikingly fine structure is sometimes observed in the latitude range $\pm 20°$ geomagnetic, especially in the Atlantic area. Here the He^+ and H^+ ions are found to disappear and there is strong evidence for the presence of ions Fe^+. The explanation of these Fe^+ ions at heights near 1000 km raises certain problems. It is expected that if they come from below the chief difficulty will be to understand their transport up to about 180 km.

Another interesting feature revealed by the OGO-6 observers is the presence of 'holes' in the F region near the magnetic equator, in which the ion density is decreased by about an order of magnitude in a distance of about 1 km.

Thuiller produced some evidence that the tropical 6300 Å nightglow features which were observed with OGO 4 satellite can be explained by the joint action of winds and dissociative recombination.

It seems evident that future analyses of these and other data could provide a new and exciting picture of the equatorial F region.

4. Reaction Chemistry: Atmospheric Models*

Not surprisingly, the substantial efforts directed to understanding the chemistry and molecular/atomic interaction processes in the atmosphere seem to generate more problems than solutions. Thus the identification of more and more complex ion and neutral species, together with the rates at which these may enter reaction, illuminates growing needs for experimental data and accompanying theoretical understanding in areas of more than academic interest. The complexity of chemical processes and inter-acting species is considerable over a substantial altitude regime; at low altitudes relatively minor species – by concentration – appear important in the life cycle of pollutants and to finite climatic changes.

A concern of importance is the possible reduction of O_3 in the troposphere and stratosphere by NO_x released by the supersonic transport. Crutzen reviewed the relevant chemistry and identified the nagging questions which must be addressed. In

---

* Presented by W. Berning.

the non-SST atmosphere the observed loss of O_3 in downward diffusion is a minute fraction of the O_3 produced in the atmosphere. Model calculations indicate that the observations and calculations may be reconciled by catalytic chains involving NO_x if mixing ratios of the order of 10^{-8} at 30 km for $NO + NO_2$ exist. Confidence in this result and extrapolation to the SST – disturbed atmosphere rests on uncertainties in the sources of NO_x, their mixing ratios, and their several removal processes. Odd-hydrogen (HO_x) compounds, while not a factor in O_3 destruction below 45 km, are important in the formation of HNO_3 from NOx and model calculations indicate pronounced conversion of NO_x to HNO_3 below 35 km. There is major uncertainty in the photodissociation rates of HNO_3, and its conversion back to NO_x may be slow and is certainly altitude dependent. Oxidation of CH_4 appears important at altitudes of interest in that reactions in the troposphere lead to O_3 formation via CO and OH reactions, whereas in the stratosphere oxidation is important to the OH balance and conversion of NO_2 to HNO_3. Thus the chemistry chain controlling O_3 loss and production in the atmosphere is complex, and there remain significant uncertainties at critical points in this chain. Wayne did call attention to definitive work by Bayes and Jones at the University of California, Los Angeles, on the quantum yield of photo-dissociated NO_2, a problem identified as important by Crutzen. Similar work for HNO_3 and NO_3 is considered of great importance.

The discovery of cluster ions in the D region and the close correlation in altitude of hydrated protons and noctilucent clouds suggests aerosols either as centers of nucleation or nucleation products. Castleman carefully reviewed present knowledge of aerosol classification and nature; large aerosols of diameters 0.1 to 1 μm appear capable of formation by chemical reactions or of participating in chemical reactions. Supersaturation ratios of approximately 6 (with respect to the condensed phase) are required for stable nucleation to occur, and times for reaching such ratios must be short compared to dissipation processes. Interestingly, experiments show that hydrated protons have one of the lowest nucleation barriers. Considering only aerosols formed *in-situ*, and above the tropopause, laboratory studies indicate: (1) reactions leading to products which undergo homogeneous nucleation; (2) chemical reactions followed by hetero-molecular nucleation; (3) ion cluster switching reactions; (4) nucleation about ions; and (5) addition and polymeric reactions. Aerosol layers have been found at many altitudes, even above 100 km. Direct, rocket carried, sampling techniques have collected materials from 65 to 145 km, and procedures are valid for particles as small as 0.1 μm. Two phenomena, of particular interest in the question of aerosol participation in chemical processes, are the noctilucent cloud layer and the Na layer. Considering the noctilucent clouds, direct sampling indicates particle concentrations 2 to 3 orders of magnitude less than photometric measurements during passage through the layers. The direct samples suggest a dense core surrounded by a deformable coating with no definite indication of the constituents. Aerosols formed by nucleation about ions may be a preferable explanation of the layer. Certainly theoretical work is needed for calculating supersaturation required for nucleation about specific ions. Explanation of the Na layer in terms of an aerosol model is quite difficult. Nucleation of Na compounds or accretion on surfaces of aerosols is followed by ever more stabilizing reactions so that a reversible cycle producing free Na at concentrations required $(10^3 - 10^4 \, cm^{-3})$ is very unlikely.

General characteristics of the Na distribution in the atmosphere were addressed by Kvifte. Although additional data are accumulating on column number densities and altitudes at which the peak density occurs, scarcity of such data and substantial differences in the measurements obtained does suggest a larger effort is needed if the Na problem is to be considered important. As things stand: (1) there appears to be a seasonal variation in column density and altitude of maximum density – more evident in the twilight than in the daytime observations, (2) observations are too few for a positive indication of a latitudinal effect, and (3) the generally accepted diurnal variation is now in question. Recent laser measurements by Sandford and Gibson support the last point. Derblom reported the interesting correlation of the Na layer altitude, with the sunspot cycle from observations in Sweden for the years 1961–1970. The altitude rose as the sunspot number decreased; this was interpreted as correlation with density and hence with the altitude of maximum deposition of meteoric material. Visconti noted an enhancement of 2:1 in emission intensity three days after near passes of comet tails in 1970 and 1971. Progress in chemistry related to the Na layer variations and location has not been spectacular. The central question of the origin of free Na remains unanswered, and concentrated work on Na chemistry will likely be paced by answers to this question. Another question, probably as important, is that of the large ratio of neutral to ionized Na. Most recent measurements by Narcisi indicate peak densities of 50 to 100 cm^{-3}. Heimerl called attention to the suggestion by Keller and Beyer whereby clustering of CO_2 and N_2 to Na^+, followed by the expected rapid recombination with electrons, might well explain the observations.

Although not a model in the sense of atmospheric chemistry, a model of the mid-latitude thermosphere was developed by Kockarts. More specifically, the phenomena contributing to heat balance – solar input, absorption, reradiation, and conduction – were developed, and the variation in thermospheric temperature at approximately 500 km was computed from a modeling of the heat balance. Not included in the model was mechanical transport (advection) which certainly is important. Without a nocturnal heat source, predicted diurnal temperature variations were larger than those observed. Adding a heat source to simulate transport, displaced 12 hr from the solar input and amounting to approximately 5% of the solar UV input, resulted in calculated temperatures in good agreement with observations. Considering the real uncertainties in UV absorption cross sections, particularly for N_2, is how the absorbed energy is distributed, and what the effective heat sinks are, the model calculation provides a surprisingly good result. Similar model calculations at the higher latitudes would probably not be good in view of substantial heat sources not considered in modeling and the very large atmospheric motions believed to occur in the auroral regions. For either high or low latitudes, substantial experimental and theoretical work is needed for confidence in model calculations under varying conditions of solar input and magnetic disturbances.

5. Physical Processes*

This summary covers two 'tutorial' reviews, by Hunten and Rishbeth, and four other papers which are mostly about energy input to the atmosphere.

Hunten presented a general theory of minor constituent diffusion. According to this theory the flux φ of a particular minor constituent (concentration n, scale height H) is given by

$$\varphi = -Dn[\ldots] \text{ for molecular diffusion (coefficient } D)$$
$$\varphi = -Kn[\ldots] \text{ for eddy diffusion (coefficient } K)$$

where the brackets contain 'concentration gradient', 'temperature gradient' and 'gravity' terms that are practically the same in each case. If diffusion dominates (i.e., production and loss are unimportant) then the steady-state equation for n (as a function of height h, say) is $d\varphi/dh = 0$, which has two independent solutions:

$$n(h) = [\text{hydrostatic distribution}] + [\text{flow term}].$$

We may expect the steady-state distributions to be established within a time of order H^2/D or H^2/K, as the case may be; but if the chemical time constants are shorter than both these, then the diffusive distributions will not become established.

Among the various applications of the theory are the following:

(a) A model with two regimes (smaller K below, greater K above) can reasonably well explain the vertical O_3 distribution in the troposphere and stratosphere.

(b) It is useful to write the eddy diffusion equation in terms of 'total hydrogen', so long as K is the same for all hydrogenous constituents.

(c) We can obtain useful expressions to describe the 'smoothing out' of minor constituent profiles around the homopause.

(d) From the observed turbopause for Ar at 99 to 102 km, we find

$$K = 3.4 \times 10^5 \text{ cm}^2 \text{ s}^{-1}.$$

(e) For atomic oxygen we find $K \sim 6 \times 10^5 \text{ cm}^2 \text{ s}^{-1}$, in contrast to older values of

$$K \sim 4 \times 10^6 \text{ cm}^2 \text{ s}^{-1}.$$

(f) N_2O and CH_4 are good tracers for the stratosphere; the former gives

$$K \sim 3 \times 10^3 \text{ cm}^2 \text{ s}^{-1}.$$

(g) For 'total hydrogen' we can consider the limiting upflux that diffusion will support in the thermosphere, and deduce lower limits for K at lower heights (10^3 at 30 km, 10^6 at 90 km). Thus H is a good tracer for eddy diffusion.

Rishbeth reviewed the theory of electric fields and currents in the ionosphere in a physical way. The equations of motion for an electron or ion are represented by a

---

* Presented by A. Egeland and H. Rishbeth.

triangle of forces comprising (a) driving force (wind or electric field), (b) friction due to collisions with air molecules, and (c) Lorentz $\mathbf{V} \times \mathbf{B}$ force. The angle θ between the driving force and the particle's drift velocity is given by:

$$\tan\theta = (\text{gyrofrequency/collision frequency})$$
$$= 1 \text{ for electrons at 70 km and for ions at 125 km.}$$

If we combine the motions of ions and electrons in one equation, and replace a neutral air wind $\mathbf{U}$ by a dynamo electric field $\mathbf{U} \times \mathbf{B}$, we obtain Ohm's Law for the ionosphere:

$$(\text{current}) = N_e(\mathbf{V}_i - \mathbf{V}_e) = (\text{tensor conductivity}) \, (\text{dynamo} + \text{polarization}$$
$$\text{fields}).$$

The quiet day Sq ionospheric currents are due *either* to tidal atmospheric winds generating dynamo fields, which also set up polarization fields, *or* to magnetospheric processes setting up electric potentials and polarization fields. Storm current systems must be mainly due to magnetospheric processes. The equatorial electrojet is driven by strong polarization fields arising from a freak geometrical situation.

In the ionosphere, an electric polarization field $\mathbf{E}$ produces a plasma drift $\mathbf{W} = \mathbf{E} \times \mathbf{B}/B^2$, which influences the electron density N only if it has a component parallel to grad N. As for the effects of $\mathbf{W}$:

In the E layer: $\mathbf{W}$ produces only small perturbations because production and loss dominate.

In the F layer: important in low latitudes where $\mathbf{W}$ has a vertical component (this causing the equatorial anomaly); less important at other latitudes, where $\mathbf{W}$ just 'convects' the plasma horizontally.

The electric fields drive winds because the drifting ions set the air into *horizontal* motion (they are not powerful enough to cause much *vertical* air motion) within a time constant:

$$\tau = (\text{collision frequency of } \textit{neutrals} \text{ with } \textit{ions})^{-1}.$$

In the F layer $\tau \sim 1$ hr (or less) so this 'ion-drag' is effective in setting up thermospheric winds, particularly in the auroral zone, which add to winds due to solar heating. In the E layer 'ion-drag' is ineffective because the time constant is too long ($\tau \gg 1$ hr).

Few parameters of the dynamo theory are really well known, and there are two outstanding topical problems:

(a) Are the Sq currents due to tidal winds or magnetospheric processes?

(b) To what extent are magnetic field lines in the ionosphere and magnetosphere electric equipotentials, as the theory assumes?

Ackerman (whose paper was summarized by Kockarts) reviewed the solar spectrum from the visible to the end of the Schumann-Runge continuum, from the point of view of the radiation coming from the whole solar disk. There are almost no major problems in the visible but the difficulties start around 3000 Å. Since the balloon observations are made at heights of only 35 to 40 km it is not possible to measure the solar flux accurately unless the absorption by the atmospheric constituents is known. Detailed computations of the O_2 absorption cross section in the Schumann-Runge bands were presented in the paper. At shorter wavelengths, in the Schumann-Runge continuum, there is still a big problem: is there a variation with the solar cycle and if so, what is its amplitude?

The major problem in the whole study of solar UV measurements is that of absolute calibration: different experiments may not agree to within even a factor of 2.

Ching suggested that the semi-annual variation in upper atmosphere density could arise from purely geometrical variations in EUV heat input. This is a very interesting suggestion, and if it is correct it might provide a simple explanation of a puzzling phenomenon. Previously it seemed that only the seasonal variation of temperature could be accounted for in this way. The detailed analysis, in which the heating rate at 150 km is found to vary semi-annually, is in course of publication in *Planetary and Space Science*.

Berthelier spoke about the transauroral ionosphere and its connections with the magnetosphere, and presented some detailed satellite data. The transauroral region is a belt extending about 8° poleward of low-latitude edge of the auroral oval; across this region there seems to be a change in the convection pattern, so it separates two completely different parts of the upper atmosphere as far as the motions of the plasma are concerned. In the polar cap the convection velocity, i.e., the circulation patterns for the plasma, is not aligned with the noon-midnight direction but is more symmetrical about the 10 to 22 LT meridian. He presented data on fluxes of particles, mostly in the keV range, and on the total energy inputs.

In discussing disturbances he gave some thought to the question of how the magnetosphere acts as a source of energetic particles: does the energy come from hot plasma in the plasma sheath or from interactions between the magnetosphere and the interplanetary medium?

6. Experimental Results and Their Interpretation* (1)

In this session the role of optical methods to determine atmospheric composition, and its possible variations, was discussed in some detail by the various authors and I shall attempt to briefly review some of the more significant results presented.

The IR instrumentation that has been used to identify the species present in the stratosphere and mesosphere was described by Schurin who emphasized that the low radiance levels make detection extremely difficult. These instrumentation problems have been particularly significant for H_2O determinations where some of the published results indicate mixing ratios that cannot apply to the atmosphere. However, new instrumentation methods were described by Murcray and it is to be hoped that these significant measurements will be extended to the ceiling for large balloons. There has also been important progress in the identification of other minor components in the

---

* Presented by E. J. Llewellyn.

lower stratosphere but there is an urgent need for the various experimenters to make direct inter-comparisons.

There is little doubt from the optical measurements that the CH_4 concentration decreases above the tropopause and contributes to the H_2O mixing ratio in the stratosphere, in agreement with the direct sampling techniques described by Martell. The new observations of HNO_3 emission at 11.3 μm (Murcray) are extremely interesting and have indicated an apparent seasonal and latitudinal variation, with a maximum mixing ratio near 25 km. These observations should be extended as the measurements are significant in the understanding of the important stratospheric chemistry (Crutzen, 1971.)

The importance of optical measurements to atmospheric models was demonstrated in the Ly-α observations discussed by Mange. Unfortunately there is at present some difficulty with absolute calibration for these measurements. However, the observations indicate that the present H models, which have a base height of 100 km, should be extensively revised to include a diurnal variation of the H concentrations, a decreased H atom concentration at 100 km, and a latitudinal asymmetry. Independent Ly-α observations by other experimenters (Bertaux) support the detected temporal variations which indicate a diurnal change of 70% with a maximum at 0600 and a minimum at 1600 LT. The reduction in the H atom concentration at lower altitudes has also been suggested from extensive Hα ground based observations (Tinsley, 1970) and inferred from the rocket observations of the $O_2(^1\varDelta)$ and the OH Meinel emissions (Evans and Llewellyn, 1972). If the total H content is unchanged from that adopted in present models the reduction in the H concentration requires an increased H_2 concentration. Hence it is possible that the photodissociation of H_2O leads to H_2 or that the mesospheric H atoms are effectively recombined to the stable molecular form.

The recent measurements of the EUV He emissions were also discussed and can lead to information on the He bulge and the solar flux. New information on the bulge has been derived from observations of the He I (10830 Å) emission which indicate that the alignment is with geomagnetic latitude. At present the solar fluxes calculated from the EUV emissions are widely discrepant and suggest that the use of thin film filters for spectral discrimination may not be sufficient. It is to be hoped that future measurements with a spectrometer may permit an improved comparison between theory and experiment. Observations of emissions at the He wavelength (5876 Å) were briefly discussed by Stoffregen but the problems of possible blending from other auroral features cannot be dismissed. This fact was illustrated by Gattinger with a series of synthetic spectra of the aurora in the wavelength range 7000 Å to 9000 Å.

The use of auroral observations to determine parameters associated with energy transfer in the atmosphere was discussed in some detail by Vallance Jones and it was shown that the transfer from $O(^1D)$ to O_2 occurred at a rate in essential agreement with the laboratory value. New observations of auroral spectra were also described and it was noted that the O^+ line at 7319 Å is apparently present in normal aurora;

this fact has probably been missed in previous studies due to the presence of OH Meinel vibration-rotation lines in the same region. Observations of the 8446 Å O$_I$ line were also presented but the excitation mechanism, which could be related to energy transfer was rather uncertain.

Other auroral observations, the $N(^2D)$ emission at 5200 Å, were discussed by Gerard but it was noted by Strobel that the inferred column concentration was probably not a constraint on the NO chemistry. In particular a column concentration of 2.5×10^{12} cm^{-2} can be readily satisfied from F region sources and does not explain the large abundance measured at 120 km.

The value of optical techniques for the study of atmospheric composition in the thermosphere was discussed by Noxon and it was shown that in combination with other techniques (e.g., incoherent backscatter) new information on the composition can be derived. It was noted that during red auroras the ionic mass at 200 km changes from a value 16 and that composition changes must therefore occur. Other compositional changes have been suggested from twilight observations of the 6300 Å emission which indicate an increase in the lower thermospheric O_2 during the summer without an associated N_2 increase. This may be particularly important for attempts to determine the O/O_2 ratio at 120 km as recent rocket observations of the oxygen green line indicate low O concentrations. Perhaps the most unusual application of optical techniques has been with the horizon scanning satellite photometer which provides electron densities below the F_2 peak. However, the analysis of these observations by Donahue and Thomas (1972) has indicated that for the F region green line emission the rate controlling step proceeds at a rate only one-half of that inferred from laboratory studies.

7. Experimental Results and Interpretation* (2)

In my opening remarks I pointed out the active interaction between the models, the laboratory kinetic measurements, and atmospheric observations. The first paper of this session really was, in part, a complaint by Wayne that the session entitled experimental methods really had very little time devoted to the critical discussion of the laboratory rate constant data. I think from the last summary there would have been particular interest in a critical analysis of some recent work on HO_2 kinetics, and on the rather messy system of NO_2, N_2O_5, and the other nitrogen oxides. This section really had most to do with airglow observations and the attempts to infer concentrations and height profiles from emissions which is a potentially useful technique, but one which requires careful interaction of laboratory data regarding both the excitation and the quenching mechanisms. Perhaps the most dramatic example of this interaction concerned excitation, by the Chapman mechanism of $O(^1S)$ where a simple redetermination in the laboratory of a quenching rate for O_2, not only resulted in a different rate of excitation, but also removed the concern that we were getting three times as many O_2 at the 100 km region than could be explained by the models.

* Presented by H. I. Schiff.

Additional measurements are obviously required on temperature coefficients for quenching. Work by Welge recently has indicated there is an appreciable activation of energy for the quenching of $O(^1S)$ by O_2.

We have seen from some of the satellite measurements the behavior of $O(^1D)$ emission at higher levels which undoubtedly is due in part to the dissociative recombination with O_2^+ although here again laboratory data are urgently needed to determine the branching ratios for ^1S and ^1D for dissociative recombination both of O_2^+ and that of NO^+. It was pointed out that there is still too much $O(^1S)$ observed at the higher levels and requirements were for photoelectron sources. In addition there have been suggestions that there were local sources (Shepherd in some of his recent work on the ISIS data sees evidence for conjugate point electrons). Wayne in reviewing the O_3 photochemistry pointed out that there was general agreement now that, in the wavelength region 2500 to 3100 Å, the products are almost exclusively $O(^1D)$ and $O_2(^1\Delta_g)$. In the Huggins bands, at wavelengths greater than 3100 Å, singlet oxygen can only be produced by violation of spin conservation. Wayne pointed out that this does not necessarily preclude their formation if the absorption coefficients are small and presented evidence for singlet O_2 formation in this spectral region.

What is urgently needed is definitive, direct measurements of the quantum yield for $O(^1D)$ in about the 3100 Å region. We know that quantum yield for ^1D drops off rapidly beyond 3100 Å but the exact dependence on wavelength is extremely important in view of the rapid increase in solar intensity with wavelength. Most of the complex reaction mechanisms proposed for the stratosphere and troposphere start with $O(^1D)$ reactions with N_2O, H_2O, CH_4, etc.

We also need to extend the measurements of O_3 photochemistry to short wavelengths to see whether excited upper states of O_2 and $O(^1S)$ can be produced at the short wavelength end of the Hartley band. Still a great deal of uncertainty remains on the reactions between OH and O_3, the effect of vibrational temperature on the rate and even the identification of the products.

Llewellyn gave us a very comprehensive review of a large number of $O_2(^1\Delta_g)$ measurements from rockets and balloons and gave us convincing evidence that O_3 and $O_2(^1\Delta)$ are well correlated during the dayglow. The evidence for a second layer or trough has now been well established and it looks as though this can be fitted to the model by the increased H catalyzed destruction of O_3 at the trough.

At the nightglow we still have a problem. There is a persistence of about 100 kR during the nightglow and a new observation made by Llewellyn's group of a second layer. A possible mechanism was rather strongly suggested but a thermodynamic argument was presented which threw doubt on that mechanism.

Noxon reported on some $O_2(^1\Delta_g)$ measurements from an aircraft. Some of these measurements could only be reconciled with O_3 photolysis if, during that time, O_3 was not being controlled by H_2O chemistry. He asked whether $O_2(^1\Delta_g)$ could be a product of the reaction between NO and O_3.

There was some discussion on the enhancement of $O_2(^1\Delta_g)$ in certain types of aurora and speculation on the mechanism. Although there seemed to be some degree of agreement that the enhancement could indeed be very large, the excitation mechanism still appears to be an open question.

We had a description of some OH height profiles from the measurements from Gattinger, including vibrational distributions, and another interesting suggestion in a short paper by Evans in which the observations could be shown to be reconcilable with the laboratory production distribution in which most of the energy from the $H + O_3$ reaction goes into the maximum vibrational levels possible. Figures were 45% into $v=9$, 45% into $v=8$. and 10% into $v=7$, but the reconciliation required the intriguing condition that the reaction rate coefficient for atomic oxygen plus OH decreases as the vibrational energy of the OH molecule increases. This would certainly be the first evidence for such behavior.

Other reactions in which vibrational energy has been considered to play an important role were discussed by Walker including the classic example of the $O^+ + N_2$ rate increasing very markedly with N_2 vibrational energy. Walker went on to give us a very comprehensive review of the types of vibrational energy exchange that could be occurring, pointing out that the conversion of vibration energy to kinetic energy in the case of N_2 has a low probablity of about 10^{-8} while the intermolecular conversion of vibrational energy is much more likely with a probability of about 10^{-2}. He pointed out the importance of the accidental resonance between vibrational levels of N_2 and CO_2 and N_2O which could then be dissipated in the form of IR radiation. Also of interest was the reverse process of kinetic energy being converted to vibrational energy by collisions of electrons with N_2 by the route of an intermediate short lived N_2 state. In O_2 vibrational relaxation by atom-atom interchange was indicated to be a rapid process.

Of considerable interest to this meeting was the quenching of ^1D oxygen atoms with N_2 to produce vibrationally excited N_2 and ground state O which has been proposed earlier. Theoretical work of Bauer indicated the probablility to be less than 1%, a view which Walker did not find completely convincing. The question of whether N_2 vibrationally excited is rapidly quenched with O was raised and McNeal's recent work indicates that it has a rate constant of about 10^{-14}.

Finally, Walker discussed mechanisms which produce non-Boltzmann distributions and pointed out that these were not likely to be important at altitudes less than 300 km and even there would be small.

To conclude, I feel that the airglow indeed looks like a promising way to get information on atmospheric processes, but I think we have to be very careful that we know all of the excitation and the quenching mechanisms very accurately.

8. Atmosphere of Other Planets*

This report must cover three planets, in contrast to the preceding six reports devoted to a single planet. The papers by McConnell, Prinn, and Strobel are already condensed

---

* Presented by D. M. Hunten.

reviews in themselves, and further condensation would be unprofitable. I shall therefore concentrate on a few highlights, general problems of current interest, and prospects for the near future. A list of general references is also included: two review papers, several conference proceedings, and collections from the various Mariner spacecraft.

It is obvious from all three papers that one cannot make even a good start on planetary aeronomy without some quantitative means of describing the vertical transport. The method is to adopt a profile of the eddy mixing (or diffusion) coefficient, a shorthand for all kinds of mechanical transport in the vertical, averaged over the whole planet. The procedure is arbitrary and unsatisfactory, and it is important to obtain a quantitative description of eddy diffusion, and insights on how to relate this rather ill-defined process to more fundamental ones such as winds and general atmospheric circulation. For most height ranges on the earth, eddy coefficients are determined empirically by following various tracers, such as CO, CH_4, N_2O, O, O_2 and Ar, and until we can do better we can have little confidence in the values postulated for the other planets. Similarly, any criticism of values adopted by others is based only on an intuition that is not well educated.

For Mars we have rather more information than we do for the other planets. Gierasch and Goody (1968) calculated temperature profiles for the troposphere and lower stratosphere on the basis of convection and radiative heat transport in the atmosphere; the input was the diurnal variation of the ground temperature. Eddy coefficients for heat transport were also derived; the values for matter transport would be of the same order, though problably not identical. More recently Stone (1972) has shown that the dominant mode in the Martian troposphere is not that described by Gierasch and Goody (local convection driven by a hot surface) but rather a baroclinic instability. Probably this is what drives the general circulation of the earth; it draws on the latitude variation of temperature such that the atmosphere is more extended near the equator and more compact at higher latitudes. The slope of the constant-pressure surfaces gives a source of potential energy on which motions can draw. This mechanism accounts for the characteristic mid-latitude circulation on the earth with its eddies, storms, and even hurricanes. However, in either model heat must be mechanically transported to the tropopause from the warm lowest layer of the atmosphere; the gas is far too opaque in the IR for radiation to be effective. To transport a given amount of heat in a given time across a given distance requires the same transport coefficient no matter what process the transport is attributed to; thus, Stone's model must give very nearly the same high eddy heat conductivity as the Gierasch-Goody model. However, above 15 or 20 km there is no compelling reason for high eddy coefficients. This region is a stratosphere, probably in radiative equilibrium with a stable temperature gradient; it resists vertical mixing. A small eddy coefficient (as on earth) seems more likely than a large one. Much higher, in the region of the mesopause, there is a well-established mechanism for producing true turbulence: essentially the breaking of gravity waves. There is a close analogy with ocean waves coming up to a beach and encountering a thin layer of water such that their amplitude grows: finally they break and create all sorts of small scale motion.

Uncertain as the situation is for Mars, it is far worse for Venus and Jupiter, where we simply have to guess. The only exception is the mesopause region on Jupiter, where H can be used as a tracer.

Next, I want to discuss the adequacy of data on temperature profiles for planetary atmospheres, important because the whole structure is defined by the composition and the temperature. Generally speaking we must rely on occultation results, from either spacecraft or stars. The physics of the two phenomena is quite similar, and they run into similar troubles as low densities (cf. Hunten, 1971). The observed quantity, integrated density, can be converted readily to local density by the equivalent of differentiation. The noise level is raised, but not unduly so because of the exponential height variation. Temperature is then obtained from the ratio of integrated density (pressure) to local density. The results are good except in top couple of scale heights. An exactly similar difficulty is found if the temperature is obtained from the logarithmic derivative of the density. It will be illuminating to compare the results from the occultation and IR experiments aboard Mariner 9. The recent stellar occultation by Jupiter seemed to show indications of a thermosphere at the very top of the profile, but this result must be regarded with deep suspicion.

From the aeronomical standpoint, Mars and Venus have a good deal in common because the major constituent is CO_2 for both. The ionospheres are presumably similar, because minor constituents and surface conditions are not important. But lower down there are large differences, and the aeronomy of Mars is much more highly developed. The reason is the presence of a visible surface, which limits the penetration of sunlight. In a cloudy atmosphere we cannot even be sure that absorption lines of various gases are formed above the clouds, rather than inside them. The slow rotation of Venus is another problem; diurnal averages, which are useful for earth and Mars, are much less so for Venus.

The stability of CO_2 has received a great deal of attention in the past few years. For both planets the problem divides into two parts:

(a) Why are CO and O so much rarer in the thermospheres than is O on earth?

(b) How is CO oxidized in the lower atmospheres to maintain the low observed abundances?

Problem (a) can be solved by fast downward transport, or in other words, rapid vertical mixing. In principle, the mixing can be on a small scale (eddies) or a large scale (circulation). Mars requires eddy coefficients of 10^8 cm^2 s^{-1} or greater at the base of the thermosphere, compared with 10^6 on earth. There may be more O and CO on Venus, but the available data are limited. Dickinson has proposed a circulation model that might sweep the thermosphere fast enough, but it is not universally accepted. In any case, no acceptable alternative to rapid transport has been suggested.

Problem (b) has seen some progress recently; indeed, two possible solutions for Mars have been offered, both involving catalysis by odd hydrogen. McConnell's comparison is so clear that it is hardly necessary to look at the original papers. The model of Parkinson and Hunten (1972) promotes the reaction of CO and CO_2 through an intermediate, H_2O_2; rather large amounts of odd hydrogen are needed. McElroy and

Donahue (1972) avoid the production of too much O_2 in the first by means of rapid stratospheric eddy transport; thus, the reaction scheme oxidizes CO with O. Each model has an unattractive feature, high concentrations of odd hydrogen or large eddy coefficients.

For Venus, Prinn has suggested that problem (b) can be solved with catalysis by products of HCl photolysis. The scheme is plausible, but many important reaction rates have not been measured.

Ionospheric models for both planets bear a considerable resemblance to the observed electron densities, except that they fall short by a factor of 2. The solar flux is uncertain, but if it were multiplied by the required factor of 3 or 4, the calculated exospheric temperatures would be far too high. The only explanation in sight is the suggestion by O'Malley that O_2^+ recombines slowly when vibrationally excited, as discussed in McConnell's paper. The ion composition is drastically affected by traces of O, and better information is necessary, especially for Venus.

Jupiter has a cloudy and very deep atmosphere; in fact there may be no real surface at all. A central problem for the ionosphere is the path by which protons recombine. If radiative recombination were dominant, the electron densities would be large; but there is a strong suspicion that some kind of ion chemistry will be found, as for O^+ on earth. Strobel gave an account of the photochemistry of CH_4 directed at the problem of identifying the principal IR radiator. He finds no molecules as important as CH_4 itself. There seems likely to be a slow production of heavier compounds, which will eventually be destroyed as they are mixed down to lower, hotter regions. We seem to have a good basic description of the thermal structure. The stratospheric temperature, from the recent stellar occultation, is about 150 K, and the temperature rise across the thermosphere is 10 or 20 K.

There has been a recent flurry of activity concerning the atmosphere of Titan, the largest satellite of Saturn (reviewed by Hunten, 1972). The presence of CH_4 has been known for decades; small amounts of NH_3 are probable, and H_2 and N_2 are suspected. If photochemistry produces heavier molecules, they should rapidly freeze out on the surface, whose temperature seems to be 140 K. This 'deep-freeze' should be an interesting thing for direct study in a possible future.

Finally, a few words may be of interest on current happenings and future prospects. Mariner 9 has been orbiting Mars for almost a year and producing a huge amount of exciting data. The UV spectrometer and the occultation experiment are telling us about the ionosphere, the occultation and IR experiments about the lower atmosphere. The great dust storm perturbed the thermal structure to great heights; for example the stratospheric temperature rose above 220 K in places. The television results are spectacular and will continue to be studied for years. For the future, the U.S.S.R. has a continuing program of landers, and the U.S.A. will launch a pair of Viking landers in 1976.

Flybys and orbiters are less productive on Venus because of the cloud cover and deep atmosphere. On the other hand, entry probe missions are aided by the atmosphere and suitable for direct measurements. The U.S.S.R. Venera series has been technically successful, though carrying rather limited instrumentation. NASA is seriously studying a program of its own (Pioneer-Venus) and we may hope for a successful and continuing effort.

From the earth's surface we may still hope to learn about time variations on Mars. Ozone is known to be highly variable in time and space, from Mariner 9. Substantial, but much slower, changes in CO and O_2 are likely to occur, because their abundances seem to depend on minor factors such as mixing rates and H_2O abundances. Even in the space age, there are still worthwhile things to be done with a telescope.

References

SECTION 6

Crutzen, P.: 1971, *J. Geophys, Res.* **76**, 7311.
Donahue, T. and Thomas, G.: 1972, *J. Geophys. Res.*
Evans, W. F. J. and Llewellyn, E. J.: 1972, to be published.
Tinsley, B.: 1970, in T. M. Donahue, R. L. Smith, and G. Thomas (eds.), *Space Res.* **10**, North-Holland Publishing Company, Amsterdam, p. 582.

SECTION 8

Gierasch, P. and Goody, R.: 1968, *Planetary Space Sci.* **16**, 615.
Hunten, D. M.: 1972, *Comments Astrophys. Space Phys.* **4**, in press.
McElroy, M. B. and Donahue, T. M.: 1972, *Science* **117**, 986.
Parkinson, T. D. and Hunten, D. M.: 1972, *J. Atmospheric Sci.* **29**, in press.
Stone, P. H.: 1972, *J. Atmospheric Sci.* **29**, 405.

GENERAL REFERENCES

Hunten, D. M.: 1971, *Space Sci. Rev.* **12**, 539.
Ingersoll, A. P. and Leovy, C. B.: 1971, *Ann. Rev. Astron. Astrophys.* **9**, 147.
The Atmospheres of Mars and Venus: 1968 (ed. by J. C. Brandt and M. B. McElroy), Gordon and Breach, New York.
(Various Authors): 1968, *J. Atmospheric Sci.* **25**, 533.
(Various Authors): 1969, *J. Atmospheric Sci.* **26**, 795.
(Various Authors): 1970, *J. Atmospheric Sci.* **27**, 523.
(Various Authors): 1971, *J. Atmospheric Sci.* **28**, 833.
(Various Authors): 1969, *Icarus* **10**, 353.
(Various Authors): 1970, *Radio Sci.* **5**, 121.
(Various Authors): 1967, *Science* **158**, 1665.
(Various Authors): 1972, *Science* **175**, 293.
'Planetary Atmospheres': 1971, (*IAU Symp.* **40**) (ed. by C. Sagan, T. C. Owen, and H. J. Smith), D. Reidel Publishing Company, Dordrecht-Holland.

CONCLUSIONS

BILLY M. McCORMAC

Lockheed Palo Alto Research Laboratory, Palo Alto, Calif., U.S.A.

1. Structure and Composition of the Atmosphere

A comprehensive survey of the current state of knowledge of the atmosphere was presented. The number of interlocking variables is so great that the only feasible method for correlating them has been to make 'model atmospheres' and then to test observations for their compatibility with current models. New findings are then often presented in the form of the deviations from models. For much of the earth's atmosphere, models exist and the data are immediately tested against the models. For atmospheres of other planets, each new set of observations serves as the basis for constructing a model.

The idea that 120 km was an 'invariant altitude' and thus had intrinsic merit as a reference altitude was repudiated. This idea may have originated with the 1965 Jacchia model that the densities of O and O_2 became equal at this level. There is no particular aerodynamic or chemical reason for the existence of an invariant altitude. In addition, von Zahn gave the ratio of $[O]/[O_2]$ at 120 km altitude as 3. An older idea that the ratio $[O]/[N_2]$ became unity at 200 km altitude is now felt to be attained more nearly at 165 km.

The region from 150 to 400 km altitude is generally regarded as the domain of diffusive equilibrium. However, using data at 150 km altitude a temperature based on the equation of diffusive equilibrium yields a value that is considerably too high at 400 km altitude.

2. Physical Processes

The mathematical treatment of the turbulent transport of minor constituents has been aided by the procedure called 'eddy diffusion.' The procedure is applicable without verification as to the type of wind or eddy or turbulence that is acting. The equations of molecular and eddy diffusion can be combined, as well as the equation of continuity, to yield the total flow rate.

By a step-by-step analysis of the contributions of molecular diffusion, eddy diffusion, and gravitational separation, and observations on densities and rates of production if applicable, Hunten was able to present a picture of the flux characteristics of numerous minor species of the atmosphere. Argon has proved to be an effective minor constituent for such studies at least in part because of its chemical inertness. The flux of H, regardless of molecular state, is upward at all heights. Typically of several species, it shows the molecular and the eddy diffusion coefficients to become equal, $D = K = 3.45 \times 10^5$ cm^2 s^{-1} at 100 ± 2 km.

B. M. McCormac (ed.), Physics and Chemistry of Upper Atmospheres, 378–381. All Rights Reserved.
Copyright © 1973 by D. Reidel Publishing Company, Dordrecht-Holland.

A careful analysis of solar UV radiation as a primary source of incident heat to the atmosphere shows that there is a factor of 2 to 3 uncertainty in incident EUV from 1400 to 2000 Å. There is also a question about solar cycle variations. The energy balance cannot be adequately solved at this time. Models only include UV absorption and IR loss (by O). There are other sources and sinks, such as joule heating and tidal dissipation, but there is insufficient information to include them at this time.

3. Chemical Processes and Models

Certain key ideas emerge. One is the almost total absence of nitrogen as a chemical reagent, whether atomic or molecular, neutral or ionized, in atmospheric chemistry. Oxygen in its various forms, O, O_2, O_3, O^+, O^-, O_2^+, O_2^-, O_3^+, O_3^-, $O(^1D)$, $O(^1S)$, $O_2(^1\Delta_g)$, $O_2(^1\Sigma_g^+)$, by contrast, is all important. While far less abundant, NO and H_2O play roles greatly exceeding their abundances. In the E and F layers, the metal vapors, largely ionized, are of interest, and at very high altitudes, H and He dominate.

Laboratory rate constants for formation and destruction of O_3 and of O have only recently become available. The rate constants predict about 50% more O_3 than is formed, and they do not account for the distribution for example, the maximum O_3 density over the winter pole.

Thoughts that O_2^- was the only negative ion in the D region have been greatly modified to include O^-, O_3^-, NO_2^-, OH^-, and NO_3^-. The last is believed to be a terminal ion in a chain of formations and charge exchanges. The positive ions in the D region are strongly dominated by H_3O^+, $H_5O_2^+$, and $H_7O_3^+$. It is known that NO^+ is produced, but laboratory studies of NO with slight traces of H_2O produce the hydrated ion too slowly. A solution to this problem was proposed that a 'cluster switching' process occurs, although it has not yet been successfully verified experimentally.

The E region is characterized by free electrons and the positive ions O_2^+, N_2^+, NO^+ and with increasing altitude, O^+, with NO^+ as the end-product ion. Metallic ions that have been observed include Na^+, Mg^+, Ca^+, Fe^+ and Si^+ as well as the oxide SiO^+. These ions are characterized further by extremely low recombination coefficients by comparison with molecular ions and hence with long lifetimes and redistribution by winds.

In auroras, there is a large amount of NO and O^+, no O_2^+, and to date, modeling attempts to characterize the observed densities fall very short of 'explaining' the largeness of the densities of NO and O^+.

An extensive study of the N_2 and CH_4 chemistry of the atmosphere was presented, motivated by considerations of SST contamination. The work is essentially modeling, based on the most recently available laboratory reaction rate date. The current concern regarding large scale SST operation is that it may cause a reduction of O_3 density in the lower stratosphere. Since O_3 absorbs UV solar radiation (peak absorption at 2700 Å), its destruction to any significant degree in the stratosphere would raise the flux of UV at the earth's surface. The supposed mechanism by which SST operation would destroy O_3 is its decomposition by NO:

$$O_3 + NO \pm O_2 + NO_2.$$

The reaction is called a catalytic one because the NO_2 is restored by sunlight to NO by photolysis of NO_2 so that the NO continues to decompose O_3 repeatedly, out of molecular proportion to its abundance. It then becomes of interest to assess the speed by which still other processes will remove NO from this cycle of reactions.

Crutzen takes note of the steady-state existence of NO and NO_2 in the stratosphere and recognizes its role in the O_3 balance. Loss of NO and NO_2 is largely through conversion to HNO_3 and N_2O, but this may be only temporary. The chemistry of odd H reactions then influences the HNO_3 production. Methane reactions further aid in removing NO_x. The conclusion is that definitive statements on hazardous levels of O_3 destruction cannot yet be made.

IR emission measurements on HNO_3 from balloons showed the maximum mixing ratio at 20 to 25 km altitude with a definite decrease above 30 km. The peak value of the mixing ratio was $\sim 10^{-8}$ g HNO_3 g^{-1} air.

Other Planets

Martian aeronomy is mainly the problem of the interaction of solar UV radiation with the low-density atmosphere of the planet. The problem of heat transfer and thermal equilibrium calls for further study. There is some uncertainty in the solar UV constant and also in the seemingly rapid transfer of heat from the atmosphere close to the surface to higher altitudes. Calculations using the eddy diffusion coefficient yield a value that seems excessive. If the heat transfer is entirely produced by turbulent gas motions, then speeds of 200 to 500 m s^{-1} are indicated, and these would be unreasonably high.

Measurements of the martian atmospheric composition disclose very little CO and O, which is difficult to explain. Since the atmosphere is transparent, down to the ground in the visible and almost to the ground in UV, an appreciable degree of dissociation of the CO_2 into CO and O is predicted. Various reaction mechanisms have been tested, but confirmation of any of them depends on a more accurate measurement of the concentration of O as a function of altitude.

Venus, like Mars, has an atmosphere that is mostly CO_2. The presence of halogen compounds is considered surprising and there is much speculation on the role of HCl in the chemical balance of Venus. Oxygen has been much sought but not found. The major point to be explained, as on Mars, is the abundance of undissociated CO_2 at higher altitudes, and the associated low levels of CO and O_2.

Temperature measurements have not been carried higher than 90 km. The nighttime profile differs only slightly from the daytime curve, and this observation combined with many others seems to establish a picture of a "super rotating" atmosphere on Venus. The atmosphere would appear to make one complete rotation in 4 days, in the retrograde direction as contrasted with reflected radar measurements from the surface that give a rotation period of 244 days, also retrograde. Such a high speed rotation necessitates wind speeds of 100 to 150 m s^{-1}.

Much work is required on the fundamental physical and chemical processes that may occur in the Jovian atmosphere. Current models are quite rough.

5. Future Institute

It was concluded that the rapid evolution of upper atmospheric physics and chemistry justified another two-week Advanced Study Institute in 1974. The 'Physics and Chemistry of Atmospheres' 1974 Institute Program Committee members are: Drs Alv Egeland, Hans Bolle, Donald M. Hunten, Günther Lange-Hesse, M. Nicolet, J. Ortner, Harold I. Schiff, Lance Thomas, Alister Vallance Jones, Richard Wayne, and Gilbert Weill.

GLOSSARY

Aerosol. Finely divided particles, liquid or solid, suspended in the atmosphere. The exact size range is arbitrary.

Auroral Absorption (AA). Measure of cosmic noise absorption through the ionosphere in the auroral oval.

Auroral Oval. Locus of auroras in latitude as a function of time, which has an oval shape.

C9.   An index number from O to 9 characterizing the intensity of a magnetic storm.

Chemical symbols in brackets, e.g., $[N_2]$. Represents the concentration of the substance indicated. In this report, the units are frequently molecules cm^{-3}.

CIRA Model. COSPAR International Reference Atmosphere, published by North-Holland Publishing Co., Amsterdam (1965).

Cosmic Noise Absorption (CNA). Surface measurement of the absorption of cosmic noise, usually about 30 MHz, passing through the ionosphere.

D Region. Altitude region in which the density of ions rises steeply, from $\sim 10^2$ to $\sim 10^6$ ions cm^{-3}. The altitude is usually 50–90 km.

E Region. Altitude region next above *D* region with ion densities $\sim 10^6$ cm^{-3}. The altitude varies from day to night, usually 90–160 km in daytime.

Eddy Diffusion. A method of treating turbulent transport of minority constituents in the atmosphere in the mathematical formulation of classical molecular diffusion.

Electrojet. Refers to a current of electrons, on occasions reaching $\sim 10^6$ A, moving horizontally from E to W in the early morning and much more weakly from W to E in the evening, at ~ 120 km altitude and $\sim 65°$ INLT.

ELF. Extremely low frequency and is from 30 to 3000 Hz.

Equatorial Electrojet. A current system in the ionosphere, flowing generally along the earth's equator.

EUV. Extreme ultraviolet, defined as starting below 1040 Å (the cutoff of the best window) and going down to 2 Å.

F Region. Altitude region next above the *E* region, characterized by a new rise in ion density.

FP. Fabry-Pérot.

Hall Currents. Current flow perpendicular to both the electric and magnetic fields.

IMF. Interplanetary magnetic field.

IN Lat. Invariant latitude, Φ.

IN LT. Invariant local time.

IN Pole. Invariant pole, where $\Phi = 90°$.

Invariant Coordinate System. McIlwain's *B, L* space magnetic coordinates.

Ionosphere. Altitude region where solar radiation is strongly effective in ionizing the atmosphere. Altitude limits usually given as 50–400 km, including the *D*, *E*, and *F* regions.

IR. Infrared radiation covering from about 7800 Å to 1000 μ.

IS. Ionospheric scatter.

Jacchia Model. Model atmosphere computed by L. G. Jacchia and staff of the Smithsonian Institute Astrophysical Observatory.

K_p. Quasi-logarithmic scale, from 0 to 9, measuring the range of activity of the most active component of the magnetic field within a 3 hr interval.

L. McIlwain's invariant shell paramenter, whose units are expressed in R_E at the magnetic equator.

LBH. Lyman-Birge-Hopfield band system.

LT. Local time.

Ly-α. Stands for Lyman-alpha. Radiation (or absorption) of atomic H at 1216 Å, involving transition of the atomic electron between $n = 1$ to 2.

Magnetosphere. Region inside a surface surrounding the earth at ~ 10 earth radii inside of which the earth's magnetic field exceeds external fields.

Mesopause. A narrow altitude region at the top of the mesosphere, often between 70 and 85 km altitude, in which the atmosphere is at its lowest temperature.

Mixing Ratio. Ratio of molecular density of a species to the total molecular density, sometimes alternatively the mass ratio or the specific volume ratio. Applies to atmospheric gases with good trubulent mixing.

MLT. Magnetic local time.

Molecular Diffusion. A term referring to interdiffusion of one gas into another in accordance with basic gas-kinetic theory.

$O(^1S)$ and $O(^1D)$. Two excited states of O atoms, ~ 4 and ~ 2 eV resp. above the ground energy states. Transitions between these states or to the ground state are quantum mechanically forbidden, hence the states are called metastable and have long lives, ~ 1 s for the first and ~ 145 s for the latter. These states only differ from each other and ground state by electron spin directions and the m quantum number.

$O_2(^1\Sigma_g^+)$ *and* $O_2(^1\Delta_g)$. Two excited states of O_2 molecules having ~ 1.6 and 1.0 eV energy resp. Transitions to the ground state are quantum mechanically forbidden leading to lifetimes ~ 12 s and ~ 3900 s resp. Basically, the states only differ from each other and from the ground state in electron spins.

Occultation. Important observations on planetary atmospheres are made during the brief period that a source of radiation is just disappearing or reappearing from behind the planet – hence occulting – and the radiation reaching the earth observers passes through the planet atmosphere tangentially.

Oxygen Green Line. Radiation of wavelength 5577 Å resulting from transition in atomic O from ^1S to ^1D states.

PCA. Stands for Polar Cap Absorption but refers to an *event*, now usually written SPE for solar proton event. During the event, solar protons increase the ion density in the earth's polar cap causing increased absorption of radio frequencies.

Pedersen Current. Current flow along electric field which is perpendicular to the magnetic field.

Photochemical. Term referring to a chemical reaction caused directly by solar photons acting on the reagent.

Photoionization. Ionization of an atmospheric constituent by solar radiation.

Polar Cap. Region inside the auroral oval.

ppmv. Parts per million volume.

Pre-dawn Enhancement. Enhanced optical emission produced before normal sunrise behavior as a result of charge particles from the sunlit conjugate region.

Quenching. Metastable atoms and molecules 'normally' de-excite to a lower energy state spontaneously in a relatively long characteristic time interval after creation or excitation. Admixture of a different gas sometimes shortens this time. If so, the foreign gas is said to quench the metastable.

Rayleigh (R). Unit of brightness or radiative intensity, for example of an aurora or airglow, in simplest form amounting to 10^6 photons $cm^{-2} s^{-1}$ but complicated by thickness considerations.

Red Arc. Airglow at 6300 Å wavelength arising from the atomic O transition ^1D to ^3P (ground state).

Retrograde Motion. Most of the planets and satellites of the solar system rotate around the sun in the same direction with orbits nearly in the same plane (the ecliptic). They also revolve on their axes in the same directional sense. If either the orbit direction or the revolution is opposite to this sense, it is called retrograde motion.

Scale Height. Boltzmann's law of atmospheres for gases in equilibirium is $p=p_0$ $\exp(-mgh/kT)$. p is the pressure at height h for a gas of molecular weight m and constant temperature T. The quantity kT/mg is the scale height. It is thus also the value of the height h at which $p=p_0 e^{-1}$.

Schumann-Runge Band and Continuum. UV absorption by O_2 with a band system between ~ 2200 and 1759 Å. The continuum begins at 1759 Å and extends toward shorter wavelengths. It arises from dissociation of the O_2. The band system arises from transitions from the $X^3\Sigma_g^-$ ground state to the $B^3\Sigma_u^-$ excited state with lowest energy 6.1 eV.

SID. Sudden ionospheric disturbance.

Solar Wind. A steady flow or flux of particles escaping from the sun with high velocity up to 10^3 km s^{-1}. The particles are charged, ions and electrons, but the net flux is neutral. The flow can rise by a factor as large 10^2 during a solar storm.

SPA. Sudden phase anomaly in reflected radio waves.

Sporadic E. Irregular fluctuation in ion and electron density in the E region sometimes yielding very high densities. It occurs commonly in a thin layer at ~ 120 km altitude, leading to the term, sporadic E layer.

Stratopause. A narrow altitude band between the stratosphere and the mesophere marking the top of the constant and low temperature domain of the stratosphere, frequently at ~ 25 km altitude in midlatitudes.

Superrotation. A condition where the atmosphere of Earth or other planets is moving in the same direction but faster than the planet itself is rotating.

UT. Universal time.

UV. Ultraviolet radiation and extends from 100 to 3800 Å.

Vibrational Temperature. The mean energy of molecular vibration, divided by Boltzmann's constant. It frequently differs from ordinary or 'translational' temperature because exchange of energy between vibration and translational motion may be extremely slow.

VK. Vegard-Kaplan band system.

VLF. Very low frequency and is from 3 to 30 kHz.

1N. First negative band system.

1P. First positive band system.

2P. Second positive band system.

ASTROPHYSICS AND SPACE SCIENCE LIBRARY

Edited by

J. E. Blamont, R. L. F. Boyd, L. Goldberg, C. de Jager, Z. Kopal, G. H. Ludwig, R. Lüst,
B. M. McCormac, H. E. Newell, L. I. Sedov, Z. Švestka, and W. de Graaff

1. C. de Jager (ed.), *The Solar Spectrum. Proceedings of the Symposium held at the University of Utrecht, 26–31 August, 1963*. 1965, XIV + 417 pp.
2. J. Ortner and H. Maseland (eds.), *Introduction to Solar Terrestrial Relations. Proceedings of the Summer School in Space Physics held in Alpbach, Austria, July 15–August 10, 1963 and Organized by the European Preparatory Commission for Space Research*. 1965, IX + 506 pp.
3. C. C. Chang and S. S. Huang (eds.), *Proceedings of the Plasma Space Science Symposium, Held at the Catholic University of America, Washington, D.C., June 11–14, 1963*. 1965, IX + 377 pp.
4. Zdeněk Kopal, *An Introduction to the Study of the Moon*. 1966, XII + 464 pp.
5. Billy M. McCormac (ed.), *Radiation Trapped in the Earth's Magnetic Field. Proceedings of the Advanced Study Institute, Held at the Chr. Michelsen Institute, Bergen, Norway, August 16–September 3, 1965*. 1966, XII + 901 pp.
6. A. B. Underhill, *The Early Type Stars*. 1966, XIII + 282 pp.
7. Jean Kovalevsky, *Introduction to Celestial Mechanics*. 1967, VIII + 427 pp.
8. Zdeněk Kopal and Constantine L. Goudas (eds.), *Measure of the Moon. Proceedings of the Second International Conference on Seleiodesy and Lunar Topography held in the University of Manchester, England, May 30–June 4, 1966*. 1967, XVIII + 479 pp.
9. J. G. Emming (ed.), *Electromagnetic Radiation in Space. Proceedings of the Third ESRO Summer School in Space Physics, held in Alpbach, Austria, from 19 July to 13 August, 1965*. 1968, VIII + 307 pp.
10. R. L. Carovillano, John F. McClay, and Henry R. Radoski (eds.), *Physics of the Magnetosphere. Based upon the Proceedings of the Conference held at Boston College, June 19–28, 1967*. 1968, X + 686 pp.
11. Syun-Ichi Akasofu, *Polar and Magnetospheric Substorms*. 1968, XVIII + 280 pp.
12. Peter M. Millman (ed.), *Meteorite Research. Proceedings of a Symposium on Meteorite Research held in Vienna, Austria, 7–13 August, 1968*. 1969, XV + 941 pp.
13. Margherita Hack (ed.), *Mass Loss from Stars. Proceedings of the Second Trieste Colloquium on Astrophysics, 12–17 September, 1968*. 1969, XII + 345 pp.
14. N. D'Angelo (ed.), *Low-Frequency Waves and Irregularities in the Ionosphere. Proceedings of the znd ESRIN-ESLAB Symposium, held in Frascati, Italy, 23–27 September, 1968*. 1969, VII + 218 pp.
15. G. A. Partel (ed.), *Space Engineering. Proceedings of the Second International Conference on Space Engineering, held at the Fondazione Giorgio Cini, Isola di San Giorgio, Venice, Italy, May 7–10, 1969*. 1970, XI + 728 pp.
16. S. Fred Singer (ed.), *Manned Laboratories in Space. Second International Orbital Laboratory Symposium*. 1969, XIII + 133 pp.
17. B. M. McCormac (ed.), *Particles and Fields in the Magnetosphere. Symposium Organized by the Summer Advanced Study Institute, held at the University of California, Santa Barbara, Calif., August 4–15, 1969*. 1970, XI + 450 pp.
18. Jean-Claude Pecker, *Experimental Astronomy*. 1970, X + 105 pp.
19. V. Manno and D. E. Page (eds.), *Intercorrelated Satellite Observations related to Solar Events. Proceedings of the Third ESLAB/ESRIN Symposium held in Noordwijk, The Netherlands, September 16–19, 1969*. 1970, XVI + 627 pp.
20. L. Mansinha, D. E. Smylie and A. E. Beck, *Earthquake Displacement Fields and the Rotation o, the Earth. A NATO Advanced Study Institute Conference Organized by the Department of Geophysics, University of Western Ontario, London, Canada, June 22–28, 1969*. 1970, XI + 308 pp.
21. Jean-Claude Pecker, *Space Observatories*. 1970, XI + 120 pp.
22. L. N. Mavridis (ed.), *Structure and Evolution of the Galaxy. Proceedings of the NATO Advanced Study Institute, held in Athens, September 8–19, 1969*. 1971, VII + 312 pp.

23. A. Muller (ed.), *The Magellanic Clouds. A European Southern Observatory Presentation: Principal Prospects, Current Observational and Theoretical Approaches, and Prospects for Future Research. Based on the Symposium on the Magellanic Clouds, held in Santiago de Chile, March 1969, on the Occasion of the Dedication of the European Southern Observatory.* 1971, XII + 189 pp.

24. B. M. McCormac (ed.), *The Radiating Atmosphere. Proceedings of a Symposium Organized by the Summer Advanced Study Institute, held at Queen's University, Kingston, Ontario, August 3–14, 1970.* 1971, XI + 455 pp.

25. G. Fiocco (ed.), *Mesospheric Models and Related Experiments. Proceedings of the 4th ESRIN-ESLAB Symposium, held at Frascati, Italy, July 6–10, 1970.* 1971, VIII + 298 pp.

26. I. Atanasijević, *Selected Exercises in Galactic Astronomy.* 1971, XII + 144 pp.

27. C. J. Macris (ed.), *Physics of the Solar Corona. Proceedings of NATO Advanced Study Institute on Physics of the Solar Corona, held at Cavouri-Vouliagmeni, Athens, Greece, 6–17 September 1970.* 1971, XII + 345 pp.

28. F. Delobeau, *The Environment of the Earth.* 1971, IX + 113 pp.

29. E. R. Dyer (general ed.), *Solar-Terrestrial Physics/1970. Proceedings of the International Symposium on Solar-Terrestrial Physics, held in Leningrad, U.S.S.R., 12–19 May 1970.* 1972, VIII + 938 pp.

30. V. Manno and J. Ring (eds.), *Infrared Detection Techniques for Space Research, Proceedings of the Fifth ESLAB-ESRIN Symposium, held in Noordwijk, The Netherlands, June 8–11, 1971.* 1972. XII + 344 pp.

31. M. Lecar (ed.), *Gravitational N-Body Problem, Proceedings of IAU Colloquium No. 10, held in Cambridge, England, August 12–15, 1970.* 1972, XI + 441 pp.

32. B. M. McCormac (ed.), *Earth's Magnetospheric Processes. Proceedings of a Symposium Organized by the Summer Advanced Study Institute and Ninth ESRO Summer School, held in Cortina, Italy, August 30–September 10, 1971.* 1972, VIII + 417 pp.

33. Antonín Rükl, *Maps of Lunar Hemispheres.* 1972, V + 24 pp.

34. V. Kourganoff, *Introduction to the Physics of Stellar Interiors.* 1973, XI + 113 pp.